A. E. NORDENSKIÖLD

FACSIMILE-ATLAS

TO

THE EARLY HISTORY OF CARTOGRAPHY

WITH

REPRODUCTIONS OF THE MOST IMPORTANT MAPS
PRINTED IN THE XV AND XVI CENTURIES.

TRANSLATED FROM THE SWEDISH ORIGINAL
BY
JOHAN ADOLF EKELÖF
ROY. SWED. NAVY
AND
CLEMENTS R. MARKHAM
C. B., F. R. S.

WITH A NEW INTRODUCTION BY
J. B. POST
MAP LIBRARIAN
FREE LIBRARY OF PHILADELPHIA

DOVER PUBLICATIONS, INC.
NEW YORK

PUBLISHER'S NOTE

Wherever possible all maps within the text and all plates have been presented at the size at which they originally appeared in the Stockholm, 1889, edition. Some plates, however, were beyond the press sizes available to us, and had to be reduced by small amounts, generally from two to five per cent:

XVI, XIX–XX, XXII–XXIV, XXVI, XXIX–XXXVI, XXXVIII–XLII, XLIV–XLV, XLIX–LI.

A glossary of cities of publication that occur in the list of maps is found on p. x.

THE EDITORS

New York, 1973

Published in Canada by General Publishing Company, Ltd., 30 Lesmill Road, Don Mills, Toronto, Ontario.
Published in the United Kingdom by Constable and Company, Ltd., 10 Orange Street, London WC 2.

This Dover edition, first published in 1973, is an unabridged and corrected republication of the English edition originally published in Stockholm in 1889. A new Introduction was written by J. B. Post especially for the Dover edition, and a Publisher's Note and Glossary of Cities of Publication have also been added.

International Standard Book Number: 0-486-22964-5
Library of Congress Catalog Card Number: 72-83741

Manufactured in the United States of America
Dover Publications, Inc.
180 Varick Street
New York, N. Y. 10014

INTRODUCTION
TO THE DOVER EDITION

Adolf Erik Nordenskiöld (1832–1901) was one of those remarkable men who are able to make a mark in more than one field of endeavor. Born in Helsinki of Swedish parents, he appeared destined to follow in his father's footsteps as a mineralogist. Nordenskiöld twice, however, incurred the displeasure of the Imperial Russian bureaucracy which governed Finland at the time: he was involved in a student group which gave vocal support to the French and British during the Crimean War, and he made a public speech lamenting the loss of freedom in Finland under the Russians. Discretion being the better part of valor, he left for Sweden in 1857. Hoping tensions had decreased, he returned to Finland the next year, but was ordered to leave by no less a personage than the Governor-General when he wouldn't recant. Settling in Sweden, Nordenskiöld became chief of mineralogy at the Stockholm Riksmuseum and there created one of the great mineralogical collections in Europe. Some men, perhaps, would consider this enough and rest on their laurels.

In 1858 Nordenskiöld began the first of his ten arctic expeditions. His greatest popular fame rests on these polar journeys. That each one was a success, even the one that threatened disaster, is due in large measure to Nordenskiöld's organizing abilities. Each phase of an expedition was planned with meticulous care. At first the Swedish government sponsored Nordenskiöld's expeditions, but soon Oscar Dickson, a wealthy Göteborg merchant, became his patron.

Nordenskiöld's arctic expeditions were made to Spitsbergen in 1858, 1861 and 1864, and each revealed more and more about these islands. These expeditions were staffed with a full range of scientific personnel, as were all of Nordenskiöld's explorations. Nordenskiöld himself discovered paleontological evidence of changing climates in the Arctic. In 1868 the steamer *Sofia* attempted to go as far north as possible, reaching 81°42′ North before being stopped by the ice. The 1870 trip to Greenland was to determine if dogs could be used as beasts of burden in arctic exploration. While in Greenland, Nordenskiöld was able to penetrate thirty miles inland on a side trip, the farthest inland anyone had gone up to that time.

The expedition to Spitsbergen in 1872–1873 was the ill-fated one. The two supply ships accompanying Nordenskiöld's ship, *Polhem,* were frozen in, and sixty-five men had to survive on provisions intended for twenty-two. Thirty-nine of the forty reindeer brought along as draft animals escaped. It was a difficult time. Still, Nordenskiöld did not remain idle. Using the frozen ships as a base, he made one side trip which produced important glaciological data.

In addition, he had brought with him a set of the *Transactions* of the Swedish Academy of Sciences and proceeded to research and write a history of Swedish science for the years 1718–1772.

Nordenskiöld's interest, after returning from this expedition, now shifted to finding the Northeast Passage, the water route around Northern Asia. In 1875 he sailed from Norway to the Yenesi River in Siberia. Critics claimed the ice was unusually favorable. In 1876 he repeated the voyage, delivered a commercial cargo and proved that the Kara Sea was navigable.

The most famous of his arctic journeys was in 1878–1880 when his ship, the whaler *Vega,* after being frozen in the ice for nine months, went on to make the first circumnavigation of Asia. As always, the trip was a scientific expedition, and five volumes of scientific reports and a two-volume popular narrative were published as a result of the observations made on the trip. In 1880 Nordenskiöld was made a baron by the Swedish King. The government of the Netherlands decided that Nordenskiöld was the discoverer of the long-sought Northeast Passage and awarded him the prize offered in 1611 by the States General for that feat.

The last arctic expedition was in 1883 to Greenland. Two Lapp ski experts managed to penetrate 145 miles into the interior and return in 57 hours. This figure was challenged and once again Nordenskiöld set about proving himself right. In a controlled experiment he held a cross-country ski race and one of the Lapps established a world record.

Most explorers, and even some casual travelers, wonder about who has been to a place before them and what maps were used. Nordenskiöld, in the course of researching for his own journeys, became an authority on earlier polar explorations. As he did so, he became aware of the deplorable state the history of cartography was in. There had always been an interest in old maps and early geographical theories, and quite often facsimiles of these maps were produced. What upset Nordenskiöld was the lack of order in it all. To an essentially organized man like himself, this unsystematic approach to the history of cartography was shocking. Even before the publication of *Facsimile-Atlas* he had begun to write articles on early cartography.

One can say with little exaggeration, however, that it is with the publication of his *Facsimile-Atlas to the Early History of Cartography with Reproductions of the Most Important Maps Printed in the XV and XVI Centuries* (Stockholm: 1889) that the modern historical study of cartography can be said to begin. Before *Facsimile-Atlas* there existed knowledge and interest but after *Facsimile-Atlas* there existed a systematic approach to cartography. No scholarly work escapes the tarnishment of time and *Facsimile-Atlas* is no exception. New information causes us to modify our assumptions as well as our conclusions. Nordenskiöld realized he was not writing the last word on the subject and concludes his preface to *Facsimile-Atlas* by saying "he will feel himself richly rewarded if it should contribute not only to supply a defect in the geographical literature of the day, but also to promote new discoveries in the recesses of libraries and map collections."

Facsimile-Atlas was published in both Swedish and English editions. Linguistic jingoists will be happy to observe that English has been a major language of communication in the study of cartography for quite a while. A few years after *Facsimile-Atlas* was published, there appeared Nordenskiöld's *Periplus: An Essay on the Early History of Charts and Sailing Directions* (Stockholm: 1897), which was also issued in both Swedish and English editions. *Periplus* is in many ways a second volume and a continuation of *Facsimile-Atlas,* though time has dealt more harshly with it than with *Facsimile-Atlas.*

The geographies of Ptolemy were for long the major source of geographic knowledge. Nordenskiöld begins *Facsimile-Atlas* by examining the various editions and pseudo-editions of, and additions to, Ptolemy. Other broad headings into which he divides his study include non-Ptolemaic ancient maps, early mapping of the Age of Discovery (including maps of the New World), early terrestrial globes, map projections, and the beginnings of modern cartography.

Both in *Facsimile-Atlas* (p. 53) and in articles published in learned journals Nordenskiöld places quite a bit of reliance on the Zeno Map. This map was first published in 1558 at Venice in a work purporting to be the adventure of two Venetian brothers in the North Atlantic about 1380. Research since Nordenskiöld's time has shown this document to be somewhat suspect. This is not a reflection on Nordenskiöld since any researcher can only be expected to know the past, not the future.

If 1889 was the "birth" of the modern study of the history of cartography, there are other dates and events worth noting. Nordenskiöld's *Periplus* was published, as we indicated, in 1897. In 1935 Leo Bagrow issued the first number of *Imago Mundi,* an annual journal devoted to the study of early maps. The year 1960 saw the founding of the Society for the History of Discoveries, though the society did not issue its annals, *Terra Incognita,* until 1969. In 1963, two events occurred: R. V. Tooley started to publish *The Map Collectors' Circle,* a ten-times-a-year publication devoted to illustrated cartobibliographies and studies in the field of cartography; and Theatrum Orbis Terrarum, Ltd., of Amsterdam began issuing magnificent facsimiles of early atlases, complete with bibliographical and historical introductions by leading authorities. The Newberry Library in Chicago dedicated its Center for the Study of the History of Cartography in 1972.

The Dover reprint of *Facsimile-Atlas* presents the first opportunity for most readers to see Nordenskiöld's great work. The book was reprinted in 1961 for $110 a copy, and five years later the Streeter copy of the original edition was sold in auction for $450. It is doubly gratifying, therefore, to have this inexpensive unabridged facsimile made available by Dover Publications now.

J. B. Post

Philadelphia, 1973

BRIEF BIBLIOGRAPHY

Bagrow, Leo. *History of Cartography,* revised and enlarged by R. A. Skelton. London: Watts, 1964.

Kish, George. "Adolf Erik Nordenskiöld (1832–1901): Polar Explorer and Historian of Cartography," in *The Geographical Journal,* December, 1968, Vol. 134, Pt. 4, pp. 487–500.

———. *North-East Passage: Adolf Erik Nordenskiöld: His Life and Times.* Amsterdam: Nico Israel, 1972.

Morison, Samuel Eliot. *The European Discovery of America: The Northern Voyages, A.D.* 500–1600. New York: Oxford University Press, 1971.

[Obituary] *The Geographical Journal,* October, 1901, Vol. 18, No. 4, pp. 449–452.

Skelton, R. A.; Marston, Thomas E.; and Painter, George D. *The Vinland Map and the Tartar Relation.* New Haven: Yale University Press, 1965.

PREFACE

The history of geography during the era of the great geographical discoveries can hardly be fully intelligible without a comparative study of the maps which were then accessible, and on which the explorers based their schemes for new enterprises. In this respect the printed maps, owing to their wider circulation, played a part by no means inferior in importance to that of the manuscript geographical drawings, of which as a rule only a few copies existed, jealously concealed in the archives of the State or in the chests of merchant-adventurers. But even printed maps of this period have become very rare, and extensive collections of them are only to be found in a few libraries. Many of the most important of these documents are therefore not easily accessible to students — a difficulty the unfavourable influence of which may be traced even in elaborate geographical treatises of the most distinguished authors.

It is this circumstance which has induced me to publish the present systematic collection of the most important maps printed during the early period of cartography. I have endeavoured to make the work as complete as possible, in order to enable every student of historical geography to examine and consult in his own library correct copies of the most important and characteristic geographical documents published in print during the XV. and XVI. centuries. But it did not enter into the plan of this work to give a systematic reproduction of *manuscript maps,* for which the reader is referred to the celebrated Atlases of Jomard, Santarem, Theobald Fischer, and others. The necessity of confining my work within reasonable limits has furthermore compelled me to refrain from reproducing some very large printed maps, the insertion of which would have been possible only on a scale so reduced that the copies would have been almost worthless. The most interesting among these omitted maps are the newly discovered large map of Scandinavia by Olaus Magnus, the terrestrial globe in gores by Mercator, Sebastian Cabot's planisphere, Mercator's Flandria, his large map *»ad usum Navigantium,»* and an immense cordiform map by Vopel. In the text references will be given to several more or less complete descriptions and reproductions of these which have been lately published.

This work was originally written in Swedish and published under the title: *Facsimile-atlas till karto-grafiens äldsta historia, innehållande afbildningar af de vigtigaste kartor tryckta före år 1600, af A. E. Norden-sköld.* But in order to make it accessible to students not versed in the Swedish language, this English edition was undertaken and sent to press during the printing of the Swedish original, which was finished in March 1889.

While preparing the sheets for the press the author has received much important information and valuable aid from Mr E. DAHLGREN of the Royal Library at Stockholm. The translation has been executed, under the author's supervision, by Captain J. A. EKELÖF of the Royal Swedish Navy, and revised by the learned geographer and explorer Mr CLEMENTS R. MARKHAM. It is a pleasant duty for the author here to record his best thanks to these gentlemen for the assistance they afforded him during the composition and publication of this work.

He will feel himself richly rewarded if it should contribute not only to supply a defect in the geographical literature of the day, but also to promote new discoveries in the recesses of libraries and map-collections.

Dalbyö, Aug. 15th 1889.

A. E. NORDENSKIÖLD.

CONTENTS

Maps reproduced by photolithography.

All, excepting T. LI, reproduced in full size;*in the text designated with N. T. and Roman numeral.

---

[1] KB. = original in the Royal Library at Stockholm. N = original in the collection of the author.
*[See Publisher's Note.]

Maps and figures in the text.

Generally in reduced size. In the text designated by N. fig. and Arabic numeral.

ix

GLOSSARY OF CITIES OF PUBLICATION
THAT OCCUR IN THE LIST OF MAPS

(Latin forms appear in the locative case)

Antverpiae: *Latin form of* Antwerp
Argentinae, Argentorati: *Latin forms of* Strasbourg
Basileae: *Latin form of* Basel
Bononiae: *Latin form of* Bologna
Brixiae: *Latin form of* Brescia
Cracouiae: *Latin form of* Cracow
Firenze: *Italian form of* Florence
Friburgi: *Latin form of* Freiburg (im Breisgau)
Hagae-Comitis: *Latin form of* The Hague
Hanoviae: *Latin form of* Hanover
Ingolstadii: *Latin form of* Ingolstadt
Lipsiae: *Latin form of* Leipzig
Lovanii: *Latin form of* Louvain
Mediolani: *Latin form of* Milan
Norimbergae, Norinbergae, Nurembergae, Nurenbergae: *Latin forms
    of* Nuremberg
Parisiis: *Latin form of* Paris
Romae: *Latin form of* Rome
Sthlm: *abbreviation of* Stockholm
Tiguri: *Latin form of* Zurich
Ulmae: *Latin form of* Ulm
Venetia: *Italian form (old spelling) of* Venice
Venetiis: *Latin form of* Venice
Viennae: *Latin form of* Vienna
Vinegia: *Venetian form of* Venice

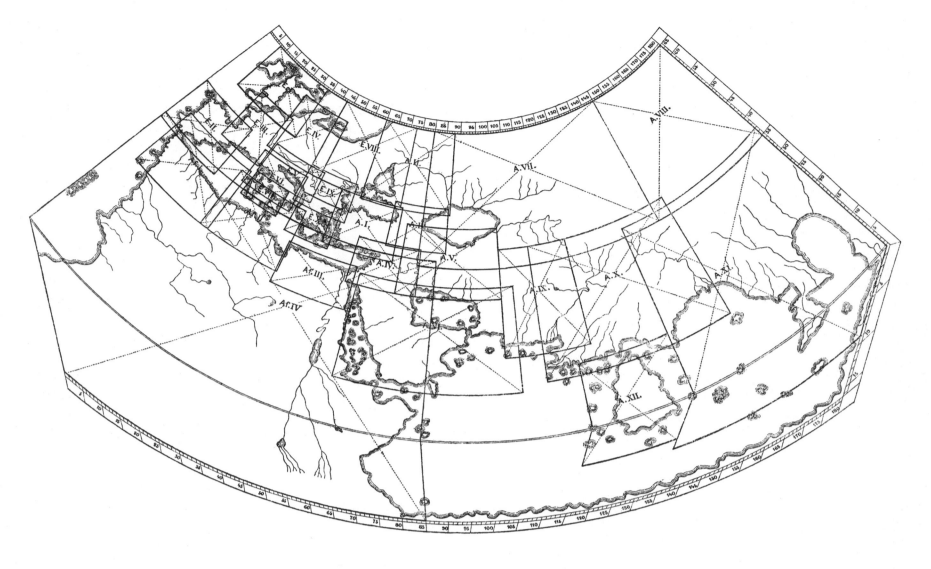

I.

The geographical atlas of Ptolemy.

The maps, connected with the oldest editions of the geography or cosmography of CLAUDIUS PTOLEMÆUS, constitute the prototype of almost all geographical atlases, published since the discovery of the art of printing. This is due not only to the circumstance, that the rules and directions, given by Ptolemy for drawing geographical maps, are still practised, in mapping continents and oceans, to which the surveyors' triangulation has not yet been extended; but also because the method of denoting boundaries between lands and seas, mountains, rivers, and towns, used in old manuscripts of Ptolemy's work and especially in its oldest printed editions, have up to this very day, with but slight variations, been followed by cartographers. They also almost always use the Ptolemaic orientation (north above, east to the right), the graduation of Ptolemy, and also very often some of his projections. The principles of geography may be said still to be published with Ptolemy's alphabet. If we compare our atlases on the one hand with the maps of Ptolemy, and on the other with maps not influenced by the work of the Alexandrian geographer — for instance with *Tabula Peutingeriana*, with Arabic maps, with the maps in *Rudimentum Novitiorum* (Lübeck 1475), of which fac-similes will be given further on, with a Japanese, or a Chinese map, or with map-sketches of savages —, this alphabet will be found to be more conventional, than may have been imagined. Symbolically speaking, we have in the former instance to do with a well known, though perhaps somewhat antiquated writing, in the latter with types, or letters, as foreign to most people, as an oriental alphabet. Considering, further, that the only, or almost the only atlases, or collections of geographical maps, from the year (1472 or 1478) when the first edition illustrated with maps of Ptolemy's geography was printed, down to

1570, when the first edition of the *Theatrum Orbis terrarum* by Ortelius was published, consisted of editions of Ptolemy, generally augmented by some *tabulæ novæ*, drawn in his manner and provided with addenda, in which the latest discoveries had been cautiously noted, it seems proper, that the present work should commence with copies of Ptolemy's oldest printed maps, as well as with a brief exposition of their origin and influence on the development of cartography.

As regards the biography of Ptolemy, scarcely anything is known of his life. We do not know either the place, or year of his birth, or the year of his death. In some latin translations of his works he is called CLAUDIUS PTOLEMAEUS PELUSIENSIS or PHELUSIENSIS, from which it has been deduced, that he may have been born in one or other of the cities bearing the name of Pelusium. There are two places of that name known from antiquity, — one, a considerable city on the Delta of the Nile and the other a seaport in Thessaly. Later criticism however has shown, that the epithet »Phelusiensis» probably arose from a faulty latinizing of a faulty way of writing the Arabic version of the name *Claudius*. The last observations in his great astronomical work, μαθηματικὴ σύνταξις, which is generally known by the hybrid, Græco-Arabic name of *Almagest*, are for the year A. D. 141, which shows, that he lived during the first part of the second century of our era.

At the end of the 2:d book of the Almagest Ptolemy promises a work, which was to give the longitude from *Alexandria*, and the distance from the equator, of the most important places on the earth. Probably this promise was redeemed by the geographical work now under discussion, in which case that work was finished later than the year A. D. 141. At the end of Bk. V, Ch. 12 of the Almagest Ptolemy says that he made his observations on the parallel of Alexandria. On the other hand it is stated in a fragment from OLYMPIODORUS, that Ptolemy resided forty years and made his observations in »what was called the Pteron», which was probably intended to indicate the side-building of the temple in Canopus, a city situated 19 kilometers to the N. E. of Alexandria. The difference of latitude between these two places, being, according to Ptolemy's Geography, only $\frac{1}{12}°$, the expression »parallel of Alexandria» seems more applicable to Canopus, as the word »parallel» formerly had a signification different from that accepted in modern geographical literature. The ancients understood by this word principally the circles parallel to the equator, which were supposed to separate the *climates*. Still Letronne is of opinion, that the assertion of Olympiodorus depends on a confounding of the temple of Serapis in Canopus with the Serapeum in Alexandria, and that Ptolemy's observations were made in the last mentioned place.[1]

Accordingly the only events of Ptolemy's life which are known with certainty are that he lived during the first part of the second century of our era, and that he, during the greatest part of his life, resided at *Alexandria*, or in its vicinity. This is all, but slight as it is, it will be found to be of no small importance in the biography of a geographer. For Alexandria was at that time not only the richest city in the world with regard to learned institutions and treasures of scholarship, but also the wealthiest commercial place on the earth, a place where seafaring people and caravans from all parts of the then known world used to meet, and where, in consequence, better opportunities were offered for collecting knowledge respecting distant lands and seas than anywhere else.

The particulars that have been added by other biographers, for instance regarding his travels, his death A. D. 147, or 165, at an age of 78 years, and his social position appear to proceed from traditions without evidence, or from evident fables. Even his name is but seldom and only incidentally met with in the writings of his contemporaries or immediate successors.[2]

We should consequently have scarcely any notion of the most remarkable geographer of antiquity, had not some of his works been saved. The most important of these are his above mentioned manual of spherical and theoretical astronomy (Almagest) and his extensive geographical work. In Greek this last work bears the title of γεωγραφικὴ ὑφήγησις, which nearly corresponds to »Geographical guide», but latin translators have sometimes abbreviated the name to *Geographia* and sometimes, though quite incorrectly, amplified it to *Cosmographia*. According to the mode of naming such works in our time, its most appropriate title would have been *Atlas of the world*.

This atlas consists of 27 maps with an extensive text, evidently written hand in hand with the drawing of the maps. In order to understand and duly appreciate the only cartographical work which escaped from the general destruction of ancient literature, a short analysis of this text will be necessary.

It is divided into eight books.

In the first chapter of the first book Ptolemy explains the difference between *geography* and *chorography*. Geography has for its object to describe the habitable, known world, or at least its more important parts; chorography depicts particular localities, such as harbours, country-seats, villages, or rivers. Geography occupies itself with that which

is great and important, chorography with less important geographical details. Geography may be symbolized by the image of a head, chorography by that of an ear or an eye.

This chapter is only of interest to us, inasmuch as it seems, in preference to other parts of Ptolemy's geography, to have captivated the geographers of the sixteenth century, such as STOBNICZA, GLAREANUS, APIANUS, and others, who generally begin their works with a more or less complete recital of these definitions and distinctions, illustrated by naïve drawings, intended further to explain the meaning of the Alexandrian Geographer.

In the next chapter necessary directions are given how to collect materials for geographical maps, and how to make use of the material collected. In the first place Ptolemy says, it is advisable to procure access to the journals of intelligent travellers, who have visited distant countries. Their observations may be either of a geometrical character — i. e. only giving distances between different places —, or they may be founded on observations made of celestial bodies by means of instruments for the measurement of the altitude of stars, and the length of the shadow of the gnomon. *It is only with the aid of such instruments, that the bearings and distances between different places can be determined.*

Moreover, the distances, given by travellers, only become serviceable, when they have been corrected with reference to the unavoidable circuits, the difficulties met with, and the changes in direction and strength of the winds, during the voyages at sea. Finally it is essential, before inserting in the maps, thus reasonably reduced, the distances obtained from travellers, duly to consider that the surface of earth and sea is spherical. Even when the terrestrial distances are exactly known, the data are still insufficient for calculating the latitudes and the proportions between the traversed distances and the circumference of the earth. This, as well as the length of the circumference of the earth and the distances between the meridians at different latitudes, can only be ascertained by astronomical observations, the nature of which is explained in the L. I. C. III.

In the fourth and fifth chapters of the first book Ptolemy says, that a perfectly reliable map of the inhabited world would be obtained, if travellers made the observations mentioned in the foregoing chapter. But only HIPPARCHUS has determined the latitudes of a few places in the northern hemisphere, and mentioned the distances from the pole, at which some other places are situated.

Others had, with the help of voyages at sea, made with the wind either from the south or from the north, enumerated some places on the other side of the equator as situated on the same meridian. The distances east and west, depend, for want of actual observations, mainly on traditional suppositions. As for the determination of the difference of time between two places, it had only been possible to observe a few eclipses of the moon, simultaneously at different places of the earth, as was the case with the eclipse, seen in Arbela at the 5:th and in Carthago at the 2:d hour. Anyone, who wishes to draw a geographical map, should found his work on trustworthy observations and then insert the remaining less reliable materials with as much accuracy as possible. As changes and variations often occur, the latest itineraries should

---

[1] Compare: G. M. RAIDELIUS, *Commentatio critico-literaria de Claudii Ptolemaei Geographia ejusque codicibus tam manuscriptis quam typis expressis.* Norimbergae 1737;

K. MANNERT, *Geographie der Griechen und Römer aus ihren Schriften dargestellt.* T. I. Nürnberg 1799, p. 135;

A. J. LETRONNE, *Composition mathématique de Claude Ptolémée;* and *Examen critique des prolégomènes de la géographie de Ptolémée* in: *Oeuvres choisies,* 2:e Sér. I. Paris 1883. P. 95 and 127 (First published 1818 and 1831);

F. A. UKERT, *Ueber Marinus Tyrius und Ptolemaeus, die Geografen,* in *Rheinisches Museum für Philologie,* VI. Bonn 1839, p. 173;

A. FORBIGER, *Handbuch der alten Geographie,* I. Leipzig 1842. P. 402.

[2] Such places in the ancient authors, where Ptolemy is mentioned, are to be found collected in the above cited works of UKERT, LETRONNE, etc., as well as in NOBBE, *Claudii Ptolemaei Geographia.* Ed. stereotypa. Lipsiae 1881. I. p. XX.

principally be relied upon, but with due criticism and selection (Ch. V).

Chapters VI—XX are devoted by Ptolemy, to a detailed, but amicable criticism of the works of his nearest predecessor, MARINUS OF TYRE, and to an explanation of the desirability of a new geographical atlas. Marinus receives the credit of having devoted himself with the utmost care to his work. He had extended his investigations far beyond what was known before. For he had made an assiduous study of all itineraries within his reach, and thus corrected not only the

however, induced Ptolemy to reproduce the atlas of Marinus, with a view to making it more systematic and useful.

Then Ptolemy begins to examine those itineraries, the accounts of which had caused Marinus to give to the known world too great an extent in length and breadth. ›It is with good reason we name the extent from east to west, length, and that from north to south, breadth. The same nomenclature is used to mark the positions of heavenly bodies. The greatest extent is generally called length, and we all agree that the extent from east to west of the inhabited world is greater, than that

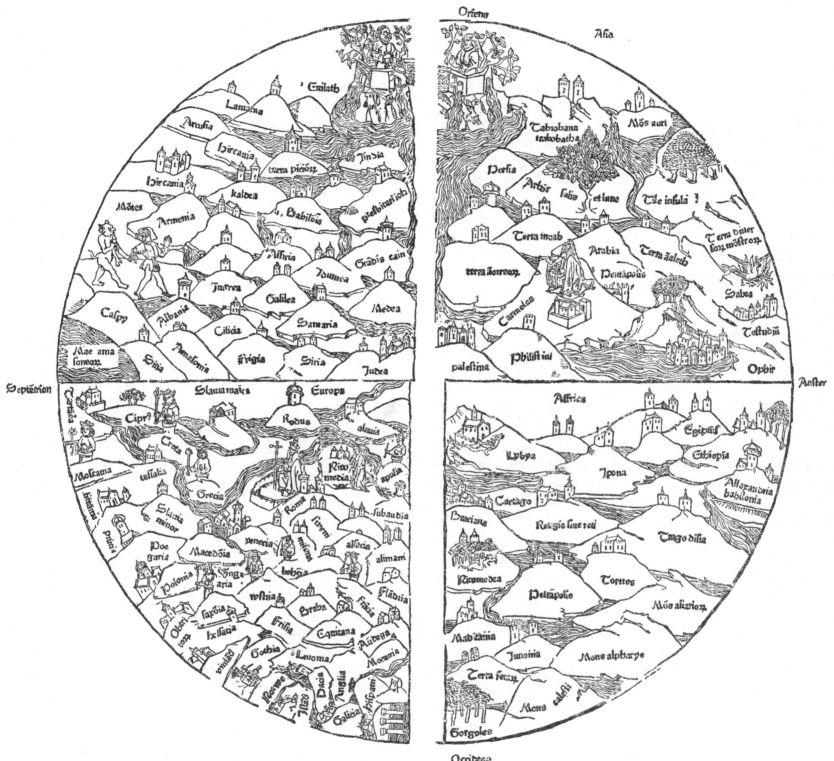

2. Map of the world from *Rudimentum Novitiorum*, Lübeck 1475. (Diam. of circle in original. = 382 m. m.).

mistakes committed by his predecessors, but also the errors in his own earlier works. This was easily perceived in comparing the different editions of his geographical tables. [1]

If the last work of Marinus had been faultless, there would have been no difficulty in producing a map of the inhabited world by its guidance. But Marinus had fallen into error partly from having adopted several itineraries which were not altogether trustworthy, and partly owing to the method of drawing, adopted by him, not having been well selected. Of course these errors were made in good faith. They,

from north to south. As for the breadth, Marinus places the northern limits of the inhabited world at the island of Thule, in lat. N. 63°, or at a distance from the equator of 31,500 stadia. The southern limits are formed by that part of Aethiopia, which is named *Agysimba* and *Cape Prasum* and placed by Marinus at the southern tropic. Hence the breadth of the whole inhabited world would be 87° or 43,500 stadia‹.

For the northern limit, or for the position of Thule, Marinus gives, according to Ptolemy, sufficient reasons. On the other hand Ptolemy is of opinion, that Marinus has

---

[1] In Greek Διόρθωσις τοῦ γεωγραφικοῦ πίνακος. Critics seem not fully to agree as to the explanation of these words. LETRONNE has »*Correction de la table géographique*». Marinus appears to have composed several different editions of geographical maps with text, or rather of sea-charts with sailing directions. Of the last edition the text had been finished, but not the corresponding pictures.

placed the southern limits too far from the equator. He bases this opinion on three different reasons, namely on astronomical observations, on journeys from northern Africa towards the south, and on voyages at sea along the east-coast. During a military expedition in Africa, Septimius Flaccus spent three months on a march from the Garamantes to the Aethiopes. On a similar expedition with the king of the Garamantes, Julius Maternus spent four months on the way from Garame to Agysimba, that is to the land where the rhinoceros is found. Then there was an Indian traveller, Diogenes, who, during a voyage at sea, was drifted by a norther along the coast of the Troglodytes (the eastern coast of Africa); and, after 25 days he arrived at Rapta, a point of land, projecting from the territory where the swampy sources of the Nile are situated. Finally, a certain Theophilus, during a voyage from Azania, ran under sail, before a southerly wind in 20 days from Rapta to Aromata.

These reasons are carefully examined by Ptolemy, who shows that the astronomical observations had been incomplete or wrongly interpreted, and that the distances traversed during travels by land, or by sea, had been overestimated, due regard not having been given for circuits, days of rest, and baffling winds under the equator. If the data given by Marinus were accepted without further examination, these countries, where Aethiopes (negroes) and rhinoceri dwell, would have to be removed as far as to the cold zone of the southern hemisphere. This would be an absurdity, for animals and plants should resemble each other in countries of the same temperature and the same atmospherical conditions, or in other words, under parallels with the same polar-distance. The truth of this objection was also perceived by Marinus himself, who, in consequence, arbitrarily removed the southern limits of the »antoikumenæ» to the southern tropic. But Ptolemy is not satisfied even with this reduction, for black people and rhinoceri, or elephants, are not met with in our hemisphere at the northern tropic; it is only farther south, that the human complexion darkens. Ptolemy therefore placed Agysimba and Cape Prasum at the same distance to the south of the equator, as Meroe is to the north, or on $16° \frac{1}{3} \frac{1}{12}$ (= 16° 25′). It is remarkable that Ptolemy, although his map in many places is said to embrace the whole inhabited part of the globe, nowhere produces any facts to prove that Africa is uninhabited beyond Agysimba. On the contrary, one might, as a corollary to his reflections on the influence of the climates, arrive at the conclusion, that the zone between the southern tropic and the south pole were inhabited by races of white people and by European species of animals.

This part of Ptolemy's criticism of the geographical work of Marinus is of special interest, because it shows that the mapping of distant countries by the former geographer rests on actual accounts from travellers, and is not, as some people have imagined, made up of fancy-sketches. Here we find, that the marshes whence the Nile springs, were very well known at the time of Ptolemy; and that voyages had been made from the Red sea, along the eastern coast of Africa, towards the south.

In chapters XI—XIV Ptolemy discusses in detail the length, or extent from east to west, of 15 hours or 225°, that Marinus had assumed for the inhabited part of the earth. He divides the space between the eastern and western limits of land into two sections: the first from the *Insulæ Fortunatæ* to the crossing of the Euphrates at *Hieropolis*, and the second one from thence to *Sera*, *Sinæ* and *Cattigara*.

Ptolemy approves of the length, given by Marinus for the first section, along the parallel of Rhodes, as based upon long experience. He says that Marinus, in forming his estimate of the distance, appears to have taken into proper consideration the reductions rendered necessary by circuits and stoppages during the voyages, and that he, assuming the length of the equatorial degree to be 500 stadia, quite properly supposes the length of a degree of the parallel at Rhodes, or in a latitude of 36°, to be 400 stadia.[1]

On the other hand Ptolemy considers the extent of land to the east of Hieropolis, as conjectured by his predecessor, to be a great deal too large, which is to be accounted for partly from the circumstance, that, passing from Hieropolis to Sera, a long circuit northwards had to be made, as far as to the *Stone-tower* (*λιθίνος πύργος*, not a tower or a town, but a mountain, situated, according to Ptolemy's maps, near to the western part of Altai) and partly from intentional exaggerations in the accounts of merchants. One merchant, a Macedonian, had communicated some notes on such a journey. Yet he had not made the journey himself, but sent one of his clerks there, and the only remarkable circumstance in his account was the extreme length of the journey, which lasted seven months.

Finally Ptolemy (Ch. XII) adopts the following differences of longitudes:

	Difference of longitude according to: Ptolemy.	modern maps.
Insulæ Fortunatæ to Sacrum Promontorium (Cape St. Vincent)	$2\frac{1}{2}°$	9°
Sacrum Prom. to the mouth of Baetis (Guadalquivir)	$2\frac{1}{2}°$	$2\frac{2}{3}°$
Mouth of Baetis to Calpe (Gibraltar)	$2\frac{1}{2}°$	1°
Calpe to Caralis (Cagliari) on Sardinia	25°	$14\frac{1}{2}°$
Caralis to Lilybæum on Sicily (Marsala)	$4\frac{1}{2}°$	$3\frac{1}{3}°$
Lilybæum to Pachynum Prom. (C. Passaro on Sicily)	3°	$2\frac{2}{3}°$
Pachynum Prom. to Taenarum Prom. (C. Matapan)	10°	7° 23′
Taenarum to Rhodes	$8\frac{1}{4}°$	$5\frac{1}{2}°$
Rhodes to Issus	$11\frac{1}{4}°$	8°
Issus to Euphrates	$2\frac{1}{2}°$	$1\frac{2}{3}°$
Euphrates to Turris lapidea	60°	$78\frac{1}{3}°$?
Turris lapidea to Sera	$45\frac{1}{4}°$	
Sum	$177\frac{1}{4}°$	

Out of these data, the first ten — the distances between Insulæ Fortunatæ and the Euphrates — agree with the distances assumed by Marinus. As for the considerable reduction Ptolemy adopts for the two last distances, he gives further reasons in a critical examination of the accounts from navigators, communicated by Marinus.

It is not known for certain what modern town in eastern Asia corresponds with Ptolemy's *Sera*. The latitude of 38° 35′, stated by him, possibly indicates the present Peking (lat. 40° 36′).[2] Whilst Ptolemy in this case assumes 62°, instead of $41\frac{1}{2}°$, for the well known distance between Calpe and Issus, he assumes $107\frac{3}{4}°$, instead of 80°, for the distance between Issus and Sera. In the former instance, he differs from the real distance by 50 %, in the latter, by 35 %. This considerable error in the maps of the part of the globe which was best known to Greeks and Romans, arose from the circumstance, that the sailors reported their distances in stadia, and that Ptolemy assumed only 500 instead of 700 stadia for each degree of longitude, at the equator.

In the chapters XV—XVIII the work of Marinus is examined from other points of view. There are incorrect and contradictory statements respecting the positions of many places. The limits of land and the directions of the coast-lines are often erroneously given. According to the accounts of merchants, who had travelled from Arabia to Aromata, the marshes, which form the sources of the Nile, are situated

---

[1] The 36:th parallel actually crosses Rhodes. The length of the degree of latitude being assumed = 500 stadia, one degree of the parallel of Rhodes is = 404,5 stadia. The difference (4,5 stadia) is obviously of no material influence on maps so rude as those of the old Alexandrian Geographer.

[2] F. von Richthofen identifies Sera with Hsi-ngan-fu in lat. 30° 6′ and long. 109° 22′. The diff. of lat., $8\frac{1}{2}°$, however seems to contradict this assumption.

— 5 —

rather far in the interior, and not on the coast: at Promontorium Rapta a river of the same name, on which a town is situated, disembogues; the eastern coast of Africa here makes a considerable bend to the west. Maps were entirely wanting in the last edition of the works of Marinus, but to make them with the assistance of his tables, would have been impossible on account of their contradictory and confused arrangement. In one place the latitudes are given, in another the longitudes. For several localities Marinus only supplies one of these important data, and for others, especially for those in the interior of the continents, neither of them. These circumstances induced Ptolemy to compose a geographical work, in which the statements of Marinus, which were not in need of correction, were to be preserved, inexact data

correspond as nearly as possible with reality. Understanding this, Marinus had criticized all the projections of maps on a plane, but nevertheless he employed exactly that method of projection, which most distorted the proportions. He drew the parallels and meridians as straight lines, forming right angles with each other, maintaining all over the map the same relation between the degrees of latitude and longitude as on the parallel of Rhodes. He consequently disregarded the attainment of proper proportions in the other parts of the map, as well as the spherical aspect of the whole. The distances between the meridians to the north of Rhodes becoming too large and those to the south too small, considerable errors arose.

To avoid, or at least to lessen these errors, Ptolemy proposed to employ, what is now called a *conical* projection,

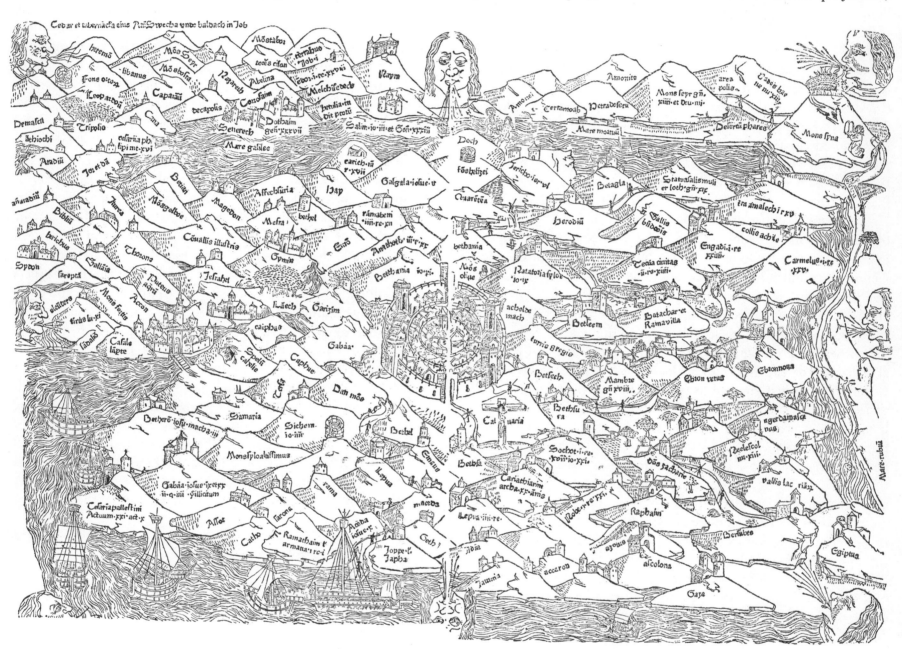

3. Map of Palestina from *Rudimentum Novitiorum*, Lübeck 1475. (Orig. size 580 × 400 m. m.).

corrected, and that which had been unknown to Marinus, added by the aid of the best maps and the most reliable itineraries. He also promised a more practical and convenient arrangement of the work.

In the remaining chapters (XX—XXIV) of Book I Ptolemy treats of the construction of the maps. The easiest and most appropriate method would have been to have delineated the map on a sphere, for which purpose directions are given in Ch. XXII. But it would have been difficult to give to a sphere a size, sufficient for the insertion of the names of all the more important places. Nor would it have been possible to get such a view, as would comprehend the whole, if the map were drawn on a sphere. These difficulties would be avoided, if the map was projected on a plane surface. To do this, however, some special method was required in order to preserve the resemblance to the sphere, and to make the distances on the plane

i. e. to project the map, with equidistant parallels, on a conical surface developed round the axis of the earth, and passing through the parallels of Rhodes and Thule. When such a conical surface is extended on a plane, a network with circular parallels and rectilinear, converging, meridians arises. Lest the proportions of certain parts of the mapped territory should be too much deformed, only the northern or the southern hemispheres should be laid down on the *same map* by this projection, which is consequently inconvenient for maps, embracing the whole earth. Ptolemy only rigorously applies the conical projection to the northern part of his map of the world (N. T. I). To represent the known parts of the southern hemisphere on the same sheet, he describes an arc of a circle parallel to the equator, and at the same distance to the south of it, as Meroe is to the north, and then divides this arc in parts of the same number and size, as on the parallel of Meroe. The network is then obtained by joining the intersections to corresponding points on the equator.

At the end of the last chapter of the first book Ptolemy describes a still more correct method of projecting the map of the earth on a plane. Half the equator is here represented by the arc of a circle, the chord of which is to its height as 180: 23 $\frac{4}{5}$ (23,833), i. e. as the length of the half of the equator (180°) is to the latitude of this point, Syene, where the mean-meridian of the known or inhabited world was supposed to cross the mean-parallel. The parallels are then laid out as equidistant arcs of circles, concentric to the equator, and placed at such distances, that the middle of the chord of the equatorial arc obtains a latitude of 23° 50'. The mean-meridian will then form a straight line, and the other meridional curves are drawn in such a manner, that a proper relation will be maintained for the parts of parallels between neighbouring meridians and the scale of latitudes. Ptolemy himself never made use of this projection, but it is used for the general map in several printed editions of Ptolemy, and after the discovery of the new world it was sometimes employed for maps embracing the whole world on a single sheet, as, for instance, by Bernardus Sylvanus 1511 (N. T. XXXIII) and Apianus 1520 (N. T. XXXVIII). Stobnicza also employed this projection in his correctly constructed, but badly executed maps of the old and new hemispheres (N. T. XXXIV).

Ptolemy's exhaustive criticism of the imperfect methods of drawing maps, adopted by Marinus, would lead to the expectation that he himself would have used some of the projections recommended by him as more correct. But such was not the case. For his first general map Ptolemy certainly employs his conical projection, but for the remaining 26 special maps he uses the rectangular projection of Marinus with due observance of the ratio between the longitude and latitude at the base of the map. This inconsistency of Ptolemy seems to have astonished his publishers in the 15:th and 16:th centuries. With one exception only (Berlinghieri's Italian translation in terza-rima, of which a description will be given further on), every editor of Ptolemy's geography has published, not the original maps, but a modification of them by Nicolaus Germanus (Donis), who, with praiseworthy exactness and without any further alterations, reproduced the originals on a projection with rectilinear, equidistant parallels and meridians converging towards the pole. For want of a better name, this projection will henceforth be called *Donis' projection*. In one of the oldest, or perhaps the oldest printed edition of Ptolemy's maps — the one which bears the incorrect date of »Bononiæ 1462» — these are reproduced on a *conical projection*, a circumstance which hitherto seems to have escaped his numerous commentators.

———

In connection with his description of different map-projections, Ptolemy gives in B. I, Ch. XXIII, an exposition of the division of the surface of the earth into *climates* and *parallel zones*. The importance of this division to cartography, induces me here to devote a few lines to the subject.

The latitude, or the distance from the equator, was generally calculated from the length of the longest and the shortest day. The earth was accordingly divided into a number of zones, parallel to the equator and within which these days had a certain length, for instance of 12 to 13, or 15 to 16 hours. These zones were termed *climates*, from the Greek word κλίμα = inclination. Different from what is now usual, this word in the old maps had a purely geographical, not a meteorological signification, though the ancient geographers early perceived that, what is now termed the climate of a place, depended to a considerable extent on its distance from the equator, i. e. on the »klima» in the old sense of that word. The lines that separated the climates were termed *parallels*.

This division into climates, however, suffered from the defect of not permitting a rigorous application without causing a considerable difference in the breadth of the zones, into which the surface of the earth was divided. That zone, for instance, within which the longest day is from 14 to 15 hours, had a breadth of 10° 32' or 632 miles, while the zone with the longest day from 19 to 20 hours only had a breadth of 2° 53' or 173 miles. It was probably this circumstance, that made the different writers vary to such an extent as to the limits of the zones. Several passages in Strabo, cited by Forbiger (I p. 199) show, that Hipparchus had divided the surface of the earth into eight lines parallel to the equator, the northernmost of which crossed the mouth of the Borysthenes. Following his predecessors, Pliny divided that part of the earth, which was well known to the Greeks and Romans, or the surface of the earth from the southern limit of the known world to 46° Lat. N., into seven »segments», parallel to the equator. For the wilderness yet further north three others were added, the total number of »segments», accepted by Pliny, thus amounting to ten. This number was further augmented by Ptolemy, who divided the northern hemisphere into 21 parallels, noted in the margin of his maps, but not drawn across them.

The following table, the last column of which is calculated by Wilberg (*Cl. Ptolemaei Geographiae libri octo*, Essendiæ 1838, I, p. 70) for an inclination of the ecliptic of 23° 50', gives the positions of the climates and parallels, enumerated by Ptolemy in Lib. I, Cap. 23, as well as the errors committed by him in the evaluation of the latitude, corresponding to a certain length of day during the summer-solstice.

	Parallels.	The longest day.	Corresp. polar alt:de according to Ptolemy.	corrected.	
Klima I	1	12$\frac{1}{4}$	4°$\frac{1}{4}$	4° 14'	These parallels also exist in the southern hemisphere.
	2	12$\frac{1}{2}$	8°$\frac{1}{3}$$\frac{1}{12}$	8° 25'	
	3	12$\frac{3}{4}$	12°$\frac{1}{2}$	12° 31'	
	4 (through Meroe) [1]	13	16°$\frac{1}{3}$$\frac{1}{12}$	16° 28'	
Klima II	5	13$\frac{1}{4}$	20°$\frac{1}{4}$	20° 15'	
	6 (through Syene)	13$\frac{1}{2}$	23°$\frac{1}{2}$$\frac{1}{3}$	23° 50'	
Klima III	7	13$\frac{3}{4}$	27°$\frac{1}{6}$	27° 12'	
	8 (through Alexandria)	14	30°$\frac{1}{3}$	30° 22'	
Klima IV	9	14$\frac{1}{4}$	33°$\frac{1}{3}$	33° 18'	
	10 (through Rhodes)	14$\frac{1}{2}$	36°	36° 2'	
Klima V	11	14$\frac{3}{4}$	38°$\frac{1}{2}$$\frac{1}{12}$	38° 34'	
	12 (through Rome)	15	40°$\frac{1}{4}$$\frac{1}{12}$	40° 54'	
Klima VI	13	15$\frac{1}{4}$	43°$\frac{1}{12}$	43° 3'	
	14 (through Pontus)	15$\frac{1}{2}$	45°	45° 2'	
Klima VII	15 (through the mouth of Borysthenes)	16	48°$\frac{1}{2}$	48° 32'	
	16 (through Montes Riphæi)	16$\frac{1}{4}$	51°$\frac{1}{4}$	51° 31'	
Klima VIII	17	17	54°	54° 2'	
	18	17$\frac{1}{2}$	56°$\frac{1}{4}$	56° 11'	
	19	18	58°	58° 0'	
	20	19	61°	60° 53'	
	21 (through Thule)	20	63°	62° 58'	

———

[1] When a parallel is said to pass through Meroe, Syene, etc. it only signifies, that it passes near to one of these places. The latitudes of these places, stated in the text of Ptolemy's atlas (Lib. II—VII), consequently depart a little from corresponding parallels, viz.

	Ptolemy.	Corresponding data from Philips' Library Atlas.
Meroe	16° 25'	—
Syene	23° 50'	24° 7'
Alexandria	31° 0'	31° 11'
Rome	41° 40'	41° 54'

As Ptolemy himself, in his geography, exclusively employs the system still in use for defining geographical positions by means of latitude and longitude, this division of the surface of the earth into a number of parallel zones seems to have been already obsolete in Ptolemy's time, and to have been continued by him for the same reason that sometimes causes antiquated standards of length to be employed in modern maps. His practice of inserting the ancient climates and parallels in the margins of the maps was, however, adhered to for a long time. It is, for instance, used in the map of CLAUDIUS CLAVUS of 1427; in the map of RUYSCH of 1508 (N. T. XXXII) — with 30 parallels, of which the

rivers. It is an exception when geographical or descriptive remarks are added to this bare enumeration of names.[1]

The fifth chapter of book VII contains a description of the map of the world, together with an enumeration of the oceans and of the more important bays and islands. The Indian ocean, which is assumed to be bordered on the south by an unknown continent, uniting southern Africa with eastern Asia, is stated to be the largest sea surrounded by land. The Atlantic *ocean* is not mentioned among the seas. It is remarkable that such questions never seem to have occurred to the Alexandrian geographer, as: What is there to be found beyond Serica and Sinarum Situs? What, to the north of Thule,

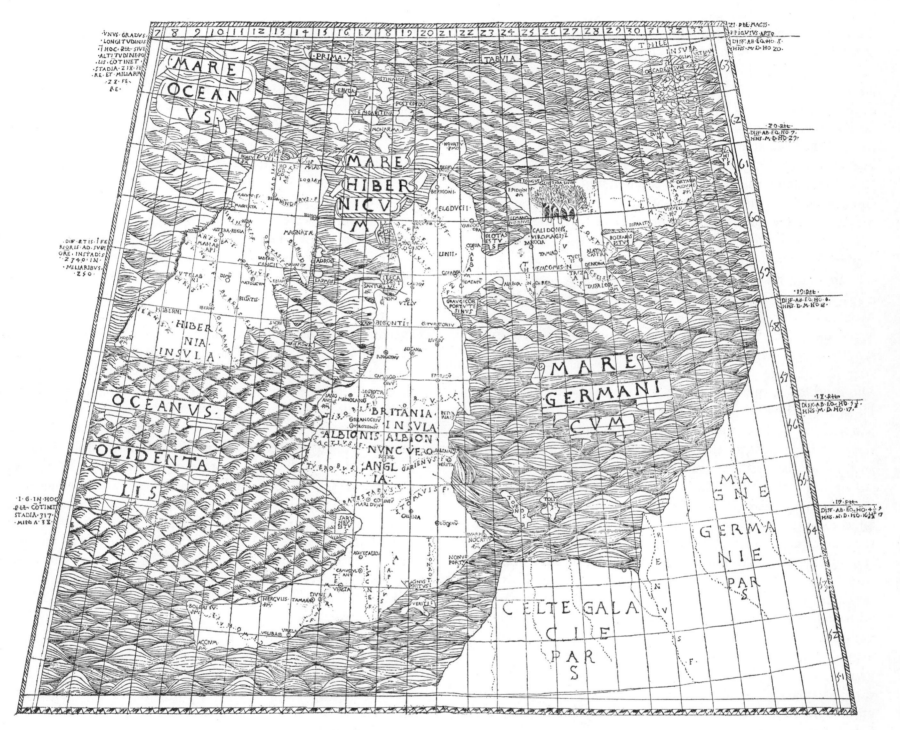

4. Map of Britannia from *Ptolemy*, Bononiæ 1462 (1472). (Orig. size 505 × 405 m. m.).

northernmost one has a summer day of four mouths —; in the map of APIANUS of 1520 (N. T. XXXVIII); in the large map of OLAUS MAGNUS of 1539 (extending to the pole, with 34 parallels and 16 climates); in the Basel-edition of the map of OLAUS MAGNUS of 1567, and in some maps of still later data.

Books II—VI and the first four chapters of book VII are devoted to a complete catalogue of all the places laid down in the 26 special maps of the geography. For every place, mentioned in the catalogue, latitudes and longitudes are given, and also for the more important capes and mouths of

or to the south of Agysimba and Cape Prasum? Where would you come to, if you sailed on, westward, from the Insulæ Fortunatæ? It seems, as if the consequences of the spherical shape of the earth were not clearly seized by Ptolemy himself. Even Strabo (for instance in Lib. II, cap. 4) only touches questions, regarding the inhabitants and physical aspects of countries beyond the Greek or Roman world, incidentally and with the same scornful dismissal, with which a modern savant would probably treat questions about organisms on Jupiter.

In the last two chapters of book VII a description is given of a projection of the inhabited hemisphere on a plane,

[1] MANNERT (cited work p. 137 & 193) calls in question, whether not Ptolemy, beside his γεωγραφικὴ ὑφήγησις, the text of which mainly consists of name-lists, also has written a descriptive geographical work, now lost. As a support for this supposition, he quotes a passage of EUSTATHIUS, in which Ptolemy is cited regarding the shape of the Caspian sea, which passage however is not to be found in the geography.

by which it would retain its circular outline, or globular aspect. Ptolemy himself never employed this manner of projection, which has since, though more or less modified, been preferred by geographers for maps, representing one of the hemispheres. The words, with which Ptolemy concludes this book, are characteristic of the idea of the ancients concerning the extension of the inhabited world. He there says, that the inhabited part of the surface of the earth is nowhere bordered by oceans, excepting at the northwestern parts of Africa and Europe (Comp. N. T. I, or the metallo-type p. 1).

The eighth and last book begins with a short preface, in which further explanations are given as to the manner of drawing the maps. Thus, for instance, the scale may vary, according to the number of names and other details, to be inserted; the parallels and meridians may, without inconvenience (on maps only embracing less extensive regions) form straight, parallel lines (as in the maps of Marinus, which had been so severely censured). Then there follow short legends for each of the special maps — 10 for Europe, 4 for Africa and 12 for Asia — mentioning the countries laid down on each plate, describing the limits and enumerating the tribes of each country and its most important towns. It is these legends which, in some editions, have been placed on the reverse of the maps, and they really appear to have been originally intended for that purpose.

In some manuscript-codices of Ptolemy's geography, the maps are preceded by the remark, that they were drawn by AGATHODÆMON of Alexandria. Hence some commentators have concluded, that Ptolemy's geography was not originally supplied with maps, but that the maps were drawn for the text after it had been finished by the above mentioned draftsman, who is otherwise quite unknown. But, it has been established by UKERT and FORBIGER, that Ptolemy's own text and the whole plan of his work completely refute this suggestion. It is on the contrary evident, that the text, excepting the first book and some part of the seventh, is composed either by the aid of maps already finished, or hand in hand with the map-drawing, and that Ptolemy's original maps had exactly the same extent, were drawn on the same projection and contain the same names, as the copies ascribed to Agathodæmon. It is possible that AGATHODÆMON was the name of some draftsman, who assisted Ptolemy in his map-making, but it is more probable that some copyist, living centuries after Ptolemy, immortalized his name by adding it to his carefully and faithfully drawn copies of the maps of the old master. To identify him only on account of identity of name with the grammarian, AGATHODÆMON, who flourished in the fifth century, is plainly as absurd, as to array Ptolemy in a Royal crown and robe, because his name is the same as that of the Greek dynasty of Egypt.[1]

The metallotype on page 1 gives a diagram of the parts of the earth, embraced by each special map of Ptolemy. It shows, that the known world had been systematically divided into 26 fields, each of which is mapped on a separate sheet. Generally these sheets are of about the same size, but the scales vary according to the space required for the legends. As the diagram shows, each special map embraces, besides its own proper territory, some parts of the neighbouring countries. But, as is also usual in modern atlases, these parts of the map are only roughly sketched: for instance, the regions of north-eastern Gallia, which appear on the map of Britannia, or the part of Britannia, which enters the map of Gallia. Ptolemy's maps are never disfigured by such sketches of animals, mon-

sters, savages, ships, kings etc. as adorn the manuscript-maps of the middle-ages, and many of the printed maps of the 16:th century. This and the manner of denoting boundaries between land and sea, rivers and towns, almost give a modern appearance to Ptolemy's maps.

Ptolemy's 27 maps seem to be the same in all complete manuscripts of his geography. Important as these maps are to the history and geography of antiquity, yet they have not been subjected to any exhaustive critical analysis, with reference to their geographical details, their division into countries and provinces, and the names of nations and towns found on them. It cannot be expected that this defect in literature should be filled up here. The task belongs rather to the historian and the ethnographer, than to the geographer. We only profess to deal with Ptolemy's work, as the prototype of modern cartography and *from this point of view*, it is, in doubtful cases, of little importance by what name any particular town or nation is indicated. Notwithstanding all their defects, those editions of Ptolemy's geography which were published at a time when his work was still the canonical book on cartography among all civilized nations, are of the greatest interest to us. This was the case during almost the whole period of the great geographical discoveries.

Ptolemy's work is the only geographical atlas still extant which has come down to us from the ancients, and it is doubtful, if any other, so complete and so systematic as this, was ever composed during that period. Incidentally a map of the world, or a map used as a wall-ornament, or a map of a country, are mentioned in ancient literature, but, with the exception of what Ptolemy himself says of the work of Marinus, no other collection of maps, such as we at present term an ›Atlas‹, is ever alluded to. It may however be possible that such works existed, for Ptolemy's great geographical work itself is but seldom mentioned during the succeeding centuries. But, if that was the case, such works must have been very little known and used, for it is almost exclusively from Ptolemy, that geographers derived their learning when they, during the succeeding centuries, compiled their descriptions of the world. With the exception of some personal experience, acquired during journeys in Africa and Syria, Ptolemy's work was the sole source of information in the description of the world in chapter 2 of OROSIUS, *De Miseria Mundi* (from the beginning of 5th century), and in the leading chapters of JORDANES, *De origine actibusque Getarum*, Ptolemy is named: *orbis terræ descriptor egregius*. Nor is there any other Roman or Greek collection of maps, mentioned by the Arabic geographers, than that of Ptolemy and Marinus of Tyre, and, among the excellent geographical treatises of the Arabs, there is not a single map or collection of maps extant, which is comparable to that of Ptolemy.

During the darkness of the middle-ages even Ptolemy and his method of map-drawing were forgotten, at least in the west. Instead of his clear and intelligible maps, drawn in proper proportion and based on astronomical observations, maps were produced, drawn without a vestige of proportion, covered with figures of princes in mantle and crown, of monsters and fantastic legends, borrowed from the mythic world of christian and heathen legends. The only exceptions are some charts, made in the beginning of the fourteenth century in Italy and at the Balearic Islands, exclusively for the use of mariners and ship-owners, and from materials collected by them. These ›Portolanos‹ or ›loxodromic charts‹, to which I will return in a future chapter, often represented the coasts better than the old maps of Ptolemy. But gene-

---

[1] In his above mentioned work RAIDEL gives a copy of the title-page of a splendid Greek manuscript of Ptolemy, belonging to the Bibliotheca Marciana in Venice. Ptolemy is here seen standing arrayed in royal crown and mantle, in the open court of a splendid palace.

rally they represented only the coasts of the Mediterranean and Black seas, with the western and north-western parts of Europe. They have in our days been studied with deep interest and have become the subject of an extensive literature, but from the 14th to the 18th century they were but little cared for by learned geographers, and scarcely acknowledged by them as any real contribution to geographical literature. When Ptolemy's geography, with its systematically and clearly drawn maps of the *whole world* then known, was divulged in the west by means of manuscripts, imported at the beginning of the 15:th century from the expiring Byzantine empire, it had the effect of an important discovery, which seized on men's minds, at first with even more force than the

rediscovery of the New World by Columbus. Not a new world, but the very world in which one was living, had been extricated from the darkness, in which it had been hidden during a whole millennium. Characteristic in this respect is the circumstance that, while Ptolemy's geography, even before the end of the 15th century, was published in not less than seven bulky folio-editions, expensively illustrated, and liberally provided with maps, the works printed during the same period on the voyages of the Portuguese along the coasts of Africa and on the discovery of the New World by the Spaniards, only formed insignificant and scantily illustrated pamphlets, whose combined contents would easily find room in a few folio-pages of the editions of Ptolemy.

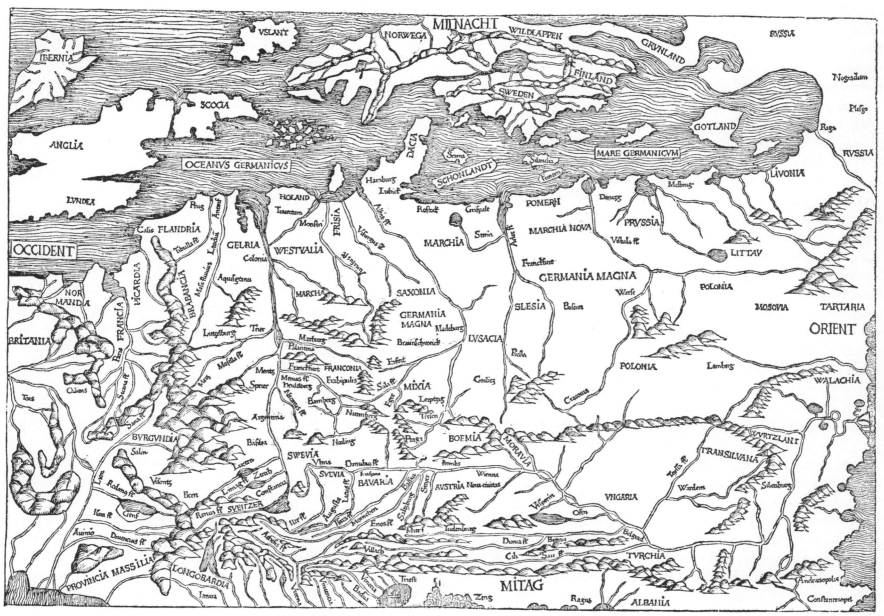

5. Germania from SCHEDEL, *Liber cronicarum*, Norembergæ 1493. (Orig. size 595 × 397 m. m.).

II.

Editions of Ptolemy's geography.

At the commencement of the fifteenth century the knowledge of the Greek language was very slight, even among learned men in the west. The immense influence that Ptolemy's great geographical work exercised, thus commenced at the time, when its spread had been facilitated by translation into Latin.[1] Such a translation had already been begun by EMANUEL CHRYSOLORAS, a Byzantine scholar whose

merit in promoting the spread of Greek literature in the west was very great. He had been employed by his sovereign as an ambassador to the other European courts during the last period of the Oriental Empire, and he then settled in Italy as a teacher of Greek. He died in 1415 at Constance, at the age of sixty. The translation was finished some years earlier by one of his pupils, JACOBUS ANGELUS, under whose name the

[1] It is stated that Ptolemy's geography already had been translated into Latin at the beginning of the 6th century by the celebrated philosopher and statesman BOËTIUS. The letter by CASSIODORUS (*Variorum liber* I: XLV), on which this statement is founded, contains only: »*Translationibus tuis* (sc. Boëtii) *Pythagoras Musicus, Ptolemaeus astronomus leguntur Itali*», and is consequently rather to be referred to Ptolemy's astronomical, than to his geographical works.

oldest latin manuscripts of the Latin version of Ptolemy's geography are usually known. In this translation, on which all the editions from the 15th century, seem to be mainly based, the word *Cosmographia* is used for the title, quite arbitrarily; instead of *Geographia*.

Nothing appears to be known with certainty respecting the life of Jacobus Angelus, judging at least from the contradictory[1] and trivial notices of him, which RAIDEL succeeded in collecting. That learned scholar possessed a special and intimate knowledge of the literature connected with Ptolemy's translators and publishers. The date (1410), when the translation was finished, can only be determined with exactness from the circumstance, that ALEXANDER V, to whom it is dedicated, was pope from 1409 to 1410. The translation was soon spread by means of numerous copies, of which several are yet extant in the public libraries of Europe. Such a codex, in the from of a small quarto, provided with maps, executed for cardinal FILIASTRUS and finished 1427, is preserved in the town-library at Nancy. It is bibliographically described by Mr. BLAU in *Mémoires de la Société Royale des sciences de Nancy*, 1835. In the same memoir (p. 95) Mr. Blau also mentions another manuscript in quarto of the translation of Angelus without maps, which was presented in 1417 by the same cardinal to the library at Reims. That the cardinal PETRUS ALLIACUS, who died in about 1420, also had access to a copy dating from the early part of the fifteenth century, is evident from the extensive and partly verbal extract from Ptolemy's »Cosmographia», which he gives in his work »Compendium Cosmographicum». In a codex, preserved in the Library of the Imperial Staff at St. Petersburg, and to which I (p. 12) shall refer again, the usual dedication of Jacobus Angelus to Alexander V begins with an A, surrounded by a miniature, representing Jacobus Angelus on his knee, delivering his manuscript to the Pope seated between two cardinals. This is evidently a reproduction of the original copy presented to the pope, *and as this codex is provided with maps*, it is a proof, that Jacobus Angelus not only translated the text, but also latinized the maps.

The critics have held various opinions respecting the merits of the work of JACOBUS ANGELUS. Some praise his learning, and his knowledge of Greek, for the study of which he had visited Constantinople and profited by the teaching of Chrysoloras. But CONRAD GESNER finds fault with his translation, and PIRCKHEIMER says of him in the preface to the edition of Ptolemy, printed in Strassburg 1525: »*Licet græca aliquantulum calluisse videri possit, disciplinas tamen mathematicas ita ignoravit, ut plerumque nec semetipsum intellexerit.*»

Dominus NICOLAUS GERMANUS, another scholar of the fifteenth century, generally cited under the name of NICOLAUS DONIS, also deserves great credit for the share he had in spreading a knowledge of Ptolemy's geography. His merit in this respect has however been misunderstood. It is often asserted that Jacobus Angelus only translated the text without supplying it with any maps provided with latin legends, and that the merit of having first translated the maps is due to Donis. This is not correct. For, as above mentioned, the latin manu-

script of Ptolemy of 1427, which is preserved at Nancy, is provided with handsome Latin maps,[2] and before the year 1480 there were already two translations of Ptolemy's geography published in Italy, and supplied with Latin maps, entirely different from the maps of Donis. These are: 1st a latin edition printed in Bologna, with the inexact date of 1462, giving 26 maps *on a conical projection and not on the characteristic projection of Donis;* and 2d BERLINGHIERIS' rhyming Italian translation with maps, all of which, excepting that of the world, are drawn on the original, equidistant, cylindrical projection. The real merit of Donis consists in the fact that he, before any printed edition of Ptolemy's geography was published, I suppose with the manual aid of several other scribes, edited a manuscript edition, based on the translation of Jacobus Angelus, and that he provided it with splendidly illuminated maps, generally drawn on an improved projection, but in other respects strictly following the Greek originals. To the old maps of the Alexandrian geographer he also added some new ones of the greatest importance to the history of geography. He dedicated his work to PAUL II, who wore the tiara from 1464 to 1471, and the initial letter of the dedication forms, as in the manuscripts of Jacobus Angelus, a richly ornamented miniature, representing Donis on his knees, and presenting his book to the Pope seated on a throne in full canonicals.[3] This initial letter is reproduced in the editions printed at Ulm in 1482 and 1486, from which it has been inferred, that Donis himself was the publisher of these editions and that they were *the first editions of his translation*. This seems however not to be the case. At least the maps, already engraved on copper in 1478, and employed for the editions of 1478, 1490, 1507 and 1508, evidently have, as is shown by their characteristic projection, Donis for their author and translator.[4] It is this collection of maps which, as to method of drawing, constitutes the real prototype of all subsequent atlases.

The only facts known with certainty respecting the life of Donis are, that he was a Benedictine from Reichenbach, and that he lived during the time of the pope Paul II.[5] MURR says (*Dipl. Geschichte des Ritters Behaims*, Gotha 1801, p. 13), but without giving his authority, that he died in 1471. The similarity of names has since caused him to be confounded with the celebrated cardinal NICOLAUS CUSA, or with a supposed printer in Rome, NICOLAUS HAHN. The former died in 1464; consequently before the pope, to whom the work of Donis had been dedicated, had ever donned his tiara; and Raidel proves (p. 46) that the identification of Donis with Hahn was owing to a *falsification* at the sale of a parchment-copy of the edition Ulmæ 1482. In this copy the name of »Donis Nicolaus Germanus» had been replaced by that of *Nicolaus Gallus*, and the copy was wrongly stated to have been printed in Rome. Later *Nicolaus Gallus* was by retranslation into German changed into *Nicolaus Hahn*. Although this error was corrected in 1737 by Raidel, who had the opportunity of examining the falsified copy, the name of Hahn is still often mentioned in geographical literature, in connection with the work of Nicolaus Germanus, whose work besides (even by HUMBOLDT and D'AVEZAC) is erroneously stated to be the first edition with

[1] Jacobus Angelus is sometimes indicated by the surname of *de Scarparia* or *Florentinus*. It is however uncertain, whether these surnames belong to him, or not, as several authors of the same name as Ptolemy's translators, appear to have existed at the beginning of the 15:th century.

[2] In the work: *Studi biografici e bibliografici sulla storia della geografia in Italia*, Vol. II: *Mappamondi, Carte Nautiche, Portolani dei secoli XIII—XVII* per G. UZIELLI e P. AMAT DI S. FILIPPO (Roma 1882, p. 5) published by *Società Geografica Italiana*, it is stated, that FRANCESCO DI LAPACINO and DOMENICO DI LEONARDO BUONINSEGNI were the first who copied and provided Ptolemy's maps with Latin text.

[3] Some latin codices of Donis' type are dedicated to other princes; thus, such a codex preserved at the *Biblioteca Estense* in Modena, is dedicated to Prince BORSO D'ESTE, who probably 1466 renumerated Donis for it with 100 florins in gold (RUELENS, *Carte de l'Europe 1480—1485*. Texte explicatif. Bruxelles 1887). One Donis-codex, dedicated to BORSO D'ESTE, is also preserved at the *Bibliothèque Nationale* of Paris. An illuminated facsimile of the initial miniature of Donis is delivered in JOHN RUSSELL BARTLETT's catalogue of the CARTER-BROWN Library.

[4] This is confirmed by the »*Codex splendidissimus latinus Ebnerianus*», described by RAIDEL. According to him the text of this codex is an unaltered copy of the translation of Jacobus Angelus, but the maps are at the end of the 2d chapter of Book VIII, expressly attributed to Donis or Nicolaus Germanus by the words: »*Nunc sequuntur tabulae per Nicolaum Germanum*». Probably a codex of this kind has served for the edition *Romae 1478*.

[5] JOHANNES TRITHEMIUS, *De scriptoribus ecclesiasticis* (RAIDEL, p. 31). Tritheim, abbot in the convent of the Benedictines at Spannheim, a prolific author on various topics, was born 1462 and died 1516. He was accordingly almost contemporary with Donis.

modern maps.[1] A curious uncertainty has also prevailed as to the proper name of »Donis». It is written »Donnus Nicolaus Germanus» in the edition Ulmæ 1482 and in the codex of the Zamoisky library at Warsaw, and sometimes »Donis Nicolaus Germanus» (edit. Ulmæ 1486). In consequence of this, the editor of this edition of Ptolemy's geography is generally called »Donis», of which name I am also going to make use for the sake of brevity and clearness, and more particularly as it has become familiar in geographical literature. But it is not a correct one. At the end of Ptolemy's text in the edition Ulmæ 1486 is expressly written: *Opus Domini Nicolai Germani*, showing that *Donnus* is a rather unlucky abbreviation, and *Donis* a perversion of *Dominus*, and that the man's real name is *(Dominus)* Nicolaus Germanus.

ted in the 15:th, and thirty-three in the following century. These editions are of extraordinary importance to the history of cartography, and have therefore often been bibliographically described, quite lately by Mr. Justin Winsor in a very meritorious manner, with much learning and an extensive knowledge of the geographical literature.[2] But even his work contains some inaccuracies. He has, for instance, like the majority of other bibliographers, omitted the only edition, which renders Ptolemy's maps on an unaltered projection and he has, on the other hand, among editions of Ptolemy cited Stobnicza's *Introductio in Phtolomei Cosmographiam* of 1512 and 1519, and Wytfliet's *Descriptionis Ptolemaicae Augmentum* of 1597, 1598, 1603, 1607 and 1611, which works do not contain a single passage from Ptolemy, nor a single one of

TABVIA NOVA HIBERNIE ANGLIE ET SCOTIE

6. Britannia from *Ptolemaeus*, Argentinæ 1513. (Orig. size 516 × 316 m. m.).

The latin translation of Ptolemy's geography, or (according to Jacobus Angelus) cosmography, was at first disseminated in numerous, often splendidly decorated copies, and after the discovery of copper-plate and wood-engraving Ptolemy's atlas became one of the first great works, for the reproduction of which these arts were employed. Seven editions, of which six were provided with large maps in folio, were prin-

his 27 maps. However zealously American bibliophiles have collected editions of Ptolemy, Mr. Winsor has been obliged to cite several without having seen them himself. Moreover, he has been induced, by Santarem's carelessly written and uncritical contributions to the Ptolemaic literature,[3] to register in his catalogue several editions which do not exist. As I have had an opportunity, at my own writing-table, of examining at

[1] — — — »mais l'Allemagne, avec ses cartes simplement taillées dans le bois sans grande habileté, par Jacques Schnitzer d'Arnsheim, pour les éditions d'Ulm 1482 et 1486, pouvait se prévaloir à son tour d'une refonte critique des dessins traditionnels d'Agathodémon, et de l'addition de cinq cartes modernes, dressées par le savant bénédictin dom Nicolas Hahn de Reichenbach» etc. *Martin Hylacomylus Waltzemüller . . . par un géographe bibliophile* (d'Avezac). Paris 1867, p. 23.
[2] *A Bibliography of Ptolemy's Geography.* Cambridge, Mass. 1884 (*Library of Harvard University. Bibliographical Contributions*, No. 18). For bibliographical data to the editions of Ptolemy, see also the before mentioned work of Raidel; Joachim Lelewel, *Géographie du Moyen Age*, Bruxelles 1852—57; Ioannis Alberti Fabricii *Bibliotheca Graeca*, Edit. 3:a, Vol. V, Hamburg 1796, p. 270; Henry Harrisse, *Bibliotheca Americana Vetustissima*, New-York 1866, and *Bibl. Americana Vetustissima. Additions*, Paris 1872; John Russell Bartlett, *Bibliographical Notices of rare and curious books, relating to America in the Library of the late John Carter Brown*, Providence 1875; *Catalogue of the printed maps, plans, and charts in the British Museum*, London 1885.
[3] In a memoir on the voyages of Amerigo Vespucci. *Bulletin de la Société de Géographie*, VIII, Paris 1837, p. 175.

leisure almost every authentic edition of Ptolemy's geography and thus have been able more completely and minutely, than my predecessors, to compare them especially with regard to the cartography, I think it useful and proper to give a new catalogue in this place. In doing this, I shall pay special attention to the *maps*, of which the *Tabulæ novæ* or *modernæ*, added in increasing number to every new edition, are of the greatest importance to the history of cartography, at least down to 1570, when the work of Ptolemy was, as a practical atlas, for ever superseded by a more perfect, modern literature of maps, headed by the *Theatrum Orbis Terrarum* of Ortelius, and Mercator's *Atlas*.

Catalogue of editions of Ptolemy's geography.

1. Bononiae 1462 (1472). Fol. Colophon (on the last sheet but one): *Hic finit Cosmographia Ptolemei impressa opera dominici de lapis ciuis Bononiensis Anno MCCCCLXII, Mense Iunii XXIII. Bononie.* On the last page of the text: *Tabulas Cosmographiae secundum dimensiones Ptolomei impressas tibi … commendo … Accedit mirifica imprimendi tales tabulas ratio. Cuius inuentoris laus nihil illorum laude inferior. Qui primi literarum imprimendarum artem pepererunt in admirationem sui studiosissimum quemque facilime convertere potest. Opus utrunque summa adhibita diligentia duo Astrologiae peritissimi castigauerunt Hieronimus Mamfredus et Petrus bonus. Nec minus curiose correxerunt summa eruditione prediti Galleottus Martius et Colla montanus. Extremam emendationis manum imposuit philippus broaldus, qui plinii Strabonis reliquorumque id genus scriptorum Geographiam cum Ptolomeo conferens, ut esset quam emendatissimus elaborauit.*

Notwithstanding repeated assurances of the assiduity with which the publication of this edition was overlooked by several learned men, yet it is full of gross misprints. This often happens even in the manner of spelling Ptolemy's name and also, what is of more consequence to the history of printing, in the date when the edition was published, which certainly could not have been 1462. This may be deduced, first from the fact, that Dominicus de Lapis, so far as is known, had not been established as a printer in Bologna as early as 1462; secondly from signatures having been employed in this work to denote the order of the sheets, which practice is not supposed to have been introduced so early as 1462; and thirdly, from Philippus Broaldus (or Beroaldus) being mentioned as one of the correctors of the work. He is said to have been born in 1453, and was thus only nine years old in 1462. In about 1472 he was, at the age of 19, rector of a school at Bologna.

From reasons, adduced by Bartolommeo Gamba in his work: *Osservazioni su la edizione della Geografia di Tolomeo fatta in Bologna colla data del MCCCCLXII*, Bassano 1796, written with learning and critical acuteness, it seems evident to me that an X has been omitted in the date of this edition. It should then have been printed MCCCCLXXII (1472), and not, as some bibliographers suppose, 1482 or 1492. At least the crude execution of the Bologna maps shows, that the copper-printer was altogether unacquainted with the excellent maps of Schweinheim-Buckinck, published in Rome in 1478. Moreover, a comparison of the text of the Bologna-edition with that of the edition »Vicenciæ 1475», seems to show, that the latter is an improved reprint of the former.[1] The maps are coloured, evidently with an illuminated manuscript as a model, the sea green (now brownish green), the mountains ultramarine-blue, and the land partly yellow, partly carmine-red. Some of the legends are richly gilded. The *Tabula secunda* and *tertia Asiæ* being united, the number of maps is only 26. It is remarkable, that the maps in so early an edition as this, are reproduced on a conical projection with the latitudes and longitudes[2] drawn out across the map. To furnish the reader with an idea of these early maps, I give a reduced fac-simile of the map of Britain. The copy I had at my disposal, belongs to the Library of the University of Upsala. As I have seen this edition in several libraries on the continent, it may not be exceedingly rare, though Gamba says that he only saw two complete copies in the Italian libraries. The assertion that one codex of Ptolemy at the library of the Imperial General-Staff at St. Petersburg is actually the original of the Bologna-edition, rests upon a mistake. To judge from the description of Edw. v. Muralt in *Bulletin scient. publ. par l'Acad. Impér. de St. Pétersbourg*, X, 1842, p. 97, the codex in question, which belonged to Prince Labanow-Rostowsky and, before him, to the French geographer Mac-Carthy, is a splendid copy of the translation of Jacobus Angelus. But several such manuscripts are to be found in the libraries of Europe, and there is no special reason why this one should be assumed to be the original of the edition of 1472. Indeed, the very opposite is proved by a careful comparison of von Muralt's extracts from this codex with the text of the editions printed during the 15th century.

2. Vicenciae 1475. Fol. Colophon: *En tibi lector Cosmographia Ptolemaei ab Hermano leuilapide Coloniensi Vicenciae accuratissime impressa. Benedicto Triuisano & Angelo Michaele praesidibus. MCCCCLXXV. Idi. Sept.* The text commences with a dedicatory letter of Jacobus Angelus to Alexander V, in which the writer among other things mentions, that he, following the example of Pliny and other Latin authors in the title of the book, changed the word *geography* to *cosmography*. This edition contains no maps, and, as in the Bologna-edition, here also the division into chapters is wanting in Lib. II—VIII.

3. Firenze 1478? Fol. *Geographia di Francesco Berlinghieri Fiorentino, in terza rima & lingua toscana distincta con le sue tavole in varii siti & provincie, secondo la geographia et distinctione dele tavole di Ptolomeo. Cum gratia et privilegio.* In some copies this title is printed in red on the first page of the first leaf; in others this first page is blank, while the contents of the book are given on the inside of the leaf by the following words: *In qvesto volume si contengono Septe Giornate della Geographia di Francesco Berlingeri*

---

[1] The date of this edition is of very great interest to the history of cartography, as well as to that of copper-engraving. If these maps were printed in 1472, they are not only the first ones published in print, but also among the oldest known specimens of copper-plate. This question has given rise to an extensive literature, enumerated in the works of Gamba and Winsor.

[2] When J. van Raemdonck in *Les sphères terrestre et céleste de Gérard Mercator*, p. 298, says: »Un dernier perfectionnement apporté par Mercator à la sphère terrestre, est la graduation. A son époque, les topographes et chorographes en général, ne se souciaient nullement ni de longitudes ni de latitudes géographiques, et leurs cartes étaient, pour la plupart, sans aucune graduation», he is completety mistaken. From the beginning of the printing of maps the graduation of latitudes and longitudes were, on the contrary, marked down on most printed maps, at least in the margin. A similar graduation is also found on the globes from the beginning of the 16th century, as well as on the map of »*Lotheringia*» in *Ptolemaeus, Argentinæ 1513*. On the contrary Sebastian Münster may be reproached with having often omitted the graduation.

Fiorentino allo illustrissimo Federigo Duca Durbino. Probably the above mentioned red title was only printed in some of the copies on the title page, originally left blank.

The Colophon at the foot of the last leaf[1] of the book is as follows: *Impresso in Firenze per Nicolo Todescho et emendato con somma diligentia dallo auctore.* Thus the year of printing is not given, but it may be approximately determined from the circumstances that the Duke of Urbino, to whom the book was dedicated and to whom it had already been presented in manuscript,[2] died in 1482, and that the *Apologus* of MARSILIUS FICINUS, inserted in the work of Berlinghieri, was printed among the letters of the former from 1481—1482 (MURR, p. 7). LIBRI gives to this book an

is a translation not from the original Greek, but from the Latin version of Jacobus Angelus. We here meet Ptolemy's chapter on the difference between chorography and geography, his warning as to a credulous and uncritical use of the reports of travellers by land and sea, his analysis of the work of Marinus of Tyre, his chapter on the navigation from Aurea Chersonesus to Cattigara, his directions as to map- and globe-making, etc. At the beginning of the 2d, 3d, 5th, 6th and 7th *libro* or *giorno*, corresponding to the 2d, 3d, 5th, 6th and 7th books of Ptolemy, Berlinghieri does not speak any longer in his own name, but in that of Ptolemy (for instance *Claudio Ptolemeo il tertio giorno*) and the customary *Feliciter incipit* has slipped into the beginning of the different sections of the

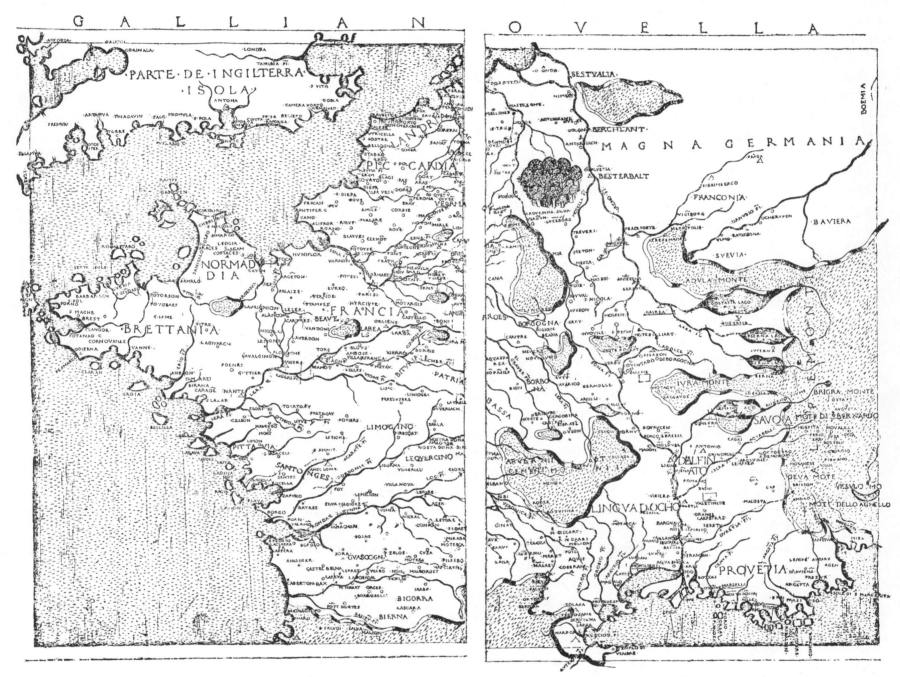

7. Gallia novella from *Geographia di Francesco Berlinghieri*, Firenze 1478 (?).   (Orig. size 493 × 373 m. m.).

earlier date than 1478 and considers it to contain the first maps printed with copper-plate. This seems to be confirmed by the want of finish and by the deficiencies in the copper-engraving, which would not have occurred, if the artist had known the fine maps of the Rome-edition of 1478.

Berlinghieri's work is a tolerably faithful translation into Italian of Ptolemy's geography or rather cosmography, for I suppose that this remarkable sample of geographical rhyming

book without translation. Ptolemy's catalogues of geographical longitudes and latitudes are also to be found in Berlinghieri's work, but alphabetically arranged, and not, as in the original, geographically.

The text is, of course, at present of slight interest at least to the geographer, but the plates are of importance, as the only copies of Ptolemys maps printed on their original projection, with equidistant parallels and meridians. I have here

---

[1] The last leaf of the book fails in my copy as well as in that of the Library of the University at Upsala, both of which also want the title printed in red. The last leaf did also not exist in the copy belonging to the Library at Nuremberg, described by von MURR in: *Notitia libri rarissimi Geographiae Francisci Berlinghieri*, Norimbergæ 1790. I have copied the colophon from the catalogue of LA VALLIÈRE (Vol. II, p. 514, Paris 1783), where also some bibliographical notes regarding the copy of La Vallière are communicated. It is indicated by the words: »*Première édition fort rare imprimée vers 1478*». There are different variations in the work, described by BRUNET.

[2] RAIDEL (in his often mentioned work, p. 75) describes a magnificent manuscript of the geography of Berlinghieri, which belonged to Count PERTUSATTI. Raidel supposes it to have been destined for the Duke of Urbino. It only contained 28 maps.

reproduced one of these maps (N. T. XXVIII). If the plate XV (from the edition of 1478 or 1490), is compared with this facsimile we shall be able to appreciate the care and accurary, with which »Donis» followed the old originals, when redrawing the maps on a new projection.

Berlinghieri's work however derives its greatest value from the modern maps it contains, »Novella Italia», »Hispania novella», »Gallia novella» (N. Fig. 7) and »Palestina moderna et Terra Sancta». With the exception of the coarse drawing of the Holy land (N. fig. 3) published in *Rudimentum novitiorum* (Lübeck 1475), these »Tabulæ novellæ» are the first printed modern maps, the first germ of modern cartography. Notwithstanding this, they have, so far as I know, never been the subject of a monograph, or even mentioned in cartographical literature. Neither HUMBOLDT, nor PESCHEL, nor LELEWEL appear to have seen these Tabulæ novellæ, and they assume, that the first modern maps were published in Ptolemaeus Ulmæ 1482 or Romæ 1508. In his excellent *Histoire de la Géographie*, VIVIEN DE SAINT MARTIN mentions the »*Précis de géographie en vers italiens, publié par Berlinghieri en 1480*», but he does not know that this work is a translation of Ptolemy, and seems not to have perceived, that it contains the first printed modern map of France. Nor have I found any statements regarding the source or origin of these maps. Probably they belonged to that latin manuscript of Ptolemy, which Berlinghieri used for his translation. If they are compared with the corresponding *Tabulæ Novæ* in the editions of Ptolemy, printed in Ulm 1482 and 1486, and with the maps of the codex in Brussels, of which the »tabulæ novæ» lately were reproduced in facsimile by M. CH. RUELENS, considerable differences will certainly be found. But probably all these maps ultimately derive from the same original, the age of which ought not to be difficult to ascertain by the aid of the numerous names of provinces and towns they contain. Such a determination would be of a certain interest even as regards the cartography of the North, for it would perhaps be possible here to get a clue for discovering the source of the remarkable new map of Scandinavia and Greenland in the Donis' codices of Ptolemys geography (N. T. XXX).

4. Romæ 1478. Fol.

Colophon: *Claudii Ptolemæi Alexandrini philosophi Geographiam Arnoldus Buckinck e Germania Romae tabulis æneis in picturis formatam impressit. Sempiterno ingenii artificiique monumento. Anno dominici natalis MCCCCLXXVIII. VI Idus Octobris Sedente Sixto, IIII. Pont. Max. anno eius VIII.* The dedication to the Pope, which introduces the work, contains the following notice interesting as regards the history of map-printing, *Ne librariorum inscitia tuæ sanctitatis aures offenderet, Domitius Calderinus, Veronensis, cui huius emendationis provintia demandata fuerat, cum vetustissimo Graeco, manu Gemisthi Philosophi emendato, latino Codices se collocaturum Magister vero Conradus Suueynheim, Germanus, a quo formandorum Romæ librorum ars primum profecta est, occasione hinc sumpta, posteritati consulens, animum primum ad hanc doctrinam capessendam adplicuit, subinde mathematicis adhibitis viris, quemadmodum tabulis æneis imprimerentur, edocuit, triennioque in hac cura consumpto, diem obiit. In cuius vigiliarum laborumque partem non inferiori ingenio ac studio Arnoldus Buckinck e Germania, vir apprime eruditus, ad imperfectum opus succedeus, ne, Domitii Conradique obitu, eorum vigiliae emendationesque sine testimonio perirent, neve virorum eruditorum censuram fugerent, immensae subtilitatis machinamenta ex amussim ad unum perfecit.* These words seem to indicate that the publisher was DOMITIUS CALDERINUS, and the printer CONRAD SCHWEINHEIM and

that the latter, by whom the art of printing was introduced into Rome, for this publication, learned (»*edocuit*»), with the assistance of skilled mathematicians (mechanics), to print with copperplate. In this he was assisted by ARNOLD BUCKINCK, who also finished the work after the death of Domitius and Conradus. The confidence, pride and satisfaction expressed in the above quoted introduction as well as in the colophon, are fully justified, for the maps of this edition were, as masterpieces of copper-plate printing, not surpassed for centuries, and they will still take the first place among the maps of all the numerous editions of Ptolemy's geography hitherto published. They are based on the copies by Donis of Ptolemy's maps, but the style of the drawing is here greatly improved. Whilst, for instance, a very practical and convenient manner of indicating the mountains is used in these maps, Donis, as expressly indicated in his dedication to Paul II, only marked the mountain ranges by encircling lines. In uncoloured maps one was thus likely to confound mountains and seas, but this is no longer to be apprehended with the plain and handsome topographical dressing used in the maps of Schweinheim-Buckinck.

I have compared the editions of 1490, 1507 and 1508 with two copies of the very rare edition of 1478, the one lent to me from the Library of Göttingen through the kind assistance of its Librarian Dr. DZIATZKO, and the other in my private collection. From this comparison it results, *that the maps of the edition 1490 only consist of new, perfectly unaltered impressions from the plates, employed for the edition 1478, and that the same plates also have been used for the original maps in the editions 1507 and 1508. But before the plates were employed for the last two editions a few slight corrections or repairs of damaged parts had been made.* In Tab. VI Europæ various emendations had thus been made in the representation of the tributaries of the Po, and in Tab. IX Europæ the name »Bosphorus Thracius» has been added.

For the facsimiles plates I—XXVII, which are thus entitled as well to the date 1478 as to that of 1490, I had access to five copies of the copper-engravings of Schweinheim-Buckinck, namely to two of the edition of 1478, and to one of each of the editions 1490, 1507 and 1508, and I thus anticipate that the photolithographic copies here given will represent the original maps not only faithfully, but, by the aid of the many originals maps to which I had access, also *completely*.

5. Ulmæ 1482. Fol.

Colophon on a leaf inserted after the maps: *Opus Donni Nicolai Germani secundum Ptolomeum finit. Anno MCCCCLXXXII, Augusti vero kalendas XVII. Impressum Ulme per ingeniosum virum Leonardum Hol prefati oppidi civem.* The work begins with a letter, *Beatissimo patri Paulo secundo pontifici maximo Donis Nicolaus Germanus*, which contains, besides the usual dedicatorial flowers, an explanation why a deviation from Ptolemy's originals has occurred in the reproduction of the maps. This principally resulted from the new improved projection, with which »Dominus Nicolaus» seems to be not a little proud.... *Cogitare cepimus quo pacto nos aliquid glorie comparemus. Rati enim nobis oblatam esse occasionem uti aliquid industrie nostre monimentum extaret et ingenii vires ducescere possent statim picturam orbis proper aratione aggressi sumus.* Donis then mentions, that he, in the maps, has represented mountains, islands etc. with more care than before, and that he has added new maps of Italy, Spain and the countries of the North, among which also *Gronelandia* is enumerated. The dedication begins with a handsomely decorated initial *(N)* representing Donis on his knees presenting a book to a Pope in full canonicals.

After the dedication follows Ptolemy's text, divided as usual into eight books, but without any subdivisions into chapters. In accordance with what seems to have been the original intention of the author, that part of the eighth book, which contains the explanatory remarks regarding the maps, is printed on the reverse of each corresponding map. These are all in double folio, excepting the last one of Taprobane. Besides the new maps of Spain (N. fig. 11), Italy (N. fig. 12) and the countries of the North, mentioned in the preface, the edition contains a new map of Gallia and one of Palestine. There are no explanatory remarks to these maps, excepting on the reverse of the map of Italy, where a short text is printed, beginning with the following highflown praise of the country: *Plurime*

This is the first printed map which is *signed*. It is further remarkable, because the geographer for the first time has ventured to make some changes in Ptolemy's picture of the world, though as yet only for the distant north, almost unknown to the ancients, but whose inhabitants since then had so often, weapon in hand, spread knowledge of their country among the southern nations. Already during the 14th century these countries were laid down in Catalan and Italian sea-charts; and manuscript maps of them in Ptolemy's style were added to his classical work from the time that it was translated into Latin. I give Plate XXX, a facsimile of the most complete among these præcolumbian maps of the North, and farther on (Ch. V) a review of the oldest cartography of

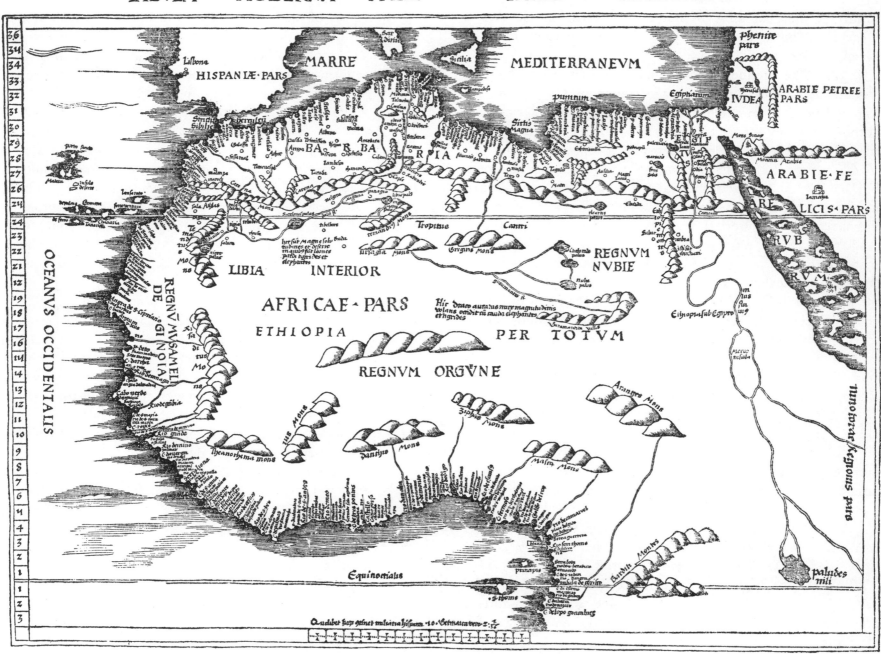

8. Prima pars Africæ from *Ptolemaeus*, Argentinæ 1513. (Orig. size 580 × 416 m. m.).

sunt regiones que quidem singule singulis rebus excellere videntur. Nam thus sola arabia gignit. Balsamum nusquam nisi in Iudea legitur: ex India ebur prouenit. Item aliud ab aliis nationibus accepimus. Verum si exactissime cuncta iudices, inuenies perfecto Italiam omnibus esse iure preferendam etc. From this we may conclude to what nationality the author of at least this *tabula nova* belonged.

The maps are rough, though clear and distinct wood-cuts, according to the legend on the upper border of the map of the world (N. T. XXIX) sculptured by JOHANNES SCHNITZER DE ARMSSHEIM (not Arnissheim). Certainly *Schnitzer* should here be understood not as a surname, but as a professional title, and the superscription should be interpreted: »Engraved by Johannes, sculptor from Armsheim».

that part of the world. It is a remarkable circumstance that in 1482, when the Ulm-edition was published, or fifteen years earlier, when Nicolaus Germanus probably finished the manuscript from which it was printed, neither in Germany nor in Italy people seem to have had any idea about the Portuguese having penetrated to the south far beyond Ptolemy's »oikumene». At least no additions to the old maps of Africa are made in the editions of 1482 and 1486.

With regard to these editions, students of the history of cartography have been led into some mistakes, alluded to before, but to which it perhaps would be convenient here shortly to revert:

1st. To indicate these editions by the name of Donis, as has often been done, is incorrect. The text of the edition follows the translation of Jacobus Angelus. As original for the maps has served the reproduction of Ptolemy's maps on a new projection, first executed by Dominus NICOLAUS GERMANUS. But these very maps were already engraved on copper by ARNOLD BUCKINCK in 1478, and this with a finish by no means to be claimed for the wood-cuts printed in Ulm. The edition Ulmæ 1482 contains nothing particularly belonging to Donis which is not found in the Rome edition, except the dedication to the Pope and principally its initial N, faithfully copied from the manuscript of Donis.

2d. The name NICOLAUS DONIS, so often used, is, as before remarked, incorrect. The writer of the codex for the Ulm edition was called Nicolaus. He was a native of Germany and accordingly designated as *Germanus*, and to him belonged the title of *Dominus*. His complete name and title were therefore *Dominus Nicolaus Germanus*. The publisher of the geography of Berlinghieri calls himself »*Nicolo Todescho*», but was probably a different person from this Nicolaus Germanus.

3d. The frequently repeated assertion that the first *Tabulæ novæ* were printed in the edition of Ulm 1482, and the supposition that Nicolaus Germanus or »Donis» was their author, is doubly misleading with regard to a question particularly important to the history of geography. The first *Tabulæ novæ* were drawn, not on the projection of Donis, but on an equidistant cylindrical projection, and were printed in about 1478, in the *Septe giornate della Geographia* of Berlinghieri, of which work I have given a short analysis above. And that Donis was not their author is shown by the fact that for the new maps of Spain, France and the Holy Land, the same original has served for the edition Firenze 1478 and for that of Ulm 1482 and 1486. But, if the merit of having communicated the first *tabulæ novæ* in print does not belong to the editions Ulmæ 1482 and 1486, these however are, in an other respect, very remarkable and important. In the former edition, printed ten years before the first voyage of Columbus, a map is given, embracing *not only the North of the Old World, but also that part of the New world (Greenland), which, half a millennium before Columbus, was discovered by the Scandinavians*. The latter edition contains a reimpression of the same map and a catalogue of geographical positions of a number of places in Scandinavia *and Greenland*.

6. Ulmæ 1486. Fol. Colophon: *Impressum Vlme opera et expensis Iusti de Albano de Venetiis per provisorem suum Iohannem Reger. Anno Domini MCCCCLXXXVI. XII Kalendas Augusti.* This sentence does not terminate the geography of Ptolemy but an extensive addition, entitled: *De Locis ac Mirabilibus mundi*. At the end of the text of Ptolemy is written: *Opus Domini Germani secundum Ptolomeum finit*. These lines correspond to the colophon in the edition Ulmæ 1482, excepting that *Donni* is corrected to *Domini*. Had this correction been observed before, the discussions about the name of »Donis» might have been spared.

The edition of 1486 is a reprint of the edition 1482, but with important additions. In the new edition the division into chapters is introduced and various smaller corrections made in the text. But in other respects the older edition is faithfully adhered to, often line for line, through whole sheets.

The most important additions are the following:

1) An introduction of forty-one folio leaves: *Registrum alphabeticum super octo libros Ptolomei*. A very great number of places, among which there are many not mentioned in the text of Ptolemy, are here enumerated, often with additional remarks, chiefly referring to the history of the martyrs and the saints. Under *Scennig* there are, for instance, allusions to the attempts made at the Council of Constance by the Swedes, to obtain canonization for a couple of Scandinavians.

No reference to the voyages of the Portuguese occurs here, although the *Fortunatæ insulæ* are mentioned with the addition: *Hic Brandianus magne abstinentie vir de scocia natus pater 1000 monachorum cum beato Maclouio has insulas septennio perlustrat* etc. Here we have the first (?) printed notice of the mythical voyages of St. BRANDON or BRENDAN. The alphabetical order in the *Registrum* is far from correct.

2) To the tenth chapter of the second book is added a *Tabula Moderna extra Ptolomeum posita*, and to the fifth chapter of the third book an appendix almost two folio-pages long: *Tabula Moderna Prussie, Svecie, Norbegie, Gotcie et Russie, extra Ptolomeum posita*. These additions contain, in the form of a text to the new map of the North, a geographical description as well of the extreme north-west of the old world, as of the extreme north-east of the new one, printed six years before the first voyage of Columbus across the Atlantic. Among other things the latitudes and longitudes are here given of 183 places in Scandinavia, Greenland, north-eastern Germany and northwestern Russia.

3) The above mentioned work printed on 24 folios: *De Locis ac Mirabilibus Mundi*, follows the maps and ends the book. It is, at least to the geographer, an almost worthless compilation from other authors about countries, towns, seas, rivers, mountains, monsters, giants, floods, land-survey, prophecies of weather, etc. That it is not written by a German is evident from the scarcity of notices about Germany. Nor are there found any traces of such a knowledge of the North, as is testified by the before mentioned addition to Ptolemy's text. Every thing shows that, under the name of Donis, a mediæval work has here been added to the edition of 1482 by the *editor* of the edition 1486 »JUSTUS DE ALBANO *de Venetiis per provisorem suum* JOHANNEM REGER», in order to make it more attractive to the buyer. This is also confirmed by the fact, that the sheets in the »De locis ac mirabilibus mundi» are marked with special signatures, from which it appears, that the original intention had been to publish it as a separate work. It must have been very popular, as it is also inserted into the editions Romæ 1507 and 1508.

The maps in the edition of 1486 are the same as those in the edition Ulmæ 1482, and printed *from the same blocks*.

7. Romæ 1490. Fol. *Claudii Ptolemaei Geographiae libri VIII.*

Colophon: *Hoc opus Ptholomei memorabile quidem et insigne exactissima diligentia castigatum iucondo quodam caractere impressum fuit et completum Rome anno a nativitate domini MCCCCLXXXX, die IV. Novembris. Arte ac impensis Petri de Turre.* This is a beautiful reprint of the edition Romæ 1478. The maps are 27 in number and are printed from the same plates as those of the above mentioned edition. Raidel's assertion (p. 49) that the maps of this edition were engraved in a somewhat rougher manner is consequently incorrect, as presupposing new copper plates. For the origin of these remarkable maps, see above (edit. Romæ 1478).

The text in the edition 1490 contains, besides the eight books of Ptolemy's Geography, the »*Registrum alphabeticum*» and »*De Locis ac Mirabilibus Mundi*», from the edition Ulmæ 1486.

8. Romæ 1507. Fol. *In hoc operæ hæc continentur: Geographia Cl. Ptholemaei a plurimis viris utriusque linguæ doctiss. emendata: & cum Archetypo græco ab ipsis collata.*

Schemata cum demonstrationibus suis correcta a Marco monacho Cælestino Beneuentano: & Ioanne Cota Veronensi viris Mathematicis consultissimis.

Figura de proiectione Sphæræ in plano quæ in libro octavo desiderabantur ab ipsis nondum instaurata sed fere adinventa: ejus n. vestigia in nullo etiam græco codice extabant.

Sex tabulæ noviter confectæ vz Hispaniæ: Galliæ: Livoniæ: Germaniæ: Poloniæ: Vngariæ: Russiæ: & Lituaniæ: Italiæ: & Iudeæ.

Maxima qvantitas dierum civitatum: & distantiæ locorum ab Alexandria Aegypti cuiusque civitatis quæ in aliis codicibus non erant.

Planisphærium Cl. Ptholemæi noviter recognitum & diligentiss. emendatum a Marco monacho Cælestino Beneventano.

Cautum est edicto Iulii II. Pont. Max. ne quis imprimere aut imprimi facere audeat hoc ipsum opus pena

The work contains:

1:o. A dedication, *R:mo patri Cardinali Nannatensi*, in which the author, EVANGELISTA TOSINUS, enumerates his assistants, emendators and correctors, expressing the conviction, that the new edition had now attained a perfection never to be surpassed. He also commends the importance of the addenda, especially of the planispherium and the *Tabulæ novæ*, enumerated on the title-page.

2:o. *Registrum Alphabeticum* etc. A slightly altered reprint from the edition Ulmæ 1486.

3:o. The text of Ptolemy's geography without the important addenda to the tenth chapter of the 2d book and the fifth chapter of the 3d book, inserted in the edition of Ulm 1486.

TABVLA MODERNA SECVNDE ⟨ PORCIONIS APHRICE

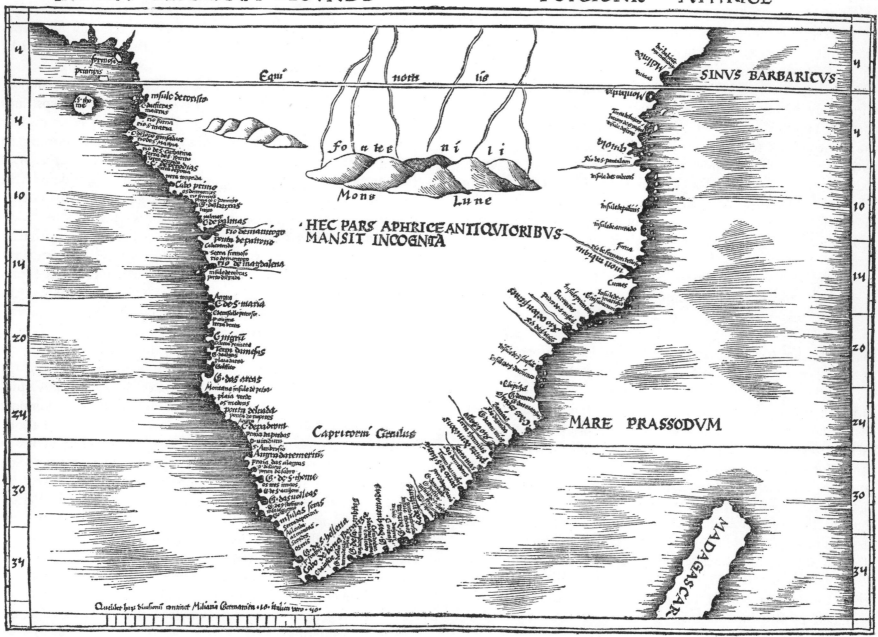

9. Secunda Pars Africa from *Ptolemaeus*, Argentinæ 1513. (Orig. size 513 × 363 m. m.).

excommunicationis latae sententiae his qui contra mandatum iussumque conari audebunt.

Colophon: *Explicit Planisphærium Ptholemaei recognitum deligentissime a Marco Beneventano Monacho Cælestinorum quod antea in multis etiam antiquis exemplaribus latinis corruptissimum reperiebantur. Nec non Claudii Ptholemæi a plurimis viris utriusque linguæ doctiss. emendatum cum multis additionibus Romæ Noviter impressum per Bernardinum Venetum de Vitalibus. Expensis Evangelista Tosino Brixiano Bibliopola. Imperante Iulio II. Pont. Max. Anno III. Pontificatus sui. Die VIII. Septembr. MDVII.*

4:o. *Planisphærium Ptholemaei*, with a dedication to the Venetian patrician JOHANNES BADUARIUS, who had presented a manuscript of the Planisphere to MARCUS BENEVENTANUS.

5:o. A letter from the Pope JULIUS II, of the 28th of July 1506, giving to the editor a patent for six years to the exclusive sale of the book, on condition that the price should be fixed by the Pope's Librarian.

6:o. 33 maps, i. e. the 27 of Ptolemy, printed from the same plates, as the maps of the editions 1478 and 1490,[1] and six new maps, viz:

Tabula Moderna Prussiæ, Livoniæ, Norvegiæ et Gothiæ.

---

[1] The assertion, that this edition should contain a new map of Benevent, is incorrect, at least to judge from the copies I had occasion to examine. I suppose that the mistake arose from some blunder in the translation of a period containing the name of MARCUS BENEVENTANUS.

Tabula Moderna Hispaniæ.

Tabula Moderna Galliæ.

Tabula Moderna Poloniæ, Ungariæ, Boemiæ, Germaniæ, Russiæ, Lithuaniæ.

Tabula Nova Italiæ.

Tabula Moderna Terræ Sanctæ.

Of these six maps the three first and the two last ones are of least interest, as only being new, slightly altered editions of corresponding maps in the work of Berlinghieri and in the editions, printed in Ulm 1482 and 1486.

Tabula Moderna Poloniæ, Ungariæ etc. is more important. It is evidently based on the same original, as the map in the *Liber Cronicarum* of HARTMANNUS SCHEDEL, Nurembergæ 1493 (N. Fig. 5) but may, on account of the richness of new cartographical details, be regarded as an original work. It is the first copper-plate map of Germany, Austria, Hungaria, Polonia and Russia, or of the whole of central Europe from »Ostende» in the west to »Neper» in the east and, for the period of incunabula uncommonly rich in details. But, so far as I know, no attention has hitherto been paid to it in literary works on cartography, and it has never been subjected to critical investigation with regard to the legends, and the age and sources of the original. It does not enter into the plan of the present work to supply this deficiency. I can here (Fig. 13) only give a faithful, although reduced, fac-simile of the map. Farther on (Fig. 14) I also give a fac-simile of the *Tabula Moderna Prussiæ, Livoniæ, Norvegiæ et Gothiæ.* This *tabula* is only a copy of the corresponding map in the edition 1482 and 1486, but yet of interest as the first copper-plate map of Scandinavia. The celebrated map of John Ruysch is also found inserted in some copies of the edition of 1507, but as it is not enumerated among the maps on the title-page, I suppose that it did not originally belong to this edition.

All the new maps are engraved on copper and with the same finish as the old ones. The explanatory remarks on each map in book VIII of Ptolemy are not here, as in the Ulm-editions, printed on the reverse of the maps, but with the other text. To facilitate the binding, the map-sheets have often been placed by twos, one inside the other, in such a way, as that the first page is left blank, the 2d contains, for instance, one half of the map of the world, the 3d (the 1st page of the 2d sheet) the other half of the map of the world, the 4th and 5th are blank, the 6th contains the western half of Britain, the 7th the eastern half of Britain, the 8th is blank. If the sheets are separated, one of them will in consequence embrace the western half of the map of the world and the eastern half of Britain, and the other, the eastern half of the map of the world and the western half of Britain.

9. Romæ 1508. Fol.
The title-page, on the left decorated with a handsome border, runs:

In hoc opere hæc continentur.

Geographia Cl. Ptolemæi a plurimis viris utriusque linguæ doctiss. emendata: et cum archetypo græco ab ipsis collata.

Schemata cum demonstrationibus suis correcta a Marco Beneventano Monacho cælestino, et Ioanne Cotta Veronensi viris Mathematicis consultissimis.

— — — — — — — — — — — — — — —

Nova orbis descriptio ac nova Oceani navigatio qua Lisbona ad Indicum pervenitur pelagus Marco Beneventano monacho cælestino ædita.

Nova et universalior Orbis cogniti tabula Ioa. Ruysch Germano elaborata.

Sex Tabulæ noviter confectæ videlicet: Hyspaniæ, Galliæ, Germaniæ Italiæ, & Iudeæ.

— — — — — — — — — — — —

Anno Virginei Partus, MDVIII. Rome.

Excepting a new title-page, this edition is identical with the previous one, but with the addition of a few verses to the dedication; of an extensive but not very important supplement *Nova Orbis descriptio* by MARCUS BENEVENTANUS, and of a new map of the world by JOHANNES RUYSCH, on which the discoveries of the Spaniards and Portuguese were, for the first time, registered in the literature of cartography. Of this map, which is also sometimes inserted in the edition of 1507, a fac-simile is given on Plate XXXII. I will return further on to the same, as well as to the edition of Ptolemy in which it was published.

10. Venetiis 1511. Fol.
Claudii Ptholemaei Alexandrini liber geographiæ cum tabulis et universali figura et cum additione locorum quæ a recentioribus reperta sunt diligenti cura emendatus et impressus. (Colophon): *Venetiis per Iacobum Pentium de leucho. Anno domini MDXI Die XX Mensis Martii.*

In his dedication to ANDREAS MATHEUS AQUÆVIVUS, Duke of Adria, Lord of Eboli etc., BERNARDUS SYLVANUS EBOLENSIS declares himself to be the author. With evident pride, as the author, he here speaks of the corrections, introduced into the new edition to make the maps of Ptolemy agree with the accounts of later navigators. »When I considered that Ptolemy, with more care than any other geographer, had determined the relative positions and distances of the places, I was astonished», Sylvanus says, »that his maps only occasionally corresponded to the experience of the mariners of our time. This seemed to me so much the more remarkable as I presumed that Ptolemy himself mostly based his work on the reports of the navigators». When comparing the several Greek and Latin manuscripts, brought together from various places, Sylvanus found considerable differences, especially as to the figures *quibus locorum signantur intervalla.* On account of this he reexamined the text with great attention, and found that this corresponded well with the reports of the mariners, but not so the numerical data. These he corrected arbitrarily and without any regard to the readings in the manuscripts. The principles of these corrections are developed in a special introduction, *»Bernardi Sylvani Ebolensis annotationes in Ptholemaei geographiam cur nostræ tabulæ ab iis quæ ante nos ab aliis descriptæ sunt differant»* etc., filling four pages in folio and containing principally an apology for the modifications Sylvanus ventured to make in the old maps. Here, however, the remarks more concern less important modifications in the maps of the anciently known world than the radical changes in the whole cartography, necessitated by the new discoveries of the Scandinavians, Portuguese, Spaniards and Britons. While almost half a page is dedicated to corrections of the old map of Sicily, the circumnavigation of Africa is despatched by the following words: *Ausi nam se Lusitani, cum loca illa ignota essent, fortunæ credere, et incognita explorare maria, plurima invenere, quibus illi et æternam sibi gloriam, et nobis ac posteris omnibus jucundam novarum rerum cognitionen peperere.* Nor is there to be found a word, immediately referring to the discovery of the new world. Evidently even at Venice, which at that time possessed a richer store of commercial experience and political insight, than any other city of the world, no general idea prevailed as to the importance of the voyages of Columbus across the Atlantic or of Vasco da Gama round Caput Bonae Spei to India. In these *»annotationes»* Sylvanus also inserts a long passage *»Adversus Marcum beneventanum Monachum»*, (one of the editors of Edit. Romae 1507 and 1508) who is charged with ›*inscitia atque negligentia*‹. His mathematical demon-strations alone read tolerably well, but this is due to the merit of JOHANNES COTTA. After the introduction first follows the text of Ptolemy without the usual addenda, then 28 maps, of which 27 slightly retouched ones of Ptolemy, and a new map of the world.

As an edition of Ptolemy, the work of Sylvanus is quite worthless on account of the arbitrary alteration of Ptolemy's data for longitude and latitude. Nor has his attempt to trans-form the work of the Alexandrian geographer into a modern atlas been attended with better success. Sylvanus for instance

In typographical respects the edition of Sylvanus is dis-tinguished by a handsome outfit. For the first time we here meet with maps printed in two colours, and, contrary to what generally was and yet is the custom, both sides of the paper are used for the map-print; excepting for the new map of the world, printed on a sheet, where the reverse is left blank. The maps are from woodcuts, for which the legends in black are produced by types fitted into the blocks.

11. Argentinæ 1513. Fol. *Claudii Ptolemei … Geo-graphiæ opus novissima traductione e Græcorum archetypis*

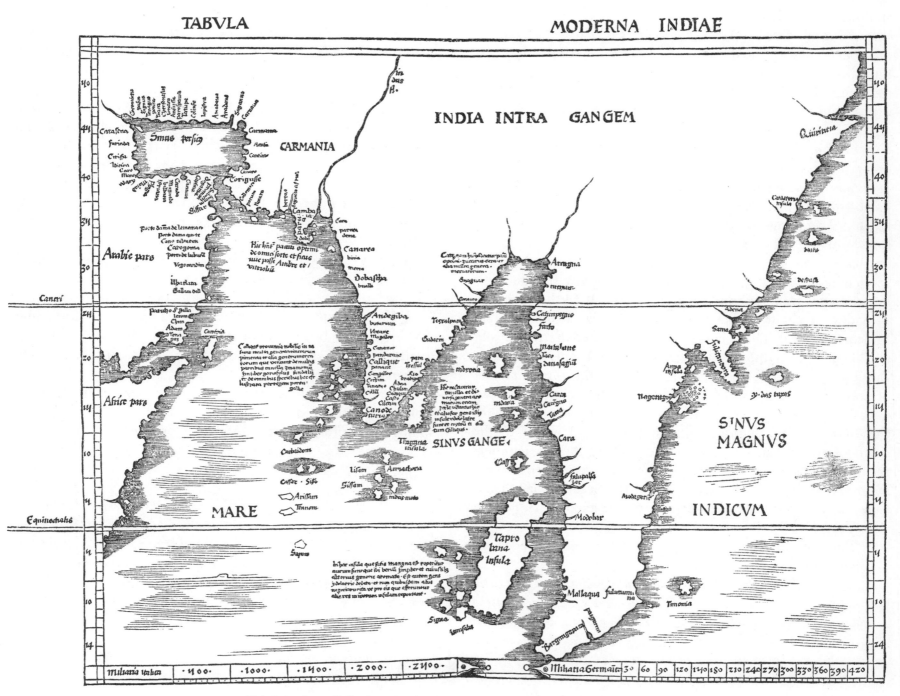

TABVLA   MODERNA INDIAE

10. Tabula moderna Indiæ from *Ptolemaeus*, Argentinæ 1513. (Orig. size 510 × 403 m. m.).

leaves the old names unchanged in maps, pretended to be modernized. The merit, however, must be conceded to Syl-vanus, that he was the first to break with the blind confidence that almost every scholar in the beginning of the 16th cen-tury had in the atlas of the old Alexandrian geographer. Sometimes also the corrections and modifications by Sylvanus were real improvements. Following the sketches of the nor-thern regions in some old portolanos, he thus published the first map of England, in which the enormous elongation to the east, which Ptolemy gives to the northern part of Scot-land, is left out. But the greatest importance of this edition to the history of cartography, consists in the cordiform map of the world, of which I have given a fac-simile on Pl XXXIII. To the analysis of this map I will return further on.

castigatissime pressum: cæteris ante lucubratorum multo præstantius.

Pro Prima parte continens Cl. Ptolemaei Geographiam per octo libros partitam, ad antiquitatem suam, integre & sine ulla corruptione, …

Pars Secunda moderniorum lustrationum Viginti ta-bulis, veluti supplementum quoddam antiquitatis obsoletæ, suo loco quæ vel abstrusa, vel erronea videbantur resolu-tissime pandit.

Colophon after Ptolemy's text: *Anno Christi Opt. Max. MDXIII. Marcii XII. Pressus hic Ptolemaeus Ar-gentinæ vigilantissima castigatione, industriaque Ioannis Schotti urbis indigenæ. Regnante Maximilano Cæsare semper Augusto.*

The book is dedicated to the Emperor MAXIMILIAN by JACOBUS AESCHLER and GEORGIUS ÜBELIN, »curiarum ecclesiasticarum Argentinarum causarum patroni». Eschler and Übelin were, accordingly, the editors of the work. JOHANNES SCHOTTUS is mentioned in the colophon as the printer, and from the »Ad lectorem» printed on the last leaf before the maps of the first part of the work, we learn, that the translation into latin was effected or revised by PHILESIUS or MATHIAS RINGMAN. This man (born 1482, † 1511) was for a time a professor at the Gymnasium of S:t Dié, celebrated in the history of geography for attaching Amerigo's name to the New World, discovered by Columbus. As we learn from a letter of Ioannes Franciscus Picus Mirandula to »Jacobo Aeschler I. U. Doctori & complicibus S. D.» dated »Novi, Quarto Calendas Septembris MDVIII» and printed on the reverse of the title-page, Philesius was sent to the learned Italian Prince for critical researches for the new edition. In this letter Mirandula writes: *Ne tamen aut vos poeniteat voti aut Philesium pigeat itineris, illi ipsi ad vos data est Græca Ptolemaei Geographia, quam si latinæ variis typis excusæ composueritis, cognoscetis forsitan & hanc ab illa tantum abesse quantum Ptolemaeus abfuit a regio stemmate.* It is this greek manuscript from which Philesius made his new translation. According to Raidel it does not, however, quite correspond to the great pretensions, with which it was published. In this edition not only the Latin, but also the original Greek, geographical names are for the first time given (in the catalogue of geogr. positions, Book II—VIII). It further contains an index of about 7,000 geographical names to be found in Ptolemy's text. Then comes a letter, dated Ferrara 1508, to Philesius from the famous Italian writer LILIUS GREGORIUS ZIRALDUS (GIRALDI), in which he explains the Greek system of indicating integers and fractions. Next follow the 27 tables of Ptolemy, now and then slightly emendated. In the map of the world for instance, the land which to the south of the Indian ocean unites Africa and Asia is left out; the river Thames is inserted on the map of *Albion Insula Britannica;* and the southern part of the Scandinavian peninsula on the north-western corner of the map of *Sarmatia Europea.*

Then follows the second part of the work, preceded by a second title-page:

In Claudii Ptolemei Supplementum modernior lustratio terrae marisque singula positionibus certissimis regulatius tradens ad sæculi nostri peragrationes. Pars Secunda.

Præmissus Canon brevis a tergo Chartæ Lectorem resolutissime ab errorum ambagibus absoluit.

Post tabulæ numero Viginti sequentes sese ordine suo sic locantur.

1. Hydrographia, sive Charta marina: continens typum Orbis universalem iuxta Hydrographorum traditionen.

Decem particulares tabulæ Europæ.

2. Oceani occidentalis, seu Terræ novæ.

3. Iberniæ, Angliæ, & Scotiæ.

4. Ispaniæ.

5. Galliæ.

6. Germaniæ.

7. Hungariæ, Poloniæ, Russiæ, Prussiæ, et Vualachiæ.

8. Norbegiæ, et Gotthiæ.

9. Italiæ, Siciliæ, et Sardiniæ.

10. Italiæ et Siciliæ alteræ.

11. Bossinæ, Serviæ, Greciæ et Sclavoniæ.

Due particulares tabulæ Aphricæ, ex Chartis Portugalensium sumptæ.

12. Primæ portionis Aphricæ.

13. Secundæ portionis Aphricæ.

Tres particulares tabulæ Asiæ.

14. Asiæ minoris prima, sive maior Turcia.

15. Terræ sanctæ, sive Iudeæ, Palestinæ etc.

16. Indici maris accolas hæc habet.

Quattuor deinde Chorograpiæ, particulariores et magis extensæ prædictis tabulæ.

17. Chorographia eremi Elvetiorum.

18. Chorographia provinciæ Rheni.

19. Chorographia Cretæ.

20. Chorographia Lotharingiæ.

Tractatus de variis moribus et ritibus gentium, nominumque localium et gentium declarationibus per 61. capita destributus finem operi ponit: lectu dignissimus, admiratione plenus, utilitatis simul ac oblectationis præcipuæ.

I shall return further on to this extremely important part of the work, which may be regarded as the first *modern atlas* of the world. Among the many important maps it contains, one has, though incorrectly, been ascribed to COLUMBUS himself and been the subject of extensive commentaries; and two others, evidently based on surveys undertaken during the two first voyages of VASCO DA GAMA, have hitherto been completely overlooked.

12. Nurenbergæ 1514. Fol. Incomplete but admirably printed edition by JOHANNES WERNER; without maps. It has, however, been of no small importance to cartography on account of the treatise it contains: *De quatuor terrarum orbis in plano figurationibus.*

13. Argentorati 1520. Fol. The title is surrounded by a splendid broad woodcut border, in which there is to be read: *Ptolemaeus auctus, restitutus, emaculatus cum tabulis veteribus ac novis.*

The last leaf of the text is decorated with the same splendid border as the title-page Within it are two dogs, biting each other's backs. Above is printed:

Caroli V Imperii Anno I.

and below:

Ioannes Scotus, Argentorati litteris excæpit 1520.

This edition is a slightly altered reprint of Ed. 1513, excluding the letter of PICUS MIRANDOLA, the dedication of AESCHLER and ÜBELIN, and the second title-page with the short, but interesting preface, printed on its reverse. There are the same number of plates in both editions, and they are *printed from the same blocks,* with the exception of *Tabula Nova Eremi Helvetiorum,* which, in the copy I examined, is different and reproduced on a somewhat smaller scale than the corresponding map of Ed. 1513.

14. Argentorati 1522. Fol. *Claudii Ptolemaei Alexandrini, Mathematicorum principis, opus Geographiæ, noviter castigatum et emaculatum additionibus raris et invisis, necnon cum tabularum in dorso jucunda explanatione. Registro quoque totius operis, tam Geographico, quam etiam historiali, facillimum introitum prebenti.* On these words follows, on the title-page, a long *Ordo contentorum in hoc libro totali,* in which the contents of the work are more particularised. Colophon: *Ioannes Grieninger civis Argentoratensis opera et expensis propriis id opus insigne, æreis notulis excepit, laudabilique fine perfecit XII. die Marcii Anno MDXXII.*

The long *ordo contentorum,* here excluded, promises, besides a faithful translation of the text, and the 27 maps of Ptolemy, a complete index and 20 new maps provided on their reverse with extremely interesting *(jucundissimae)* legends about the customs and manners of the inhabitants. Moreover, the edition contains a general map by Laurentius Frisius (N. T. XXXIX), also alluded to in the title, though in obscure and scarcely comprehensible words, and two new maps of eastern Asia, not mentioned on the title-page. The number of the new maps is consequently 23 (including the map of *Lotharingia*, which, in the copy I examined, was printed on the reverse of the map of the northern countries). The 27 maps of Ptolemy are (with the exception of *Tab. V Asiæ*, printed from the same block as the corresponding map of Ed. 1513 and 1520) badly executed copies, on a reduced scale, of the maps in Ed. 1513. Also 20 of the *tabulæ novæ* are copied from Ed. 1513, most of them on a reduced, a few on an unaltered scale. These (for instance the map of Britain) seem to have been transported

tion 1522, no traces to be found, showing that the learned men who superintended its publication, had any notion of, or paid any attention to the geographical discoveries made during the fourteen or fifteen years that had elapsed since the maps in the edition of 1513 were finished.

This edition with its 23 new maps thus had but slight influence on the progress of cartography. A retrograde step may on the contrary be traced, reaching perhaps its maximum in the ugly maps of Münster, and manifesting itself, amongst other ways, by the revival of the mediæval custom of decking continents and oceans with figures of castles and sailing ships, of monsters and demons, cannibals cutting their fellow-creatures into pieces and roasting them on spits, naked savages, kings with crown and sceptre, and other similar fantastic ornaments.

11. Hispania nova from *Ptolemaeus*, Ulmæ 1482. (Orig. size 530 × 389 m. m.).

to the new blocks by mechanical means, and reproduce the originals with tolerable accuracy, which is by no means the case with the plates, that give the maps of 1513 on a reduced scale. Out of the three maps, not borrowed from the edition of 1513, the above mentioned map of the world by Laurentius Frisius is certainly an original work, but bad beyond all criticism, as well from a geographical as from a xylographical point of view.

The other two maps are the *Tabula Moderna Indiæ orientalis* and *Tabula superioris Indiæ et Tartariæ Majoris* (N. fig. 62, 63). A closer examination of these maps shows that they are still almost exclusively based on traditions from Marco Polo, and the globe of Behaim. They may indeed, to a certain extent, be considered as the first reproduction of this globe, so often referred to in modern geographical literature. Consequently there are, in the tabulæ novæ of the Edi-

As regards the text to the edition of 1522 the following remarks may be of some geographical interest. The publisher of the edition, Thomas Aucuparius, praises Amerigo Vespucci in his preface, but does not seem to know Columbus. On the contrary, the honor of the discovery of the New World is exclusively attributed to the latter by the famous inscription on the *Tabula terræ novæ* (copied from the ed. 1513), as well as by the explanatory remarks to this map printed on its reverse.

Verso on folio 100:

Et ne nobis decor alterius elationem inferre videatur, has tabulas e novo a Martino Ilacomylo pie defuncto constructas et imminorem, quam prius unquam fuere, formam redactas notificamus. But it seems to be a mistake when these words are translated in the sense that Waldseemüller was the author of the maps in the Ptolemy of 1513. Evidently they only indicate, that he copied the maps of 1513 on a reduced scale for the Ptolemy of the year 1522.

In the legend to *Tabula moderna terræ Sanctæ* is the following passage: ... *Scias tamen, lector optime, iniuria aut jactantia pura, tantam huic terræ bonitatem fuisse adscriptam, eo quod ipsa experientia mercatorum et peregre proficiscentium, hanc incultam, sterilem, omni dulcedine carentem depromit. Quare promissam terram pollicitam et non vernacula lingua laudantem pronuncies.*

This phrase was repeated in the edition 1525, without attracting the attention of the religious fanatics of the time. But when they were copied by SERVETUS, in the edition of 1535, they created a great scandal, and became one of the main charges against this unfortunate sceptic, who was burned through the machinations of CALVIN.[1]

The superscription of the last of the new maps is ›*Tabu Gran Russie*‹, which, as is shown by the legend on the reverse of the map, is an abbreviation of *Tabula Gronlandie et Russie.* By this strangely abbreviated name the editor indicates an almost unaltered copy of the inner part (*Europa, Asia* and *Aphrica*) of *Orbis typus universalis juxta hydrographorum traditionem* in Ed. 1513 (N. T. XXXV). This is one of the many proofs of the carelessness with which the printing of this splendid edition was superintended.

15. Argentorati 1525. Fol. *Claudii Ptolemaei Geographicæ enarrationis libri octo Bilibaldo Pirckeimhero interprete. Annotationes Ioannis de Regiomonte in errores commissos a Iacobo Angelo in translatione sua.*

Colophon: *Argentoragi, Iohannes Grieningerus communibus Iohannis Koberger impensis excudebat. Anno a Christi Nativitate MDXXV. Tertio Kal. Apriles.*

This edition, which, as regards its typography, is richly illustrated, is mentioned by Raidel as: ›*omnium, quas vidi, editionum Ptolemaei splendidissima est*‹, and contains 50 maps, printed on 49 double folio-leaves. They are all, with the exception of the *Tabula V Asiæ*, printed from the same blocks as the maps of Ed. 1522, and, like these, are almost unaltered copies, on a reduced scale, of the maps in the edition of 1513. They, therefore, present nothing new and of special interest to the history of geography. The same may be said of the paper of REGIOMONTANUS on the errors in the translation of Jacobus Angelus, and of the extensive and singularly arranged index, provided with a separate title-page. But the following passage in PIRCKHEIMER's dedication to *Sebastianus Episcopus Brixinensis*, may be of importance as to the history of Mercator's projection.

Ego quidem si Deus permiserit, novas aliquando tabulas edere constitui, meridianis æquidistantibus, ut Ptolemaeus jubet, et haud quaquam inclinatis, quo longitudo recte ex utraque tabulæ extremitate, cum latitudine conveniat; conservabitur et certa parallelorum ratio, non solum cum meridianis, sed in vera quoque ab æquinoctiali distantia, ac quantitate diei, iis locis pro fundamento positis, quæ nostro etiam tempore diligentiori observatione sunt rectificata.

Pirckheimer here promises, or seems to promise, for his expressions are not quite clear, a modern atlas with equidistant, not converging meridians, where the degrees of longitude should be in a proper ratio to the degress of latitude and in which such places should be selected, as a basis for the delineation of the maps, as had had their positions determined by modern observations. Unfortunately Pirckheimer never re-

deemed his promise, but the above quoted words seem to indicate, that the learned humanist and patrician of Nuremberg had the intention of publishing a map on a projection resembling that, to which Mercator afterwards attached his name and which, properly understood, has become of such immense benefit to navigation.

16. Argentorati 1532. Fol. *Ptolomei Tabulæ Geographicæ cum Eandaui annotationibus eggregie illustratæ.*

Colophon: *Argentorati apud Petrum Opilionem MDXXXII.*

I have not seen this edition, which only contains eight maps in double-folio. Raidel does not mention it, and Winsor cites it after BRUNET. A more complete description is to be found in HARRISSE: *Bibliotheca Amer. Vetust.*, p. 285. The short notice here given of one of the maps induces me to suppose that they are printed from the same blocks as the eight maps in ZIEGLERS: *Quae intus continentur Syria, Palestina, Arabia, Aegyptus, Schondia* etc. *Argentorati apud Petrum Opilionem* 1532.

17. Basiliæ 1533. 4:o. The first complete Greek edition of the text to Ptolemy's geography, published by ERASMUS. No maps. A complete description of this edition is given by RAIDEL (p. 34). As before mentioned the local names had already been given in the editions of 1513 and 1520, in Greek as well as in Latin.

18. Ingolstadii 1533. *Introductio geographica Petri Apiani in doctissimas Verneri Annotationes ... Huic Accedit Translatio nova primi libri Geographiæ Cl. Ptolemaei ... Authore Vernero ... Locus etiam pulcherrimus desumptus ex fine septimi libri ejusdem Geographiæ Claudii Ptolemaei ... Cum gratia et privilegio Imperiali. Ingolstadii An. MDXXXIII.*

WERNER's edition of Ptolemy, printed in Nuremberg in 1514 with a new title-page, a new very extensive introduction, and an appendix describing some instruments invented by APIANUS. Of the actual text in the work of Werner there are only a couple of leaves reprinted at the beginning, and one at the end of that part of the work. The rest is inserted into this work from the old edition. No maps.

19. Lugduni 1535. Fol. *Claudii Ptolemaei Alexandrini Geographicæ enarrationis libri octo. Ex Bilibaldi Perckeymheri translatione, sed ad Græca & prisca exemplaria a Michaële Villanovano iam primum recogniti. Adiecta insuper ab eodem Scholia, quibus exoleta urbium nomina ad nostri seculi morem exponuntur. Quinquaginta illæ quoque cum veterum tum recentium tabulæ adnectuntur, variique incolentium ritus & mores explicantur. Lugduni ex officina Melchioris et Gasparis Trechsel fratrum MDXXXV.*

The maps of this edition are impressions from the blocks employed for the editions of 1522 and 1525, without any additions, or emendations. They are, therefore, as bad as their prototypes, and still less on a level with the geographical discoveries of the time when the edition left the press.

As seen by the preface *(Michael Villanovanus lectori)*, printed on the reverse of the title-page, the text of this edition, or rather of its commentaries, was edited by the celebrated martyr for freedom of thought, MICHAEL SERVETUS or VILLANOVANUS. The edition is, therefore, designated as ›*editio prima Serveti*‹. Remarks are often inserted in the margin of the text, displaying the vast learning of the editor. Here he also gives the modern names of the towns and

---

[1] Compare RAIDEL, p. 60, note. Raidel did not know the edition 1522.

countries mentioned in Ptolemy's text; an innovation, expressly mentioned by Villanovanus as a recommendation to his edition. In the margin of Lib. III cap. 5 there is, for instance: *Sinus Finnonicus*, i Lib. VII cap. 3: *Sinum Magnum dixi esse hodie mare de Sur, ad quem Hispani versus occidentem navigando perveniunt, terrestri itinere duorum dierum interjecto.* In Lib. VII cap. 2—5 there are numerous marginal notes respecting the discoveries of the Portuguese in Africa and southeastern Asia. Several of the legends of the *Tabulæ novæ* are also very remarkable. Here we find the above cited legend about Palestine, which, though only reprinted from the editions of 1522 and 1525, was used among the pretexts for sentencing Servetus to be burned alive. The quarrelsome enthusiast probably drew down upon himself a good deal of ill will by the way in which he here descri-

cupari contendunt, cum Americus multo post Columbum eandem terram adierit, nec cum Hispanis ille, sed cum Portugallensibus, ut suas merces commutaret, eo se contulit. Yet the map of LAURENTIUS FRISIUS is to be found in this edition, reproduced from the same block as in the edition 1525, and without the name of »America» having been erased.

20. Lisboa 1537. Fol. A translation of Ptolemy's first book is inserted in: *Tratado da Esphera com a Theorica da Sol e da Lua, e o primeiro Livro da Geographia de Ptolemeu, e duos Tratados da Carta de marear. Com muitas notas;* by the distinguished Portuguese mathematician PEDRO NUÑES. As I have not had access to this work, the reader is referred for further information to: FABRICIUS, *Bibl. Graeca,*

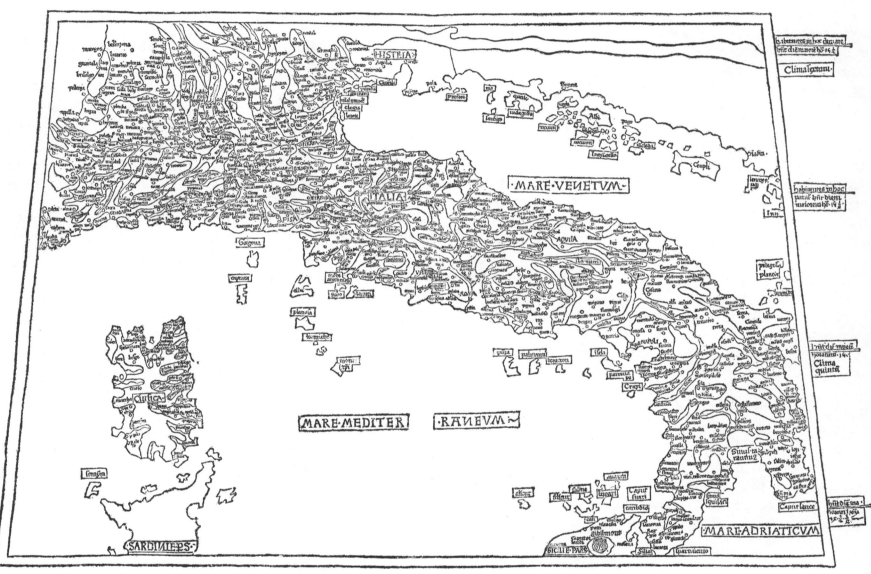

12. Italia nova from *Ptolemaeus*, Ulmæ 1482. (Orig. size 567 × 370 m. m.).

bes the national character, the habits and customs of the several peoples. For instance in the legend to »Germania» there is this passage: *Producit Hungaria boves, Bavaria sues, Franconia cepas rapas & glycerhisiam, Svevia meretrices, Boemia hereticos, Bavaria iterum fures, Helvetia carnifices bubsequas, Vestfalia fallaces, tota denique Germania ac totus Septentrio gulones et potatores;* and his estimates of the characteristics of the Spaniards and French are scarcely more complimentary.

The text to the »*Oceani occidentalis seu terræ novæ tabula*» deviates considerably from the corresponding passage in the edition of 1525. Like that passage, Servetus begins with a description of the outward appearance of Columbus, but the *procera statura* is here changed into *mediocri statura*. The text, printed on the reverse of this map, ends with the following protest against the newly adopted name for the New World, derived from that of Amerigo: *Toto itaque, quod ajunt, aberrant coelo qui hanc continentem Americam nun-*

V: 278, and POGGENDORFF, *Handwörterbuch zur Geschichte der exacten Wissenschaften*, Leipzig 1863.

21. Coloniæ 1540. 8:o. *Claudii Ptolemaei Alexandrini Philosophi et Mathematici præstantissimi libri VIII de Geographia e Græco denuo traducti ... Ioannis Noviomagi opera. Coloniae, excudebat Ioannes Ruremundanus, Anno MDXL.* A new Latin translation by the distinguished mathematician and philosopher NOVIOMAGUS or BRONCHHORST; with Greek names of places, besides the Latin ones, and an ample register. No maps.

22. Basileae 1540. Fol. *Geographia universalis, vetus et nova, complectens Claudii Ptolemaei Alexandrini enarrationis Libros VIII ... Basileæ, apud Henricum Petrum. Mense Martio Anno MDXL.* There is also a long passage on the title-page giving the information that this edition, besides a corrected translation of Ptolemy's text, also contains the modern names of towns, mountains, rivers etc., as well as a short description of the manners and customs of different peoples, and a complete index. It includes not only Ptolemy's old maps, *opera Sebastiani Munsteri novo paratæ modo*, but also several new ones, *modernam orbis faciem literis et pictura explicantes.* From the introduction we learn that the editor was SEBASTIAN MÜNSTER, Professor in Hebrew at the University of Basel, and an eminent geographer and mathematician.

This edition contains 48 maps; 27 of Ptolemy's (pl. 2—28) and 21 new ones, namely: Pl. 1. *Typus Universalis*, (N. T. XLIV); 29. *Europa;* 30. *Anglia;* 31. *Hispania;* 32. *Gallia;* 33. *Helvetia, prima Rheni tabula;* 34. *Germania;* 35. *Alsatia et Brisgoia, secunda Rheni tabula;* 36. *Tertia Rheni tabula;* 37. *Quarta Rheni tabula;* 38. *Brabantia, quinta Rheni tabula;* 39. *Svevia et Bavaria;* 40. *Franconia;* 41. *Schonlandia;* 42. *Polonia et Ungaria;* 43. *Italia;* 44. *Terra Sancta;* 45. *Novæ Insulæ Tabula* (America; with the names of *Mare Pacificum* and *Fretum Magaliani*); 46. *Africa Nova;* 47. *India Extrema* (Asia, excepting its westernmost part); 48. *Lacus Constantiensis.*

The maps are tolerably good wood-cuts, but from a chartographical point of view they are by no means comparable to the maps of Schweinheim-Buckinck, Armsheim or Sylvanus. For the longitudes are never indicated in the »tabulæ novæ» and in most of them there is no graduation at all, neither for longitude nor for latitude. Without the excuse of typographical necessity Münster often deviates from the orientation (north above, east to the right) of Ptolemy. The manner of indicating the variations of the ground is by no means fortunate, and several maps, especially of distant parts of Asia and America, are extremely poor in geographical details. They are instead disfigured by large figures of savages and monsters, entirely foreign to the cartography and also often to the nature of the country, on which they are laid down. But, on the other hand, Münster has the merit of having, in composing his »tabulæ novæ», not altogether neglected to take notice of the latest geographical discoveries. In this respect his work is in advance of the editions of 1522, 1525 and 1535. Münster is the first to give general maps of the four parts of the earth then known; his new map of »Anglia» is far more rich in details and more correct, than the new map of Britain in the preceding editions of Ptolemy; and his maps of the territories about the Rhine, of Suabia, Bavaria etc. may to this day be referred to with advantage by students of medieval geography. To Münster the merit is also due of having, in his dedication to the Bishop of Basel PHILIPPUS A GUNDELSHEIM, given references to the sources used by him for his tabulæ novæ: *Proinde viri quorum opera ego in novis meis tabulis usus sum, hi sunt.*

In descriptione Galliæ consului Oruntii excellentis mathematici topographiam. In tabula Norvegiæ, Sveciæ, Gothiæ etc. usus sum opera praestantis viri Iacobi Ziegleri, quem et pro magna parte imitatus sum in descriptione terræ sanctæ. Helvetiam et Rhætiam jampridem ministravit eximius vir Aegidius Tschudus. Alsatiam et Brisgoiam ego observavi, usus tamen in quibusdam consilio et subsidio ornatiss. viri Beati Rhenani. Alias duas Rheni tabulas ego quoque descripsi, sed cui non vulgares suppetias tulit illustrissimus princeps D. Ioannes Palatinus Rheni, dux Bavariæ et Comes Sponheimensis, omnium bonorum studiorum amator et studiosorum singularis patronus, qui mirabili suo ingenio descripsit Vastum regnum et Hunorum stationem, vulgo Hunesruck dictam. Cooperatus est denique mihi in quarta Rheni tabula humaniss. vir Iohannes Dryander pro sua Hassonia. Quintam Rheni tabulam ministrarunt Brabantini. Angli quoque et Poloni sua suppeditarunt

regna. Franconiam a Sebastiano Rotenhan nobili viro descriptam, ego mea peregrinatione nonnihil auxi. Idem feci in Bavaria, quam primum Iohannes Aventinus omnis vetustatis Germanicæ amator descripsit. Sveviam vero, fontes Danubii et Nigram sylvam ego mea lustratione et observatione in tabulam coëgi. Porro lacum Podamicum exhibuerunt Constantinenses, nempe insignes viri Iohannes Zuickius et Thomas Blaurerus.

In his introduction Münster further declares that he changed the old Ptolemaic manner of denoting geographical latitude and longitude, so far as to replace the fractions of degrees by minutes and seconds; as for instance 40°½ or 38°½⅓, by 40° 30' and 38° 50'. This very useful reform had already been introduced for astronomical data in manuscripts of the Almagest; but so far as I know, it is first employed for the indication of geographical latitudes and longitudes in the text to the map of Scandinavia of 1427 by CLAUDIUS CLAVUS (N. fig. 27).

23. Lugduni-Viennae 1541. Fol. *Claudii Ptolemaei Alexandrini Geographicæ Enarrationis, Libri Octo. Ex Bilibaldi Pirckeymheri tralatione sed ad Græca & prisca exemplaria a Michaële Villanovano secundo recogniti, & locis innumeris denuo castigati. Adiecta insuper ab eodem Scholia, quibus & difficilis ille Primus Liber nunc primum explicatur, & exoleta Vrbium nomina ad nostri seculi morem exponuntur. Quinquaginta illæ quoque cum veterum tum recentium Tabulæ adnectuntur, variique incolentium ritus & mores explicantur. Accedit Index locupletissimus hactenus non visus. Prostant Lugduni apud Hugonem à Porta. MDXLI.*

Colophon printed on separate leaves as well at the end of the whole work as between the maps and the text: *Excudebat Gaspar Trechsel, Viennæ MDXLI.*

Here we have a new edition of the Ptolemy of SERVETUS, *(Ed. secunda Serveti)* very inferior to the first one, owing to the omission of the interesting, although often offensive legends to the new maps, in the edition of 1535. Text and maps are otherwise unaltered. By *Lugduni* and *Viennæ*, recorded as the places where the book had been printed and the wood-cuts made, is here meant Lyons and Vienne, a small adjacent town on the Rhone. But, although the assertion that the maps had been engraved in *Vienne MDXLI* constitutes the alfa and omega of the work, we have here only to deal with reprints from the same blocks which had been used for the editions of 1522, 1525 and 1535. The handsome wood-cuts, which form the borders of the explanations, printed on the reverse of the maps in the editions of 1525 and 1535, have not been employed for this edition.

24. Basileæ 1541. Fol. I have not seen, this Basel-edition, but RAIDEL cites (p. 62) an edition of Münster's Ptolemy with a title-page bearing this date.

25. Basileæ 1542. Fol. The long title is the same as that on the title-page of the edition of 1540, excepting that the two last lines (*Mense Martio anno MDXL*) are excluded. The year of printing is here given in the colophon: *Basileæ apud Henricum Petrum Mense Martio, An. MDXLII.* This edition is supposed by Raidel to be the same as the edition of 1540, only provided with a new title-page. But this is not the case. It is a reprint of the edition of 1540, repeated page by page with the utmost care, but with some slight corrections, alterations of the punctuation, of the types in the rubrics, etc. Regarding the editions of 1540 and 1542 it must further be mentioned, that the part of the book paged 1—195, does not only contain Lib. II of Ptolemy, as stated by Harrisse, but the whole of his geography, excep-

ting the first book, which is printed in the commencement of the volume on unnumbered leaves, and the map-legends of the eighth. It contains also various addenda, for instance an extensive *Appendix Geographicum* (p. 157—195), by Münster; tables (p. 153) for calculating the degrees of longitude at different latitudes; a table for finding the squares etc. The book is often bound in such a manner, that the maps are placed between p. 156 and p. 157. Every map occupies a double folio-sheet, on the reverse of which the legends are printed, surrounded by wood-cut borders often executed in a masterly way, though worn, and probably originally intended for other works. They seem to have been taken at random from the printer's main stock of wood-cuts, and are generally arranged in quite an arbitrary and careless way. Some of them are ascribed to HOLBEIN.

The edition contains 54 maps. Out of these, the 27 of Ptolemy, the new map of the world *(Typus Universalis)* and the maps of *Europa, Anglia, Hispania, Gallia, Germania, Lacus Constantiensis, Svevia et Bavaria, Franconia, Polonia et Ungaria, Italia, Terra Sancta, India, Africa* and *Novæ insulæ tabula* are printed from the same blocks as the maps of the edition of 1540. The other 12 are either amended maps from Münster's editions, or new ones. Among the former may be mentioned the maps of the Rhine and Switzerland, which had probably been corrected by observations and communications collected since 1540 in Basel, and the map of Scandinavia, clumsily and badly reproduced from Olaus Magnus. The new maps are *Valesia 1:ma; Valesia 2:da; Nigra Sylva; Slesia; Sclavonia; Transylvania; Græcia* and *Bohemia.* The last one (N. fig. 69)

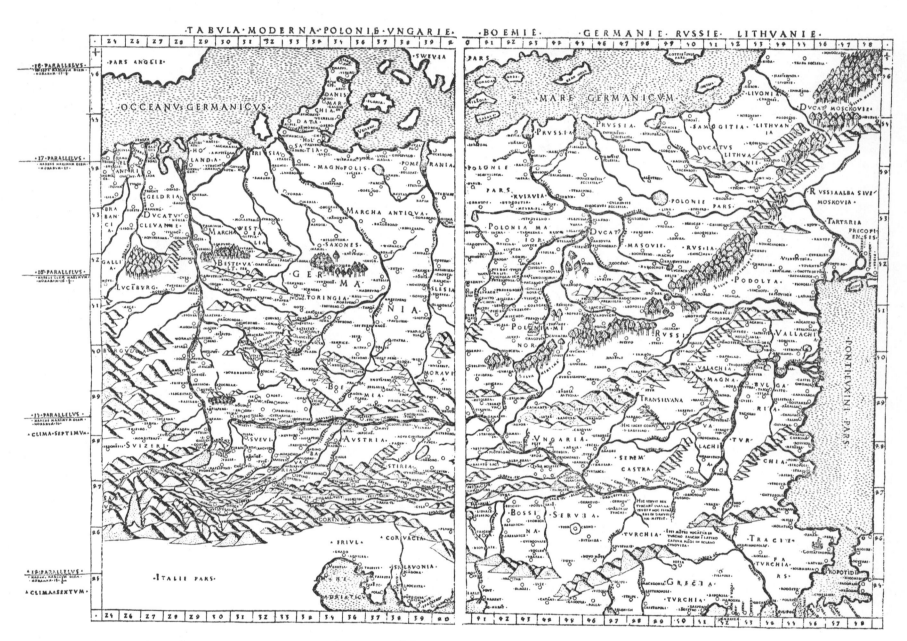

13. The first map of central Europe in copper-print, probably a copy of the map of NICOLAUS A CUSA. From *Ptolemaeus*, Romæ 1507. (Orig. size 515 × 382 m. m.).

26. Basileæ 1545. Fol. Here the long title-page also corresponds perfectly with the title of ed. 1540, with the exception of two additional lines, containing: *Adjectæ sunt huic posteriori editioni novæ quædam tabulæ, quæ hactenus apud nullam Ptolemaicam impressuram visæ sunt,* and the words *Apud Henricum Petrum Mense Martio MDXL,* are changed into *Per Henrichum Petrum Anno MDXLV.* The new edition has a woodcut on the reverse of its title-page, representing Ptolemy occupied with astronomical observations. Münster's letter of dedication to the Bishop of Basel is left out. This edition is moreover a reprint of Münster's former editions, and the reprint is so complete that it generally corresponds page for page and line for line. Even the irregular pagination of ed. 1540 and 1542 is followed.

is of interest as the first map on which the different religious and political conditions of a country are denoted.

27. Parisiis 1546. 4:o. First a Greek title; then: *Claudii Ptolemaei Alexandrini Philosophi cum primis eruditi, de Geographia libri octo, summa cum vigilantia excusi. Parisiis, apud Christianum Wechelum, sub Pegaso, in vico Bellovacensi, Anno MDXLVI.* Reprint of the Greek edition of Erasmus 1533; no maps.

28. Venetiis 1548. 8:o. *Ptolemeo. La Geografia di Claudio Ptolemeo Alessandrino, con alcuni comenti & aggiunte fatteui da Sebastiano munstero Alamanno, Con le tavole non solamente antiche & moderne solite di stamparsi,* *ma altre nuove aggiunteui di Messer Iacopo Gastaldo Pia-montese cosmographo, ridotta in volgare Italiano da M. Pietro Andrea Mattiolo Senese medico Eccellentissimo. Con l'aggiunta d'infiniti nomi moderni, di Città, Provincie, Castella, et altri luoghi, fatta con grandissima diligenza da esso Meser Iacopo Gastaldo, il che in nissun altro Pto-lomeo si retrova. Opera veramente non meno utile che necessaria. In Venetia, per Gioâ. Baptista Pedrezano. Co'l privilegio dell' Illustriss. Senato Veneto per anni X. MDXLVIII.*

Colophon: *In Venetia, ad Instantia di messer Giouan-battista Pedrezano libraro al segno della Torre a pie del ponte di Rialto. Stampato per Nicolo Bascarini nel Anno del Signore, 1547, del mese di Ottobre.* A small but very elegant Italian edition with plates, handsomely engraved in copper by the famous cosmographer GASTALDI. They are 60 in number, of which 26 are Ptolemy's (his map of the world is excluded) and there are 34 new ones, namely: *Inghilterra; Spagna; Francia; Brabantia, Fiandra & Holandia; Ger-mania; Schiavonia et Dalmatia; Italia; Piemonte; Marcha Trevisana; Marcha de Anchona; Sicilia et Sardegna; Prusia et Livonia; Scholandia* (after ZIEGLER); *Polonia et Ungheria; Grecia; Mauritania; Aphrica minore; Mar-maricha; Egitto; Aphrica; Natolia; Moschovia; Soria; Persia; Arabia Felice; Calecut; India Tercera; Terra nuova* (South America); *Nova Hispania* (Central America); *Terra nova de Bacalaos; Isola Cuba; Spagnola; Universale nuovo; Carta marina universale.* A whole series of plates of the New World is here met with, for the first time, and some of them are of no slight interest to the history of geography.

29. Basileæ 1552. Fol. *Geographiæ Claudii Pto-lemaei Alexandrini... Libri VIII, partim a Bilibaldo Pirckheymero translati ac commentario illustrati, partim etiam Græcorum antiquissimorumque exemplariorum colla-tione emendati atque in integrum restituti. His accesserunt*

Scholia, quibus exoleta locorum omnium nomina in Ptolemaei libris ad nostri seculi morem exponuntur.

Indices duo hactenus a multis desiderati... Conradi Lycosthenis Rubeaquensis opera adjecti.

Quibus praefixa est epistola in qua de utilitate tabu-larum Geographicarum ac duplicis indicis usu late dis-seritur.

Tabulae novae quæ hactenus in nulla Ptolemaica edi-tione visæ sunt per Sebastianum Munsterum.

Geographicæ descriptionis compendium, in quo varii gentium ac regionum ritus, mores atque consvetudines per eundem explicantur.

Cum Regiæ Majestatis Gratia et Privilegio ad sex-ennium.

The year and the place when and where the work was printed, are not directly stated, but the introduction by Mün-ster, a letter of licence for six years by the King of France, and a letter by CONRAD LYCOSTHENES to JOANNES FRISIUS »De utilitate tabularum geographicarum», are dated 1552. Here we have a new edition of MÜNSTER's Ptolemy, provided with an elaborate index. The plates are the same, and printed from the same blocks as those of the edition of 1545, with the exception of the map of *Lacus Constantinensis* being left out, instead of which a map of *Pomerania* (map 42) is inserted.

30. Venetia 1561. 4:o. *La geografia di Claudio Tolomeo Alessandrino, Nuovamente tradotta da Greco in Italiano da Girolamo Ruscelli.... In Venetia, Appresso Vincenzo Valgrisi, MDLXI.* A new translation into Italian with numerous remarks and extensive addenda by RUSCELLI. It is notified on the title-page that the work was to contain

26 old and 36 new maps, but there are in reality 27 old and 37 new plates. The maps are enlarged copies of Ga-staldi's maps in the edition of 1548, excepting: »*Universale Novo*» drawn on a new projection, and called »*Orbis descrip-tio*»; the map of Britain, for the northern part of which the author has modified the type hitherto followed of the »*Tabula nova Hiberniae, Angliae et Scotiae*» in the Ptolemy of 1513; and the map of Central America *(Nueva Hispania)*, where Yucatan is drawn as a peninsula, and not separated from the main by a strait as in the map of 1548. Four maps are added, viz: *Toscana, nova tabula;* Zeno's map or »*Septentrionalium partium tabula nova*»; *Brasil, nova tavola,* and the old map of the world by Ptolemy, excluded from the edition of 1548, is here re-inserted.

Two important innovations were introduced by this edition into cartographical literature, or at least into the literature of atlases, viz:

1st. The division of the map of the world into two hemispheres, of which the right one represents the Old World and the left the New. As may be seen by the fac-simile pl. XLV, this map of Ruscelli is extremely well designed, and engraved on copper with Italian taste and Italian skill. But the first idea of dividing the map of the world into two parts, one for the Old World and one for the New, belongs to STOBNICZA, of whose map a fac-simile is given pl. XXXIV.

2d. The drawing of the map of the Arctic Regions, in accordance with a type published in Venice three years pre-viously in: *De i commentarii del Viaggio in Persia... libri due. Et dello scoprimento dell' Isole Frislanda, Eslanda, Engrouelanda, Estotilanda, & Icaria, fatto sotto il Polo Artico, da due fratelli Zeni, M. Nicolò il K. e M. Antonio. Libro uno, con un disegno particolare di tutte le dette parte di Tramontana da lor. scoperte.... in Venetia per Francesco Marcolini MDLVIII.* If the remarkable map in this little work, had not received extensive circulation under the sanction of Ptolemy's name, it would probably have been soon forgotten. During nearly a whole century it now exercised an influence on the mapping of the northern coun-tries, to which there are few parallels to be found in the history of cartography.

According to THOMASSY (*Les Papes géographes* in *Nouv. Ann. de Voyages,* 33, 1853, p. 155), the maps of this edition of Ptolemy have served as models for the famous wall-pain-tings executed in the Vatican during the reign of Pope Pius IV.

31. Venetiis 1562. 4:o. *Geographia Cl. Ptolemaei Alexandrini olim a Bilibaldo Pirckheimherio translata, at nunc multis codicibus græcis collata, pluribusque in locis ad pristinam veritatem redacta a Iosepho Moletio Mathematico... Venetiis, apud Vincentium Valgrisium MDLXII.*

The maps in this edition are the same in number, and are all printed from the same plates, as those in the edition of 1561.

32. Venetia 1564. 4:o. I have not seen this edition. According to Winsor it is a new edition af Ruscelli-Gastaldi's Ptolemy of 1561, divided into two separate works: *La Geo-grafia di Claudio Tolomeo tradotta di Greco in Italiano da Ieronimo Ruscelli,* and *Expositioni et introduttioni uni-versali Di Ieronimo Ruscelli sopra tutta la Geografia di Tolomeo,* both of which are printed in Venice, *appresso Gior-dano Ziletti, al segno della Stella, MDLXIIII.* It contains 64 maps, printed from the same plates as the maps in the edition of 1561.

33. Venetiis 1564. 4:o. A re-issue of the Latin edition of 1562. Only the title and the preface are reprinted. On the reverse of the last folio of the introduction *(Aloysio Cornelio Cardinali ... Iosephus Moletius)* is a woodcut, representing Ptolemy in oriental attire observing the stars.

———

After 1570, when the first edition of the *Theatrum Orbis terrarum* by Ortelius was published, the new editions of Ptolemy lost their main interest to the history of cartography. I shall, therefore, only give a summary catalogue of them with a few remarks, which suggest themselves from a cartographical point of view.

34. Venetiis 1574. 4:o. A new edition of Ruscelli-Gastaldi's Ptolemy by Gio. Malombra, printed in Venice, *Appresso Giordano Ziletti MDLXXIIII.* The maps are

36. Coloniæ Agrippinæ 1584. Fol. A new edition of Mercator's Ptolemy, but with the addition of Ptolemy's text. The maps are printed from the same plates as in the foregoing edition.

37. Venetiis 1596. 4:o. *Geographiæ Universæ tum veteris tum novæ absolutissimum opus ... Auctore Io. Ant. Magino Patavino, Mathematicarum in almo Bononiensi Gymnasio professore.* An extensive geographical work by Io. Antonius Maginus, divided into two volumes, with separate title-pages and indexes. Vol. I contains the text of Ptolemy's geography (excepting the map-legends from the VIII Book) and separately paginated commentaries; Vol. II the 27 maps of Ptolemy with his map-legends, and 37 ›Tabulæ recentiores›, all admirably engraved in copper by Hieronymus Porro. The maps are printed on single quarto-leaves, with the exception of a map of the two hemispheres

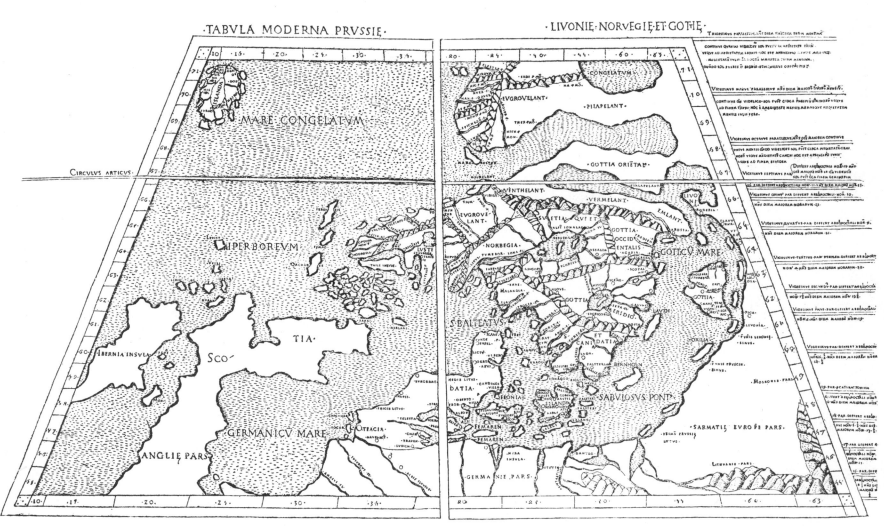

14. The first map of Scandinavia in copper-print. From *Ptolemaeus*, Romæ 1507. (Orig. size 568 × 312 m. m.).

a reprint from the same plates as those used in the editions of 1561 and 1562, with the exception of Ptolemy's map of the world, for which Malombra had returned to the original conical projection of Ptolemy. Moreover, one map is added: the map of ›*Territorio di Roma*›.

35. Coloniæ Agrippinæ 1578. Fol. *Tabulæ Geographicæ Cl. Ptolemaei ad mentem autoris restitutæ et emendatæ per Gerardum Mercatorem Illustriss: Ducis Cliviæ etc. Cosmographum.* This edition, the first with Mercator's series of maps, contains only the 27 old maps of Ptolemy in double folio, and a smaller one of the Delta of Nile. All the maps are printed from plates engraved in a masterly style by Gerard Mercator. This edition does not contain the text of Ptolemy, but extensive explanations of the modifications introduced in the accepted Ptolemaic types are printed on the reverse of the maps. An elaborate index of the 7,000 names in Ptolemy's geography concludes the work.

of the world on a stereographic projection, copied from Mercator's Atlas, and occupying two quarto-leaves. There is also an extensive text for the new maps by Maginus. A mariner's compass with the variation of the needle indicated is inserted in the map of *Forum Iulii et Histria.* Such a drawing had previously been given in 1532, in *Tabula Universalis Palestinæ* of Ziegler. In Waghenaers first sea-charts of 1584 no allusion is made to the variation of the compass, but in some of Barents' Mediterranean charts of 1595, there is, as shown by fig. 24, a difference made between the *Directorium Nauticum vulgare* (our common magnetic compass) and the *Directorium Nauticum Italicum* (the compass corrected for magnetic variation and consequently giving the true North). Diagrams showing the variation of the compass, but without combination with any map, are found in the *Cosmographiæ introductio cum quibusdam Gæometriæ ac Astronomiæ principiis ad eam rem necessariis, Ingolstadii MDXXIX* (at the end *MDXXXIII*).

38. Coloniæ Agrippinæ 1597. 4:o. A reprint of the preceding edition executed in Cologne. Even the maps are copied with extreme care from the edition Venetiis 1596 without any additions or alterations. Two somewhat dissimilar title-pages, with the same printer and year of printing, seem to exist for this edition.

39. Venetia 1598. Fol. A handsomely illuminated translation into Italian of the Ptolemy of MAGINUS (N:r 37) printed by GIO. BATTISTA & GIORGIO GALIGNANI FRATELLI. The maps (by PORRO) are of the same number and printed from the same plates as those of the Latin original.

40. Venetia 1598. 4:o. A new Italian edition by GIOSEFFO ROSACCIO. Most of the maps are faithful copies of the maps in the edition of 1561. Even the map of Zeno is here inserted with the legend of 1561 beginning: »*Il disegno, ò la descrittione di questa Tavola Settentrionale, non è stato fin quì in alcuno de gli altri Tolomei, così Latini, come Italiani*». This map had then been introduced into six previous editions (1561, 1562, 1564, 1564 bis, 1574 and 1597).

41. Venetia 1599. 4:o. Identical with the preceding edition: only with a new date on the title-page.

42. Dusseldorfii 1602. Mercator's Ptolemy with 34 maps. I have not seen this edition, indicated in *Catalogue of the printed maps in the British Museum*, II p. 3403.

43. Francofurti & Amsterodammi 1605. Fol. Edition with Greek and Latin text and 28 maps from Mercator's plates. The title-page of this edition varies. In some copies Frankfort, in some Amsterdam and in others both these towns are mentioned as the printing place (conf. RAIDEL p. 35). From the preface by I. HONDIUS we learn that PETRUS BERTIUS had assisted in the publication of this edition. Hondius here also gives us some interesting information as to the sale of the copper-plates of Mercator.

44. Coloniæ Agrippinæ 1608. 4:o. A new edition of MAGINUS with maps from the same plates as those in the edit. »Coloniæ Agrippinæ 1597».

45. Arnheimii 1617. 4:o. A new edition of the preceding one, with most of the maps printed from the same plates.

46. Lugduni Batavorum 1618. Fol. A new edition of the maps of Mercator with Greek and Latin text, published by PETRUS BERTIUS BEVERUS. To this edition are added *Gerardi Mercatoris Rupelmundani in tabulas Ptolemaicas a se delineatas Annotationes; Tabula itineraria ex illustri Peutingerorum bibliotheca, quæ Augustæ Vindelicorum est, beneficio Marci Velseri Septemviri Augustani in lucem edita; Fragmenta Tabulæ itinerariæ antiquæ* and finally *Abrahami Ortelii Geographiæ veteris tabulæ aliquot.*

47. Padua 1621. 4:o. A reprint, carefully executed by the brothers PAOLO and FRANCESCO GALIGNANI, page for page, from the edition which was published in Venice in 1598 by the brothers GIO. BATTISTA and GEORGIO GALIGNANI and having, like these, maps from the same plates as those already used in the editions Venetiis 1596 and 1598.

48. Francofurti-Amsterodammi 1624. Under this date an edition is cited in the catalogue of HENRY C. MURPHY,

with exactly the same title-page as in the edition of 1605; probably only a title-edition.

49. Franequeræ et Trajecti ad Rhenum 1698 (and 1695?). **Fol.** A reprint of the maps of Mercator. No text. The year 1695 is given in MURPHY's catalogue; but the difference possibly depends on a printer's error.

50. Trajecti ad Rhenum 1704. Fol. Reproduction of the preceding edition. (LELEWEL, II: p. 209).

51. Amstelædami 1730. Fol. Reprint of the maps of Mercator without the Ptolemaic text, but with an extensive index.

52. Paris 1828. 4:o. French translation of Ptolemy's geography by HALMA. No maps are added to the work, and only the first book of the Greek-French text is published.

53. Essendiæ 1838. Fol. A translation by Dr. FR. WILH. WILBERG of *Claudii Ptolemæi Geographiæ Libri VIII*, commenced but not finished.

54. Leipzig 1843—1845. 12:o. A stereotyped edition with Greek text, published by CARL FRIEDR. AUG. NOBBE. It contains only a single insignificant map, which is intended to represent an imitation of Ptolemy's map of the world; but in this respect it is entirely useless, following Ptolemy neither in the projection, nor in the distribution of land and sea. The preface of the first part is dated 1843, that of the third part 1845. In the introduction (*Epistola litteraria ad Fridericum Albertum de Langenn*) an account is given of the codices and earlier editions used for this work.

55. Paris 1867. Fol. Photolithographic reproduction of a manuscript of Ptolemy's geography, preserved at the convent Vatopedi on mount Athos, published by PIERRE DE SEWASTIANOFF. According to the statement in the introduction of Mr. VICTOR LANGLOIS, this manuscript dates from the end of the 12th, or the beginning of the 13th century. Of the 27 Ptolemaic maps some are lost (the map of Britannia, the eastern part of Hispania, the eastern part of Tabula sexta, and the western part of Tab. septima Asiæ). Evidently the maps in this codex were never very carefully executed and those now extant are sadly reduced by wear of time. According to present requirements, the reproduction perhaps also leaves something to be desired. But an opportunity is supplied to the enquirer by this edition, of convincing himself, in his own study, how exactly and minutely the fine maps which were published in Ptolemy's name at the end of the 15th century, correspond with manuscript maps from the beginning of the 13th, as regard the main geographical features and the legends.

56. Parisiis 1883. The first part of a new, critical, Greek-Latin edition by CARL MÜLLER. At the beginning of the work an *index codicum* is inserted, enumerating thirty-eight manuscripts consulted by the author, without however communicating any further information regarding them. The complete work will consist of text in two volumes and one volume of maps.

———

According to the above critical catalogue, from which a number of pseudo-editions, enumerated in the next chapter, are excluded, 56 editions of Ptolemy's geography have hitherto been published, one of them photolithographically and fifty-five by book-print. Many of these editions, however, are not complete; sometimes the maps, sometimes more or less of the text is wanting. Others again are only title-editions. Yet by a careful comparison it will be found that many a supposed title-edition is an actual reprint of a former edition, rigorously following the original, page for page, line for line. In consequence of the cheapness of manual labor during the 15th and 16th centuries the reprint of a ready prepared text, even of such a voluminous work as Pto-lemy's geography, might not have been connected with very heavy expenses, or difficulties. It was more costly and difficult to prepare new plates or blocks for the printing of the maps. For this reason the copper-plates and wood-cuts were often used for repeated editions, as may be perceived by the accompanying table. With regard to this it must be borne in mind that the old maps are not generally signed, and that it is often difficult to decide whether a name given on the title-page, or in the preface, in connection with the maps, belongs to the editor, to the drawer of the maps, or to the engraver.

Copper-prints and wood-cuts in the editions of Ptolemy's geography.

1. Copper-print by SCHWEINHEIM-BUCKINCK: N:o 4 1478; N:o 7 1490; N:o 8 1507; N:o 9 1508.
2. Wood-cut by JOHANNES DE ARMSHEIM: N:o 5 1482; N:o 6 1486.
3. Wood-cut, first published in an edition by AESCHLER and ÜBELIN: N:o 11 1513; N:o 13 1520.[1]
4. Wood-cuts of the preceding maps, reduced into a smaller size by WALDSEEMÜLLER: N:o 14 1522; N:o 15 1525;[2] N:o 19 1535; N:o 23 1541.
5. Wood-cut from SEBASTIAN MÜNSTER: N:o 22 1540; N:o 24 1541; N:o 25 1542; N:o 26 1545;[3] N:o 29 1552.[3]
6. Copper-print first published in an edition by RUSCELLI and MOLETIUS: N:o 30 1561; N:o 31 1562; N:o 32 1564; N:o 33 1564 bis; N:o 34 1574.[4]
7. Copper-print by MERCATOR: N:o 35 1578; N:o 36 1584; N:o 42 1602; N:o 43 1605; N:o 46 1618; N:o 48 1624; N:o 49 1698; N:o 50 1704; N:o 51 1730.
8. Copper-print by HIERONYMUS PORRO: N:o 37 1596; N:o 39 1598; N:o 47 1621.
9. Copper-print in an edition by ROSACCIO: N:o 40 1598; N:o 41 1599.
10. Copper-print by PETRUS KESCHEDT; exact copies of the previous, executed in Germany: N:o 38 1597; N:o 44 1608; N:o 45 1617.[4]

Maps only printed once: N:o 1 Maps engraved in copper for the edition of MANFREDUS and PETRUS BONUS 1472 (?); N:o 3 Maps engraved in copper for the edition of BERLINGHIERI 1478 (?); N:o 11 Wood-cut by BERNARDUS SYLVANUS 1511; N:o 28 Maps engraved in copper by GASTALDI 1548; N:o 55 Chromo-lithographe from the Athos-manuscript 1867. Without maps, or with a few only, are the editions N:o 2 1475; N:o 12 1514; N:o 16 1532; N:o 17 1533; N:o 18 1533; N:o 20 1537; N:o 21 1540; N:o 27 1546; N:o 52 1828; N:o 53 1838; N:o 54 1843—1845; N:o 56 1883.

---

[1] Exception: Tabula nova Eremi Helvetiorum. [2] Exception: Tab. V. Asiæ. [3] Exception: Some new maps of European countries. The same blocks as those employed in these editions of Ptolemy, have afterwards been used for Münster's Cosmography and for other works, printed in Germany during the latter half of the 16th century. [4] Exception: Some maps from other plates.

II.

Pseudo-editions of Ptolemy. Ptolemy's errors and merits.

The catalogue of editions of Ptolemy's geography given above is, as far as I know, *complete*. It contains a few editions never noticed before, and among them one of the oldest and, in a cartographical point of view, most important. But on the other hand I have excluded twenty six works, erroneously enumerated among editions of Ptolemy's geography, viz:

1. *Cracoviæ 1512,* and
2. *Cracoviæ 1519,* two different editions of the *Introductio in Ptolomæi cosmographiam* etc. by IOANNES DE STOBNICZA.

I shall further return to this work and to the remarkable map (N. T. XXXIV) in the edition of 1512. It need here only be observed, that the introduction of Stobnicza certainly contains some geographical statements from Ptolemy, but not one single page directly translated from him, nor any Ptolemaic maps. It is mainly composed of extracts from the works of AENEAS SYLVIUS, ISIDORUS, OROSIUS and ANSELMUS.

3. *Lovanii 1597;* 4. *Louvain 1597* (English edition); 5. *Lovanii 1598;* 6. *Duaci 1603;* 7. *Douay 1605;* 8. *Dovay 1607;* 9. *Dovay 1611.* Different editions of a very important and, as may be conceived by the numerous editions, highly appreciated work by CORNELIUS WYTFLIET, of which the latin title is: *Descriptionis Ptolemaicae Augmentum.* Here, it is true, Ptolemy's name is on the title-page. But the work, does not contain one line of Ptolemy. It describes a part of the globe entirely unknown to the ancients, and this in a manner completely different from the style of the Alexandrian geographer.

Moreover, the following »editions» will have to be excluded, their insertion in the catalogue evidently being due to a confounding of other works of Ptolemy with his geography, to errors of printing, or to other mistakes. I must especially point out that the contribution of Santarem to Ptolemy's bibliography seems to be so hastily and uncritically written that the statements in his short, but often cited paper, are not deserving of the slightest regard, when not confirmed by more reliable references.

10. *Bologna 1480* (THOMASSY, *Les papes géographes et la cartographie du Vatican,* in *Nouvelles Ann. des Voyages,* T. 32, 1852, p. 57; T. 33, 1853, p. 151; T. 34, 1853, p. 7). Probably identical with the edition Bononiæ MCCCCLXII,

for the misprinted date of which Thomassy has adopted 1480 instead of 1472.

11. *Florence 1481.* This edition is cited by Thomassy (*Nouv. Ann. des Voyages*, T. 32, p. 75). Probably Berlinghieri's *Septe giornate* is here meant, in which case this work has already been enumerated by Thomassy among the editions of Ptolemy.

12. *Cl. Ptolomaei Geographiae libri VIII. 1500.* »*Noted by Butsch, but thought to be apocryphal*» (WINSOR). I have in vain searched for this edition in other catalogues, and suppose that the date is due to an error in writing 1500 instead of 1490.

13. *Nuremberg 1524.* Cited by Santarem (*Bulletin de la Société de Géographie*, Sér. II, T. 8, 1837, p. 175). The date depends no doubt on an error in writing 1524 instead of 1514.

14. *Paris 1527;* 15. *Venice 1528.* These two editions are also cited by Santarem, but are not found either in the *Catalogue of the printed maps in the British Museum*, or in any other of the many Catalogues I have consulted. Evidently Santarem's citation of these editions depends on errors in writing, or printing.

Basileæ, ex officina Henricpetrina, 1571, most of the maps of Münster's Ptolemy are inserted. They are numbered in connexion with the other text and provided on the first page with the usual Ptolemaic legends, borrowed from the Book VIII. The supervision of the printing, however, has been done so carelessly, that the map of the world and three maps of Asia have been omitted, whereas the map of Greece and the 4th, 5th and 9th maps of Asia have been introduced twice. In the book there also appear small wood-cut maps of Euboea, Creta, Lesbos, Rhodes, Cyprus and Cephalonia. The work can not be enumerated among editions of Ptolemy.

22. *Venice 1575;* 23. *Basel 1582.* Again two editions cited by Santarem, but otherwise unknown.

24. *Bononiæ 1608* (LELEWEL, II, p. 209). Evidently the edition: »*Anno 1608. In celeberrima Agrippinensium Colonia excudebat Petrus Keschedt*». The error of stating Bononiae as the printing place has arisen from the sentence »*Mathematicarum in Almo Bononiensi Gymnasio publico professore*» being added to the name of Maginus, on the title-page.

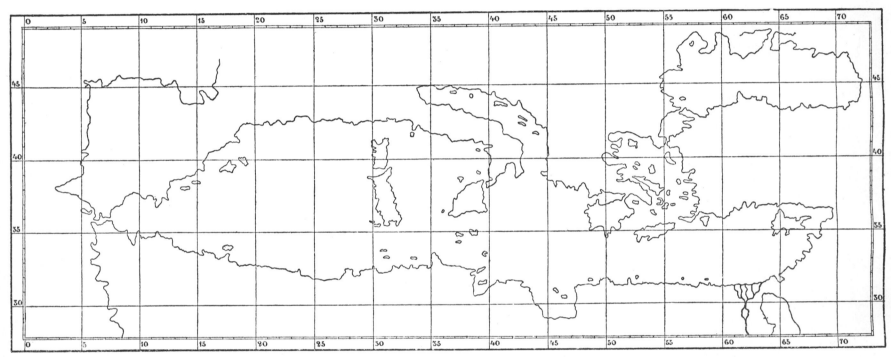

15. The Mediterranean and Black seas according to *Ptolemy* Tab. II, III, V—X Europæ, I—III Africæ, I—IV Asiæ.

16. *Basel 1538, fol.* according to Santarem. Here a confusion no doubt exists with »*Claudii Ptolemaei magnæ constructionis libri XIII. Theonis Alexandrini in eosdem commentar. libri XI*», which was published 1538 at Basel by SIMON GRYNÆUS.

17. *Venetiis 1543.* As shown by D'AVEZAC, the citation of this edition depends on a printing-error (the year 1543 in stead of 1548) in ZURLA: *Di Marco Polo et degli altri viaggiatori Veneziani... con Appendice sulle antiche mappe idrogeografiche lavorate in Venezia.* Venezia 1818, T. II, p. 368.

18. *Basileæ 1555.* The citation of this edition depends on a printing-error in FREDERIK MULLERS, *Catalogue of books etc. relating to America*, Amsterdam 1877, N:o 2626. Muller refers to HARRISSE, *Additions*, N:o 155. But there is only spoken of the edition of 1545.

19. (Place of printing not stated) *1559.* Again one of Santarem's »*Ptolémée sans cartes*», not mentioned by any other of the bibliographers of Ptolemy.

20. *Venice 1568.* Cited only by Santarem.

21. *Basileæ 1571.* In *Strabonis rerum geographicarum libri septemdecim a Guilielmo Xylandro... recogniti...*

25. *Trajecti ad Rhenum 1695.* Cited by MURPHY; probably identical with the edition: Franequeræ et Trajecti ad Rhenum 1698.

26. *Parisiis 1715.* Among editions of Ptolemy Lelewel cites (p. 209) BERNARD. DE MONTFAUCON's *Bibliotheca Coisliniana olim Segueriana*, Parisiis 1715, fol. This magnificent work does not contain any essential part of Ptolemy's geography, but, p. 611—768, an extensive comparison between the text of a Greek codex of Ptolemy *(Codex Coislinianus)* and the edition Lugduni Batavorum 1618.

In the critical catalogue, which I have given here, the number of the different editions of Ptolemy's geography is much reduced, 26 spurious editions being excluded. But there yet remain 56 authentic editions, most of them provided with maps. Thirty-three were issued before 1570, 26 of which contain about 700 old Ptolemaic maps and about 400 »tabulæ novæ». By comparing this number with the small number of maps printed before 1570 without any connection with Ptolemy's geography, we get an idea of the great influence that the Alexandrian geographer, after fourteen hundred years, still exercised not only upon the history of geography, but upon the whole history of civilisation. Under such circumstances it would be of no small interest to have a reliable answer to the questions: how did the first printed maps of Ptolemy agree with the manuscripts? to what extent can these latter claim to be faithful reproductions of Ptolemy's own maps, and what was the main source of Ptolemy's extensive geographical knowledge?

The already very extensive literature on Ptolemy's geography does not, so far as I know, contain any answer to the first question, founded on a careful collation between the printed maps and the manuscript, and I had myself only the opportunity, during short visits to libraries, of examining a few of the many codices of Ptolemy's geography still extant. I have, however, been able fully to convince myself:

1st. That the original maps of Ptolemy are faithfully reproduced in Berlinghieri's *Septe Giornate della Geographia*,

the maps here reproduced with the fac-simile of the Athos codex, convince himself that no essential change has been introduced into Ptolemy's maps through these errors. To judge from the reproduction, published in Paris 1867, and from the introduction by M. Victor Langlois, the Athos manuscript itself is not very well preserved. It is sadly worn, and some of the 27 original maps have now disappeared. In many places the maps are so injured by moisture and age, that the outlines of the countries and the names are often in such a state that they cannot be deciphered. But in this manuscript, of the 12th or 13th century, one finds exactly the same maps, with the same territorial limits, the rivers following the same courses, the same mountains-ranges, and the same legends, the same division into climates etc., as in the maps of 1478 or 1490. If, for instance, the *Tabula V Europæ* in the edition of 1478 is compared with the corresponding map in the manuscript of Athos, it will be found that both these maps embrace exactly the same parts of the earth, and that exactly the same parts in the north,

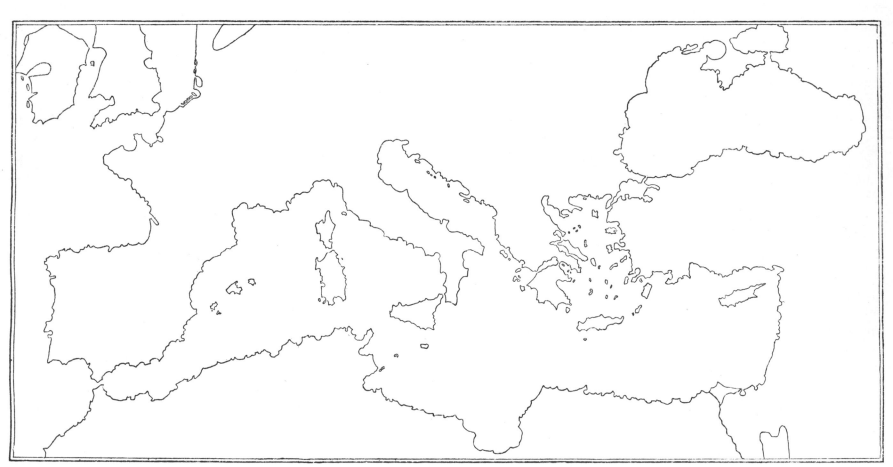

16. The Mediterranean and Black seas according to a portolano by *Dulcert* 1339.

Firenze c. 1478. But this reproduction is full of gross typographical errors, and the copper-engravings have serious technical defects.

2d. That the manuscript-maps, redrawn on a kind of conical projection, are very faithfully reproduced in the edition of Bologna c. 1472. Here the technical execution of the maps, perhaps the first ones printed from copper-plate, is also very defective.

3d. That the manuscript maps of Nicolaus Germanus carefully reproduce the original Greek maps without any other modification than that of Ptolemy's equidistant cylindrical projection being changed into a projection with rectilinear, converging meridians.

4th. That the maps printed on copper by Schweinheim-Buckinck, Romæ 1478, 1490, 1507 and 1508, here reproduced in facsimile (T. I—XXVII), are faithful and unaltered copies of the original maps of Nicolaus Germanus.

It was inevitable that many errors, through negligence and caprice of the copyists, should have slipped in during the repeated transcriptions, but any one may, by comparing

east and south, in both these maps, have been left blank, without any territorial detail. Here are, within the frame of the map, portions of countries, of which complete maps are given on other sheets. On these territories are written: *Italiæ pars, Magnæ Germaniæ pars, Iazigum Metanastarum pars, Daciæ pars* etc. which names correspond to Ἰταλι.. μερος; μεγαλης γερμανιας μερος; ιαζυγων μηταναστων μερος, and δακιας μερος, in the Athos codex. The same is the case with the other legends on both maps. In both the Danube forms the northern border to the finished part of the maps, and this river makes, in both, the same bends, receives from the south the same tributaries, which again have exactly the same courses, and rise from the same mountains etc. If the manuscript had been drawn with more care and artistic skill and was not damaged, I have no doubt that name after name, river after river, and town after town, with very few exceptions, would be found on the maps of the old manuscript and on the corresponding printed maps; and although, so to speak, several generations of copyists separate these two atlases, yet even the direction in which the names are written (for instance from right to left, or from above downwards) indicates an affinity between the two atlases, which incontestably proves that both are not only copies, or copies of copies from the same original, but also present a faithful, though more or less roughly drawn reproduction of the prototype.

No older manuscript than that from Vatopedi seems to be known. The fact that these maps exhibit no trace of christian influence, makes it all the more improbable that they should have been much altered or modified between A. D. 200 and 1200. All this seems to prove that the 27 maps, given here in fac-simile, furnish us with a very faithful copy of the atlas composed in the middle of the second century after Christ.[1] When the maps were engraved on copper, however, some changes as to the manner of drawing were introduced, mainly with the view of reproducing in black that which had been rendered by colours on the old manuscripts. These changes are of less importance in the oldest editions, printed in Florence and Bologna. Here the engraver has not been able to emancipate himself from the style of drawing, employed for the manuscript-maps.

With regard to the sources of Ptolemy's atlas Dr. N. H. Brehmer in a detailed, but perhaps not sufficiently critical work (*Entdeckungen im Alterthum*, Weimar 1822) has put forward the opinion, that the maps were not of Greek or Roman, but of Tyrian origin, and that their principal sources were the experiences gained during sea and land voyages from the Phenician towns. Professor A. H. L. Heeren, on the other hand, has in *Commentatio de fontibus Geographicorum Ptolemaei tabularumque iis annexarum* etc., Gottingæ 1827, made an attempt to prove that the atlases of Marinus and of Ptolemy rest, not upon old Tyrian sources, but upon contemparary Greek and Roman writings and itineraries. But the arguments of Heeren are not convincing, and I do not hesitate to adopt the opinion of Brehmer with regard to this question, which is of such great importance to the history of geography. An atlas so exten-

sive and comparatively so correct as that of Ptolemy cannot have been the work of a few years by some few geographers. Experience, collected during centuries, has evidently been necessary for its production. We have no evidence, that such collections of maps as those of Ptolemy, or such works, as are now termed atlases, existed in ancient Rome or Greece. On the contrary all the passages regarding maps in Pliny, Strabo and other old authors, only speak of isolated drawings of the earth, of which not a single one is still extant. It is in no wise proved, that they had any resemblance to maps in a Ptolemaic or modern sense. Many of the numerous names in Ptolemy suggest a Phenician origin; even according to Heeren, the oikumenæ of Ptolemy extend to the north, south, east and west far beyond the limits of the military expeditions of the Romans and Greeks, and on the very maps we may often trace the pathway of the caravan, by which the knowledge of the distant countries arrived on the Mediterranean coast. As the *tabulæ novæ* added to Ptolemy's geography in the 15th century are founded on sea-charts or portolanos, originally drawn for practical use and attaining a rare perfection by being improved generation after generation; so the maps of Ptolemy are, as is expressly pointed out in his own text, in the first place founded on the Tyrian charts of Marinus, and these again have, as proved by Ptolemy's criticism of the works of Marinus, only been the last, the most complete and correct of the Phenician »portolanos» during Ptolemy's time. This does not exclude the belief that Ptolemy, in compiling his work, added some corrections founded on observations by Hipparchus, Eratosthenes, Posidonius, Strabo, Pliny etc., or inserted some new observations collected directly from European, Asiatic or African military commanders, mariners or merchant adventurers. But these corrections and additions were probably not very extensive, and the main part of the work, which after the discovery of the art of printing formed the prototype of all modern atlases, is thus, as Brehmer has supposed, most likely to be of a Tyrian or Phenician origin.

Errors and peculiarities in Ptolemy's geography.

High as Ptolemy's atlas stands above all other similar works either of ancient times or of the middle ages, yet a glance at his maps is sufficient to show that his geographical ideas and his notions of the distribution of continents and seas were often, not only very incomplete, but also quite erroneous. This is easy to explain and is excusable, for an atlas is only capable of representing the geographical knowledge of the age, when the work was composed. But in consequence of the well deserved reputation of the author and the unlimited faith in antiquity still prevailing in the 15th and 16th centuries, these errors and defects long exercised a retarding influence on the development of cartography, and induced learned cartographers to adhere for a long time to antiquated representations of countries, of which new and more correct maps had already been long published by illiterate mariners and travellers.

The most conspicuous of these errors are the following:

Ptolemy gave, in his maps, too great a longitudinal extention to the Mediterranean sea, and generally to the whole

world known to the ancients. This error partly arose from his adopting a length of only 500 stadia for the degree of latitude instead of 700, in consequence of which all astronomically measured distances became too short. This error principally affected the distances from north to south; for it was then only possible to determine longitudes astronomically under exceptional circumstances. There must necessarily have been great uncertainty in the determinations of longitudes, before chronometers were invented. The error, as regards the length of the Mediterranean, was early discovered by mariners, as is shown by the more correct dimensions given to the Mediterranean and Black Seas on some portolanos of the 14th century. But in maps drawn by learned scholars, and even by learned mariners, the old error was adhered to far into the 17th century. This has led some geographers to censure Ptolemy and those who revived the study of his geography in the 15th and 16th century, unjustly. The injustice of this censure may be deduced from a comparison of the maps in the Roma-edition of 1478 with maps of

---

[1] Mannert shows (cit. work I. p. 180) that the very order of the names in Ptolemy's text (Book II—VII) renders extensive interpolations improbable and difficult.

Scandinavia or of America from the first part of the 16th century. Such a comparison will show that Ptolemy's atlas, notwithstanding its deficiencies, occupied the first place among cartographical works, even far into the 16th century, at least, if the Portolanos are excepted. To the quite exceptional position in the history of cartography held by these I shall return further on. When making this comparison it should be remembered, that it was certainly possible to construct good charts of regions so well known and so many thousand times

2:0. Ptolemy made an unexplored continent, *Terra Incognita*, connect southern Africa with eastern Asia, thus forming an inland-sea of the Indian Ocean. It is probable that this theory regarding the extension of the Indian Ocean originally arose from accounts of continents and islands far in the South, among the inhabitants of the Eastern and Western Indian peninsulas, but certainly also from a tendency of the older geographers to apply the contours of known localities to unknown lands and seas. With the rejection of all reports

17. Map of the world from the 12th (?) century from a manuscript in the library of Turin. (From JOMARD).

traversed by mariners, as the Mediterranean and Black Seas, without a network of latitudes and longitudes, founded only upon guessed distances, and probably with Ptolemy's map as the point of departure. But if the same principles had been followed at the mapping of other parts of the globe, the mariners of the age of the great geographical discoveries would, during their passages to new countries and seas, for a long time have had to be guided, not by maps of Ptolemy's model, but by fancy maps in the Arabian style, or by representations of the earth resembling the map from the 12th (?) century, of which a facsimile is given fig. 17, or the maps in the *Rudimentum Novitiorum* N. fig. 2 & 3.

of the circumnavigation of Africa, the Indian Ocean was thus supposed to be an inland-sea like the Mediterranean, Black and Caspian Seas. It is here to be remembered that formerly there was a difference made between *Oceanus* and *Mare*. The Atlantic Ocean was generally not enumerated among the seas of the globe.

After the voyage round the southern point of Africa by BARTHOLOMEUS DIAS and especially after the return of VASCO DA GAMA 1499 from his first voyage to India, the error of this theory regarding the distribution of land in the Old hemisphere was obvious. Maps of the world of the old type, however, were printed not merely as representations of the geographical conception of antiquity, but also as real maps, far into the 16th century, f. i. in the *Margarita Philosophica* by REISCH, 1503 (N. T. XXXI), in *Cosmographia Pii papæ* etc. Parrhisiis 1509. The first printed maps giving to Africa a tolerably exact extension to the south are the map of RUYSCH in Ptolemy, Romæ 1508, a map on the title-page of the *Itinerarium Portugallensium*, Mediolani 1508 (N. fig. 37), one in the *Globus mundi*, Argentinæ 1509 (N. fig. 22), and the new maps of Africa in Ptolemy, Argentinæ 1513 (N. fig. 8 & 9). I shall give a more particular account of these maps further on.

3:0. On Ptolemy's maps the Indian peninsula is drawn as a slightly protruding projection from the south coast of Asia, whereas Ceylon *(Taprobane)* has an enormous extension, so as to make it the largest island of the world. This delineation was owing partly to statements of older Greek geographers, for instance Eratosthenes, who expressly declares the southern point of India to be situated on the latitude of Meroe (extract from Eratosthenes in the beginning of 2d book of STRABO) and partly to the lands beyond Ganges being known to Ptolemy only through doubtful and misunderstood reports. One millennium later, MARCO POLO first acquainted western Europe with the existence of a number of large islands in that part of the world, unknown to Ptolemy, and after the voyages of the Portuguese to India and the East-Indian Archipelago the first tolerably true maps of these regions were obtained. Here RUYSCH again takes the initiative in the reformation of the cartography. But it was during the former half of the 18th century that a thorough knowledge of the eastern boundaries of Asia was first obtained through the voyages of the Dutch in the Japanese waters and the Russian expeditions under Bering, Spangberg etc. in the northern parts of the Pacific Ocean.

4:0. Ptolemy did not know the northern limits of the old hemisphere. The large Scandinavian peninsula is only represented by two islands, *Scandia* and *Thule*,[1] and the northern coast of Asia is not indicated at all. The existence of populous Christian countries in the northern ocean was made known in Italy through the invasion of the south by Scandinavian freebooters, and by the reports of northern ecclesiastics to Rome, long before Ptolemy's atlas was generally made known through the translation of Jacobus Angelus. So that the Scandinavian peninsula had already been laid down on a number of old portolanos as a land of considerable extent, as for instance on the portolanos of DULCERT (1339), of PIZZIGANI (1367) and of ANDREA BIANCO (1436). Among the Tabulæ novæ, which in the 15th century were added to the Latin codices of Ptolemy's geography, a large map representing the Scandinavian peninsula, Iceland and Greenland is early met with, and this map was added to one of the earliest printed editions.

5:0. Ptolemy gave to the northern part of Scotland an enormous extension towards the east. This error had already been corrected on the earliest known portolanos, f. i. that of DULCERT of 1339, but on printed maps the error of Ptolemy

was long adhered to, a reform being first introduced on the maps of RUYSCH 1508 and of BERNARDUS SYLVANUS 1511.

6:0. Ptolemy reduced the distance between the Baltic and the sea of Azof to one third of its actual length. So far as I have been able to ascertain, this error was first corrected on the small *Carta marina nova*, in MATTIOLO-GASTALDI's Ptolemy of 1548 (N. T. XLV). How defective Ptolemy's knowledge was of the interior of Russia in Europe may also be conceived from his letting mountain ranges cross the *Sarmatia Europae*. Appealing to the protection of his superiors, MATTHIAS DE MIECHOW first ventured a protest against the existence of such mountains in a work, *Tractatus de duabus Sarmatiis*, printed Cracoviae 1517. It is to be observed that the short distance between the Black Sea and the Baltic in Ptolemy's maps not only depended on the coast of the present Pomerania and Livonia, almost unknown to the Romans, being placed too far to the south-east, but also on the Propontis and the Black Sea having got too northerly a position, and the sea of Azof too large an extension. That »APIANUS, MERCATOR, ORTELIUS, MAGINUS, and other geographers,» following the example of Ptolemy, gave to Constantinople or Byzanz a latitude of 43° 5' instead of 41° 6', was first 1686 (!) pointed out by JOHN GREAVES (*Philosoph. Transact.*, Vol. XV, p. 1295).

Ptolemy's maps of the »oikumenæ» may also be charged with some other more or less important errors, which space will not allow me to point out here. But his atlas became an unsurpassed master-piece for almost 1500 years, and owing to its richness of detail, it still constitutes an inestimable source of knowledge for the student of ancient history and geography. This is perhaps chiefly the case in the study of the earliest history of the tribes that encompassed the Roman empire in the first century of the Christian era, which were then barbarous, but whose descendants have since become the bearers of civilization. Finally Ptolemy's maps may, in the form and manner in which they were published by Schweinheim-Buckinck, be said to constitute the technical type or model for the whole modern literature of atlases. I therefore hope to do the students of geography a service by affording them an opportunity of consulting in their own studies reliable and exact fac-similes of these very rare maps, to which they would otherwise only have access in the few libraries containing one of the very rare editions of 1478, 1490, 1507, or 1508. The maps in these editions, which are, as before mentioned, printed from the same plates, seem to me to be the only serviceable ones for the study of Ptolemy as an authority for ancient geography and a model for modern cartography. The maps in the older editions of Bologna 1472 (?) and Firenze 1478 (?) are still too rude, incomplete and incorrect for this purpose, and the maps in all the later editions, excepting the ed. Paris 1867, are too much corrected or »improved,» or, as Mercator expresses himself, »ad mentem auctoris restitutæ et emendatæ.» Again, the original maps in the old codex of Mount Athos, published in fac-simile in Paris 1867, are drawn with too little artistic skill and are too worn to supply the want here indicated.

---

[1] Various opinions exist about the country designated by the ancients with the name of *Thule*, and it is possible that different lands or islands were at different times known by that name. But what *Ptolemy* or rather the mariners in Ptolemy's time meant by it, seems to me to be clearly deduced from his map of the world and his *Prima Europæ tabula*. Thule here corresponds to the south-western part of Norway, where the name *Telemarken* still reminds us of its ancient appellation.

IV.

Ancient, not Ptolemaic atlases.

HERODOTUS begins the 49th chapter of his fifth book *Terpsichore* with the following words: »When Cleomenes was king in Sparta, Aristagoras, the ruler of Miletus, arrived there. And when the latter went to speak with Cleomenes, he brought with him — so the Lacedemonians say — a copper-plate, on which the circle of the whole earth was engraved, and the ocean, and all the rivers.» It is expressly said that ARISTAGORAS referred to this drawing in his attempts to make the king of Sparta assist the Ionians against the Persians. This happened in about the year 500 B. C. We have no further information about this map, and the conjectures that it was constructed by HECATÆUS, a countryman and contemporary of Aristagoras, or by the Miletian ANAXIMANDER, who was living one generation earlier, and who, according to a quotation by Strabo (lib. I. cap. 1) from Eratosthenes, made the first map of the world, are without any foundation. But it is in this passage of Herodotus that a map is mentioned in literature for the first time, and that not merely as a valuable curiosity, but as the means of explaining the relative position and extension of the dominions of the king of Persia as regards the Ionians and Greeks, at a negotiation of extreme importance to the independence of the whole Grecian nationality.

There are a few other instances of maps being mentioned in Greek and Roman literature, and then only incidentally, and in a way which denoted that they were scarcely so generally used and so important in antiquity as in our time. CLAUDIUS AELIANUS, for instance, relates in *Variæ Historiæ*, III: 28, that Socrates, in order to check the vanity of Alcibiades regarding his large estates, brought him to a place, where a map of the world was exhibited, and asked him to point out on the same where his estates were situated. DIOGENES LAERTIUS mentions (lib. V cap. II) that THEOPHRASTUS had bequeathed a portico in the vicinity of the Lyceum at Athens, for the suspension of a map of the world. Pliny speaks in lib. III cap. II of a map, whose construction and mounting in a portico at Rome »*ex commentariis M. Agrippa*» was commenced by his sister and finished by AUGUSTUS. In the geography of STRABO (f. i. lib. II cap. IV) some very insufficient rules are given as to the drawing of maps, while in the work of the great geographer references to maps in order to facilitate the understanding of the text, are exceedingly rare and scanty. Other similar, but still more meagre notices of maps having existed in antiquity may be found in the writings of the philosophers and poets. Bur, with the exception of Ptolemy's atlas (and Ptolemy's description of the charts of Marinus of Tyre) and the Itinerarium of CASTORIUS (?) *(Tabula Peutingeriana)*, there is no ancient atlas extant, nor the description of any sufficiently intelligible to make it possible that a reconstruction could be effected, at least of its main features. With regard, for instance, to one of the maps of antiquity, the above mentioned map of Agrippa, most often referred to and long ago commented upon more than enough, we have not arrived farther than that it was declared by MOMMSEN to be round, by MÜLLENHOFF to be an oval, by DETLEFSEN to be square, by MANNERT and others to be oblong (compare KONRAD MILLER, *Die Weltkarte des Castorius*, Ravensburg 1888, p. 69). This is not to be wondered at, as all we actually know of it is contained in the words, »*orbis terrarum urbi spectandum expositus.*»

The cause of this deficiency in classical literature evidently depends on the fact, that the geographers and philosophers of antiquity were principally occupied with theoretical schemes of the »Orbis terrarum,» considering it beneath their dignity to work out really practical, detailed maps. In the middle ages the same predilection for generalization still prevailed. At least I suspect that allusion to the existence of the excellent Italian or Catalan portolanos of the 14th and 15th centuries will seldom be found in the writings of the learned men of that time.

At all events, the above quoted passages show, that practically serviceable maps existed in Greece and Rome several centuries before the days of Marinus of Tyre and Ptolemy. But not one of these maps is still extant, and owing to the scanty descriptions given of them, it is impossible to form any idea of the manner in which they were drawn. It is, however, possible that the maps of which fac-similes are given on Tab. XXXI from old editions of MACROBIUS and SACROBOSCO, were ultimately founded on præ-Ptolemaic originals. They give us at least a faithful, although roughly sketched representation of the notion prevailing in antiquity respecting the relative positions of continents, islands, seas, and oceans of the earth. It is also possible that the remarkable medieval charts, known under the name of portolanos, compass or loxodromic maps, were originally founded on the maps of Marinus of Tyre, described in Ptolemy's geography. At least the portolanos have played the same part with regard to the first *tabulæ novæ* in Ptolemy's geography, as the maps of Marinus did with reference to Ptolemy's own original maps.

––––––––

The following Catalogue of the few and insignificant maps printed before 1520, without any direct connection with the different editions of Ptolemy's geography, will give an idea of the predominance of the old Alexandrian geographer even during the greater part of the first century of geographical discoveries.

1 and **2**. A wood-cut circular map of the world, and a wood-cut map of Palestine in the *Rudimentum Novitiorum*, Lübeck 1475 (N. f. 2 and 3). These maps are the first ever published in print with an unquestionable date. The map of the world is inserted fol:s LXXIIII and LXXV, the map of Palestine fol:s CLXII and CLXIII of this bulky folio volume, printed on vellum, or on uncommonly thick paper. In the postscriptum or long colophon it is stated, that the book »with the aid of the art of printing newly invented by the special grace of God to the redemption of the faithful,» was published to serve as a manual to students, and to dispense the poorer of them with the necessity for buying other books. Here it is further mentioned, that the work was finished in Lübeck the day of Oswald MCCCCLXXV and printed by *Magister* LUCAS BRANDIS DE SCHASS. The work is at present very rare. The Library of the University of Upsala possesses a copy printed on vellum, which is not quite complete, and a perfect copy printed on thick paper is preserved to the town library at Lübeck. The fac-similes here are from that copy, to which I had access through the courtesy of the Librarian, Dr. C. Curtius.

3. A map of the world in *Pomponii Mellae Cosmographi Geographia: Prisciani quoque ex Dionysio Thessascoliensi de situ orbis interpretatio.* (Colophon): *Erhardun*

Votum ponitur bene merenti humano viro aspiranti flores novellæ ætati necessarios ad vermiculatos calles geographiæ.

4. *Isolario di Bartolomeo da li sonetti, s. l. e. a.* 8:o. In about 1477 a description of the islands of the Greek Archipelago, written in sonnets and accompanied by a number of small maps, resembling the small maps in the *Isolario* of Bordone was printed in Venice under this title. No

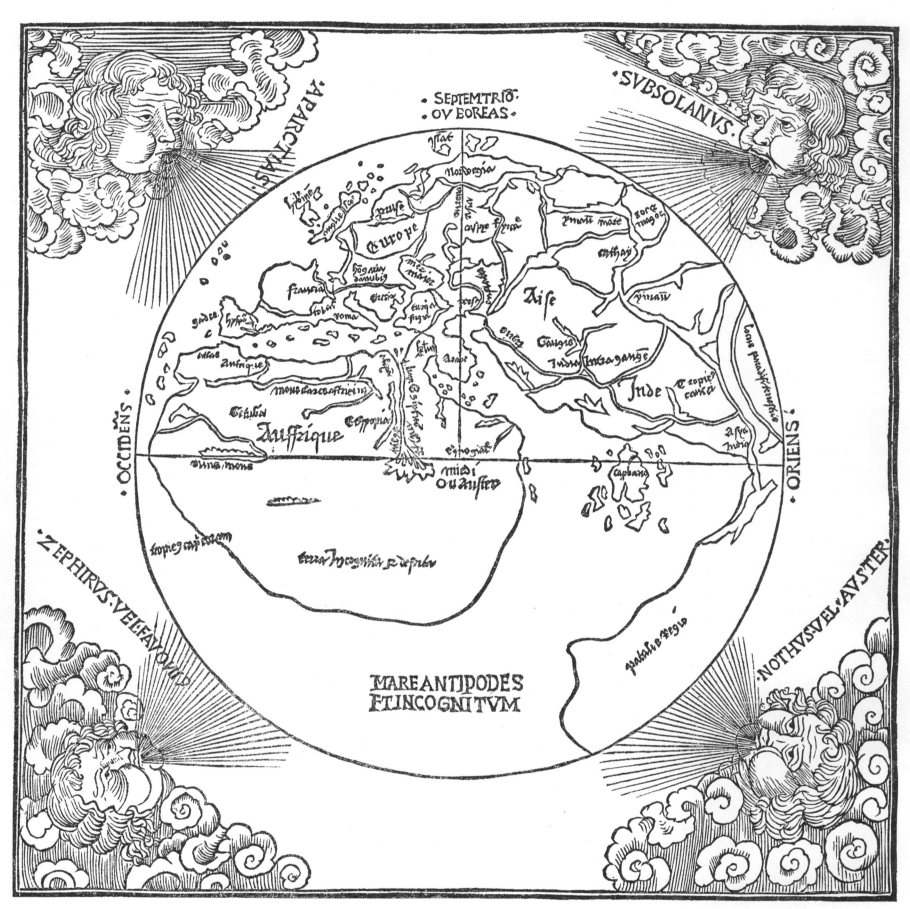

18. Map of the world from *La Salade nouvellement imprimée*, Paris 1522. (Orig. size 254 × 257 m. m.).

ratdolt Augustensis impressit Venetiis 1482. Laus Deo. 4:o. (N. T. XXXI). The map which occupies the first leaf of the book, is generally wanting, but that it really belongs to the work is shown by the water-mark corresponding to the water-mark of the text.

If we except the more correct form given to Scandinavia, it is a reproduction of Ptolemy's map of the world on a reduced scale. The curious inscription may have to be read:

large map is to be found in this edition, but in a later one of 1532 a map of the world, resembling the oval map of Bordone (N. T. XXXIX), was inserted. As I only had the opportunity of examining this rare work superficially in the Biblioteca Marciana, I must refer to Carlo Castellani, *Catalogo ragionato delle più rare o più importanti opere geografiche a stampa che si conservano nella biblioteca del Collegio Romano*, Roma 1876, p. 66, for a fuller description. In one

---

of the sonnets the author names himself BARTOLOMEO DA LI SONETTI. He states himself to have been a skilled sailor, who

>*Quindece volte intrireme son statto*
officiale e poi patrone in nave,

and who >*ho piu volte ogninsula chalchatta*
e porti e vale e scogli i sporchi e i netti
col bosolo per venti ho i capi retti
col stilo in charte ciaschuna segnatta.

Although the maps in the work of Sonetti are very insignificant, yet they are of a certain interest as being the

AURELIUS MACROBIUS lived at the end of the 4th and the beginning of the 5th century. His *Interpretatio in Somnium Scipionis* contains, besides essays on different metaphysical and cosmographical topics, some curious geographical speculations in the 2d book, the meaning of which was explained, when the work was printed, by means of a map, probably copied from a sketch in some old manuscript. A map is also referred to in the text, in the following words: *Omnia haec ante oculos locare potest descriptio substituta, ex qua et nostri maris originem, quæ totius una est, et*

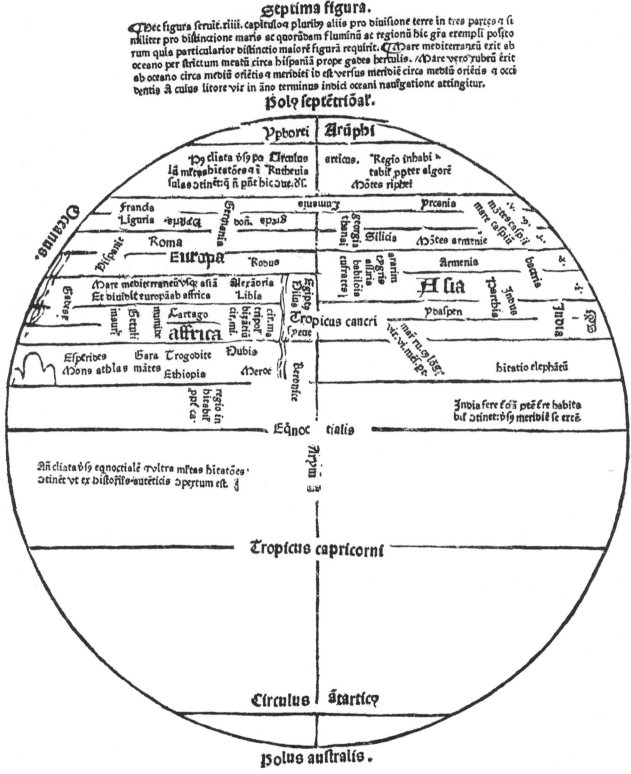

19. Map of the world from *Ymago Mundi*, by PETRUS DE ALIACO, c. 1483. (Orig. diameter 202 m. m.).

first printed maps of which it is expressly stated that they are founded on actual measurements.

5. The edition of MACROBIUS, published at Brescia (Brixiæ) in 1483, under the title >*In Somnium Scipionis Expositio*,< contains a map of the world of which a fac-simile is given on T. XXXI. The same map is reproduced almost unaltered in several later editions f. i. Brixiæ 1501, Venetiis 1521, Venetiis 1574, and others. In the edition Basileæ 1535 this sketch of the earth is replaced by a map of the old hemisphere, on which Africa is drawn in accordance with the later discoveries, but even here the inscriptions *Aethiopia perusta* at the equator, and *Frigida* at Africa's southern extremity, still remind us of the old prototype.

Rubri atque Indici ortum videbis. Caspiumque mare unde oriatur invenies, licet non ignores esse nullos qui ei de oceano ingressum negent. On this map the Caspian Sea is drawn, not as an inland sea, but as a large gulf of the eastern ocean. The passage in which Macrobius treats of the shape and extension of the inhabited world, is quoted by HUMBOLDT (*Kritische Untersuchungen* etc., I p. 166). It is scarcely quite intelligible, but it may be conjectured, that Macrobius considered the earth to be divided by the ocean-currents into four large islands, of which two were in the northern and two in the southern hemisphere. The equatorial zone, which separates the northern and southern islands, was supposed to be a sea impassable from heat, but Macrobius speaks of no obstacle to navigation between the two islands in the northern hemisphere. Humboldt remarks that this remarkable passage was not referred to by Columbus in his letters, although several editions of Macrobius had been published before 1492, and Columbus, through his own studies and the assistance of his friends, seems to have been familiar with the *dicta geographica* of the classical authors and to have placed great reliance on them. According to LETRONNE (HUMBOLDT, *Krit. Unter-suchungen*, II p. 82—91) the geographical theories of Macrobius have probably been borrowed from some old commentaries on the description of the currents of the ocean by Homer.

The map from Macrobius, here given, is the first printed map on which the currents of the sea are denoted.

6. Maps of the world in various editions of SACROBOSCO's *Opusculum Sphericum*, from the 15th and 16th centuries. As may be perceived from the fac-simile on pl. XXXI, these maps bear a strong resemblance to the map of Macrobius.

JOHANNES DE SACROBOSCO, or HOLYWOOD, was a distin-guished and learned scholar, living in the first part of the 13th century. Though born in England he was a teacher of Mathe-matics and Astrology at the university of Paris. Few medieval literary works have enjoyed such extensive credit as his *Opus-culum Sphericum*, or *De Sphera*, of which 60 to 70 editions, generally provided with extensive commentaries, have been published. The work is a manual of the principles of astronomy and cosmography. The author here adopts the

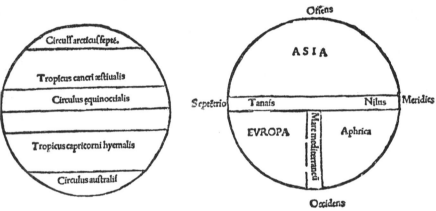

20. From *Orbis Breviarium*, by ZACHARIAS LILIUS.

spherical form of the earth, and the existence of antipodes. At least he divides the surface of the earth into five zones, to which he applies the verses of Ovid (Metamorph. I: 49):

»*Quarum quae media est non habitabilis aestu,*
Nix tegit alta duas; totidem inter utrasque locavit
Temperiemque dedit, mixta cum frigore flamma».

With but slight modifications the map, which is here repro-duced, is inserted into almost all editions of Sacrobosco's Sphera. During the 13th, 14th, and 15th centuries, it was probably the only accessible general map of the world to a number of students.

7. Frame to a map of the world, published in a work probably printed at Louvain in 1483, containing *Ymago Mundi* by PETRUS DE ALIACO, besides eleven smaller tracts by the same author and four by JOHANNES GERSON. PETRUS DE ALIACO, or PIERRE D'AILLY, was a French prelate and cardinal, celebrated for his learning. He had great influence on the theological controversies of the age, especially during the council at Constance, where he, amongst others, took part in the sentence of death on HUSS. He died in about 1422, and the *Imago Mundi* was, according to a statement at the end of the work, finished in 1410.

As this work has a certain importance in the history of the discovery of the new world, it may here be remarked that the place of printing and the date (»Paris c. 1490») generally given to it are not exact. It should be *Louvain before 1487*, as may be deduced by comparing the types with which it was printed, with those employed for a work of Petrus de Cre-scentiis, printed by Johannes de Westfalia at Louvain, and

by the following annotation, in the handwriting of the 15th century, on the 1st sheet of a copy taken at Olmütz during the Thirty Years' War, and at present belonging to the Royal Library at Stockholm: *Petrus Auliacus. Liber Philippi de Penczicz, emptus Parisiis die Octobris undecima Anno Domini etc. octuagesimo septimo, manu propria ligatus in Wyssaw et signatus in Budesin, die Aprilis 28. Anno etc. 89.* HUMBOLDT (*Cosmos*, II. p. 286) ascribes to this work an influence on the mind of COLUMBUS even greater than his correspondence with TOSCANELLI. Every one having access to this old and rare work may convince himself of the truth of the assertion by perusing the chapters in it treating of cosmographical and geographical questions, the trouble of which will be richly rewarded by the insight he obtains here into the geographical theories and reasonings of the middle ages. D'Ailly seldom pronounces any positive opinion of his own. But the learned prelate, otherwise so orthodox, often allows a doubt to glimpse forth as to the geographical dogmas of the church. He brings doubtful points to view, and gives as guidance to their answer, citations from ARISTOTELES, PTO-LEMY, SENECA, PLINY, AUGUSTINUS, ESRA, AVERROES etc. For instance, in the 8th chapter, mostly borrowed from ROGER BACON's *Opus Majus* (p. 183 in the ed. London 1733), d'Ailly says, that Ptolemy supposed the continents to occupy only the sixth part of the surface of the earth, and later in the same chapter he seems to accept this theory himself: *Ex præ-missis igitur et ex dicendis inferius apparet, quod terra habitabilis non est rotunda ad modum circuli sicut dicit Aristoteles, sed est velut quarta superficiei unius sperae, cuius quartae due partes aliquantulum extremae rescindun-tur, scilicet illae quae non habitantur propter nimium ca-lorem aut nimium frigus.* But in the same chapter, as well as in other parts of his works, he mentions on the contrary, without disavowal, that Esra in his fourth book assumes six parts of the surface of the earth to be inhabited and the seventh covered with water. He also cites Aristoteles as a supporter of the view that the sea separating India from *His-pania ulterior* (=western Africa) has not a very great extension.

Columbus' own copy of the work is still preserved at the Bibliotheca Colombina in Seville. It is supplied with marginal annotations, but not, as formerly supposed, by the great explorer himself, but probably by his brother BARTHO-LOMAEUS (HARRISSE, *Christophe Colomb*, Paris 1884, II p. 190). All this contributes to give some interest to d'Ailly's very rudimentary map, of which fig. 19 is a fac-simile.

8. ZACHARIAS LILIUS, *Orbis breviarium*, Florentinæ, An-tonius Miscominus 1493. 4:o. On the reverse of the fourth leaf the above geographical sketches (fig. 20) are inserted. They reproduce a rudimentary map of the world often met with in old manuscripts, and in printed works from the 15th and 16th centuries. Lilius was an Italian geographer from the end of the 15th century. His Orbis Breviarium seems not to contain one line founded on original investigations, or a single expression alluding to the great geographical discoveries during the century in which he lived. It begins with these characteristic words: *Terrarum orbis universus in quinque distinguitur partes, quas vocant zonas. Media solis torretur flammis. Vltimas æternum infestat gelu. Duae habitabiles inter exustam et rigentes. Altera a quibus incolitur, teste Macrobio, non licuit unquam nobis nec licebit agnoscere.* When he wrote this sentence, the Portuguese mariners had long ago sailed across the sunburnt zone, and before the first edition was printed, Columbus had already discovered the New World. The work of Lilius, however, for a long time remained a popular manual of geography and cosmography.

9. and **10.** Two maps in HARTMANN SCHEDEL's *Liber cronicarum*, of which two large folio editions, the one in Latin

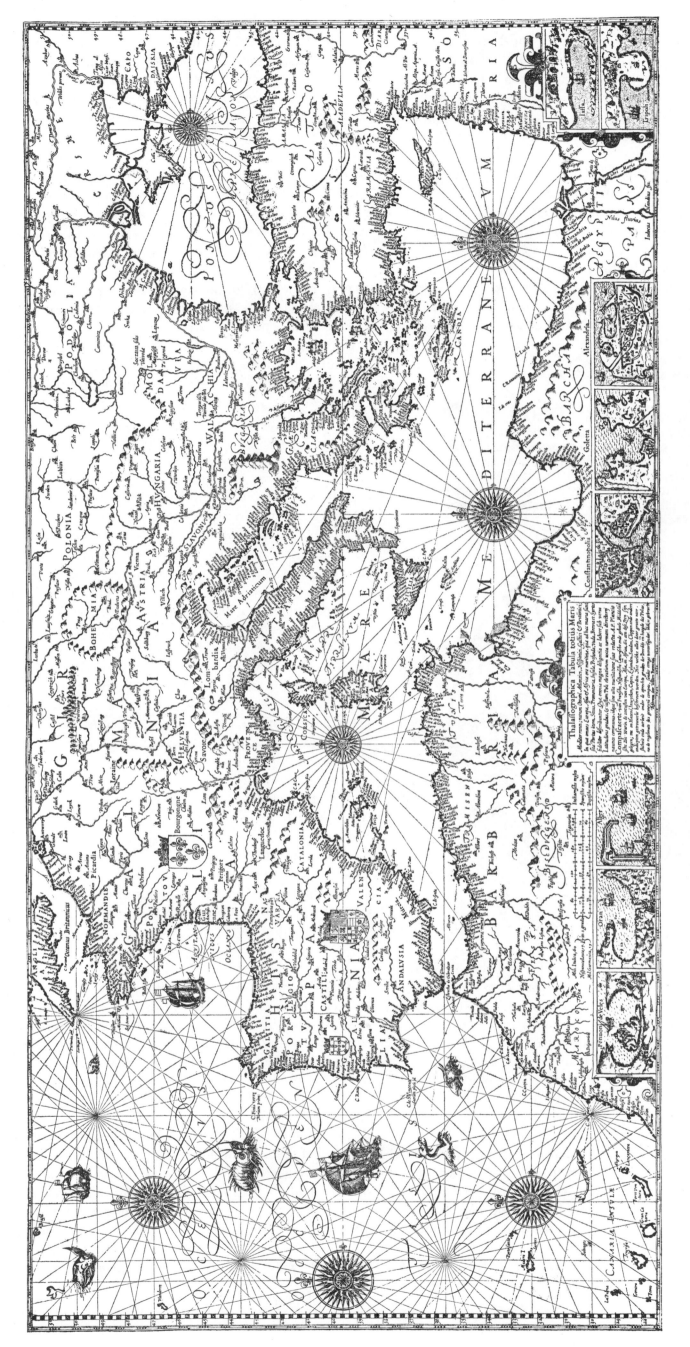

21. Chart of the Mediterranean Sea by WILLEM BARENTSZOON. Engraved in copper 1595.
Almost unaltered copy of a Portolano from the 14th century.

(Orig. size 418 × 855 m.m.).

and the other in German, were printed at Augsburg in 1493. The Latin edition commences with an elaborate index, on the first page of which is printed, in large richly ornamented characters: *Registrum hujus operis libri cronicarum cum figuris et imaginibus ab initio mundi.* The place and year of printing are given on fol. cclxvi by the words: *Completo in famosissima Nurembergensi urbe Operi de hystoriis etatum mundi, de descriptione urbium, felix imponitur finis. Collectum brevi tempore auxilio doctoris Hartmanni Schedel, qua fieri potuit diligentia. Anno Christi Millesimo quadringentesimo nonagesimo tercio, die quarto mensis Junii.* Probably the editor originally intended to finish the work here. But before it was published, two extensive appendices were added. The first occupies fol. cclxvii—ccc, and ends with a new colophon, in which Schedel is not mentioned, but the editors SEBALDUS SCHREYER and SEBASTIANUS KAMER-

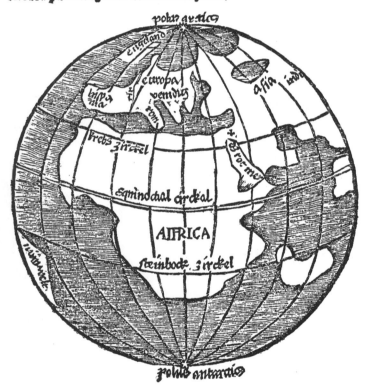

22. The title-page of *Globus Mundi*, Argentorati 1509.

MAISTER, the printer ANTHONIUS KOBERGER, and the drawers MICHAEL WOLGEMUT and WILHELM PLEYDENWURFF — *viri mathematicis, pingendique arte peritissimi.* The second, rubricated »*De Sarmacia regione Europe,*» occupies five unnumbered folio leaves. The work contains two maps. The first (fol. xii and xiii) is a copy of Ptolemy's map of the world. It presents nothing of special interest from a geographical point of view, but is a vigorous wood-cut, decorated on its right side with a frame of human monsters and on three of its corners with drawings of *Shem, Ham,* and *Japhet.* The other map, of which fig. 5 gives a reduced fac-simile, is of more importance. It is the first modern map of Central Europe published in print and, as such, of so much the more interest as it seems to be based on the lost map of the celebrated Cardinal NICOLAUS A CUSA.[1] In the German edition of 1493 the same maps are to be found as in the Latin one. Of later editions (1496, 1497, 1500) I have only seen the edition

[1] The proofs of this will be given in the last chapter of this work.

published at Augsburg in 1497. It contains bad and much reduced wood-cuts of the editio princeps. There is still to be read in this edition (fol. xiv): »*extra tres partes orbis, quarta est pars trans oceanum interiorem in meridie, quae solis ardoribus nobis incognita est, in cujus finibus Antipodes fabulose habitare dicuntur*».

11. *Venetiis 1489.* On T. XXXI I give the fac-simile of an extraordinary modification of the map of Macrobius, from: (colophon) *Summa astrologiae iudicialis de accidentibus mundi quae Anglicana vulgo nuncupatur. Iohannis Eschuid viri Anglici eiusdem scientiae astrologiae peritissim. Finis hic imponitur faustus* etc. *Anno salutis 1489... Venetiis.* The work begins with an *Ad lectorem,* in which it is said, that a unique copy of the famous astrologer JOHANNES ESCHUID's work, called *Summa Anglicana,* having been saved by the special grace of the Divine providence, this one had now brought forth such a numerous offspring that its total loss was no more to be apprehended. ESCHUID or EASTWOOD was a celebrated English physician and astrologer, who lived in the middle of the 14th century. The map is only an unsuccessful, reversely drawn copy from that of Macrobius.

12. *Margarita Philosophica,* Friburgi 1503. GEORG REISCH, the confessor of the Emperor Maximilian and prior of the Carthusian convent at Freiburg, wrote an encyclopedia called Margarita Philosophica, which seems, in a very satisfactory way, to have provided for the general want of knowledge of the educated classes during the 16th century. This may be deduced from the number of editions published. The manuscript was already finished in 1496, but it was not printed before 1503. Amongst various other subjects the work contains, in its 7th book, a few chapters on cosmography and geography, but they are of slight interest to the history of geography. It is strange to compare the self-sufficient positiveness of Reisch as a geographer with the incessant seeking for truth of Pierre d'Ailly. Evidently the geographical learning of the former was very insignificant, but he fancied he knew everything, and he doubted not a single one of the geographical fables of the middle ages. The knowledge of the latter was wonderful for his time, but he was very sceptical as a geographer; and he sought in vain to find out the truth by comparing the works of the most prominent authorities and scholars of bygone times. Thus in the text of the edition of the Margarita Philosophica of 1503 there is, as far as I have been able to ascertain, not a single sentence to be found indicating that the author or his collaborators at the university of Heidelberg, had the faintest idea of the geographical discoveries of the 15th century. But on the map which the publisher, JOHANNES SCHOTTUS, added to the work (N. T. XXXI), there is to be found a slight hint of it in the following legends at the southern limits of Africa, and on Ptolemy's fabulous territory to the south of the Indian Ocean: »*Hic Africæ terra longius protenditur ad quadragesimum ferme gradum*» — »*Hic non terra sed mare est: in quo miræ magnitudinis insulæ, sed Ptolemaeo fuerunt incognitæ*». The outlines of the Scandinavian peninsula are copied from the map in the edition of Ptolemy Ulmae 1482.

13. JACOBI PHILIPPI BERGOMENSIS *Supplementum Supplementi Chronicarum,* Venetiis 1503. On fol. 7 there is a ⊕-shaped map of the world. The work also contains a number of views of towns, of which a few are real likenesses and not free-hand-drawings. On fol. 230 verso is, for instance, a view of the place of St. Mark in Venice, still quite easy to recognize. Such a view, though ruder, and less correct, had previously been inserted in a former edition (Venetiis 1486) of the same work.

14. *Itinerarium Portugallensium e Lusitania in Indiam et inde in occidentem et demum ad aquilonem*

(Mediolani 1508). The greater part of the title-page is occupied by a map of Africa (N. fig. 37). I shall in a subsequent chapter return to this map, which is the first or second published in print, on which the outlines of Africa are laid down with tolerable exactness.

15. *Cosmographia Pii Papæ in Asiae et Europae eleganti descriptione* (Per Henricum Stephanum. Parrhisiis . . . MDIX). Under this title the Pope Pius II, or Aenaeas Sylvius Piccolomini (born 1405, dead 1464), published a work containing geographical, historical, and ethnographical descriptions of a part of Asia and Europe. That of Asia is introduced by a few chapters on the form of the earth, the division of the mainland into four islands separated by the currents of the ocean, the possibility of circumnavigating the

book. It is an almost unmodified, but badly executed copy of the map in the Margarita of Reisch of 1503 (N. T. XXXI). The legends near the southern border of the map are copied from Reisch. The map is only of interest as being the first (?) map printed in France, and a new instance how slowly the knowledge of the voyages of the Portuguese and Spaniards was spread over the rest of Europe.

16. *Argentorati 1509.* Fig. 22 is a fac-simile of the title-page of a small print, published at Strasburg in 1509. It is illustrated by a globe, on which a small corner of South America, designated as the *Niew Welt*, is laid down. Insignificant as this map or drawing may be, yet it is of interest, as being the second printed map on which a part of the New World is represented. I have not seen the original, and

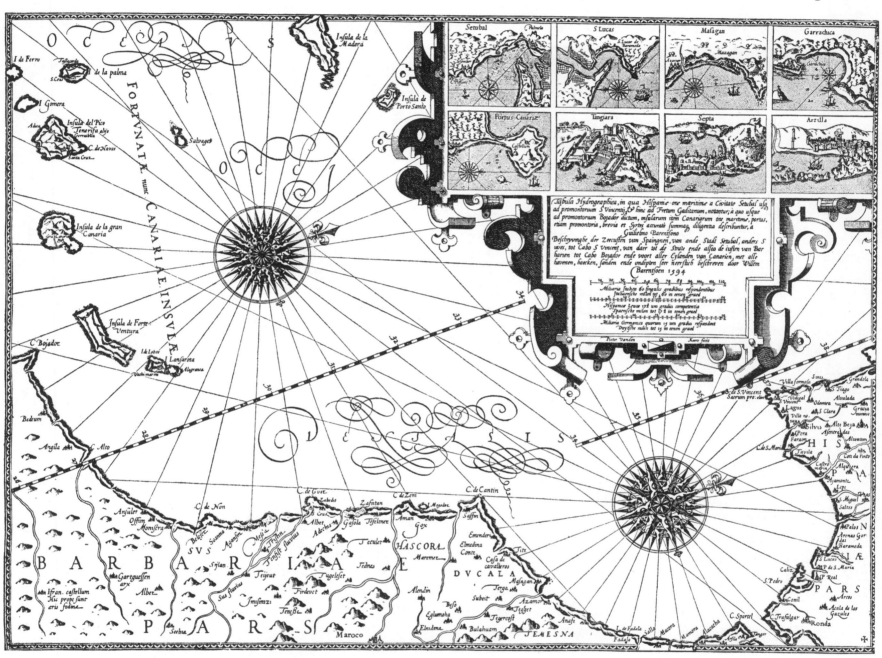

23. Chart of the sea westward of the Straits of Gibraltar by Willem Barentszoon 1594. (Orig. size. 548 × 394 m. m.).

large island formed by Europe, Asia, and Africa, and the suitability of the tropics and the arctic regions for the abode of man. These preliminary chapters are written in the same style of uncertainty and searching for truth, as the chapters in the *Ymago Mundi* of Pierre d'Ailly. Considering the wide circulation and authority enjoyed by the works of the learned Pope, these chapters ought to have exercised an immense influence on the general disposition for exploring voyages, which characterized the century of the great geographical discoveries. I have not seen the first edition printed in Venice 1477 in folio. It seems not to have contained any map. Nor is there any map to be found in the edition Venetiis 1503, described by Castellani in his *Catalogo delle . . . opere geografiche del Collegio Romano,* Roma 1876, p. 68. But the edition printed in Paris in 1509 contains a wood-cut map of the world printed on a folio-leaf, bearing the same water-mark as the other paper in the

the copy of its title-page, which is given here, is borrowed from Winsor, *Critical History of America*, II p. 172. Judging from the analysis given by d'Avezac (*Martin Hylacomylus Waltzemüller*, Paris 1867, p. 114), the text contains, on its 14 leaves in small quarto, very little of interest and, in spite of the promise on the title-page, scarcely anything of the *»quarta orbis terrarum pars nuper ab Americo reperta.»* A copy of this map, but without any legends, is inserted into the work mentioned below under No. 21.

17. *Hispali 1511.* The map of the West Indies by Petrus Martyr (N. fig. 38); and

18. *Cracoviæ 1512.* Map of the world by Stobnicza (N. T. XXXIV). In a subsequent chapter, analysing the first maps of the New World, I shall have occasion to return to these two important maps.

19. A map in *Meteorologia Aristotelis, eleganti Iacobi Fabri Stapulensis Paraphrasi explanata, commentarioque Ioannis Coclaei Norici declarata* etc., Norinbergæ 1512 (N. T. XXXI). On fol. lx verso there is a map of the world showing that the circumnavigation of Africa was known to the author.

In the commentaries of the German theologist Coclæus there is on fol. lxii written: »*Nam et Germaniæ imensas esse insulas non pridem compertas cognitum habeo ... Et revera longe maior est habitatae nunc terrae mensura, quam veteres isti Geographi descripsere. Nam ultra Gangem Indiæ immensa protenditur terra, cum maxima orientis insula Zipangri. Africa quoque ultra tropicum Capricorni longe fertur extensa. Ultra Tanais item ostia plurima habitatur terra, ad mare usque Glaciale. Quidque nova illa Americi terra admodum nuper inventa, vel tota Europa maior esse dicitur.*» These words deserve to be quoted as forming a contrast to the way in which the great geographical discoveries were almost completely ignored in Reisch's Margarita. They are also of interest as one of the very first instances of the New World being designated by Amerigo's name.

20. *Margarita Philosophica Nova, cui annexa sunt sequentia ... charta universalis terræ marisque formam neoterica descriptione indicans*, Argentorati 1515. The first edition of *Margarita Philosophica* by Reisch was, as mentioned above, printed at Friburg in 1503. Later new editions were published at Friburg 1504, Strasburg 1504, 1509, 1512, 1513, 1515, 1520, 1565, Basel 1508, 1535, 1583, Venice 1599. If the folded map of the world is not totally wanting, these editions generally contain copies of the antiquated map in the edition of 1503, or of the worthless map of GEMMA FRISIUS of 1540 (N. T. XLIV), which, however, was much admired and often reproduced in the 16th century. The only exception in this respect is the edition of 1515, which, when complete, contains an original map, *Typus universalis terrae iuxta modernorum distinctionem et extensionem per regna et provincias* (N. T. XXXVIII), on which a part of the New World is laid down. I shall further on, in my account of the first maps of the lands discovered on the other side of the Atlantic, return to the subject. How little account the publishers of this work made of the great geographical discoveries, and what exclusive preponderance the classical and theological learning had in the schools and universities of the 16th century, is shown by the Margarita Philosophica, published at Basel in 1583. By different additions it is now increased to thrice its original volume, and it was, according to a statement on the title-page, »*ab Orontio Finaeo Delphinate, Regio Parisiensi Mathematico, necessariis aliquot auctariis locupletata.*» However, it still contains (page 346) an old Ptolemaic map of the world, and the geographical discoveries of the previous century are only touched upon in a few words, in this encyclopedia of 1403 quarto pages. Pages 1347—1349 a chapter, »*Nova terræ descriptio secundum neotericorum observantiam,*» is inserted. It is borrowed from the text on the reverse of the map of the edition of 1515. But it neither contains any allusion to the new geographical discoveries, nor any explanation to the cordiform map of Gemma Frisius inserted in this edition from the cosmography of Apianus, the publisher very likely disposing of the old block of 1540. In one of the last addenda (HONTER's *Rudimentorum Cosmographiæ Liber unus*) there are for the first time enumerated among the Oceanic islands: »*In Australi (Oceano): Iona, Taprobana, Scoyra, Madagascar, Zanzibar. In occiduo: Dorcades, Hesperides, Fortunatae, America, Parias, Isabella, Spagnolla & Gades.*» The words here cited are already found in the first edition, Cracoviae 1530, of Honter's small and often reprinted Cosmography. Five or six

bare names from a fifty years' old work, are all that Reisch's Margarita, revised by Orontius Finæus, A. D. 1583, contains regarding the New World or the parts of the Old newly discovered.

21. *Opusculum de Sphæra clarissimi philosophi Ioannis de Sacro busto* etc., Viennæ 1518. In this edition of Sacrobosco a copy of the map in the *Opusculum Geographicum*, Argentorati 1509, is inserted, but without any legends. As in the original, the outlines of Africa are here laid down tolerably well, and in the south-western corner of the map a small part of South America is shown. A similar insignificant map, omitting, however, the corner of South America, is also inserted into IOANNES SCHÖNER's *Opusculum Geographicum*, s. a. et l. (sed Norimbergæ 1533).

The synopsis here communicated of maps not destined for any edition of Ptolemy's geography, and printed before 1520, is by no means complete. I suppose e. g. that the old theological literature will contain several maps of the Holy Land not mentioned here, and I have intentionally omitted some worthless copies of the maps of Ptolemy inserted in editions of other classical authors, as the maps of Spain, Gallia, and Germania in editions of Julius Cæsar. Some few maps mentioned in the literature as still existing I have never seen, f. i. a *map of Germania of 1491* mentioned in *Catal. of the Printed Maps in the Brit. Museum*, p. 1535, and two maps mentioned by CARL HARADAUER among the treasures of the Hauslab Library (*Mittheilungen der k. k. Geographischen Gesellschaft in Wien*, XXIX, 1886, p. 392) viz. an itinerary of the Roman empire, printed by Georg Glogkendon in Nuremberg 1501, and a map of Italy by Valvassor of 1516. I, however, suppose that few maps of importance have escaped my investigations, and that the above catalogue thus furnishes us with an adequate idea of the extreme poverty in the literature of printed maps prevailing far into the 16th century, if the old and new maps in the different editions of Ptolemy's geography are excepted.

Among the maps here enumerated only the following have a size of at least a common quarto page (18 × 12 c. m.):

1) 1475. The map of the world in the Rudimentum Novitiorum.

2) 1475. The map of Palestine in the same work.

3) 1482. The map of the world in Pomponius Mela of that year.

7) 1483. The map sketch in d'Ailly's Ymago mundi.

9) 1493. The map of the world in Schedel's Liber cronicarum.

10) 1493. The map of Germania in the same work.

12) 1503. The map of the world in Reisch's Margarita.

14) 1508. The map of Africa in the Itinerarium Portugallensium.

15) 1509. The map in the Paris edition of the cosmography of Aeneas Sylvius.

17) 1511. The map of the West-Indies by Petrus Martyr.

18) 1512. The map of the world by Stobnicza.

20) 1515. The map of the world in the Margarita Philosophica of that year.

Nos. 1, 2, and 7 among these are scarcely to be considered as maps in a modern sense. The Nos. 3, 9, 12, and 15 are bad copies of Ptolemy's map of the world. Consequently only the following five maps, printed before 1520 »*extra Ptolemæum*» and based on real independent geographical investigations, remain, viz.: No. 10, the first new map of central Europe; No. 14, one of the first printed modern maps of Africa; No. 17, the first map of the West-Indies; No. 18, the first map in which the earth's surface was divided into two hemispheres, and the first, on which the *Novus Orbis* is placed on a par with Europe, Asia, and Africa, or constitutes a part of the world, and No. 20, the map in the Margarita of 1515, which has become remarkable through the legend »*Zoana Mela.*»

It is possible that this number may hereafter be somewhat increased by new bibliographical discoveries, but there is very little probability that these discoveries will change the general results of the exposition here given, showing that I have by no means over-estimated the importance of the different, more or less modernized editions of Ptolemy's geography during the 15th and the first part of the 16th century.

all the countries to the south of the Black and Mediterranean Seas, from the Indus to the coasts of the Atlantic ocean. Already, during the course of the invasion, this people not only appropriated but also further developed certain branches of the Græco-Roman civilization. They especially devoted themselves with success to the mathematical, astronomical, and geographical sciences. The circumstance that countries so distant as Spain, the shores of the Indus, and the oases of the African deserts were, for a short time at least, united under the same banner, tended to facilitate the spread of geographical and ethnographical knowledge. This was further promoted by the yearly meetings at Mecca of pilgrims from the most remote parts of the world, and by the inclination for wandering inherited by the Arabs from their ancient nomadic life, which,

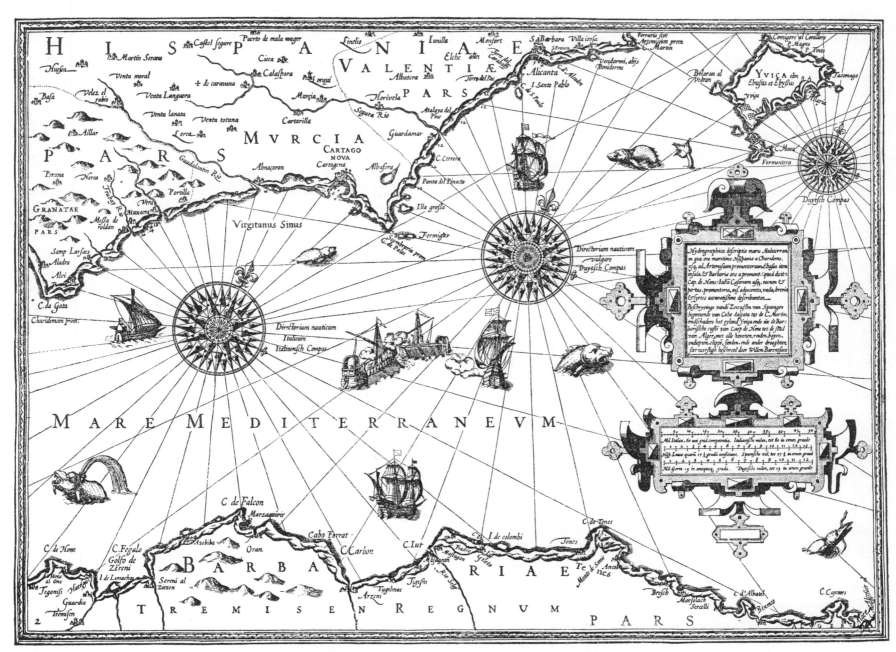

24. Chart of a part of the Mediterranean by WILLEM BARENTSZOON c. 1595. (Orig. size 532 × 375 m. m.).

As before mentioned, the main object of this work is to enumerate and briefly to describe the maps printed during the period of cartographical incunabula, and it is accordingly such maps that have almost exclusively come under discussion in the foregoing chapters. But, before proceeding further, I am induced by the close connexion between the maps published in print and the hand-drawn delineations on which these are based, to devote a few pages to a review of the manuscript maps of the middle ages, especially with regard to the influence they have exercised on the printed literature.

The number of medieval manuscript maps, scattered through the libraries of Europe, is very great. They may be classified into four different groups, viz.:

I. Arabian maps. History teaches us that a people from the deserts of Arabia, formerly little known and not very numerous, during the 7th century, invaded and conquered

in connection with a desire to propagate the doctrines of the Prophet all over the earth, induced them to undertake extensive journeys far beyond the dominions of their own Caliphs. All this produced a flourishing period for the geographical sciences in several Arabian countries. Various admirable descriptions of distant lands and of extensive voyages, written by Arabian scholars and far surpassing the geographical productions of the same period among the Christians, are also still extant. But similar perfection was never attained by the Arabian maps, which, if they were original drawings and not, as the planisphere of EDRISI, mere copies or reproductions from Ptolemy, are not only far inferior to the maps of the Alexandrian geographer, but not even comparable to the Esquimau-sketches brought home by English and Danish polar travellers from the icy deserts of the polar regions. »Il est impossible de rien imaginer de plus informe. Pas de projection, pas de graduation, rien qui ressemble à une image régulière où l'on a eu égard à la vérité des formes, des positions et des distances. On ne comprend pas comment les Arabes ont pu descendre à de pareilles productions, ayant sous les yeux les cartes graduées de Ptolémée» (Vivien de St. Martin, *Histoire de la Géographie*, Paris 1873, p. 263). They have not exercised any influence upon the development of cartography, and none on the map-printing of the 15th and 16th century. For further details about the Arabian geographical works I may refer to Lelewel, *Géographie du Moyen-Age*, Bruxelles 1852; A. F. Mehren, *De Islamitiske Folks geographiske Kundskaber* (*Annal. for Nordisk Oldkyndighed*, Kjöbenhavn 1857); Vivien de St. Martin, *Histoire de la Géographie*, Paris 1873; Peschel, *Geschichte der Erdkunde*, München 1865, and to the numerous monographies (by Reinaud, Jaubert, Wüstenfeld etc.) cited by these authors.

II. European maps from the middle ages, founded neither on Ptolemy nor on portolanos. During the next millennium after Ptolemy the art of drawing maps had become almost extinct among learned men and scholars in Europe. Yet some passages in writings from this long period may be cited, showing that maps, of which a few are still to be found inserted in old manuscripts, were then in use, even for educational purposes. Thus Santarem, in his great work, »*Essai sur l'histoire de la cosmographie et de la cartographie pendant le Moyen-Age*,» enumerates a map by Cosmas of the 6th century; a ⊕-formed planisphere from a manuscript in the Library at Strasburg of the 9th century; a ⊕-formed planisphere from a manuscript by Sallustius in the Biblioteca Laurentiana; a similar planisphere of the 10th century in a manuscript by Isidorus from Seville. But a glance at the first leaves in his atlas will show, that not a single one of these »Mappemondes» deserves the name of map, and they have scarcely exercised any other influence on the development of cartography than that the wind-heads, so often employed as ornaments on maps of the world in the 15th and 16th centuries, probably derive their origin from them,[1] and that some of these maps evidently have served as models to the figure of the earth in the works of Macrobius, Lilius, and Sacrobosco (N. T. XXXI and Fig. 20). From the 12th century the medieval maps first became of general interest in the history of civilisation through their greater fullness of detail, although they were, with the exception of the portolanos, in every respect inferior to the old work of Ptolemy. Yet their only influence on the art of map-making was the introduction of the custom prevailing to the end of the 16th century, of adorning maps with drawings of towers and temples, of kings sitting on their thrones in full attire, of monsters and ethnographic details, and with inscriptions of a doubtful geographical character, borrowed from the heathen mythology or Christian martyrology.

To show that the above estimate of maps of this group has not been unjust, I have here given fac-similes on a reduced scale of a couple of them (fig. 17 and 18). The reader will find numerous fac-similes and further information respecting these maps in:

Vicomte de Santarem, *Essai sur l'histoire de la cosmographie et de la cartographie pendant le Moyen-Age, et sur les progrès de la géographie après les grandes découvertes du XV:e siècle*, 3 vols. 8:o. Paris 1849—52, and the *Atlas composé de mappemondes et de cartes hydrographiques et historiques du XI:e au XVII:e siècle pour la plupart inédites* etc. by the same author. Complete copies of this

work are scarce, and the work has besides been so irregularly published, that the well-known bookseller Mr. Quaritch in London was compelled to print a special title-page and an *Index of Maps* to some copies at his disposal. This index is almost indispensable to the collation of the work. It is introduced by the following bibliographical data.

»The Vicomte de Santarem published originally, in 1842, a work entitled ›Recherches sur la priorité de la Découverte de la Côte Occidentale de l'Afrique,› with an Atlas consisting of 30 plates. He afterwards made this Atlas (which was in fact unfinished at the time) the foundation of the present great work, which contains 78 plates, and was published at the expense of the Portuguese Government. It was not, however, completed, in consequence of his death in the year 1855. The Maps in this last Atlas are not numbered, except those belonging to the original work, the numeration of which is no longer appropriate. There are frequently several Maps on one sheet or page of the new series, and these have been selected without any principle of sequence or order. There is no list or index for the arrangement of the sheets; but M. de Santarem communicated to the »Nouvelles Annales des Voyages» (1855, vol. 2) shortly before his death, a classified list or catalogue of the several Maps.... M. de Santarem speaks of all the Maps enumerated in his list as ›published;› but there are a few which cannot be traced. Many inexactitudes will be observed in his descriptions and notices — two or three of the Maps in the list in the ›Nouvelles Annales» are repeated more than once. The headings to the engraved Maps, and the corresponding titles in the list in the »Nouvelles Annales» are sometimes very dissimilar. Occasionally the dates assigned on the maps and in the list differ by a century. It would appear that the list in the ›Nouvelles Annales» was very carelessly drawn up (probably in consequence of illness), but it is nevertheless the only guide afforded to the intention of the author.›

Jomard, *Les monuments de la géographie ou recueil d'anciennes cartes européennes et orientales publiés en fac-simile de la grandeur des originaux*, Paris (s. a.). Even this important atlas was never finished, and there is no other text to it than an *Introduction à l'Atlas des Monuments de la Géographie par feu M. Jomard*, a short but excellent memoir by M. E. Cortambert, published in 1879 in the *Bulletin de la Société de Géographie*.

Among the larger and more important maps of the world of the middle ages only one has, as far as I know, been reproduced in full size: that of Richard of Haldingham, of A. D. 1300, a fac-simile of which was published in London 1869 by Rev. T. T. Havergal.

III. Sea-charts of the middle ages. Portolanos, compass or loxodromic charts.

Besides the contributions to what may be called the mythic cartography mentioned above, there are also preserved from the middle ages two other kinds of maps, namely charts of the Black and Mediterranean Seas, most nearly corresponding to the mariners' maps of our time, and maps of the world, for which these old charts have served as a basis. The former have never had for object to illustrate the cosmographical speculations of some classical author, or some learned prelate, or the legends and dreams of feats and chivalric deeds within the court-circle of some more or less lettered feudal lord. They have only been intended to serve as guides to mariners and merchants in the Mediterranean sea-ports. They have also seldom had learned men for authors, and slight was

---

[1] The first known map of that kind is a map of Cosmas from the 6th century. The division of the horizon has played no small part in the cartographical and geographical literature of old, and often, especially on maps of the world, the different wind-directions were marked by blowing heads, placed on the border of the map (e. g. the maps here given on pl. XXIX, XXXI, XXXIII, XXXVIII, XLIV and XLIX). For further particulars, the reader is referred to a very interesting letter from d'Avezac to M. Henri Narducci inserted in: *Bolletino della Società Geografica Italiana*, XII, Roma 1875, p. 379.

the attention paid to them in the 15th and 16th centuries by the learned geographers. Thus Münster seems to have totally overlooked them, and in the first edition of his *Theatrum Orbis Terrarum* Ortelius does not mention a single drawer of portolanos among the cartographical authors enumerated in his *Catalogus Auctorum*. At present the investigator into the history of geography acknowledges them as unsurpassed masterpieces, and reckons them among the most important contributions to cartography during the middle ages. They have long ago become a subject of inquiry, embraced with a special predilection by modern geographers. They have more frequently than other maps been reproduced in fac-simile, and the literature regarding them has already grown to such an extent, that the space at my disposal would not even allow an enumeration of the different works and memoirs. However, one of

though the loxodromes drawn obliquely across the map are left out and some modifications of the legends were introduced. Moreover, extensive parts of portolanos of the middle of the 16th century, and perhaps earlier, have often been printed in Italy, and one portolano from the beginning of the 14th century has been without any essential modifications engraved on copper in 1595, to guide the Dutch mariners, then the first in the world, on their commercial voyages. Afterwards this chart of 1595 served as a model to other similar publications during at least a whole century. Quite in contradiction to the theory generally adopted, the portolanos have excercised an immense influence on the printed cartography. They cannot, therefore, be entirely overlooked in this work on the oldest printed cartography, even if want of space, as well as the scarcity of original drawings of this

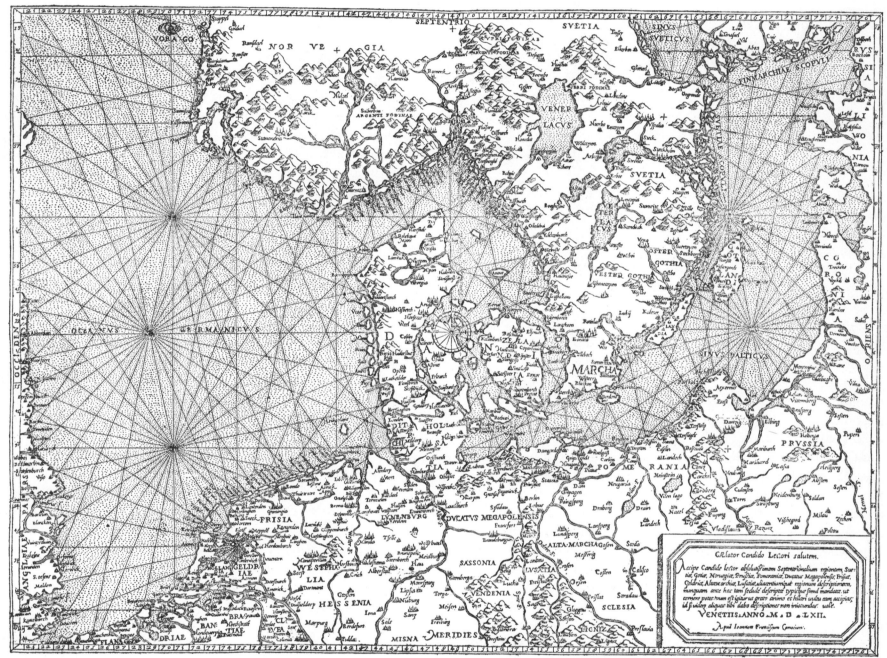

25. Chart of the Baltic and North-sea by JOANNES FRANCISCUS CAMOCIUS, Venetiis 1562. (Orig. size 524 × 382 m. m.).

the most important features with regard to the influence of these maps on cartography, seems altogether to have escaped the investigators who have as yet occupied themselves with this subject. It has been asserted that the portolanos had never been published in print, with the exception of the insignificant reproductions on a reduced scale in the works of Mattiolo-Gastaldi and Medina. This, however, is so far from being correct, that the delineations of the coasts on most of the *tabulæ novæ* in editions of Ptolemy up to the middle of the 16th century are principally founded on the portolanos, al-

class in our Libraries,[1] prevent me from publishing an extensive treatise on this interesting subject.

The portolanos are almost always drawn on parchment on a scale varying from one millionth to a ten millionth; usually the scale amounts to a five millionth. They are generally richly ornamented in gold and colours, and the same colouring has often been maintained for centuries in otherwise very dissimilar editions or reproductions. The typical portolanos only embrace the coast-lines and the towns situated in the immediate vicinity of the sea, or near the mouths of navi-

---

[1] As far as I know only three portolanos exist in Sweden: 1:0 (in the Library of Upsala) an undated portolano of the Black Sea, the Mediterranean and the western coast of Europe as far as to Scotland (large 1076 × 650 m. m.), extremely well preserved; 2:0 (in the Royal Library at Stockholm) a magnificent, exceedingly well preserved, small (222 × 153 m. m.) Portolano-atlas of several sheets — neither the author nor the date are given, but it is indubitably a work of BATTISTA AGNESE, drawn between 1530 and 1540; and 3:0 (in my private collection) an old portolano neither dated nor signed, embracing the Black and Mediterranean Seas and Europe as far as to the middle of the Baltic (815 × 600 m. m.).

gable rivers. Only exceptionally the interior of the continents is provided with territorial details, extensive legends etc. But when this is the case, it will generally be found that the neat and correct manner of drawing, so characteristic of what may be called the portolano-style, is no longer adhered to in that part of the map, which seems to indicate that the maps of the interior were introduced on the portolanos at a later period of decadence.

The work, *Studi biografici e bibliografici sulla storia della geografia in Italia*, Vol. II. *Mappamondi, Carte Nautiche, Portolani ed altri Monumenti cartografici specialmente Italiani dei Secoli XIII—XVII per* G. UZIELLI e P. AMAT DI S. FILIPPO, Roma 1882, edited by the Italian Geographical Society, contains a review of earlier manuscript maps, preserved in Italian libraries. The catalogue embraces 524 numbers, of which about 400 are portolanos, or sea-charts, drawn before the beginning of the 16th century. Another important work on the same subject was lately published by Professor THEOBALD FISCHER of Marburg under the title: *Sammlung mittelalterlicher Welt- und Seekarten italienischen Ursprungs und aus italienischen Bibliotheken und Archiven*, Venice 1886. Besides a general historical and critical description of the maps, a detailed analysis and (in a separate atlas) photographic fac-similes are here given of eleven of the most important of them. The learned Professor thus sums up the result at which he arrives:

A) It is not proved that maps expressly drawn for the use of mariners (sea-charts), were employed by the Greeks and Romans. Probably their navigators were only directed by sailing-directions (Peripli), indicating the directions of coastlines, the harbours, the distances between halting-places, prominent capes etc., with great exactness not only in the Black Sea and the Mediterranean, but also in the north-western part of the Indian Ocean. Nor did the Arabs delineate any sea-charts; the only sea-chart known with Arabic legends being a copy of an Italian original.

B) Map-sketches of the western Mediterranean for the use of mariners were already drawn in Italy during the beginning of the 12th century. The compass having been generally adopted in the middle of the 13th century, at least in some of the Mediterranean countries, the first loxodromic charts (portolanos) were from that time constructed by its aid. Already in the latter part of the 13th century these charts probably attained the finish met with in the maps still extant from the beginning of the 14th century. Such maps were also made in Catalonia from Italian models (thence the name *Cartes Catalanes*).

C) In 1883 498 loxodromic charts, of which 419 are of Italian origin, were known in the Italian libraries. The one of earliest date among them is the chart of PIETRO VISCONTE of 1311, and maps of the same model were still drawn by hand long after the discovery of the art of printing. Numbers of copies of the same original were fabricated for sale by professional artisans, adhering for generations to the model once accepted. We accordingly find portolanos of the 15th and 16th centuries copied from charts from the beginning of the 14th century with astonishing exactness not only as to the cartographical details, but also as to the manner of drawing, legends, colouring etc.

D) The original charts are founded on a careful utilization of the observations of mariners. They were drawn for practical purpose and never much cared for by scientific geographers. With regard to the Black and Mediterranean Seas, northern Africa, and the western coast of Europe, as far as to the mouth of the Scheldt, the charts are founded on the experiences of thousands of mariners, collected during a long succession of commercial voyages. Here the charts have attained a wonderful perfection in comparison with other cartographical works from that remote age. But the mapping of the lands and seas beyond the extreme limits of the voyages of the Italian mariners is, on the other hand, generally very incomplete and defective, as only depending on tales and rehearsals.

————

The name ›Portolano‹ is rejected by FISCHER and BREUSING as being improper and equivocal, having originally signified not a chart, but a sailing-directory or book of courses. The name »compass-chart» is also disapproved by these eminent authorities, who propose the name ›loxodromic charts‹ for the maps formerly styled portolanos. It, however, seems to me highly inconvenient and objectionable to denote old practical maps by this newly invented Greek name. These maps are characteristic not of the learning, but of the great personal experience of their authors or draughtsmen, and of the care and want of learned prejudices, with which the necessary material was collected.[1] The name ›Compasskarten‹ is also improper and deceptive, it being by no means proved that these charts were originally constructed by compass-bearings. Many of them are evidently older than the use of the compass on board ships. I shall therefore here retain the old name of *Portolanos* or *Portolan-charts*, referring to the manner in which these charts have originated (observations while sailing from port to port). The double meaning of the word will scarcely cause any misunderstanding. The word ›chart‹ has a similar double meaning in many languages. Two course-directories without charts were, for instance, printed at Amsterdam in 1540 and 1541 by JAN JACOBSZOON under the title: *Dit is die kaerte van dye Suyd zee tot dat Ranserdyep toe etc.* and *Dit is die caerte van der zee om Oost en West te zeylen*, and the same word *Caerte* is used by LUCAS JANSZ WAGHENAER to denote some of the *sea-charts* in his celebrated *Spieghel der Zeevaerdt*, Leyden 1584.

After having carefully compared the fac-similes of the most important portolanos, which BUCHON and TASTU, SANTAREM, JOMARD, FISCHER, LESOUJEFF, and others have given, with Ptolemy's maps, with the *Tabulæ Novæ*, printed in the oldest edition of Ptolemy's geography, and with other maps of the 15th and 16th centuries, I am also obliged to differ in some other respects from the opinions pronounced by my predecessors.

When Professor Fischer wrote his above mentioned elaborate memoir, he did not know that any portolanos of the Mediterranean and Black Seas had been published in print; he expressly says (p. 85): ›*Es kann auffallen, dass man niemals dazu schritt dieselben durch Holzschnitt oder Kupferstick zu vervielfältigen*,‹ and he nowhere in his paper cites an instance of their having been used to any great extent as material for printed maps. This is correct so far as they are never or scarcely ever mentioned by learned geographers during the century of the great geographical discoveries. But it is from these very portolanos that they have gathered their materials for several of the *Tabulæ Novæ* which were added to Ptolemy's geography; and in 1595 a true portolano of the Black Sea, the Mediterranean, and the coasts from the Canaries to Antwerp was faithfully and in a masterly way engraved in copper and published in: *Nieuwe beschryvinghe ende Caert Boeck vande Midlandtsche Zee etc. door Willem Barentzoen. Tot Amstelredam Ghedruct by Cornelis Claesz. op't Water in't Schrijfboeck, by d'oude Brugghe 1595.*[2] On the title-

---

[1] It is to be remembered that the loxodromes only form straight lines on maps with parallel meridians, and that only on maps on Mercator's projection they both form straight lines and indicate the proper course to be steered in sailing from one place to an other. The name ›loxodromic chart‹ for a portolano is thus mathematically exact only under the supposition that it is constructed on Mercator's projection.

[2] With regard to the long title, and the different editions of this rare atlas, the reader is directed to: P. A. TIELE, *Nederlandsche Bibliographie van Land- en Volkenkunde*, Amsterdam 1884.

page is a view of the roadstead of Genoa, in which a naval combat is progressing. This *Caert Boeck* forms a thin folio volume with 10 charts, engraved in copper, with innumerable wood-cut coast-views. The maps are signed WILLEM BARENTS-ZOEN and dated 1593—95. On 15 folio leaves at the end of the work there is a *Haven-wyser vanden Middellandsche Zee*, said to be translated by MARTIN EVERART from an Italian book called *Portolano*. But in several places in the book the celebrated arctic explorer Willem Barentszoon is expressly given as the author of the whole. Notwithstanding this, the first double folio map is a copy of a Mediterranean portolano from the beginning of the 14th century. With regard to the general outlines of the Mediterranean and Black Seas and to the main character of the charts, any one may convince himself of the exactness of this assertion, by a comparison between the fac-simile on a reduced scale of Barents' map (fig. 21) with the portolano of Dulcert of 1339 (fig. 16), or with any other similar medieval work. As for

Dulcert 1339.	Carte Catalane (from Buchon et Tastu) 1375.	W. Barentszoon 1595.	van Keulen 1694.
Beroardo.	Beroardo.	Castel Verando.	Castro Verando.
Jaffa.	Jaffa.	Jaffa.	Jaffa.
Arzufo.	Arzuffo.	Arsuffo.	Alzulo.
Cæsaria.	Cesaria.	Spezaira.	Caesaria.
Castro Pelegri.	Castel Pelegri.	Castel Peregrino.	Castro Pelgrine.
Carmen.	Carmeni.	Carmini.	
Acri.	Acre.	Acri.	I. Juan d'Acari.
Cauo Jancho.	Cavo Iancho.	C. Blanco.	C. Blanco.
Sur.	Sur.	Sur.	Sur.
Sarafent.	Sarafent.	Sara feret.	Saraferet.
Saytos.	Saytos.	Saites.	Saita.
Damor.	Damor.	Damor.	Dantor.
Baruti.	Barut.	Barut.	Baruti.
Fluvius canis.	Flum Canis.	F. Canis.	P. Canis.
Gibeleto.	Gibellet.	Gibileto.	Gibiletto.
Bodroan.	Bodron.	Vadro.	Vadro.
Conestar.	Conestabilli.		
Nerfin.	Nofim.		
Tripoli de Suria.	Tripolli de Suria.	Tripoli.	Tripoli.
Larcha.	Larcha.	Larca.	Laraca.

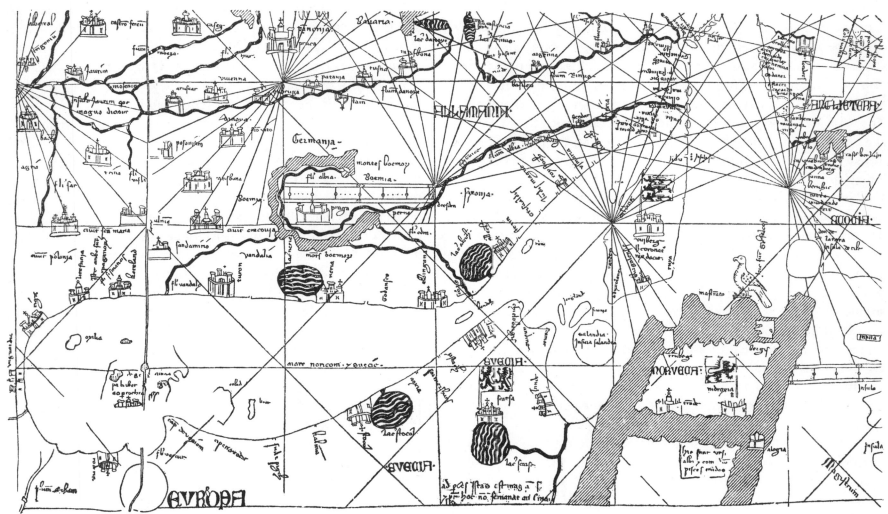

26. The Baltic with surrounding countries from the portolano of DULCERT 1339. (The part of the original here reproduced 409 × 230 m. m.).

the legends, the tables below will show the extreme care with which Barents and, one century later, VAN KEULEN, followed the old models or originals.

Comparison between the legends on portolanos and W. Barentszoon's and van Keulen's sea-charts.

Dulcert 1339.	Carte Catalane (from Buchon et Tastu) 1375.	W. Barentszoon 1595.	van Keulen[1] 1694.

1. The coast of Syria.

Damiata.	Damiat.	Damiata.	Damiata.
..nese.	Enes.	Tenes.	Tenere.
Faramia.	Faramia.	Faramia.	Faramida.
Rasalcasero.	Rasal Casero.	Raxalgagero.	
Staygnom.	Stagnom.	Stangoni.	Sangoni.
		C. Gall.	C. Gallo.
G. de Larissa.	Golfo de Larissa.	G. Larissa.	G. de Larissa.
Berto.	Berto.	P. Berton.	P. Berton.
Darom.	Darom.	Damar.	Damor.
Gazara.	Gatzara.	Galsara.	Gosara.
Excalona.	Eschalon.	Æscalona.	Escalona.

Prexon.	Prexom.	Pxini.	Proxoni.
Tortossa.	Tortosa.	Tortosa.	Tortosa.

2. The eastern coast of Tunis.

	Golfo di Tunis.	G. de Tunis.	Tunes.
Nubia.	Nubia.	Calibia.	Gallipia.
Cobin.	Cobon.	C. Bona.	C. Bona.
Maometa.	Mameta.	Mahometa.	Mahometa.
Recholia.	Rechilia.	Recolia.	Araclea.
Sussa.	Sussa.	Susa.	Susa.
Monisterie.	Monestir.	Monastir.	Monaster.
Cunie.	Conjeras.	Conieras.	Comigeras.
Affricha.	Affricha.	Africa.	Aphrica.
Capulia.	Capulia.	Capulla.	C. Capudia.
Casar Pegnatar.	Casar Pignatar.	Casar Punator.	Casar Mol.
Capisse.	Capis.	Caphis.	Zuchis.
Cæsar Nachar.	Casar Nacar.	Casar Nacur.	Casar Natur.
Muroto.	Muroto.	Marota.	Maroto.
Insula de zerbis.	Illa de Gerba.	I. de Serbi.	I. Zerby.

3. The French coast of the Mediterranean.

Porto venre.	Port Venre.	P. Veneris.	P. Veneri.
Coliuro.	Copliura.	Calibre.	Calibre.
Sasse.	Salses.	Salsas.	

[1] General chart of the Mediterranean, signed on the right corner I, and a special chart, signed VI (*De Groote Nieuwe Vermeerderde Zee-Atlas ofte Water-Werelt*, Amsterdam 1694).

Dulcert 1339.	Carte Catalane (from Buchon et Tastu) 1375.	W. Barentszoon 1595.	van Keulen 1694.
Leocata.	Leocata.	C. Leucata.	C. Leucata.
Nerbona.	Nerbona.	Narbona.	Narbona.
Sanper.	San Per.	S. Pera.	S. Pedro.
Sirignan.	Serignan.	Serignan.	Serigian.
Agde.	Agde.	Agde.	Agde.
C. de Seta.	Cap de Seta.	M. de Seuta.	C. Zeuta.
Magalona.	Magalona.	Maguelone.	Magdelena.
Lates.	Lates.	Latas.	
Mopesler.	Montpesler.	Mompeillier.	Montepelliers.
Aque Morte.	Aygues Mortes.	Aigues Mortes.	Aqua del Morte.

In some other parts of these maps, for instance in Corsica and Sardinia, there are differences, perhaps depending upon an Italian, not a Catalan original having served as original to Barentszoon's chart, or upon the latter having followed portolanos not of the 14th, but copies of copies of them of the 15th or 16th century. Yet the identity of the legends is in most parts of the maps so complete that the insignificant differences evidently can be explained by clerical errors. Should any doubt exist regarding the age of Dulcert's map, it may be remarked that the same outlines of the sea and in most cases also the same legends meet us in the chart of Pietro Visconte of 1311. Thus, notwithstanding all the progress made during the 15th and 16th centuries in the art of drawing maps with the aid of newly invented nautical instruments, there was published a chart in Holland in 1595 by one of its most expert mariners, which is only a copy, or rather a copy of copies of portolanos drawn 250 to 300 years earlier. This is an extremely remarkable fact in the history of civilisation. But moreover the principal features of the portolanos from the beginning of the 14th century are still to be found on VAN KEULEN's sea-charts of 1681—1722 (I have used the edition of 1694), and probably also on charts of a far later date. I suppose that up to the beginning of the 19th century the influence of the old portolanos may yet be traced on the charts of several parts of the Mediterranean and Black Seas. In consequence of the rather hap-hazard combinations of names, to which investigators in the history of cartography are often inclined, I may finally point out, that quite reliable scales for appreciating the variations in the map-legends, caused by repeated copying, are to be found in the above table.

Of course it would be of great interest to be able to decide with certainty the epoch when the first very rude maps of this kind were drawn, as well as the time when they reached the finish we still admire. It seems to me, that the richness in legends and the almost stereotypic tenor of the legends during centuries, should make it possible for a scholar, well versed in the history and geography of the middle ages (especially the crusades?), to give a reliable answer to the latter question by a careful analysis of the numerous names on the portolanos. With regard again to the former question, a definite answer will be involved in great difficulties and great uncertainty, until new data for its elucidation have been discovered. I suppose that these data, if yet extant, must be furnished from the lands of the old Byzantian empire. In a short appendix (»Nachträge und Verbesserungen«), appearing at the end of the index of the above cited »Sammlung mittelalterlicher Welt- und See-Karten,« Fischer seems to accept the opinion pronounced by FIORINI in Le projezioni delle carte geografiche, Bologna 1881, p. 648, that the Italians had learnt the art of drawing portolanos from Byzantium shortly after the year 1000. On the other hand, he says (p. 78): »Somit ist die Mitte des 13. Jahrhunderts auch gewiss der früheste Zeitpunkt, bis zu welchem wir die Entwerfung der ersten loxodromischen Karten hinaufrücken dürfen, in der zweiten Hälfte des 13. Jahrhunderts erlangten sie die vollendete Form, in welcher wir

sie schon zu Anfang des 14. auftreten sehen.« Fischer's assertion that the Greeks and Romans probably never employed nautical charts (Itineraria maritima) for practical use seems scarcely to be quite correct. The description and criticism of the works of Marinus of Tyre by Ptolemy shows on the contrary that the Διόρθωσις of Marinus was a real portolano, provided with a text, which seems to have had much in common with the »Opera chiamata Portolano« of the middle ages, and of which, as of the portolanos, several editions had been issued. If Ptolemy himself had not always spoken of Marinus as a definite personality, it could even be conjectured that the name »μαρῖνος ὁ Τύριος«, or the Tyrian sea-fish, had only been a collective name for a certain category of nautical maps, like the name Waghenaer, or Waggoner, fourteen hundred years later. The numerous editions mentioned by Ptolemy indicate that the Tyrian charts were made for a practical purpose, and the improvements, introduced according to Ptolemy in every new edition, constituted the germ for the future master-piece. Unfortunately none of the maps of Marinus has, as far as known, been preserved down to our time, and there is little hope of their recovery. But the Arabian geographer MASUDI declares in a work written A. D. 955, that he had seen maps drawn by Marinus, and that they were even better than the maps of Ptolemy (UKERT, cited work p. 195; MEHREN, cited work p. 7). Thus the Tyrian collection of maps was extant down to the 10th century. The portolanos are the only maps drawn in ancient times or in the middle ages which are comparable with those of Ptolemy, and it is scarcely too much to assume that the maps mentioned by Masudi under the name of Marinus, were a collection of portolanos. It is, besides, not difficult to restore the maps of Marinus as regards their main features. Ptolemy expressly declares that his maps are based on the work of Marinus, whose longitudes for the Mediterranean were left unchanged. I have (fig. 15) restored Ptolemy's outlines of the Mediterranean. If this map is compared with the corresponding part of a portolano, not inconsiderable differences will be found. The main features, however, are the same, or rather: the maps of the Mediterranean and Black Seas given by Ptolemy after Marinus were already so complete, that the delineations of these Seas on the portolanos might very well have proceeded from them by gradual improvements. The very manner of drawing, or what I have, on a former occasion, styled the cartographic alphabet, is the same on both, and numerous purely Ptolemaic features are clearly to be discerned on the oldest of the portolanos, for instance the course of the Tanais on the map of Dulcert of 1339. If we further consider, that the outlines of these seas on all other independent maps of the middle ages were disfigured so as not to be recognizable, it seems to me highly probable that the first origin of the portolanos is to be derived from the Tyrian charts described by Ptolemy under the name of Marinus.

No graduation appears on the original portolanos, but when they were published in print, they were generally provided with meridians and parallels on a projection of equidistant parallelograms, or on the projection on which the maps of Marinus of Tyre, according to Ptolemy, were constructed. This is criticized by BREUSING, who, supposing that these maps were based on deviating loxodromes, means that a conical net of graduation should be preferred for their graduation. I cannot unconditionally accept this opinion. In many cases, especially when the problem is to draw meridians and parallels on portolanos of great extent from north to south, the conical projection, or perhaps rather the »projection of Donis,« may be the most suitable or the one coming nearest to accuracy. But this is by no means generally the case. In fact, a highly interesting, but in no wise easily solved problem here presents itself to the cartographer. It may be stated thus: If the relative position of a number of places on the earth is fixed: a) by measuring the distances between these places, or b) by measuring the azimuths of the courses steered by vessels, or c) by giving the distances as well as the azimuths; what kind of projection on a plane surface will give the relative positions of these places either exactly, or, this not being possible, with the least possible deviation from the direct observations? An exact solution is only possible in the event b), and then we get a map on Mercator's projection.

If, again, as probably has generally been the case, the maps were based on measurement as well of the distances as of the courses of ships, the observations cannot be rendered on a

meridians and parallels, the projection of Marinus was generally preferred. But already, in 1558, on the publication of the voyage of the brothers Zeno by Marcolini, a very remarkable attempt was made to provide an old map with a conical net of graduation. The principles here followed are described in Ruscelli's edition of Ptolemy, Venetiis 1561, 1562, 1564 etc., on the reverse of *Septentrionalium partium nova tabula*. The chart of the North Sea and the Baltic by CAMOCIUS of 1562 (fig. 25), and the above mentioned maps of Barents of 1595, are portolanos graduated on the projection of Marinus.

While Ptolemy's maps, immediately after the discovery of the art of printing, were spread abroad in numerous copperprints or wood-cuts, a long time elapsed before the first complete portolano or loxodromic map was published in print.

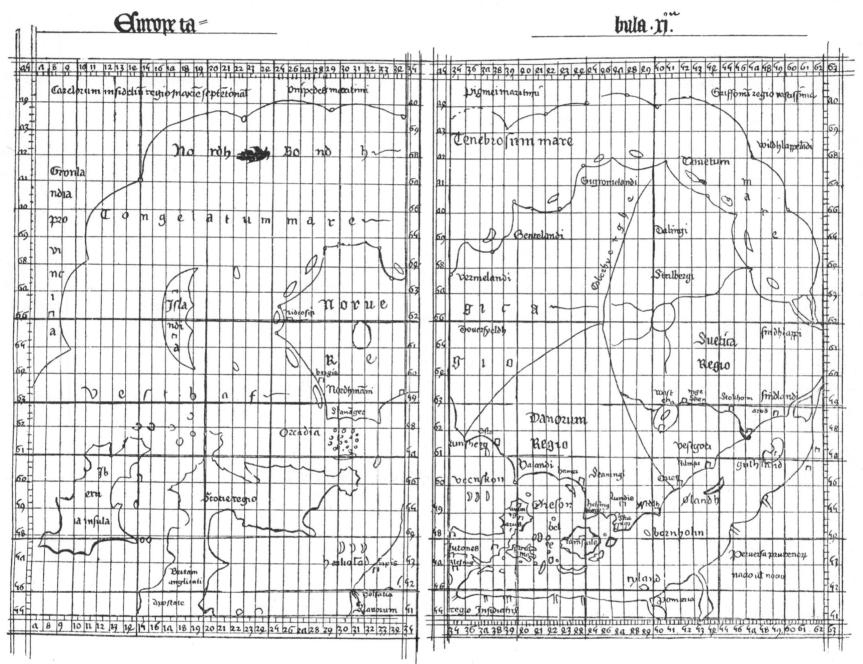

27. The map of Scandinavia by CLAUDIUS CLAVUS 1427. (Orlg. size 220 × 155 m. m.)

plane surface without a considerable adjustment. Even a mathematician of our days would not find it easy here to propose a general rule. It would be necessary by means of the method of the least square to connect observations, the relative values of which it is almost impossible to determine. Of course a mathematical treatment of the problem could not have come under consideration in the middle ages, and the problem was then, no doubt, simply solved by every map-maker trying as well as possible to introduce into the old map-model such corrections as were considered necessary by his customers. If the courses were laid out according to a deviating compass, as perhaps usually was the case after the 14th century, the problem became still more complicated. A general rule for the graduation of a medieval portolano accordingly can scarcely be fixed. When the portolanos in the 16th century were provided with

The reason for this seems partly to have been the little attention paid by the printers to these productions of unlearned mariners, partly the difficulty of reproducing the different colours of these maps by copper-print or book-press, and the cheapness at that time of manual labour, even of artists. But from the great number of portolanos still extant in Italy we may deduce that they had a greater circulation in the 16th century than printed cartographical works. The portolanos saved from destruction seem at least to be more numerous than the total number of copies still extant of the editions of Ptolemy's Atlas printed before 1500.

The following maps based on portolanos or drawn according to the principles adopted by the portolano-makers, although generally without any compass-lines or loxodromes, were published in print during the 16th century, viz.,

1. *Orbis typus universalis iuxta hydrographorum traditionem*, printed in the Ptolemy of 1513 and 1520 (N. T. XXXV) and copied, with exception of the westernmost part of the map, in the editions of Ptolemy of 1522, 1525, 1535, and Viennæ 1541.

2 and 3. Two portolanos (without compass-roses) of Africa, printed in Ptolemy of 1513 and 1520 and reproduced on a reduced scale in ed. 1522, 1525, 1535, and 1541 (N. Fig. 8 and 9). Notwithstanding the want of the usual straight lines denoting the compass bearings, these maps[1] are evidently copies of true portolanos and, through the richness and exactness of the map of the coast, quite comparable with the old charts of the Mediterranean Sea so highly admired. The legends prove them to have been drawn after the second voyage of VASCO DA GAMA. A fuller account of them will be given in a subsequent chapter.

4 and 5. *Tabula Nova Asiæ Minoris* in the Ptolemy of 1513 and 1520, reproduced in a very careless manner on a reduced scale in the editions of 1522, 1525, 1535, and 1541. This map is a rough woodcut reproduction of the corresponding part of the earth on the common portolano of the Mediterranean. Also the *Tabula moderna Indiæ* (N. fig. 10) in the above cited editions of Ptolemy is founded on a Portuguese portolano of the first years of the 16th century, as may be deduced from the want of names and topographical details in the interior of the continents and from several coast-legends.

6. A map in the *Arte de navegar* by PEDRO DE MEDINA, 1st edition, printed in Spain[2] 1545 and afterwards often translated and reprinted (according to BRUNET and CARTER BROWN: Lyons 1553, 1569, and 1576; Rouen 1573; Venice 1554 and 1555; Germany 1576; London 1581). I have had access to the edition Venetia 1554, where the map is printed in fol. xxxiii. Notwithstanding its small size, this map is remarkable for its correct delineation of the Isthmus of Panama, for the insertion of the famous papal line of demarcation between the ultramarine possessions of Spain and Portugal and, finally, on account of its original being one of the few maps printed in the Pyrenean peninsula before A. D. 1570. Only three small Spanish maps of that period are known to me, viz., the map of PETRUS MARTYR, Seville 1511 (N. fig. 38), the map of Medina in the *Arte de Navegar* and a very rude map of Spain by Medina in *Libro de grandezas y cosas memorables de España*, Sevilla 1549; Alcala de Henares 1548—1566 and 1595. This work also contains the above mentioned chart in *Arte de Navegar*. With the exception of some copies of medieval maps, which I suppose to exist in Spanish editions of classical authors, this seems to be almost the whole contribution during the earliest period of printed cartographical literature from the countries, from which the New World and the south-east passage to India were discovered, and from which hundreds of the most important voyages of discovery started during that period. Fig. 75 I give a fac-simile of Medina's original map. It is far inferior to the copy published in Venice and also, I suppose, to the copies published in France.

7. *Carta Marina Nova* (N. T. XLV), at first printed on a small scale, as a general illustration to a note on sea-charts, in Ruscelli-Gastaldi's Ptolemy, Venetiis 1548, and afterwards reprinted on a slightly enlarged scale in the editions of 1561, 1562, 1564, and 1574.

8. Various charts of parts of the Mediterranean, printed in Italy. They are generally direct reproductions from por-

tolanos. Like all old maps not inserted in books but published on separate sheets, these charts are now very rare, and of most of them, no copies are extant. But I suppose that they were, like the portolanos, once largely circulated and used by mariners and merchants. Some of these maps are inserted in the Collegio Romano copy of Atlas Lafreri (Nos. 8; 35; 115; 129; and 130, pages 240—250 in the catalogue of CASTELLANI).

9. A chart of the Baltic and the North Sea, engraved in copper in 1562 in Venice *apud Ioannem Franciscum Camocium* (N. fig. 25). When this chart was published, it was beyond all comparison the best map of the Southern Baltic and the North Sea.

10. *Spieghel der zeevaerdt* etc. *door* LUCAS JANSZ WAGHENAER *Piloot ofte Stuijrman residerende inde vermaerde Zeestadt Enchuijsen* (First edition, Leyden 1584—1585, fol.). The first part of this marine-atlas begins with a short treatise on the principles of navigation, followed by a general map of western Europe on the equidistant projection of Marinus, provided with a number of compass-roses, and intersected by compass-lines. Then follow 22 special maps of the coasts between Enkhuisen and Cadiz. The second part contains, besides the text, 21 maps of the coasts of the North Sea and the Baltic. The special maps are not graduated, but are provided with compass-roses and generally also with a scale in Spanish and German miles; one degree is stated to contain $17\frac{1}{2}$ of the former and 15 of the latter. In several places on the maps the soundings are given. This work is the first collection of sea-charts published in print. It was much appreciated among the seafaring nations of Europe, and published in a number of editions in various languages described by P. A. TIELE, *Nederlandsche Bibliographie van Land- en Volkenkunde*, Amsterdam 1884, and by F. VANDER HAEGHEN in *Bibliotheca Belgica*. Evidently the work is partly based on Italian or Catalan portolanos, and partly on ancient sailing directions for the Baltic. The same work was also printed in Leyden in 1592 and 1598 in oblong folio, with double-folded sea-, or coast-charts, of size 195 × 555. On the title-page of these editions it is spoken of »*het oude vermaerde Lees-caertboeck van Wisbuy vermeerdert, ende van ontallijcke fauten en valsche coersen ghesuyvert.*»

11. The marine atlas by WILLEM BARENTSZOON, *Caert boeck vande Midlandtsche Zee*, Amstelredam 1595. Besides the general map (N. fig. 22), the work contains eight charts of the western part of the Mediterranean and the entrance to it from the Atlantic. Whilst the edition of Waghenaer of 1592 gives *het oude vermaerde Lees-caertboeck van Wisbuy*, this work ends with an appendix introduced by the words: *Dit navolghende Boeck is ghetranslateert door Marten Everart, uyt een Italiaens Boeck, ghenaemt Portolano, twelck te segghen is Havenwyser, ofte een speciale verclaringhe van alle de havenen der Midlandsche Zee*. It is very remarkable that the authors of the sea-charts of the North Sea and the Baltic, as well as of the Mediterranean, referred to written descriptions from the middle ages, when the first marine-atlases were published in print at the end of the 16th century. This further confirms the circumstance, before pointed out, that, as Ptolemy's geography is the model of the modern atlases and was still often referred to at the end of the 16th century, so our sea-charts have been developed from medieval portolanos. On some of the special maps in Barentszoon's atlas there are two kinds of compass-roses, one (*Directorium nauticum italicum, Italiensch compas*) giving the true points, and the other (*Directorium nauticum vulgare*,

---

[1] Loxodrom or compass-lines are laid down on the large map of Olaus Magnus, although otherwise this map has not at all the character of a portolano.
[2] In Cordova, according to BRUNET, in Valladolid, according to HARRISSE. The names of three cities, Sevilla, Cordova, and Valladolid, are mentioned on the title-page.

Gemeyn Duytsch compas) the deviating ones. The special charts of the Mediterranean are not graduated, whereas a broad double line, on which the degrees of latitude are marked out, is drawn in the direction of the meridian, across the chart of the sea between Cape St. Vincent and Cape Bojador (N. fig. 23). I assume that this sea-chart is based on Portuguese maps and that we have here a reminiscence of the introduction by the men of Prince Henry the Navigator of the

of Europe and a part of Africa are laid down with tolerable exactness. The portolanos were used as models, and the map was completed in accordance with accessible data and prevailing traditions, and as well as was then possible. Necessarily it was incorrect, but nevertheless it was constructed with better judgment than that which guided the drawer of the map of the world f. i. in the *Rudimentum Novitiorum*. The most complete maps among these are:

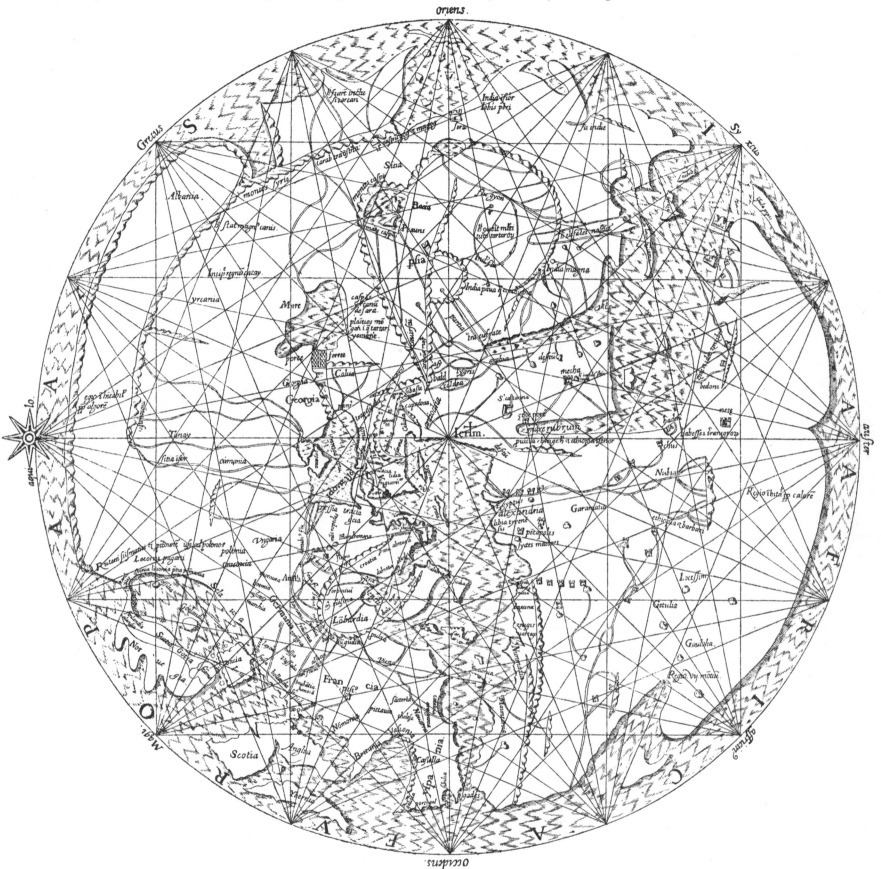

28. MARINO SANUDO'S map of the world from the beginning of the 14th century. From Bongars, *Gesta Dei per Francos*. Hanoviae 1611. (Diam. of circle in original 330 m. m.)

method of determining the ship's position by means of observations of latitude.

4. General maps of the world based on portolanos. Besides the general maps before (mom. 2) mentioned in this chapter, we also have, from the middle ages, other maps of the world, for which the portolanos served as starting points, and which, though still rude, far surpass other medieval maps or geographical drawings of the world.

On these maps Europe, Asia, and Africa form a large circular island with Jerusalem almost in the centre. In this circle the Black Sea, the Mediterranean, and the western coasts

MARINO SANUDO'S atlas of the beginning of the 14th century, composed for his great work *Liber Secretorum Fidelium Crucis*. According to UZIELLI and AMAT DI S. FILIPPO (II: p. 50) this atlas, in the most complete codex extant, consists of nine plates, among which one is a planisphere of the world. Four of these maps (Egypt, Jerusalem, Palestine, and the planisphere) were printed *Hanoviæ* 1611 in J. BONGARS' *Gesta Dei per Francos*, which, among other works, also contains Marino Sanudo's Liber Secretorum Fidelium Crucis. Copies of the planisphere have been published by Lelewel and Santarem (two variations, the one from a manuscript in Paris, the other from a codex in Brussels). For several memoirs on Sanudo's work I may further refer to the work of G. UZIELLI and AMAT DI S. FILIPPO before cited, as well as to H. SIMONSFELD, *Studien zu Marino Sanuto dem Aelteren* (*Neues Arch. der Gesellsch. für ältere deutsche Geschichtskunde*, Vol. 7, 1881—82). Sanudo's atlas appearing to be partly founded on original informations and being very remarkable through the delineation as well of the northern countries as of Africa, I here (fig. 28) give a fac-simile of its map of the world from the copy in Bongars' work.

The map of the world of ANDREA BIANCO of 1436. Of this excellent artist a map of the world or a planisphere

is still extant besides various portolanos resembling that of Marino Sanudo, but not so well executed. It has been reproduced several times, amongst others in the works of Santarem and Theobald Fischer.

FRA MAURO's map of the world of 1457. A well preserved planisphere, in the *Biblioteca Marciana*, having a diameter of 1,96 m. It is rich in details and legends and of great importance to the history of geography. It was for the first time fully described and reproduced in a work: *Il Mappamondo di Fra Mauro Camaldolese*, published in 1806 in Venice by PLACIDO ZURLA; then by SANTAREM, and on a more or less reduced scale by several other geographers.

V.

Extension of Ptolemy's Oikumene towards the north and northwest.

As may be perceived from the general map of Ptolemy, and from his Tabula 1ma, 8va Europae, and 2da, 7ma, 8va Asiae (N. T. I, II, IX, XVII, XXII and XXIII), he, following the example of Marinus, made his atlas of the known world, towards the north, terminate everywhere at lat. 63° N. This boundary-line quite arbitrarily adopted crosses, for about two thirds of its length, the Sarmatian and Scythian deserts. Until the middle of the 16th century, when the English and Dutch began their north-east voyages, the geographers took no notice of what was situated beyond lat. 63° N. Although some notices respecting northern Asia may have penetrated to the civilized countries of Europe through Marco Polo or other Asiatic travellers, and although the constructors of globes, and drawers of general maps of the world, were compelled to let the continent of Europe and Asia terminate towards the north in a coast-line, this hardly justifies us in speaking, before that time, of an extension of the maps of this part of the world beyond Ptolemy's boundary-line. The case was different in the west, where the 63d parallel was drawn by Ptolemy across the island of Thule through *Oceanus Hyperboreus* and *Deucaledonius*. Even here the maps of Ptolemy do not correspond very well with reality, but they show that the accounts of the distribution of land in these parts of the world had already reached, before A. D. 150, as far as to Egypt and Syria. Thus the northern coasts of Germany and the Cimbrian peninsula (Jutland) are laid down at least recognizably and near to their proper latitudes. But in the sea farther to the north there are, instead of the Scandinavian peninsula, only two islands, *Scandia* and *Thule*. Between Thule and the northern extremity of Scotland, which extends much too far to the east, the »*Orcades Insulæ 30*» are placed, and between Scandia and Jutland, the »*Scandiæ Insulæ 3*.» The main island *Scandia*, as well as *Thule*, had an extension from east to west of only about 150 kilometres. On Scandia we read the names of *Levonii, Chedini, Dauciones, Gutae, Phiresi, Phanone*, among which only a couple can be referred

to nations whose names are still registered on the pages of history. But the inhabitants of the Roman empire soon learned that this map of the northern part of Europe could not be correct, first through the tribes, which from the north invaded the rich countries of the south, and which mentioned with pride the large territory in the North for their home, and afterwards by the Vikings. It was also for that manifestly incorrect part of Ptolemy's atlas that the geographers of the middle ages first composed new maps completely different from the old type. Of such maps the following are yet extant:

1. Maps of the North on portolanos and general maps of the world based on portolanos of the middle ages. Regarding these works, the reader is referred to a preceding chapter. The usual manner in which the Baltic and surrounding lands were drawn on the portolanos is shown by the fac-simile of a part of one of the oldest of them yet extant, the map of DULCERT of 1339. Analogous drawings of Northern Europe are found on the famous *Carte Catalane* of 1375 (see above p. 46), on ANDREA BIANCO's map of 1436, and, as far as I have been able to ascertain, on almost all other earlier portolanos embracing the countries north-east of the mouth of the Scheldt. The most characteristical features of the northern parts of these maps appear to me to be the following.

When these maps were drawn, the Gulf of Bothnia was not yet »discovered.» Thus the Baltic got its principal extension in east and west, and a form not in the least corresponding to reality. In the centre of its eastern part there is a large island, Gotland (*y:a Codladie in qua sunt nonaginta parochie;* ANDREA BIANCO, 1436), often marked with purple and gold, probably to indicate its power and wealth. There are still about ninety parishes on Gotland, but its power and wealth long ago came to a sudden end with the destruction of Visby, in 1361, by the Danish king Valdemar.[2] The north-eastern parts of the portolanos also contain in the east the names of Novgorod and of some other cities, with which the Gotlanders had commercial intercourse. Like the Baltic

---

[1] If this map is compared with which Pliny's account of *Scandinavia, Baltia, Thule, Nerigon*, etc. (Lib. IV Cap. XIII and XVI) it will be perceived, that the Alexandrian geographer was not well aquainted with the information about the northern countries, collected during the military expeditions of the Romans to Britain, Gallia, and Germany.

[2] The importance of this event seems, however, to have been exaggerated. The trade and wealth of Gotland commenced already to decline at the end of the 13th century, with Visby ceasing to be the emporium for the commerce of the Baltic.

the Scandinavian peninsula has on these maps got its greatest extension from east to west. A number of northern names,[1] some of which may be easily identified with known localities, are here met with. The Norwegian mountains are indicated by rough contours, and several legends indicate that the draughtsman of the map knew the country personally. Even to the west of Norway the land and islands are provided with inscriptions (f. i. *y:a Rovercha, Stockfis, Stilanda* on Andrea Bianco's map), testifying that Greenland and Iceland were vaguely known to the mariners from whose reports the map was compiled. As to the style of drawing, the charts of these northern parts differ from the typical portolanos of the Me-

the limits between sea and land, and the forms of lakes and islands, this part of these medieval maps acquires a remarkable resemblance to the cartographical productions of the Arabs. I conjecture, however, that this is only owing to the same cause as that which gives a common stamp to the first scratches of children in different ages and different countries.

Professor Theobald Fischer does not assume the northern parts of the portolanos to be based on direct observations, but on narratives collected by mariners from the South, when they have met the mariners from the Baltic at the mouth of Scheldt. But the rough and shapeless character of these maps, unchanged for centuries, indicate that they

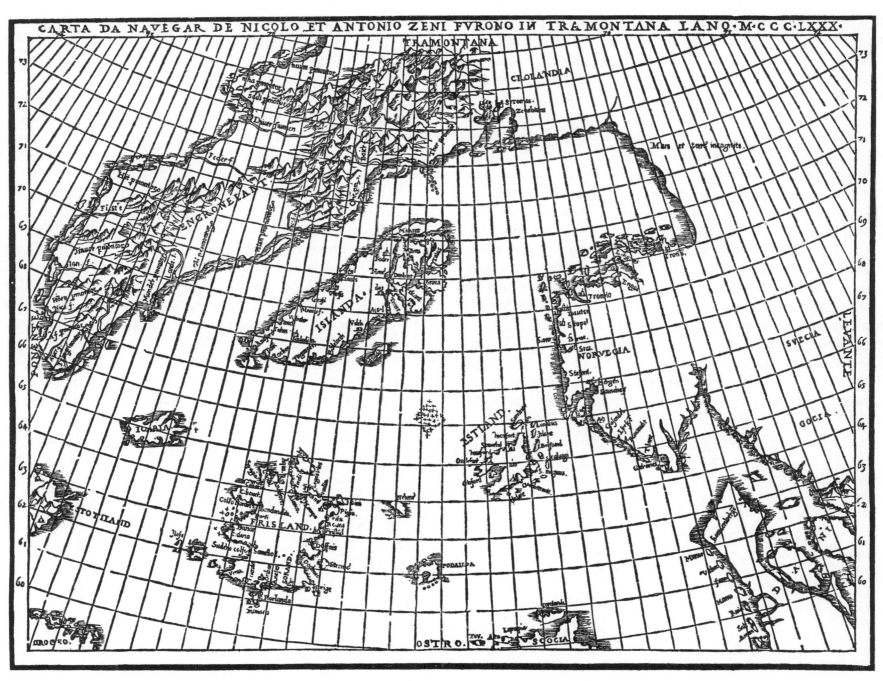

29. Map of the North of the Zenos, printed in Venice in 1558. (Orig. size 380 × 284 m. m.).

diterranean and Black Seas, and by no means to their advantage. Owing to the circle-segments here used to mark

were copies from a common prototype, only slightly modified by generations of map-makers, and various details, e. g.

---

[1] Some of the legends on the portolanos have been erroneously identified with modern names by authors, not quite familiar with our geography and history. The following collation, made by Mr. E. Dahlgren, may therefore be of use and interest to investigators into the history of cartography.

Dulcert's map of 1339.	Corresponding modern names.
Norway.	
alogia	Halogaland, Helgeland.
nidroxia tronde	Drontheim.
bergis	Bergen.
mastranto	Marstrand.
trunbeg	Tönsberg.
Sweden.	
lacus scarse	Scara lake = Venern.
scarsa	Skara.
lunde	Lund.

Dulcert's map of 1339.	Corresponding modern names.
scamor	Skanör.
andine	Nundinæ (Schanienses) = the great fair at Skanör.
chiclobergis	Trelleborg.
lundes	Lund.
ystach	Ystad.
sormershans	Simrishamn.
aoxia	Åhus.
lacus stocol	Stockholm lake = Mälaren.
stocol	Stockholm.
kalmä	Kalmar.
suderpigeh	Söderköping.
riperia roderin	Roden, Roslagen.

the sketch of a reindeer on a Catalan portolano of the beginning of the 15th century, preserved at the Biblioteca Nazionale in Florence, and reproduced in Fischer's atlas (map XIII), give them the appearance of having been drawn by a Scandinavian, or by a foreigner who had visited the Scandinavian countries.

2. A small map of the North by CLAUDIUS CLAVUS, annexed to the Latin manuscript of Ptolemy's geography mentioned above and preserved at the town-library in Nancy (N. fig. 27). The important codex[1] which contains this map was drawn, probably in Italy, for the Cardinal GUILIELMUS FILIASTRUS, and finished in 1427. Filiastrus died in 1428. He not only got the Latin translation of Ptolemy and the latinized copies of the 27 old maps copied by some skilled scribe and artist, but he also added a new map of the northern countries, besides an explanatory text in the Ptolemaic style, to the old work. Thus Filiastrus makes the following extensive addition to the original legend of Tab. VIII Europæ:

›Octava Europe tabula continet Sarmatiam Europe, vel illas regiones que sunt ab Germania ad septentrionem versus orientem, in quibus est Polonia, Pruthia, Lituania et alie ample regiones usque ad terram incognitam ad septentrionem partem Dacie et Tauricam chersonesum usque ad paludem Meotin; et ibi Thanais fluvius, qui dividit Europam ab Asia in parte septentrionali et versus orientem. — Item continet, ultra quod ponit Tholomeus, Norvegiam, Suessiam, Rossiam utramque et sinum Codanum, dividens Germaniam a Norveigia et Suessia. Item alium sinum ultra ad septentrionem, qui omni anno congelatur in tercia parte anni. Et ultra illum sinum est Grolandia, que est versus insulam Tyle magis ad orientem. Et ita tenet totam illam plagam septentrionalem usque ad terram incognitam. De quibus Tholomeus nullam fecit mencionem, et creditur de illis non habuisse noticiam. Ideo hec 8:a tabula est multo amplior describenda. Propter quod quidam Claudius Cymbricus illas septentrionales partes de- scripsit, et fecit de illis tabulam que jungitur Europe, et ita erunt 11. Et tamen nullam facit mencionem de illis duobus sinibus maris Norveigie et Grolandie. In hiis regionibus septentrionalibus sunt gentes diverse; inter quas Unipedes et Pimei, item Griffones, sicut in oriente, velut vide in tabula.›

On the tenth map of Europe there is further written: ›*Hac descriptio et tabula editae sunt a quodam Claudio Cymbrico,*› and in the geographical description of Scandinavia inserted in Ptolemy's text may be read (at *Ohdhonis insula = Fyen*): ›*in qua parte est Salinga, patria villa Claudii Clavii, Svarthonis Melis Petri Tuchonis fili*› etc. We learn from this, that a Dane, CLAUDIUS CLAVUS or CLAVIUS, was the author as well of the additions to Ptolemy's text as of the new map. The attributive *quidam* indicates that he had not been personally known by Filiastrus. Mr ERSLEV (*Jylland*, p. 118) gives good reasons for the identity of Claudius Clavus with the Danish ›mathematician› CLAUDIUS NIGER, who, according to FRANCISCUS IRENICUS (10th book, 21st chap. of *Germaniae Exegeseos volumina XII*, Hagenau 1513), probably at the beginning of the 15th century, delineated a map of Denmark at the request of the Danish king. The map in the codex of Nancy seems, however, only to be a copy on a much reduced scale.

In the text of the new map the longitudes and latitudes are given for 133 places in the Scandinavian peninsula, Iceland, and Greenland. Most of them are of course only based on estimated distances. Yet, some of the latitudes may have been calculated from actual observations. It appears from a passage in the *Liber Daticus Roscildensis* that the latitude of Roskilde[2] was already in 1274 calculated from the length of the midsummer-day, and similar calculations were probably made at several residences of the Scandinavian Bishops.

From the above cited legend on Tab. VIII Europae we may conclude that the original map of Claudius Clavus did not embrace Greenland and the northern Scandinavia. For

Dulcert's map of 1339.	Corresponding modern names.
capitulum de vexiom	Vexjö.
fluvius vettur	Vettern.
Roderin	Roden, Roslagen.
flumen Etham	?

The eastern and southern coasts of the Baltic.

Flumen Nu	Nyen, Neva.
Unguardia	Ivanogrod.
Riga	Riga.
litefanja	Lithuania.
Flumen sismaticis	?
kateland	Courland.
fl. vandalus	Vistula.
turon	Thorn.
lacus nerie } neria	Nehrung (= Kurische or Frische Haff).
Godansce	Dantzic.
elbingana	Elbing.
scorpe (? scrope)	?
lacus alech } allech	Hela (Lelewel & Tœppen).
stetin	Stettin.
Grisualdis	Greifswald.
lundis magne	Stralsund.
roystoch	Rostock.

Dulcert's map of 1339.	Corresponding modern names.
usmaria	Wismar.
lubech	Lübeck.

Denmark, Schleswig, and Holstein.

Castro Gotorp	Gottorp.
caldeng	Kolding.
randeus	Randers.
Eduxelant	?
burgalensis	Börglum.
ruya	?
vujberg hic coronatur rex dacie	Viborg.
ripis	Ribe.
Insule sce	Sild.

Islands in the Baltic.

Salandia. Insula salandia	Seeland.
finonja	Fyen.
Langland	Langeland.
eria	Aerö.
ruya	Rügen.
bondolh	Bornholm.
liter	?
eolad	Öland.
(Insula de gotlandia)	Gotland.
visbi	Visby.
oxilia	Ösel.

[1] Notices regarding this codex of Ptolemy are given by:

JEAN BLAU, *Mémoire sur deux monuments géographiques conservés à la bibliothèque publique de Nancy* (*Mémoires de la Société Royale de Nancy*, 1835, p. LIII and Supplement, p. 67), an excellent monograph, carefully written and not disfigured by any chauvinism, with enlightening notes, containing extracts from the manuscripts and a reproduction of the map of Claudius Clavus with annexed text.

RAYMOND THOMASSY, *De Guillaume Fillastre considéré comme géographe* (*Bullet. de la Société de Géographie*, T. 17, 1842, p. 144).

G. WAITZ, *Des Claudius Clavius Beschreibung des Skandinavischen Nordens* (*Nordalbingische Studien*, 1, Kiel 1844, p. 175), reproduction, principally from Jean Blau, of the map and text of Claudius Clavus.

A. E. NORDENSKIÖLD, *Om bröderna Zenos resor och de äldsta kartor öfver Norden* (*Studier och forskningar föranledda af mina resor i höga Norden*, Stockholm 1883) contains (after p. 60) a fac-simile of the map as well as the text of Claudius Clavus.

EDV. ERSLEV, *Jylland*, Kjöbenhavn 1886, p. 118. Contains important researches of the origin, age etc. of the map.

[2] JACOBUS LANGEBECK, *Scriptores rerum Danicarum Medii aevi*, Tom. III, Hafniæ 1774, p. 267. According to this remarkable passage the latitude was calculated from the time the sun was above the horizon during the longest and the shortest day of the year. The longest day in Roskilde was stated to be 17 h. 4 m., the shortest 6 h. 56 m. Roskilde is situated on 55° 38′. The inclination of the ecliptic in 1274 was 23° 32′, whence it may be deduced that the upper border of the sun in that place, observed from a height of 20 feet, including the influence of refraction, in 1274 was 17 h. 35 m. above the horizon.

this part of his new map, Filiastrus must have had access to other information, of which, unfortunately, no further account is given. But some guidance for determining their age may possibly be found in the curious manner in which the nations surrounding the native country of Claudius are characterized: *Britanni anglicati apostate — Carelorum infidelium regio maxime septentrionalis — Slavorum regio insidiatrix — Perversa Prutenorum nacio.* After the conquest by the Angles, Christianity was definitively introduced into England during the 7th century, and, after that time, the inhabitants of the country never relapsed into paganism; but in the beginning of the 13th century, during the reign of King John, the land was excommunicated by the Pope. The inhabitants of Carelia were baptized in 1296 by Thorkel Knutsson; the conversion of the Prussians commenced in the 10th century, but was not completed until the 13th, and the Sclaves to the east of the Elbe relapsed for a short time into paganism, at the end of the 12th century. Thus these discourteous legends seem to indicate that the new map in the Nancy codex was partially copied from an original of the 13th century.

That two different originals had been employed for the drawing of the map may also be deduced from its double graduation. In one of these originals the latitudes of the most important places had evidently been determined, if not by direct observations, at least by calculations from northern itineraries; in the other by the graduation of a portolano, with the aid of Ptolemy's Tab. V Europae. No other means being available to make these different statements agree, the difficulty was solved by providing the map with a double graduation. The map of Claudius Clavus is possibly the first map thus graduated.[1] The middle of the graduation on the right and left sides of the map answers tolerably well to the true latitudes, f. i.:

	According to the graduation on the right side of the map.	According to the text and the graduation on the left side of the map.	Average.	True latitude.
Drontheim	62°	66°	64°	63° 26′
Stockholm	58° 40′	62° 40′	60° 40′	59° 21′
Middle of Oeland	56°	60°	58°	57°
Northern point of Jutland	55° 30′	59° 30′	57° 30′	57° 44′
Helsingborg	55°	59°	57°	56° 3′
Ribe	53° 50′	57° 50′	55° 50′	55° 20′

It is remarkable, that the codex of Nancy does not contain any Tabulæ novæ of Italy, France, or Spain. This seems to indicate that no such maps existed at the commencement of the 15th century, or that they were then not generally known.

3. A map of the northern countries in a Latin codex from c. 1467 of Ptolemy's geography, belonging to the Zamoisky Majorat-library at Warsaw *(Biblioteka ordynancyi Zamoiskiei)* and given here in fac-simile on pl. XXX. With regard to this map I have made an exception from the adopted rule to give on the lithographed tables fac-similes only of printed maps. The reason or excuse for this is the extraordinary importance to cartography in general, and especially to the precolumbian discovery of America, of this previously unknown map. We have here the prototype, or rather a carefully executed copy on the Donis projection of the prototype as well of the maps of Scandinavia and Greenland in the editions of Ptolemy printed in 1482 and 1486, 1507 and 1508, 1513 and 1520, 1522, 1525, 1535 and 1541, in Schedel's chronicle (Nürnberg 1493), and Bordone's Isolario of 1528, as of the celebrated and much discussed map of the brothers Zeno. It belongs to a splendid codex on vellum, bound in two folio-volumes covered with red velvet. According to the Librarian, Professor JOSEF PRZYBO-ROWSKY, the following annotation, dating from the beginning of the 18th century, is written on one of the leaves of paper inserted before the text of the first volume:

Cosmographia Claudi Ptolomaei Alexandrini, Mathematicorum principis, seculo secundo, scilicet circa annum a nativitate Domini nostri Jesu Christi Centesimum trigesimum octavum, sub Antonio Pio, Imperatore Romano florenti, manu Donni Nicolai Germani, Presbyteri Secularis, descripta, Tabulisque egregie pictis adornata ac Paulo Secundo Summo Pontifice ab eodem circa annum 1467 dedicata.

It is not known for certain how the library obtained this manuscript, but a tradition says, that it was a present from a Pope to the Polish Chancellor JOHANNES ZAMOISKY, who in 1589 founded the library and some years later, in 1594, the high school in Samosc. When this school was closed, in 1810, the library was removed to Warsaw. The Chancellor Zamoiski visited Paris in 1573, as the head of the embassy which was to offer the Royal crown of Poland to the Duke of Anjou. The manuscript may thus possibly have arrived in Poland via France.

30. Fac-simile from the Zamoiski manuscript.

The manuscript is well preserved. The handwriting is clear and legible, with magnificent capitals, illuminated in gold and colours. It begins with the dedication *Beatissimo Patri Paulo secundo Pontifici Maximo Donnus Nicolaus Germanus* etc., and ends with the fac-simile lines given below:

I have not succeeded in tracing any notice as to the time, when the »scriptor» ANTONIUS VITELLENSIS lived. According to the celebrated Director of the Bibliothèque Nationale at Paris, M. LÉOPOLD DELISLE, to whom some fac-simile lines of the codex have been transmitted, the handwriting belongs to the latter part of the 15th century.

The first volume of the codex contains the text of Ptolemy's geography, excepting the map-legends of the eighth book; the second volume contains thirty maps in double folio, viz., Ptolemy's own twenty-seven maps and three Tabulæ novæ.

---

[1] As to the double graduation of maps see: A. BREUSING, *La toleta de Marteloio und die loxodromischen Karten (Zeitschrift für wissenschaftliche Geographie;* II, Lahr 1881, p. 195), and EUGEN GELCICH, *Columbus-Studien (Zeitschrift der Gesellschaft für Erdkunde zu Berlin,* Bd. 22, 1887, p. 378).

The legends on the maps of the eighth book are here inserted, on verso of each of Ptolemy's special maps, as in the edition of 1482. But there are no legends either on the general map of the world, or on the new maps. These are:

Tab. IV. A new map of Spain.
> VII. The new map reproduced here of the northern countries.
> X. A new map of Italy.

All the maps are splendidly executed and illuminated in gold, ultramarine, green, red, brown, black, etc. As may be observed from the fac-simile T. XXX the map VII is, like all the special maps of the codex, drawn on the projection of Donis. The main features of the map are almost the same as on the map in the Brussels manuscript and in the edit. Ulmæ 1482 and 1486, save that a narrow strait, extending from west to east in the vicinity of the polar-circle, here connects the North Sea with the Baltic, and thus makes Scandinavia an island[1], and that Greenland is placed, not to the north of Norway, but to the west, between the latitudes 62°½ and 71° North. The main form of Greenland is astonishingly correct, and more closely approaches reality, than the form given by all cartographers to the Scandinavian peninsula, until the publication of the map of OLAUS MAGNUS. From the topographical details one is almost led to believe that the draughtsman knew the interior of the country to be occupied by impassable masses of mountains and ice, occasionally reaching the coast. On the north-western coast of Greenland two legends are inserted, corresponding too exactly to the nature of the country not to be based on actual observations, viz., *Mare quod frequenter congelatur; Ultimus terminus terrae habitabilis*. These sentences are so much the more remarkable, as the seas generally visited then by European mariners, i. e. the Mediterranean and the Atlantic coasts of Europe as far as the North-Cape, never freeze, excepting in the Cattegat and the Baltic, the almost inland character of which was early known to geographers.[2]

If the unavoidable errors in copying are taken into consideration, the names in Iceland and Greenland on the map of the Zamoiski codex are almost identical with the corresponding names on the map printed in Ulm 1482 and 1486, and on the Zeno map. The agreement with the map in the Brussels codex, where a number of names have been omitted and a few added, or altered, is less perfect. If, on the contrary, Denmark and Scania are excepted, extremely few of the names on the Zamoiski map are found either on the map of Claudius Clavus, or in the detailed description of the same inserted in Ptolemy's text. Thus it appears to me that there cannot exist the slightest doubt but that entirely different prototypes served as originals for the delineation of the northern part of the Scandinavian peninsula, Greenland, and Iceland, on one hand on the map of Claudius Clavus, and on the other, on the map in the Zamoiski codex.

In my essay on the voyages of Nicolò and Antonio Zeno I have tried to show, that some of the names on the map of Scandinavia in the *Ptolemeus Ulmæ 1482*, and on the Zeno map, have a Scandinavian etymology. Regarding some of the analogies there pointed out, the objection might be made that they were based on purely phonetic resemblances and could not, consequently, be accepted as sufficient and convincing proofs as to the Scandinavian origin of the Zeno map. This

important question may now be considered as definitively settled by the discovery made by Mr. E. DAHLGREN, that several of the legends for rivers and islets on the Zamoiski map are unquestionably derived from the Scandinavian ordinal numbers, *första, andra, tredje, fjerde* (= the first, second, third, fourth).

On the Zamoiski map there may thus be read:

On the eastern coast of the Baltic: *fursta, auenas, trodiena, fierdis*.

On the western coast of the Baltic: *agna, trediera, fierdena*.

On the coast of Scania: *fursta, agnen*.

On the coast of Halland: *forst, āga, trodia*.

These words are certainly of purely northern origin. But inscriptions in good Latin may, on the other hand, be read on the map, for instance: *Non licet ultra ire, Ultimus terminus terrae habitabilis*. This indisputably shows, that the map here under discussion is founded on a northern original, probably compiled by somebody not versed in Latin, and that it had been adopted for Ptolemy by a good latinist but little conversant with the northern languages. In the description of the map of Claudius Clavus the northern words that are capable of translation are generally latinized, though not very correctly.

4. A map of the North, belonging to a codex of Ptolemy's geography, written between 1480 and 1485, and kept at Bibliothèque Royale in Brussels. This map has lately been reproduced in fac-simile in *Les monuments de la géographie des Bibliothèques de Belgique. Cartes de l'Europe 1480—1485. 4 cartes en 8 feuilles, texte explicatif par* CH. RUELENS, Bruxelles (1887). M. Ruelens takes it for granted, that the important maps reproduced by him from this work belong to the »Donis type.» This view, however, clearly arises from a mistake. As I have several times pointed out here, Donis was never, as far as we know, the author of any new maps. Probably aided by other scribes he produced a number of slightly improved copies from the Latin translation by Jacobus Angelus of Ptolemy's geography, and for these he redrew what may be termed a manuscript edition of the old maps on a new projection. It is this projection which constitutes the only characteristic of the maps of Donis, and which alone gives to »Donis» a particular place in the history of cartography. For the maps in the Brussels codex, the old equidistant projection of Marinus, not that of Donis, is exclusively employed, which plainly shows that the cartographer Nicolaus Germanus had had nothing to do with them. Yet the maps published by M. Ruelens obtain a higher interest through this circumstance. Evidently we have here tolerably unaltered copies of the first *Tabulæ novæ*, added to manuscripts of Ptolemy in the 15th century, and of which the Donis maps are only second-hand copies on a new, considerably improved projection.

5. A map of the North, first published in the edition of Ptolemy, printed at Ulm in 1482, and then in the editions Ulmæ 1486, Romæ 1507 and 1508, Argentinæ 1513 etc.; of these editions only that of 1486 contains some explanatory remarks on the map inserted as an appendix to Lib. II. cap. 10 and Lib. III. cap. 5. The map is generally regarded as an original work of »Nicolaus Donis,» although, as before indicated, without sufficient reason.

Of this map an exact fac-simile was published in *The voyage of the Vega round Asia and Europe* and in *Trois*

---

[1] In copying the original map for Ptolemy's geography, this strait was evidently added, in order to agree with the old conception of Scandinavia as an island in the Northern Ocean.

[2] It had already been mentioned, in ZIEGLER's *Schondia* (edition Argentorati 1532, fol. XCII) that the Baltic, the saltness of which decreased on account of the considerable influx of sweet water, freezes, whilst the Ocean between Norway and Iceland is never covered with ice.

[3] To avoid misunderstandings, I must here mention that as a standard for the resemblance of names on maps, which are considered to have been obtained by repeated copies of the same original, I take the resemblance of names between the maps of the Mediterranean by Dulcert, Barents, and van Keulen (1339—1700), or between the first printed maps of Ptolemy and the maps in the Athos manuscript. On different maps of the same country, there are always some names in common, even when no genetical connexion exists between the maps.

cartes pré-colombiennes représentant une partie de l'Amérique, Stockholm 1883. I here (Fig. 14) give a fac-simile on a reduced scale of the corresponding excellent copper-print in the edition of Ptolemæus Romæ 1507.

6. *Typus Orbis universalis juxta hydrographorum traditionem* (N. T. XXXV), first published in Ptolemaeus Argentinae 1513, and, as to the Northern regions, probably based on the same prototype as Zamoiski's map. In his atlas Lelewel publishes a much reduced copy under the name of *Charta Marina Portugalensium*, and it is possible that we here really have to do with a copy of a Portuguese original. So far as I know, this type was first met with in the *Insularium illustratum Henrici Martelli Germani*, at the end of the 15th century. The original is preserved in the British Museum, and a splendid reproduction in colours is published

was under different forms adhered to until the end of the 16th century, for instance on the globe of VOPEL of 1543 (N. T. XL), and on the map of the world by MYRITIUS of 1590 (N. T. XLIX). The latter (Ruysch's drawing of the polar basin) had the unmerited honor of serving as a model to several works of MERCATOR and his successors. It may be said to be the introduction to a mythical conception of the geography of the Polar regions prevailing down to our day, in the form of a popular belief in an open Polar sea. I will return to Ruysch's map in a succeeding chapter on the first printed maps of the New World.

8. The Zeno map, printed for the first time by Marcolini in Venice in 1558, and pretending to be a copy of a lost original of the 14th century. It was afterwards inserted, with some unimportant modifications, in RUSCELLI's Ptolemy,

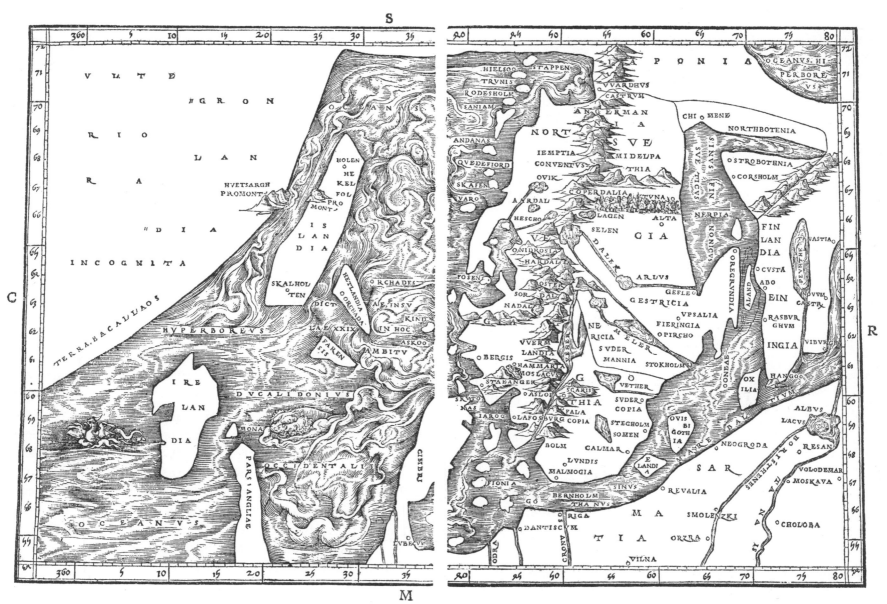

31. JACOBUS ZIEGLER's map of Scandinavia, Argentorati 1532. (Orig. size 358 × 220 m. m.).

in the *Exame das viagens do Doutor Livingstone*, por D. JOSÉ DE LACERDA, Lisboa 1867.[1]

In order here to give a complete enumeration of all the maps belonging to what may be termed the mythic age of northern cartography, two maps must be further mentioned, namely:

7. The map of Ruysch of 1507 (N. T. XXXII). This important map is, as regards the drawing of the arctic regions, characterized by two peculiarities. Firstly, Greenland here forms the point of Asia, which extends farthest to the north-east, and secondly, Ruysch makes the Polar basin contain a number of large imaginary islands. The former theory, which may be regarded as a corollary to the identification by Columbus of his newly discovered lands and islands with *India extra Gangem*,

Venetia 1561, 1564 etc. and often reprinted, or serving as a model for other maps of the northern countries. In this century it has, as an historical document, more frequently been reproduced in fac-simile than most other maps, perhaps with the greatest accuracy in my paper *Om bröderna Zenos resor och de äldsta kartor öfver Norden* (*Studier och forskningar föranledda af mina resor i höga Norden*, Stockholm 1883). I cannot here enter upon an analysis of this map, or of the Zeno question, in many respects so highly interesting to geographers. I can only refer the reader to the extensive literature enumerated by R. H. MAJOR in *The Voyages of the Venetian Brothers Nicolò & Antonio Zeno*, London (Hakluyt Soc.) 1873, PAUL BARRON WATSON, *Bibliography of the Pre-Columbian discoveries of America* (*Library Journal*, Vol. 6, No. 8,

---

[1] There is also a reproduction in the atlas of Santarem, in my paper on the voyages of the brothers Zeno and (of the northern portion of the map) in Winsor's *Narrative and critical History of America*. According to Marinelli (*Saggio di Cartografia della Regione Veneta*, Venezia 1881, p. 94) the maps in a codex of Ptolemy at the Biblioteca Magliabechiana are signed »*Henricus Martellus Germanus fecit has tabulas.*»

New York & London 1881), WINSOR, *A Bibliography of Ptolemy's geography*, p. 31. A comparison between the Zeno map and the map from the Zamoiski library in Warsaw, indisputably shows both to be copies of the same original, though various addenda and modifications have been arbitrarily introduced by ANTONIO ZENO, Jun., or by the publisher, MARCOLINI, in order to adapt the map to the text, and perhaps also to remove from it the most striking discrepancies with the map of Olaus Magnus, and other delineations of the northern countries existing in Venice at the time of its publication. As a slight contribution to the Zeno question it may here be mentioned, that in my copy of the collection of maps generally known under the name of Lafreri's Atlas and engraved in copper in Italy, from 1558 to 1572, of which a more detailed account will be given further on, there is among other maps one of »Frisland» (size 250 × 185 m. m.) and one of »Estland» (size 246 × 183 m. m.). They entirely agree with the »Frisland» and »Estland»[1] on the Zeno map, except that they exhibit a greater richness of topographical detail and some slight variations in the inscriptions. These differences are in themselves of little importance, but they nevertheless seem to indicate these copper-engravings to be independent copies of the original employed in the drawing of the Zeno map.

As far as may be inferred from the data at present extant, the development of the cartography of the northern parts of Europe and the north-western parts of the New World may be characterized as follows. At the end of the 14th and at the beginning of the 15th century there existed in Italy, if pure fancy-drawings are excluded, four different types of maps of these regions, viz.

A. Ptolemy's general map and his Tabula V Europae, which, down to the end of the 16th century, exercised no little influence on Scandinavian cartography.

B. Portolanos. With regard to the North, it is probable that these were originally based on drawings of the 13th century. They were afterwards often reproduced and slightly, although not much, modified with the aid of later notices, directly or indirectly collected from Scandinavian mariners. Of these maps numerous copies and modifications are still extant, but their influence on the first printed maps of Scandinavia has generally been insignificant. On the other hand the first *Tabulæ Novæ* of Britannia are evidently copied from portolanos, and from these also the mythical island *Brasil* (not to be confounded with the present Brazil) was introduced into the printed cartography, for instance on the *Tabula nova Hiberniæ, Angliæ et Scotiæ* in the Ptolemy of 1513 (N. fig. 6).

C. A map of the Scandinavian peninsula, Iceland, and Greenland, composed ere the northern mariners became acquainted with the use of the compass, perhaps in the beginning of the 13th century. This map is lost, but an approximate idea of it may be obtained from some more or less altered reproductions (the map in the Brussels codex, the Zamoiski map) of the end of the 15th century.

D. A large map of Denmark and southern Sweden (the ancient Denmark), by Claudius Clavus. This map is now also lost, but a detailed description of it and a reproduction of its main features on a much reduced scale, are to be found in the Ptolemy codex of Filiastrus.

By the aid of these four originals, probably still extant in the beginning of the 15th century, the following maps of the North were drawn:

1. The map, of 1427, inserted into the codex of Nancy. It is generally founded on the map of Claudius Clavus (D), but also, as regards the more distant parts of the Baltic, on the portolanos (B), and as regards the map of Greenland and the northern part of Scandinavia, on the type C. England, and the south coasts of the Baltic, are drawn from Ptolemy. The map of Filiastrus obtained little circulation and exercised no influence on later drawings of the northern countries.

2. Another *Tabula nova Septentrionis* made for Ptolemy's geography. This map is also lost, but may yet be found in some Ptolemy codex of the middle of the 15th century. It can easily be restored by redrawing the Zamoiski map on the rectangular projection of Marinus, as I have done, fig. 33.

3. A map obtained by redrawing the preceding one on the projection of Donis. This is the map which I have found in the Zamoiski library at Warsaw. The maps 2 and 3 are, as regards the countries *extra Ptolemæum*, almost exclusively founded on the prototype C. With regard to Denmark and southern Sweden, they may also, to some extent, have been influenced by the map of Claudius Clavus (D), or rather by the same originals as those used by Claudius, for not a single geographical position and exceedingly few names are, as far as I have been able to ascertain, common to the Zamoiski copy of the prototype C and to the geographical description of the map of Claudius Clavus published by Filiastrus. Considering, on the one hand, the care and exactness with which the copyists of Ptolemy, century after century, produced unaltered copies of the old prototype, and how much the portolano legends remained the same during generations of copyists, and, on the other, how much the Danish portion of the Zamoiski map and the map of Claudius Clavus differ, it seems difficult to assume the map of Claudius to have been, even regarding Denmark and southern Sweden, directly copied from the original to the *Tabula nova Septentrionis* in the Zamoiski codex.

4. When the map 2 was compared with sea-charts of the north-western coast of Britain or examined by mariners of great nautical experience, but unacquainted with the variation of the compass, it was found that no such land as that laid down on this map to the west of Ireland existed, especially if the direction was determined by the newly discovered nautical instrument (the deviating compass). But as the existence of Iceland and Greenland could not be denied, the supposed error was corrected by the introduction of such a modification on the map 2, that Greenland became situated to the north of Norway. The map in the Brussels codex belongs to this type, and I suppose that analogous maps will be found in other codices of Ptolemy of the 15th century.

5. Maps of the type 4 were then redrawn on the projection of Donis by Dominus Nicolaus Germanus, or some artist of his school. It is these maps that were printed at Ulm in 1482 and 1486, at Rome in 1507 and 1508, etc.

6. NICOLÒ ZENO Jun. finally reproduced a much worn copy of the original map C on the conical projection, introducing some free-hand corrections borrowed from more modern cartographical works, and adding some islands and mainlands wanting on the original map, but mentioned in the description of the Zeno voyages published by Marcolini in Venetia 1558. That the original map, C, was used as a model for the new delineation, and not maps of the types 2, 3, 4, and 5, may be inferred from the circumstance that none of the features cha-

---

[1] It is remarkable, that these names in the text of the voyage of the Zenos have got a more Italian form (»Frislanda», »Estlanda»). Hence the fact of no slight importance to the Zeno question may be deduced, that these maps are not freehand-drawings for the text printed by Marcolini, but copies from some independent original. These maps are not signed, but are evidently works of the same engraver, who executed two different, though closely resembling maps of *Gotlandia* in my collection, of which the one is signed FERRANDO BERTELLI. He published (1560—1568) several other maps and copper-engravings in Venice.

32. The copy engraved in Rome 1572 of Olaus Magnus' map of Scandinavia. (Orig. size 805 × 533 m. m.).

racteristic of Ptolemy's map and left unchanged on the maps 2—5 (for instance the *Insula Thule*, the *Orcades*, the long projection of the northern Scotland to the east), are found on the Zeno map, and also from the circumstance that, according to the legend to the Zeno map in the Ptolemy of 1561, no net of graduation existed on the original, which was copied for the work of Marcolini.

My view of the correlation between the maps 2, 3, 4, and 5 will be understood by their schematized representation given below (fig. 33—36).

It would be of much interest to the history of the earliest cartography both of Scandinavia and of the New World to be able, as regards the prototype C, to find out its date and origin, and the epoch when copies of it were first known in Italy.

At present no definite answer can be given to these questions. This, however, seems to be beyond doubt that the materials of the map were furnished by a Scandinavian thoroughly acquainted with the voyages of his contrymen to Iceland and Greenland, and that it was reproduced by somebody but slightly versed in the northern languages and perhaps also in the manner of writing of the Northmen during the 14th century. It is further not impossible, that here, as in so many other cases, the answer nearest at hand is the most correct one, i. e. that an old Scandinavian map, afterwards copied and adapted to the work of Ptolemy, was actually brought to Italy by the Venetian freebooters NICOLO and ANTONIO ZENO, at the end of the 14th century. But it is evidently absurd to give these brothers a conspicuous place in the history of geographical discovery because, at least in the first instance, they involuntarily followed a northern viking, sea-king, or pirate, in his enterprises to that part of the New World, which half a millennium before had been discovered by Northmen.

During a period of fifty years the map printed Ulmae 1482 constituted the only type on which the countries of the North were delineated. But in 1532 a new type was introduced through a map published by the Bavarian theologist JACOBUS ZIEGLER in a work with the following long title: *Quae intus continentur.*

Syria, ad Ptolomaici operis rationem . . .

Palestina, iisdem auctoribus . . .

Arabia Petraea . . .

Aegyptus . . .

Schondia, tradita ab auctoribus, qui in eius operis prologo memorantur.

Holmiae, civitatis regiae Svetiae, deplorabilis excidij per Christiernum Datiae cimbricæ regem, historia.

Regionum superiorum, singulae tabulae Geographicae.

Argentorati apud Petrum Opilionem MDXXXII.

According to a statement in the text, fol. lxxxv, the data for this new map were furnished by not less than four Scandinavian prelates, whom ecclesiastical and political disturbances had brought to the papal court during a visit of Ziegler in Rome. These were, the Arch-bishop of Drontheim ERIK WALKENDORF, his successor OLOF ENGELBREKTSSON, the Arch-bishop of Upsala JOHANNES MAGNUS, and the Rector PEDER MÅNSSON from Vesterås, ordained Bishop in Rome. It is remarkable that the author, for his information on »*Gronlandiæ Chersonesus et insula Tyle,*» gives references not to Walkendorf, who had worked with such energy for the rediscovery of the colonies in Greenland, but to the last named two Swedes. Ziegler, according to his own avowal, calculated the numerous latitudes and longitudes given in this work, from information respecting the reciprocal distances and azimuths of the most important places in the Scandinavian countries. The chapter on Schondia contains a tolerably extensive geographical descrip-

tion of *Gronlandia, Islandia, Hetlandia, Farensis, Laponia, Nordvegia, Svecia, Bothnia, Ostrobothnia, Gothia; Finlandia.* Under *Gronlandia* some interesting communications about »ANTONINUS» (IOANNES) CABOTUS are inserted. His report that he had encountered ice during the month of July in the Greenland Sea is dismissed with the positive assertion that this could not have been possible even at the pole, at that time of the year.

If the map (N. fig. 31) is to be considered as a faithful copy of the original, it shows that the German theologist and the Northern prelates were not particularly well skilled in the art of drawing maps. As regards the principal features of the Scandinavian peninsula their map, however, denoted considerable progress, and the legends on it, or in the text, may generally, without difficulty, be identified with known localities. Unfortunately only one name is to be found on the eastern coast of Greenland, that of a high mountain, *Hvetsargh promontorium.* Both Greenland and Iceland are, besides, far less accurately drawn here than on the map in the Zamoiski codex. In Scandinavia the direction of Kölen is tolerably well represented; several of the great Swedish lakes, Vener, Melar, Vether, Somen, Selen (= Siljan) are laid out on the map, and we have here for the first time a map, though a rough one, of Finland. With regard to further details I may refer to the original, of which there are two editions (Argentorati 1532 and 1536), and also to the reprint of *Schondia*, published by H. HILDEBRAND with a number of explanatory notes in the Journal of the Swedish Geographical Society of 1878.

Seven years after the first edition of Ziegler's work another map of the Northern countries was published in Venice by the Swedish Bishop OLAUS MAGNUS. It was printed in 1539 on nine large folio-sheets occupying together a surface of 1700 × 1250 m. m. With the map a short introduction by the same author was printed under the title of: *Opera Breve, laquale demonstra, e dechiara, ouero da il modo facile di intendere la charta, ouer delle terre frigidissime di Settentrione: oltra il mare Germanico, doue si contengono le cose mirabilissime de quelli paesi, fin'a quest'hora non cognosciute, ne da Greci, ne da Latini.* (Colophon:) *Stampata in Venetia, per Giouan Thomaso, del Reame de Neapoli, Nel anno de nostro Signore. MDXXXIX.* Of this rare brochure a fac-simile was published at Stockholm in 1887. Slightly modified translations had already been printed in German, Venetiis 1539, and as an introduction to the Latin and German editions of OLAUS MAGNUS, *Historia de Gentibus Septentrionalibus*, printed at Basel in 1567.

The large map of Olaus Magnus was considered as lost, or confounded with poor reproductions of the 16th century, until the rediscovery of the original in Munich, a few years ago, by Dr. OSCAR BRENNER, who published an excellent, though reduced fac-simile of it with a critical description in: *Die ächte Karte des Olaus Magnus vom Jahre 1539.* (*Christiania Videnskabs-Selskabs Forhandlinger*, 1886, N:o 15.) Later, through the exertions of Mr. G. E. KLEMMING in Stockholm, a few copies of a full size fac-simile of the map was published. From a cartographical point of view the large map itself is certainly not to be compared with the map of Schweinheim-Buckinck of 1478, but as regards its size and the profusion of geographical and ethnographical details on it, it stands unrivalled amongst cartographical productions of the first part of the 16th century. When the map of Olaus Magnus was published, there did not, as far as is known, exist any printed map of Italy, Spain, France, England, or Germany, the size of which was equal to that of one of the nine folio-leaves of this atlas of the far North. Nor are there, in all this literature, to be found many descriptions of the former nature of a country of the habits, customs, household-furniture, etc. of its people, so interesting, so exhaustive and, with all its naïve credulity, so important as the large work of Olaus Magnus Gothus, for which the map was originally intended.

This large map never seems to have had any considerable circulation. It was never fully described and was seldom (as f. i. by CONRAD GESSNER, *Bibliotheca universalis*, fol. 528) mentioned by contemporary authors, and the maps of the Scandinavian countries were invariably drawn according to

the large and artistically executed original, covered with innumerable legends and drawings. A copy of the original map of Olaus Magnus, carefully reduced to about half size (805 × 533 m. m.), was published in Rome, *ex typis Antonii Lafreri Seguani. Anno MDLXXII* (N. fig. 32). But even this fine copper-engraving cannot, as regards the elegance and clearness of its execution, be compared with the original, which, according to Dr. Brenner, was produced by woodcut and which, from an artistic point of view, as a woodcut-

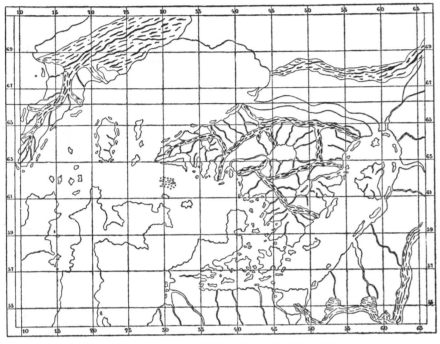

33. The Zamoiski map reconstructed on the projection of the map in the Brussels codex.

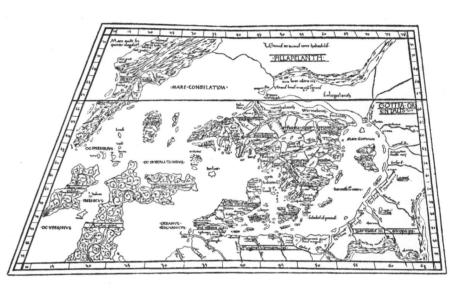

34. The map of the North in the Zamoiski codex of 1468 (N. T. XXX). (Orig. size 568 × 313 m. m.).

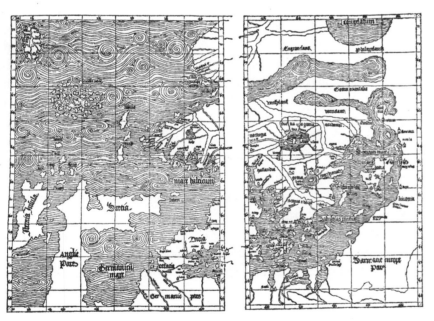

35. The map of the North in the Brussels codex. (Orig. size 651 × 450 m. m.).

36. The map of the North in Ptolemaeus Ulmæ 1482. (Orig. size. 564 × 312 m. m.).

Ziegler, until a reduced and much modified copy of the original was published first in: *De omnibus Gothorum Sveonumque regionibus Historia*, Romæ 1554, by JOHANNES MAGNUS, and shortly after in the above mentioned work *(Historia de Gentibus Septentrionalibus)* of Olaus Magnus himself. This work was first printed at Rome in 1555 and then reprinted several times, for instance Venetiis 1565 (with the map of 1555 unmodified), Basileæ 1567 (with a new, somewhat better copy of the map of 1539). Yet these reproductions of the 16th century scarcely give us an idea of

map was hardly ever surpassed by any other analogous work. Only a few copies are known of Lafreri's engraving; one in the British Museum, another, the only one in Scandinavia, in my collection of ancient maps. The drawings of monstrous animals, with which Olaus Magnus so liberally decorated his map, and of which Conrad Gessner said, *quae (animalium figurae) tamen verae aut ad vivum pictae minime videntur* (*Bibliotheca universalis*, fol. 526), were collected by Münster in a large double-folio woodcut, inserted in his cosmography.

VI.

The first maps of the New World and of the newly discovered parts of Africa and Asia.

It is generally supposed that the successful voyages of the Portuguese in the *Regio perusta*, or *Regio inhabitabilis propter nimium calorem*, and the re-discovery of the New World by Columbus must have made a great and immediate impression throughout Christendom. It seems as if statesmen and scholars at least ought to have clearly conceived the immense importance of this sudden increase of the territory adapted for the use of man. For this increase did not consist of deserts scarcely cultivable and difficult to defend, but of immense continents and islands, which, through the excellence of their climate and the fertility of a virgin soil, were capable of giving millions of human beings the means of a subsistence, more easily acquired, richer and more abundant, than in the densely populated countries of the Old World, with its soil impoverished by repeated harvests, and its social conditions fettered by thousands of traditional prejudices. Yet, this was so far from being the case, that scarcely any discovery of importance was received with so much indifference, even in circles where sufficient genius and statesmanship ought to have prevailed to appreciate the changes thus foreshadowed in the development of the economical and political conditions of mankind. The truth of this assertion will easily be perceived, if we take the trouble to study, not only the contributions to the history of geographical discovery written during the last century, but also the earliest original literature itself. With regard to the history of the discovery of America, such an investigation may now be pursued with few difficulties, thanks to the indefatigable pains taken by bibliophiles, especially on the other side of the Atlantic, in collecting »Americana,» and the care with which these collections have been examined, registered, and described by prominent scholars, most recently by Mr. HENRY HARRISSE in: *Bibliotheca Americana vetustissima. A description of Works relating to America published between the years 1492 and 1551*, New York 1866, and *Bibliotheca Americana vetustissima. A description of Works relating to America published between the years 1492 and 1551. Additions*, Paris 1872. According to the *Chronological Table* at the end of the last mentioned work, Mr. Harrisse has, in his »Bibliotheca,» registered altogether 432 works or pamphlets printed before 1551 and containing passages respecting the new world. Different editions and

unaltered reprints are here registered under separate numbers, and most of the works cited contain only slight allusions to the subject. If, in collating and making statistics on the oldest literature relating to America, due attention is paid to these circumstances, it will be found that scarcely one work containing an original communication about the New World of the length of at least one printed page, was annually published during the first fifty years after the discovery of Columbus, and that all these original communications together would be easily comprised in a single volume of very moderate size.

Regarding the early discoveries in the east of Asia and along the coasts of Africa, round the Cape of Good Hope to India, the oldest literature has been subject to no such exhaustive researches as the oldest literature relating to America. It might perhaps fill a greater number of pages, for here we have narratives of travels, rich in exciting details, and of which repeated editions were early published in print, though more as a contribution to the belles-lettres of the epoch, than as serious contributions to the knowledge of our earth. Such publications are the narratives of the travels of MARCO POLO (first edition printed in 1477), of VARTHEMA (Editio princeps Milano 1505), and CADAMOSTO (Editio princeps Vicenza 1507). But, if we except the numerous editions of these works and a few brochures of the same significance with respect to the history of the discoveries in Africa and the eastern Asia, as the letters of Columbus and Vespucci to that of the New World, the geographical literature relating to the newly discovered lands in the eastern hemisphere, during the 15th and the beginning of the 16th century, was as poor and scanty as the literature enumerated in »Bibliotheca Americana vetustissima.»[1]

Still poorer is the oldest printed literature of maps. The first drawings or inscriptions on a printed map referable to the voyages of the Portuguese is met with, as far as I know, fifty-six years after the voyages of CADAMOSTO, on the map in REISCH's *Margarita Philosophica* of 1503, which I have already mentioned above (p. 40), and of which a fac-simile is given on T. XXXI. But it was five years later, in 1508, that a map was first published in print, on which the coast of Africa discovered by the Portuguese, and the newly discovered passage

---

[1] One of the first publications containing some words about the discoveries of the Portuguese, is *Valasci Ferdinandi ... Regis Portugallie oratoris ad Innocentium octavum pontificem maximum de obedientia oratio*, s. l. et a., (first edition Romæ 1485. HAIN No. 15760). In about the middle of the small, unpaginated work we read the following remarkable words, which I give here from an edition (HAIN No. 15761) printed somewhat later (1494?): *Non desunt Beatissime pater quam plura alia eius in christi ihesu fidem et Romanam ecclesiam merita, que si sigillatim recensere velim longius quam par esset progrederer, sed duo tantum quam brevissime perstringam. Primum quod eo regnante Henrici patrui eius de quo supra meminimus industria cepta navigari Ethiopia est. Alterum vero sit quod eodem tempore in oceano athlantico decem insule vix ipsis orbis descriptoribus cognite a nostris invente sunt et in omnes Lusitanie colonie reducte. In quibus iam mira incolarum frequentia habitatis christiana fides haud minore ceremonia quam inter nos colitur. Ita ut mihi vere Alfonsus Rex gloriosus ad christianam religionem sacratissimam colendam non contentus maiorum suorum finibus videatur, nisi etiam novas provintias, nova regna, novas insulas et quasi novos et incognitos orbes christi nomini et Romane ecclesie et vobis tandem pater beatissime et successoribus vestris in posterum addiceret.*

to India were clearly laid down. To prevent any misunderstanding I may here again expressly state, that I speak of printed, not of manuscript maps. Tolerably complete map-sketches[1] drawn to illustrate the reports of explorers or adventurers were probably made for the government or the ship-owners after almost every more or less successful voyage, but they were seldom published. They generally seem to have been jealously concealed in public or private archives. Most of them have since been lost, or exhumed from the dust of libraries for the first time in the present century. They have thus often had much less influence on the development of cartography than many an insignificant printed production, compiled from hearsay reports.

Two maps of 1527 and 1529, preserved in the military library at Weimar, were considered to be the oldest manuscript maps of the New World yet discovered, even as late as 1832. But in that year Alexander von Humboldt discovered, among the literary treasures of Baron Walkenaer, a map drawn in 1500, at Puerto de Santa Maria, by the celebrated navigator Juan de la Cosa, or Juan Biscaino, one of the companions of Columbus on his second voyage. Humboldt has given a critical account of this map in Dr. F. W. Ghillany's *Geschichte des Seefahrers Ritter Martin Behaim ... Eingeleitet durch eine Abhandlung: ueber die ältesten Karten des neuen Continents und den Namen Amerika von* Alexander v. Humboldt, Nürnberg 1853, where a part of the map is also reproduced in fac-simile. It is also reproduced on a reduced scale by Ramon de la Sagra in *Histoire physique ... de l'île de Cube*, Paris 1842 (according to Harrisse, *Cabot*, Paris 1882, p. 157), by Lelewel, Sophus Ruge, H. H. Bancroft, and others. A complete fac-simile is found in Jomard's Atlas. It has given rise to an extensive literature, enumerated by Winsor (*Bibliography of Ptolemy's geography*, p. 7). After the death of Walkenaer the map was bought by the Spanish government and is now preserved at the Marine-Museum of Madrid (Harrisse, *Cabot*, p. 157). The size of the original is 1,80 × 0,97. It does not appear to have exercised any direct influence on the first printed maps of the New World, as may be concluded from a comparison with the maps of Ruysch, Sylvanus, Stobnicza, Aeschler and Übelin, and others (N. T. XXXII—XXXVI).

More important in this respect is a map sent to Hercules d'Este, the Duke of Ferrara, by Alberto Cantino, his ambassador in Lisbon, between the years 1501 and 1505. The original is at present in the *Biblioteca Estense in Modena*. A fac-simile has been published by Harrisse for his work *Les Corte-Real et leurs voyages au Nouveau Monde*, Paris 1883, where a minute analysis of the map is given in Chapt. IV (p. 69—158). This map, or copies of it, has evidently been used for the first printed maps of the New World. It is of special interest in the cartography of the North, Greenland being represented on it with tolerable exactness, although too

far to the east. Probably Cantino here followed a map of the Zamoiski type. But it seems, from a long legend on the east coast, that the Portuguese had also penetrated as far as to the vicinity of Cape Farewell.

Among manuscript maps of the New World, one map of 1503—1504 attributed to Salvat de Palestrina, and one by Pedro Reinel of 1505, require further mention. Regarding these maps, I may refer to Harrisse, *Cabot*, p. 161 and 162, and to Kunstmann, *Atlas zur Entdeckungsgeschichte Americas*, München 1859. In a Latin manuscript in the British Museum there is also the above mentioned map of the world by Henricus Martellus, on which the discoveries of the Portuguese along the African coasts down to 1489 are registered. Finally Dr. E. T. Hamy has lately, in *Notice sur une Mappemonde portugaise anonyme de 1502* (*Bulletin de géographie historique et descriptive*, No. 4, Paris 1887), published parts of a Portuguese map of the world, which, as far as I have been able to judge from the photographs published in Mr. Hamy's paper, closely resembles the maps of Africa in Ptolemy of 1513, of which I have given fac-similes on fig. 8 and 9. The original belonged to M. Alphonse Pinart, who had bought it from the English traveller Mr. King.

No other manuscript maps of the lands discovered during the 15th and in the beginning of the 16th century, and drawn before 1508 (i. e. before the year when the first printed map of the New World was published), are, as far as I am aware, at present known. A few other maps, now lost, are mentioned in old documents. But the majority, doubtless including the most important, were so well concealed that every reference to them has been suppressed. It is thus often very difficult to point out the originals[2] of the oldest printed maps of the New and of the new-discovered lands of the Old World. Here I can only cursorily refer to this question, which will, perhaps, hereafter be elucidated by new discoveries in Libraries and among Archives. I must confine myself to a chronological enumeration of the oldest printed maps and to a reproduction in facsimile of the most important, viz.

1. *Nova et universalior Orbis cogniti tabula, Ioa. Ruysch Germano elaborata*, Romæ 1508 (N. T. XXXII). This map was published among the *tabulæ novæ* in the edition of Ptolemaeus Romae 1508, and its engraving was hardly finished before that year, as may be concluded from the following legend at Trapobane: *Ad hanc Lusitani nautæ navigarunt anno salutis MDVII.*[3] Sometimes it is also inserted in the edition of 1507, without however being mentioned on the title-page, on which, according to the custom of the period, a synopsis of the contents of the work is given. But on the new title-page, with which the edition of 1508 was provided, this passage is printed: *In hoc opere haec continentur: — — — Nova et universalior Orbis cogniti tabula Ioa Ruysch Germano elaborata.*

[1] In the 31st chap. of Gaspar Correa's account of the first voyage of Vasco da Gama (edit. by Henry E. J. Stanley for the Hakluyt Society, London 1869, p. 260) it is related how several persons, during the return-passage along the eastern coast of Africa, zealously occupied themselves in collecting material for a chart of the coast by order of Da Gama. It was due to an attack of illness suffered by the clergyman Joan Figueira, at Melinda, that an account of these proceedings was inserted in Correa's *Lendas da Indias*. Figueira seems to have been a secretary to the pilots. During his illness he transmitted his notes to Gama. Several copies were afterwards taken of them. Correa saw one of these copies among old papers belonging to Albuquerque, and the maps given by me in fac-simile from Ptolemy of 1513 are certainly partly founded on this survey.
[2] The early printed maps, as well of America as of the newly discovered parts of Africa and Asia, are generally founded on Portuguese, not on Spanish originals. The reason probably is that the commercial intercourse of the Portuguese with the rest of Europe, owing to the discovery of the new way to the commercial treasures of India, was far more considerable during the 15th and the first part of the 16th century than that of Spain. The latter country only imported cargoes of the precious metals from its colonies, which were procured by immense sacrifices and at great cost; while their amount was much over-rated. The large commercial factories, through which the maps and accounts of new voyages were generally transmitted to Italy, Germany, etc. were accordingly situated chiefly in Portugal, not in Spain. Perhaps also powerful Spain was better able to protect, what may be termed the »map-secret,» than the more feeble Portugal. I have never heard of any maps printed in Portugal during the period of the incunabula of cartography.
[3] In *Kritische Untersuchungen* etc., Berlin 1852, II p. 294 note, Humboldt says: »*Reidel will selbst ... dass sie aus dem Jahre 1507 herrühre, wegen einer Angabe in calce Planisphaerii welche ich in keinem der Exemplare gefunden habe, deren ich mich in Frankreich und Deutschland bediente*». By *Planispherium* Raidel, however, did not mean, as Humboldt assumes, the Ruysch map, but the translation of *Planisphaerium Ptolemaei* inserted between the text and the maps in the editions of 1507 and 1508, and dated (in the colophon): Romae, Die VIII Septembris MDVII.

The map of Ruysch forms an epoch in the development of cartography.

It is the first printed map of the world on which the discoveries of the Portuguese along the coasts of Africa are laid down. With the exception of some small maps based on the cosmographical speculations of the ancients, and inserted in the works of MACROBIUS, SACROBOSCO, and others (N. T. XXXI),[1] it is the first printed map representing Africa as a peninsula encompassed by the ocean. The southern point of Africa moreover is here placed on a nearly correct latitude, thus giving a tolerably exact form to that part of the world. Ruysch also gives on his map a relatively correct place to the *Insule de Azores, Insula de Madera, Ins. Canarias* and *Insule de Capo Verde*.

Ruysch's map is the first published in print, on which India is drawn as a triangular peninsula projecting from the south coast of Asia and bordered on the north by the rivers Indus and Ganges. Even though it has not yet received its full extension as a peninsula, yet an important deviation from Ptolemy's geography is thus made on the map of a part of the world to which almost a *privilegium exclusivum* of knowledge was attributed to the ancients. Ceylon is also laid down by Ruysch under the name of *Prilam*, with about its proper size, and correctly as regards the southern point of India. *Taprobana alias Zoilon* is placed further towards the East Indian peninsula, in which position this geographical remnant from the time of Alexander the Great was retained, down to the middle of the 16th century.

Ruysch has given the first p r i n t e d map on which the delineation of the interior and eastern parts of Asia is no longer based exclusively on the material collected by Marinus of Tyre and Ptolemy more than a millenium previously, but on more modern reports, especially those of MARCO POLO. Various new names are here added in *Scytia intra Imaum*, as *Tartaria Magna* and *Wolha* (= Volga), and an immense, quite new territory, an *Asia extra Ptolemaeum*, or *Asia Marci Pauli Veneti*, is added beyond the eastern limits of Ptolemy's Oikumene. Here the Chinese river-system is given in a manner indicating other sources for the geography of eastern Asia, than Marco Polo's written words. In its main features the delineation of eastern Asia, to the south of lat. N. 60°, on the map of Ruysch, so nearly resembles BEHAIM's globe, that a common original might have served for both. Both deviate from Fra Mauro's map of the world, which gives us a representation of these regions much inferior to both Behaim and Ruysch.

The exaggerated extension given by Ptolemy to the Mediterranean is here much reduced, or from 62° to 53°, the actual difference of longitude between Gibraltar and the western coast of Syria amounting only to 41° or 42°. This correction had been made, centuries previously, on the portolanos. The first cartographer who adopted Ruysch's reduction, was the celebrated Gerard Mercator. On his famous map »*ad usum Navigantium*» of 1569, he gives the Mediterranean Sea a length of 52°. Ruysch's map is also the first printed map on which, in conformity with the drawings on the portolanos, a tolerably correct direction is given to the northern coast of Africa, by attending to the considerable difference of latitude between the coast-lines to the east and to the west of Syrtis, and by giving a proper form to that bay.

The map of Ruysch is the first map published in print which, following a correction made in the portolanos since the beginning of the 14th century, leaves out that excessive projection towards the east, which characterizes Ptolemy's map of the northern part of Scotland.

Ruysch is the first who gives us a map of the New World. This part of his map may be said to be a tolerably exact representation of the geographical knowledge of that part of the world in the beginning of the 16th century. Greenland is here, for the first time, drawn without being connected with Europe by a vast polar continent, bordering the northern part of the Atlantic. Instead it is connected with Asia, through New Foundland or Terra Nova; an hypothesis regarding the extension of the continents in the northern hemisphere, which was still adhered to by some geographers in the beginning of the 17th century.

———

It is evident, from what has already been said, that Ruysch deserves to be placed in the first rank among the reformers of cartography. His map is not a copy of the map of the world by Ptolemy, nor a learned master-piece composed at the writing-table, but a revision of the old maps of the known world, made on a Ptolemaic, i. e. on a scientific basis, with the aid, on the one hand of great personal experience and geographical learning and, on the other, of extensive knowledge combined with a critical use of the traditions among practical seamen of different nations. The legends on this map are consequently of very high interest, and form a more important contribution to the history of geography than many a bulky volume. Only a few of them can here be the subject of some critical remarks.

Far in the north we find an imaginary representation of the regions nearest to the pole, afterwards reproduced with some modifications by MERCATOR and, after him, by various other cartographers. Ruysch gives a reference to this extravagant theory respecting the geography of the north polar-regions in the following legends on the right limb of the map: *Legere est in libro de inventione fortunati sub polo arctico rupem esse excelsam ex lapide magnete 33 miliarium germanorum ambitu. Hanc complectitur Mare Sugenum fluidum instar vasis aquam deorsum per foramina emittentis* etc. By the *liber de inventione fortunati* no doubt is meant the same work as that thus referred to by HAKLUYT in: *The principael Navigations ... of the English Nation*, London 1589, p. 249: *A Testimonie of the learned Mathematician, maister John Dee, touching the foresaid voyage of Nicholas de Linna. Anno 1360 a frier of Oxford, being a good Astronomer, went in companie with others to the most Northern islands of the world, and there leaving his company together, he travelled alone, and purposely described all the Northern islands, with the indrawing seas: and the record therof at his returne he delivered to the king of England. The name of which booke is Inventio Fortunata (aliter fortunae) qui liber incipit a gradu 54 usq. ad polum.*[2] *Wich frier for sundry purposes after that, did five times passe from England thither, and home againe.* In connection with this story Hakluyt mentions »*privileges granted to the Fishermen of the towne of Blacknie in the said countie of Norfolke, by King Edward the third, for their exemption and freedome from his ordinary service, in respect of their trade to Island.*» Mercator, again, in the first edition of his Atlas (on the verso of *Palus*

---

[1] These maps ought rather to be considered as cosmographical schemes, than as actual representations of the earth. More important are some hand-drawn maps of the 14th and 15th centuries, the authors of which, guided perhaps by the writings of Arabian geographers, make the Atlantic Ocean in the south communicate with the Indian Ocean: for instance MARINO SANUDO's planisphere of 1306; »*Portolano della Mediceo-Laurenziana*» of 1351; »*Planisfero della Palatina*» of 1417; and FRA MAURO's planisphere of 1457. In his Lib. II, cap. III STRABO gives a tolerably detailed account of the opinions held in antiquity, regarding the possibility of circumnavigating Africa, from which it appears that this question had, even then, been already much discussed.

[2] The maps of the northern region in the Zamoiski codex, and in the editions Ulmæ 1482 and 1486 begin near that latitude.

Arcticus, ac terrarum circum jacentium descriptio) gives the Itinerarium of Jacobus Cnoxen Buscoducensis as reference for the description of the arctic regions by Nicolas de Linna (*Menorita quidam Anglus, Oxoniensis Mathematicus*).

To the north of Iceland we read the remarkable inscription: *Hic incipit Mare Sugenum. Hic compassus navium non tenet, nec naves quae ferrum tenent revertere valent.* The story of a magnetic mountain exercising a powerful attraction on vessels with iron nails is very old. It is mentioned by Ptolemy on »Undecima Asiæ tabula» in a legend referring to some islands, *Maniolae Insulae*, in the Indian Ocean and in Lib. III, cap. II of the text. Afterwards such mountains play a part in the fictions of the Arabs, and from these it found its way into the medieval literature of Europe. The existence of a magnetic rock on the east coast of Greenland was also said to be the cause of the failure of Mogens Heinessen's voyage in 1567.[1] Yet the words »Hic compassus navium non tenet,» with which this legend on Ruysch's map commences, seem to indicate an actual experience regarding the uselessness of the compass in the vicinity of the magnetic pole.

Below this inscription, there is an island between Iceland and Greenland, with the interesting legend: *Insula haec in Anno Domini 1456 fuit totaliter combusta.* The Iceland annals do not mention any such occurrence, but, considering the want of knowledge, at the end of the 15th century, of the changes occasioned by volcanic forces, and the slight interest then paid to similar occurrences in distant countries, the catastrophe here mentioned cannot have been invented. It probably refers to a volcanic eruption on the western coast of Iceland. It is, however, remarkable that the sagas of Iceland mention a small island between Iceland and Greenland, from which the coast-mountains of both were visible, although no such island at present exists in the strait between these countries.

Somewhat farther down, between *Gruenlant* and *Terra Nova*, there are two islands, with the legend: *Apud has insulas quando naute perveniunt illuduntur a demonibus ita ut sine periculo non evadunt.* This legend no doubt recalls an encounter between European sailors and Esquimaux, in which the former did not play the winning part. This may possibly have occurred during the old voyages of the Scandinavians to Greenland. There is to be seen on Andrea Bianco's map of 1436, to the west of the Straits of Gibraltar, in the Atlantic Ocean, first a large island, *y:a de Antillia*, the outlines of which are only indicated, and then farther north, at the western corner of the map, another rather large island, *y:a de la man Satanaxio*. This legend, or the narrative to which it alludes, seems to have impressed the geographers of the following centuries, the *Insula Dæmonum* being retained on manuscript and printed maps long after the rediscovery of the New World, f. i. on a map of Wytfliet of 1597.

Still farther south there is, on Ruysch's map, a large island in the middle of the Atlantic Ocean between Lat. N. 37° and 40°. It is called *Antilia Insula*, and a long legend asserts that it had been searched for in vain, but that it had been discovered long ago by the Spaniards, whose last Gothic king, Roderik, had taken refuge there from the invasion of the Barbarians. The inscription depends on a myth, which has played a certain part in the history of geography, and from which is derived the present name of the islands between Florida and the northern coast of South America. The earliest delineations of an island, *Antilia*, in the Atlantic Ocean, are found on a portolano of 1425, belonging to the

Library at Weimar, and on Andrea Bianco's map of 1436. On the Globe of Behaim to the south of the Azores an island of the same name is also represented, provided with a long inscription, corresponding, but not identical with the legend of Ruysch.

Farther to the west, on Ruysch's map there is inserted a resumé of the description by Marco Polo of a large, independent island, *Sipangus*, situated 1500 miles to the east of *Zaiton*, inhabited by idolaters and rich in gold and precious stones. »But,» adds Ruysch, »as the islands discovered by the Spanish navigators exactly occupy this place, I have not ventured to lay down this island, presuming that the land called Hispaniola by the Spaniards must be Sipangus, especially as everything written about the former is applicable to the latter, excepting the idolatry».[2] The considerable distance from the eastern coast of China adopted for Zipangu by the geographers of the first part of the 16th century depends, according to Peschel, on the distance being given by Marco Polo in Chinese *Li,* of which there are 250 on one degree of latitude. This Chinese *Li* was by the European cartographers confounded with the Italian mile (60 = 1°).

The westernmost of the legends on the Asiatic coasts declares the discoveries of the Portuguese to have proved that the Indian Ocean, which was considered by Ptolemy to be an inland-sea surrounded by land, is a part of the Ocean.

On *Taprobane alias Zoilon*, which almost corresponds to the immense island at present called Sumatra, there is a long legend, partly borrowed from Ptolemy, but with the interesting addition that Portuguese mariners arrived there in 1507. Another legend on the south-eastern parts of Asia alludes to the existence of numerous islands in that part of the ocean, of which notices from Indian merchants seem to have already reached Europe.

Ruysch's delineation of the New World seems to indicate that he was not acquainted with the latest discoveries of the Spaniards. Cuba has thus got too large an extent. Its western coast is unknown to Ruysch, and of the northern part of the New World only Greenland and Newfoundland are represented. The names applied to the West Indian islands are, at least partly, taken from the narrative of the second voyage of Columbus. South America is called *Terra Sancte Crucis sive Mundus novus*. Twenty-nine names are here given. Most of them correspond to the names on *Tabula Terrae Novae* in Ptolemy 1513 (N. T. XXXVI). Only a few are found in the letters of Vespucci, which may be owing to the scarcity of geographical names in the description of his voyages.

A legend on South America at the lower border of the map is particularly interesting. It tells us that Portuguese mariners had followed the eastern coast of the country, down to Lat. S. 50°, but without reaching its southern extremity. We here obtain notices regarding exploring-voyages undertaken before 1508, of which no other information is met with in the history of geographical discovery. In the interior of *Terra Sanctae Crucis* there is another long legend describing the inhabitants and natural productions of the country, »*quae a plerisque alter terrarum orbis existimatur*.» The main part of this inscription is taken from the letters of Vespucci, but there are differences showing that Ruysch had had access to other materials than the printed and probably corrected editions. Ruysch f. i. speaks of lions in the interior of the continent, the existence of which Vespucci expressly denies, and of the abundance of gold in the rivers and mountains of

---

[1] Regarding this compare: *Den andra Dicksonska expeditionen till Grönland*, p. 55.
[2] »*Preter idolatriam*». As the long legend in the middle of the South American continent shows, the natives of the New World were considered to live *nulla religione, nullo rege*. They were accordingly not even idolaters.

the newly discovered countries — a statement not found in Vespucci's letters, at least not in the printed version. The western coast of South America is occupied by an unfolded roll of paper, on which it is repeated that the country had been called the New World by the Spaniards on account of its great extent. They, however, had not yet fully explored it, in consequence whereof the map here has remained unfinished, especially as it was not known to them in what direction the country was extending.

This summary review of the most important features and legends on Ruysch's map, will suffice to show its immense importance to the cartography as well of the old as of the new hemisphere. It would therefore be very interesting to obtain some biographical data respecting its author. On the title-page of the edition of Ptolemy of 1508 we read his name and nationality (JOANNES RUYSCH, *Germanus*), and from the *›Nova orbis descriptio»* etc. by MARCUS BENEVENTANUS, inserted in the same work, we know that he had sailed from southern England to the eastern coast of America. From his map we may further conclude that he was both a man of some learning and a practical mariner. But nothing, save these scanty notices, is known of his life, social position, and voyages. The name belongs to an old noble family in the Netherlands, a member of which, the knight W. RUYSCH of Amsterdam, fell in war in the year 1288. Another JACOB RUYSCH took part in a pursuit of heretics at the Hague in 1512. The cartographer may have been a brother of the last mentioned. The attempts I have made to gain further information in Amsterdam through persons best informed respecting the history of the Dutch nobility, have not been crowned with success.

The editor of the work in which the map of Ruysch was published, was EVANGELISTA TOSINUS. After Ptolemy's text there is a letter to *›Cardinalis Nannetensis»* (ROBERT GUIBÉ, cardinal and bishop of Nantes) inserted, in which Tosinus says:

»When an improved edition of Ptolemy with maps and other additions was lately printed, many insisted, *ut novi etiam orbis descriptionem et inventa nuper continentis loca adiungeremus.* This, however, was long impossible, as none of those to whom he addressed himself were able to give reliable informations regarding this subject. *Interim (ut fit) novarum rerum cupiditate cum huiusce cognitionis studium increbesceret, ab iis qui inventas insulas perlustrarunt, et quem novum appellant orbem, cognoscendi industria permearunt, vere liquidoque omnia ad unum audita cognitaque sunt. Et cum nonnulli Geographiæ periti de his agerent, tum imprimis Marcus Beneventanus Monacus Celestinus noster, qui illa alia multa in maiore volumine adiecit, tanta cura et diligentia, haec ipsa etiam metitus est, ad amussimque omnia collegit, tantaque fide conscripsit, ut ad huius etiam novi orbis cognitionem nihil prope desiderandum amplius videatur. Itaque cum nova omnia placere scirem et haec potissimum cuique iucundissima fore arbitrarer.* On account of this Tosinus was able to add the map of Ruysch and *Marci Beneventani Monachi Celestine congregationis Mathematici orbis nova descriptio* to the edition of Ptolemy of 1508. The text of Beneventanus is introduced by a letter to the Roman patrician MARIANUS ALTERIUS, from which the remarkable information is obtained that Ruysch's map was printed before it was incorporated with the Ptolemy of 1508. At least this seems to be deducible from the following passage in Tosinus' letter, from which we also find that Beneventanus only wrote commentaries on the map, and had nothing to do with its authorship: *quæ dum sollicite perquirebat* (Tosinus) *factus est desiderii compos: Beneficio enim Ioannis Ruiischi Germani viri Geographi impressa est vel univer-*

salis orbis tabula, in qua tum tellus illa noviter reperta quem mundum appellant novum, tum Lusitanorum navigatio atque Brittannorum desingnatæ sunt. Quod cum vidisset iudicavit ad operis complementum satis conferre, si cum commentariis haec ederetur tabula, iuxta Cl. Ptole. sententiam cap. XVIII. Lib. I. Ideo mihi suasit ut hanc mihi adsumerem provinciam.

In *Orbis nova descriptio*, which principally consists of a commentary on the map of Ruysch, in Ptolemy's style, the following passages deserve notice. In Cap. III is written: *Ioannes vero Ruschi Germanus Geographorum meo iudicio peritissimus ac pingendo orbe diligentissimus cuius adminiculo in hac lucubratiuncula usi sumus, dixit se navigasse ab albionis australi parte, et tam diu quo ad subparallelum ab subaequatore ad boream sub gr. 53 pervenit, et in eo parallelo navigasse ad ortus littora per angulum noctis atque plures insulas lustrasse quarum inferius descriptionem assignabimus.* Cap. VIII contains some very vague and general information touching the voyages to India by the Portuguese. Cap. X contains critical remarks as to the latitudes and longitudes of Madagascar (here called *Camaroca*), Taprobane, Java, Hispaniola etc. In Cap. XI is written: *Potuit igitur Italiae nostrae reformari ex ploographia descriptio, et quo vergat*

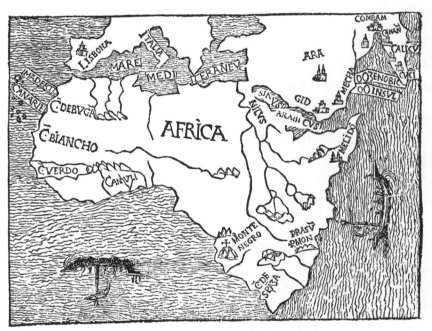

37. Map from *Itinerarium Portugallensium*, Mediolani 1508. (Orig. size 225 × 171 m. m.).

optime cognosci, similiter et quæcumque cetera littora quandoquidem una universalis facta sit tabula ex pluribus reformata navigationibus uti Archoplous Columbus nepos mihi ostendit, et Ioannes Ruysch unam condidit universalem orbis descriptionem parallelum usque 38 subgr. ad Austrum a Subcardine exordiens. Ex observationibus nam Anglorum, Gottorumque descriptum est mare a Subpolo ad parallelum per Thylem. Germani sua reformarunt littora atque mare eorum descripsere. Galli similiter sua atque Hispani. Genuenses autem nostrum mare, similiter Veneti una conscripserunt atque littora reformarunt, Insulas, Scopulos, Syrtes adverterunt. Columbus primus et nepos orientale descripserunt mare, Lusitani Meredionale, tum ex suis, tum ex Indorum navigationibus, ex quibus omnibus una confecta est (uti diximus) ploographia. Cap. XIV bears the superscription: *De tellure quam tum Lusitani tum Columbus observavere quem Mundum appellant Novum,* but does not contain anything noteworthy. In Cap. XV, again, is to be read: *quando Columbus nepos mihi dixerit quod ubiubi opus erat astrolabo, quadranteque utebatur pro elevatione polari captanda atque plures observaverit eclypses lunares . . . Et quod hoc pacto plurimorum locorum situs perbelle formavit descripsitque.*

Excepting these excerpts, the *Orbis nova descriptio* of Marcus Beneventanus only contains an exhibition of learning, which is now quite worthless, but which was perhaps necessary, at that time, as an introduction for the new world to her older sisters Asia, Europe, and Africa. It is at any rate a remarkable fact that Beneventanus, every time he deigns to descend from his pedestal of learning, communicates a fact of great importance to the history of geographical discovery. He thus incidentally informs us that the author of this map, which from a geographical point of view marks an epoch in cartography more distinctly than any other that has ever been published in print, had joined in a voyage from England to America. We also learn that Beneventanus had been personally acquainted with »Columbus Nepos.» By this name he probably designates either the illegitimate son of Columbus, Ferdinand, who sojourned in Europe till his nineteenth year (1509), or rather

be remarked that, in the XV chapter of *Nova orbis descriptio*, his »Nepos» bears strong testimony in his favour, as regards this point.

Winsor (*Bibliography of Ptol. Geogr.*, p. 7) mentions two somewhat different prints of Ruysch's map, whilst the legend *Plisacus sinus*, at the eastern coast of Asia, is wanting in some copies. On the copies to which I have had access, there are also other traces of successive corrections and emendations, e. g. on the long legend identifying Hispaniola with Sipangu. A copy of Ruysch's map, which bears no mark of having been bound and inserted in any edition of Ptolemy, is, according to Harrisse (*Cabot*, p. 164), preserved in Mr. Barlow's library at New York. Perhaps we here have a copy of the original map mentioned above.

2. Map of Africa printed at Milan in 1508 (N. fig. 37). This wood-cut map occupies the title-page of the following

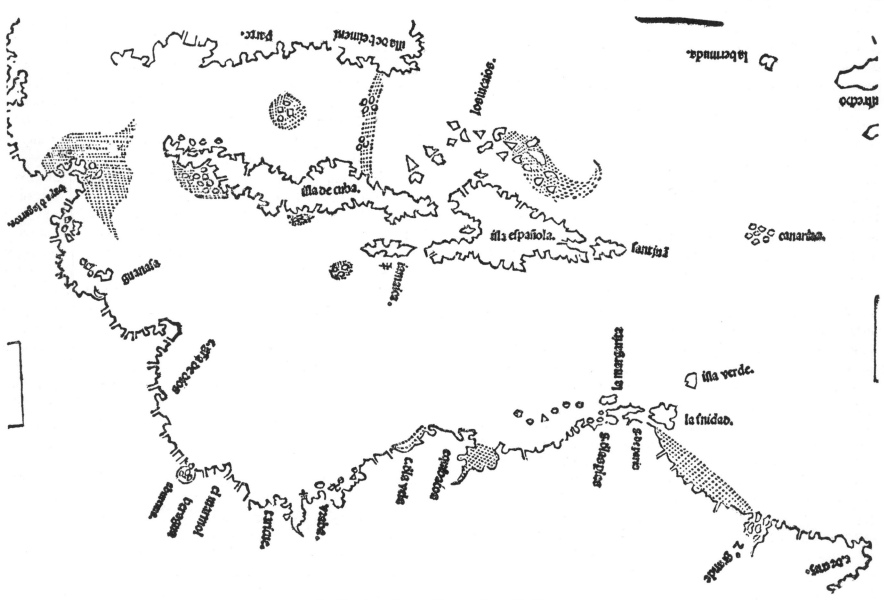

38. Map from: Petrus Martyr, *Opera*, Sevilla 1511.

the brother of Columbus, Bartholomaeus, who seems to have been an eminent cartographer. For there is an annotation on a copy of: *Paesi nouvamente retrovati* etc., Vicentia 1507, at the library Magliabechi, stating that Bartholomæus, when visiting Rome in 1505, wrote, for a canon of the church of San Giovanni di Laterano, a narrative of the first voyage across the ocean, to which a map of the new discoveries was appended. The canon presented the map to Alessandro Zorzi, *suo amico e compilatore della raccolta* (Humboldt, Krit. Unters., II. p. 343). We here probably have a notice respecting the same map, which Marcus Beneventanus had seen with »Columbus Nepos,» and which appears to have been partly copied by Ruysch, whose map consequently may be regarded as a direct illustration of the ideas prevailing in the family of Columbus as to the distribution of the continents and oceans of the globe. As the competence of Columbus in the use of nautical instruments has been doubted, it may further

rare collection of voyages: *Itinerarium Portugallensium e Lusitania in Indiam et inde in occidentem et demum ad aquilonem. Mediolani. Kalendis Iuniis MCCCCCVIII.* It is a translation of a still rarer work: *Paesi nouvamente retrovati. Et nuovo monde da Alberico Vesputio Florentino intitulato*, printed at Vicencia in 1507, but this original does not contain the remarkable map, which seems, at least for some years, to have constituted the prototype of modern maps of Africa. From the few inscriptions on the map, and from the name of the only European town *(Lisbona)* marked on it, it is evident that we here have a work of Portuguese origin.

The fac-simile fig. 37 is from a copy belonging to the *Biblioteca Marciana* in Venice. With regard to the contents of »Itinerarium Portugallensium» or »Paesi nouvamente retrovati,» I may refer the reader to: Humboldt, *Kritische Untersuchungen*, II, p. 343; Harrisse, *Bibl. Amer. Vetust.*, p. 113, and BARTLETT, *Bibliograph. Notices of books in the Library of Carter Brown*, p. 36 and 40.

3. A map printed in Strasburg in 1509 on the title-page of a small work, *Globus mundi* etc. (see p. 40). As the fac-simile (N. fig. 22) shows, the map is very insignificant, but it is the first printed to the north of the Alps on which a small part of the New World is laid down, and on which any regard is given to the circumnavigation of Africa by the Portuguese.

4. A map of the West Indies (fig. 38), inserted in some copies of: *P. Martyris angli mediolanensis opera. Legatio babylonica. Oceani decas. Poemata. Epigrammata.* (Colophon:) *Hispali cum summa diligencia per Iacobum Corumberger alemanum. Anno Millesimo quingentesimo XI.*

This map is rare. It has in vain been sought for in most copies of the very rare work for which it was intended. This circumstance is believed to be due to a suppression of the small drawing by the suspicious Spanish authorities. Notwithstanding its size and insignificant exterior, the map is of interest as the first printed Spanish map of some part of the New World, and perhaps also as the first (?) map printed in Spain. It is far more correct than other contemporary maps of the West-Indian islands, which is not astonishing, as PETRUS MARTYR D'ANGHIERA was personally acquainted with several of the great discoverers of the 15th and the beginning of the 16th century. For further bibliographical minutiae I may refer to the works of HUMBOLDT, HARRISSE, WINSOR, and CARTER-BROWN already cited. A special monograph of Petrus Martyr, with a fac-simile of the map, was published by HERMANN A. SCHUMACHER, New York 1879.

5. A map of the world printed in the Ptolemy of 1511 by BERNARDUS SYLVANUS under the title of: *De universalis habitabilis figura cum additionibus locorum nuper inventorum* (N. T. XXXIII). This is the second printed map of the world, in the delineation of which some attention has been paid to the great geographical discoveries of the preceding years. I have already (p. 18) given the full title and some bibliographical data of the work in which this map was first published. The following characteristic statement in the introduction, as to the origin of the new map and its relation to the old maps of Ptolemy, may be further quoted: *Placuit insuper universæ habitabilis figuram, cum iis omnibus quæ recentiorum navigationibus reperta, et nobis tradita sunt, ex nostro addere. Quam nulla tamen ex parte, ab universali Ptholemaei descriptione differre sentias, modo illa quæ Ptholemaeo ignota fuerunt demantur. Id vero ea tamen ratione egimus, ut videant qui Ptholemaeum damnarunt, quod nihilo nostri temporis navigationibus, ac veritati, modo neglectis prioribus numeris verba observentur, adversari videatur.* Bernardus Sylvanus here pronounces the indisputable truth, that if Ptolemy's longitudes and latitudes are duly corrected, and these new data employed as the basis of a new delineation of the world, we shall obtain a correct map, except of those parts of the earth which were unknown to Ptolemy. Sylvanus has generally followed Ptolemy, although with necessary modifications, in the delineation of Britain, the Scandinavian peninsula, south Africa, and eastern Asia, which have received a considerable prolongation beyond the bordering meridián of Ptolemy, with the addition of the recently discovered countries to the west of the Atlantic Ocean. The map of Sylvanus does not embrace the part of the earth which is situated between the 260th and 300th meridian. This map

is further remarkable as the first on which a heart-shaped homeoter-projection was employed.

We here also find the earliest allusion on a printed map to the discoveries of CORTE REAL, in the outlines of a land, *Regalis Domus* (instead of *Terra de Corte Real*), situated beyond *Terra Laboratorum*.[1] No other part of the North American continent is here laid down, and Greenland *(Gruenlant)* is transferred to eastern Asia. Cuba is drawn as a large island *(Terra Cube);* South America as an enormous continent *(Terra Sanctæ Crucis)*, of which the western limits, as well as the western limits of *Regalis Domus* and the eastern of Asia, are left unfinished. Africa has a tolerably correct shape, but the legends are copied from Ptolemy with a few exceptions, e. g. *Caput bonæ spei* and *Melinde*, which is the only town indicated on the whole African continent. The *Menutias* of Ptolemy is drawn as an insignificant island and not identified with Madagascar, which has received an enormous extension and is, under the name of *Comorbina Insula*, transferred too far to the south.

6. A small map of the world in the *Meteorologia Aristotelis*, Norinbergæ 1512, betrays a knowledge of the discoveries of the Portuguese. This is shown on the fac-simile T. XXXI. The New World is not laid down, but in the text *Nova illa Americi terra admodum nuper inventa* etc. is mentioned (compare p. 40).

7. Map of the world by JOHANNES STOBNICZA, Cracoviæ 1512 (N. T. XXXIV). This map, printed from a very rude and badly executed wood-cut, occurs in a rare work, of which the complete title is: *Introductio in Ptholomei Cosmographiam cum longitudinibus et latitudinibus regionum et civitatum celebriorum. Epitoma Europe Enee Silvii. Situs et distinctio parcium tocius Asie per brachia Tauri montis ex Asia Pij secundi. Particularior Maioris asie descriptio ex ejusdem Pij asia. Sirie compendiosa descriptio ex Isidoro. Africe brevis descriptio ex paulo orosio. Terre sancte et urbis Hierusalem apertior descriptio: fratris Anselmi ordinis Minorum de observancia* etc. (Colophon:) *Impressum Cracovie per Florianum Unglerium Anno Dni MDXII.*

The name of the author is given in a dedication, *Ioanni dei gracia Episcopo Posnaniensi.* Besides the above cited edition there are two others, viz. *Editio princeps*, for the most part identical with the preceding one, save the last leaf, which is somewhat differently printed and without date; and an edition of 1519, in which the numerous printer's errors and the faulty and capricious orthography of the previous editions are corrected. None of the many copies of these three editions, which I have examined in the libraries of Scandinavia, Russia, and Poland, contained any map. But in the Imperial library at Vienna there is a copy of the edition of 1512 with a map. The Munich Library is also said to possess a copy of the same edition, with a map. It is possible that the map has been excluded from the work, on account of the rudeness of its wood-cut, or on account of its being contrary to the old doctrines of the church. But that it was originally intended for the »Introductio» is shown by the identity of the watermark on the map with that on the paper employed for the text. The fac-simile here given on pl. XXXIV is a photolithographic copy of the original in Vienna. This map is, in spite of its rudeness, of great interest and importance to the early history of cartography, because:

1st. North and South America are here drawn, for the first time, as two large continents connected by a long and narrow isthmus. It is the earliest printed map on which the

---

[1] The origin of this strange name, which here appears for the first time on a printed map, is not certainly known. In the Library of Wolfenbüttel there is a large (2ᵐ,21 × 0ᵐ,75) map of 1534, drawn on parchment, on which there is a legend at »Tiera del Labrador,» stating the land to have been discovered by Englishmen from Bristol and that it got this name because the man who gave the first notice of it was a labourer from the Azores (HARRISSE, *Cabot*, p. 186).

newly discovered lands in the Atlantic are in their whole extent so separated from the Old World, that they may, with full reason, claim the name of *Novus Orbis*.

2d. Stobnicza's map, published one year before the 25th of September 1513, when Vasco Nuñez Balboa sighted the »Mare del Sur» from the mountains of the Isthmus of Darien, is the earliest on which the sea between Europe and Asia was divided by the newly discovered continent into two almost equal oceans, communicating only in the extreme south and the extreme north. This complete breaking with the old theory of one single Ocean, surrounding Europa, Asia and Africa, may to a certain extent be explained by the fact that coastlines are here substituted for the large unfolded rolls with legends which occupy the western coast of America on Ruysch's map. Several details, however, seem to prove that Stobnicza, or the unknown author of the map in his *Introductio*, had had access to geographical reports unknown to Ruysch. The method here employed of indicating the western coast of America by a succession of straight lines, in order to denote that the delineation was conjectural and not dependent on real observations, is worthy of note.

3d. There is no place on the map for a full extension of the new continent towards the south, but the coasts on both sides of its southern extremity are drawn in such a manner, that a southern communication between the two oceans evidently seems to have been admitted by the author.

4th. On Stobnicza's map the surface of the earth is, for the first time, divided into two hemispheres, each of which was laid down on the homeother projection of Ptolemy.

The map is based on the map of the world in Ptolemaeus 1482 (the Mediterranean Sea, northern Europe and southern Asia), on the map of Ruysch (Africa and eastern Asia), and, as regards the West-Indies, the Isthmus of Panama and North America, on data not before reproduced in printed maps. This map certainly has nothing in common with the two *tabulæ novæ* of the world in the Ptolemy of 1513. The inscriptions on Stobnicza's map are often difficult to decipher. Those on the newly discovered lands alone have some interest. They appear to be:

? de bona ventura.
Isabella.
Spagnolla.
Arcay.
Caput destado.
Gorffo Spemosa.
Caput S. Crucis.
Monte Stegoso.
Alla pego.
Terra incognita.[1]

At a considerable distance from the eastern coast of Africa two large islands are drawn, the northern designated by the name of *Mardagascar*, the southern by that of *Zinzabar*. The name of Madagascar is, though somewhat differently spelt, already found in Marco Polo, but here it is for the first time shown on a printed map. The name *Menutias* of Ptolemy is supposed to refer to this island. It is called *Camarocada* by Ruysch (1508), *Comorbina* by Sylvanus (1511), and *Madagascar* in Ptolemaeus 1513. When it was rediscovered in 1506 by Ant. Gonçalves, on the day of St. Laurence, it received the name of *San Lorenzo*, which, as far as I know, is first met with on a printed map on the

heart-shaped map of Orontius Finæus of 1531 (N. T. XLI) and on the mappemonde in gores printed at Nuremberg in about 1540 (N. T. XL).

The copy of the map given on pl. XXXIV is a carefully photolithographed fac-simile. If the inscriptions are sometimes difficult to decipher and consist of letters with indistinct outlines, this is the fault of the original, which, with all its defects, is faithfully rendered by the copy.

8—12. Five important maps among the tabulæ novæ in Ptolemaeus Argentinæ 1513. As mentioned above (p. 19) this edition contains, besides Ptolemy's text and the ordinary 27 old maps, an appendix of twenty new ones, preceded by a long second title-page: *In Claudii Ptolemei Supplementum* etc., and a short preface. On five of these new maps the discoveries of the Spaniards and Portuguese during the preceding century are represented. Of these, the *Hydrographia, sive Charta marina: continens typum Orbis universalem iuxta Hydrographorum traditionem* is given on pl. XXXV; the *Tabula oceani occidentalis seu Terræ novæ* on pl. XXXVI, and two maps of Africa and one of southern Asia, on a reduced scale, in fig. 8—10.

On the verso of the second title-page we read: *Charta autem Marina, quam Hydrographiam vocant, per Admiralem quondam serenissimi Portugaliae regis Ferdinandi, ceteros denique lustratores verissimis peragrationibus lustrata: ministerio Renati dum vixit, nunc pie mortui Ducis illustriss. Lotharingiæ liberalius praelographationi tradita est: cum certis tabulis a fronte huius chartae specificatis. Cuius item Ducis illustriss. honori cedit extensa ad finem Dominii sui tabula studiosissime pressa. Nam eius terrae latebris, Vosagi dico rupibus nobile hoc opus inceptum, licet quorundam desidia ferme sopitum, a sexennali sopore per nos tandem excitatum est.*

Portugaliæ is here probably a misprint for *Hispaniæ*. The mistake has, however, given rise to some controversies among students of the history of discovery of the New World. This passage has, furthermore, been interpreted by several prominent authors, such as Santarem, Harrisse etc., as an assertion that the map on which a portion of the new world is represented, was a copy of a drawing by the »admiral» (Columbus) himself. But the passage supposed to support this conjecture evidently only states that this sea-chart *(Charta Marina)* was based on observations made in the time of Duke Renatus, during voyages of discovery, »*per admiralem . . . ceterosque lustratores,*» and that the map had, after the death of Renatus (in 1508), together with the other maps enumerated on the title-page, been handed over with much liberality to the ducal printing-office of Lorraine. The above quoted words seem further to suggest[2] that the new edition of Ptolemy, printed at Strassburg in 1513, was commenced six years earlier in the Vosges mountains. This corresponds with the supposition that this edition of Ptolemy, so important owing to its supplement, was prepared by the same learned coterie at St. Dié, that inscribed its name on the history of geography by having unjustly given Amerigo's name to the world newly discovered by Columbus. But the argument generally quoted for regarding Waldseemüller as the author of the twenty new maps in this edition depends, as before (p. 21) pointed out, on an incorrect interpretation of a passage in the edition of 1522. There further occurs another very common error, with respect to the new maps in this edition, namely

---

[1] On fol. VII of the text of the work for which this map was originally drawn, is written: *Quarta pars orbis America. Non solum autem praedicte tres partes* (Europa, Africa, Asia), *nunc sunt lacius lustrate verum et alia quarta pars ab Americo vesputio sagacis ingenii viro inventa est, quam ab ipso Americo eius inventore amerigem, quasi americi terram sive americam appellari volunt* etc. Thus Stobnicza is one of the first geographers who adopted the name *America*, proposed by Waldseemüller.

[2] If not the map of Lotharingia is meant by »*nobile hoc opus*» in the above cited passage from the introduction to the supplement. The opening word *Nam* supports this interpretation as well as the epithet *nobile*, so appropriate for a work executed under princely patronage. But in this case much of what is written about the gymnasium of St. Dié and Waldseemüller has been a useless exhibition of learning.

that the map 2, with the inscription: *Haec terra cum adja-centibus insulis inventa est per Columbum Ianuensem ex mandato regis Castellae*, was connected with the above cited, erroneously interpreted passage in the preface . . *per Admiralem Portugaliæ regis Ferdinandi*, which map, accord-ingly, has been designated »*the Admiral's map.*» But the passage in question does not refer to the second but the first among the tabulæ novæ (N. T. XXXV), on the second title-page termed »*Hydrographia sive charta Marina.*» This map is evidently of Portuguese origin and scarcely shows any progress from the point reached by the maps of Ruysch, Sylvanus, and Stobnicza. More original and important is the second map, enumerated on the title-page, among »Particulares tabulae Europae,» under the name of *Tabula Terræ Novæ* (N. T. XXXVI). A number of names are here met with, on the American continent. It is here that we first find a group of islands, *Y. tebas*, to the south of the Equator, between Africa and the New World, roughly corresponding with Ascension or St. Helena, discovered by JUAN DE NOVA in 1501 and 1502.

The most important maps in the Ptolemy of 1513, how-ever, are the two maps of Africa in double-folio which I have reproduced on a reduced scale in fig. 8 and 9. They form admirable though hitherto completely neglected illustrations of one of the most important episodes in the history of Na-vigation, and they are evidently directly based on carefully registered observations during the Portuguese exploring voyages round Africa to India.

For their exactness, and in the richness of names along the coasts of Africa, these maps are comparable with the old portolanos of the Mediterranean Sea.

The corresponding map of Asia (N. fig. 10) shows on the contrary that the geographical notions about the Indian Peninsulas when the map was drawn (1507?) were still very vague in Europe, and dependent on hearsay.

The remaining Tabulæ novæ in the Ptolemy of 1513 are:

Tabula nova Hiberniæ, Angliæ et Scotiæ. If we except tabula prima Europae (*Albion et Ibernia*) of Ptolemy, often published in print before 1513, this is the first printed map of Great Britain. It is evidently copied from some portolano, and the interior of the country is, as in the portolanos, left as blank as the interior of Africa on maps printed half a century ago.

Four new maps, one of Spain, one of France, and two of Italy. New maps of these countries are already found in BERLINGHIERI (*Gallia*, N. fig. 7), and in the editions of Pto-lemy printed Ulmæ 1482 and 1486 (*Hispania* and *Italia* N. fig. 11 and 12) and Romæ 1507 and 1508.

A map of Scandinavia, copied from the corresponding map of the edition Ulmæ 1482.

Tabula moderna Germaniæ, roughly copied from the central part of Nicolaus a Cusa's map in the Ptolemy of TOSINUS of 1507 and 1508 (N. fig. 13).

Tabula moderna Sarmatiæ Europeæ sive Hungariæ, Poloniæ, Russiæ, Prussiæ et Walachiæ. This map of the country between the Baltic, the Oder, the Danube, and the Dnieper is a copy of the eastern part of the above mentioned map in the Ptolemy of Tosinus.

Tabula moderna Bossinæ, Serviæ, Greciæ, Sclavoniæ, and *Tabula nova Asiæ Minoris.* Here we have the first printed modern maps of Greece with the Balkan countries, and of Asia Minor. From the want of details in the interior of the countries, we may conclude that they are copies of portolanos of the 15th century.

Tabula moderna Terræ Sanctæ. The map of Palestine, or of *Idumea, Palestina, Iudea*, and *Samaria*, in the Geo-graphy of Ptolemy, is inserted on the Quarta Asiæ tabula as a part of Syria. No Christian influence can be detected on this map, and this may be considered as evidence of its prototype having been drawn in the heathen time of the Roman empire. When Ptolemy's geography, a millennium later, became the principal geographical codex of the Christian world, the insufficiency of this map was manifest. Palestine therefore became one of the first countries on which the geographers of the 15th century exercised themselves in the art of map-drawing. The first productions destined for book-print were not very successful, as may be judged from the maps in *Rudimentum Novitiorum* (N. fig. 3). But already in Berlinghieri and in the Ptolemy editions of 1482, 1486, 1507, 1508, and 1513, the new maps of the Holy Land, all founded on a common prototype, are far better.

Aeschler and Übelin, after having added one tabula ma-rina and fifteen new maps *(Tabulæ particulares)* of larger districts to the old series of maps of Ptolemy, furthermore inserted in their edition four *Chorographiæ*, i. e. maps of minor districts, but on a larger scale and more minute than the previous ones; namely, of Switzerland, of the Provinces of the Rhine, of Crete, and of Lorraine. The scale of these maps, to which I shall have occasion to return in a future chapter, varies between one millionth and 2,2 millionths.

As I have pointed out several times before, isolated new maps had already been published in several works before 1513. But in the addenda of Aeschler's and Übelin's edition of Ptolemy we for the first time obtain a modern atlas with maps of all the parts of the globe of which new geographical data could be had. To some extent this atlas may therefore be regarded as the opening chapter of the modern literature of atlases. This attaches additional interest to its publication, though its maps, as regards their execution, cannot compete either with the copper-plates of Buckinck-Schweinheim, with the woodcuts of Johannes Schnitzer von Armsheim, or with those in the Ptolemy of Bernardus Sylvanus.

13. A map in REISCH's *Margarita Philosophica*, Ar-gentorati 1515 (N. T. XXXVIII). The map in the first edition of this work, dated 1503 (N. T. XXXI) is, as before pointed out, only a badly executed copy of the mappemonde in Ptolemaeus Ulmæ 1482, and is only of secondary interest in the history of cartography, on account of an incidental allu-sion to the voyages of discovery from Portugal during the 15th century. On this map, which was printed in 1503, but inserted in several editions of a much later date, we look in vain for any lands, or islands, or inscriptions indicating some acquaintance with the voyages of Columbus.

This omission was first rectified in the edition Argen-torati 1515 by the insertion of a map on which the newly discovered lands on the other side of the Atlantic Ocean are laid down. This map is, in its main features, a rude repro-duction in wood-cut of the *Orbis typus universalis* and *Ta-bula Terræ Novæ* in Ptolemaeus 1513 (N. T. XXXV and XXXVI). Only two inscriptions on the New World are worthy of notice. On its southern part we read *Paria seu Prisilia.* It is the first time such an appellation is on a printed map applied to the large continent of which the main portion now bears that name.[1] In the northern part we read *ZOANA MELA*, an inscription of rather mysterious appearance at the first sight. In the *Zeitschrift für wissenschaftliche Geo-graphie*, Bd. 5, Wien 1885, p. 1, WIESER has shown that the

---

[1] Centuries before CABRAL this name had been employed on portolanos to designate a mythical island in the Ocean to the west of Ireland, and in this sense it is also employed on the new map of England in the Ptolemy of 1513 (N. fig. 6). In its present sense it is for the first time met with in *Copia der Newen Zeytung auss Presillg Landt*, s. l. et a., a print from 1508 or 1509 of only a few leaves, but important to the history of the discovery of America (WIESER, *Magalhães-Strasse und Austral-Continent*, Innsbruck 1881, p. 92).

derivation of this name may be deduced from the following passage respecting the first voyage of Columbus in a writing of PETRUS MARTYR at the municipal library at Ferrara: *Et in questa prima navigatione scopersono sei insule sole do delle quali de grandecia inaudita, una chiamò la Spagnola, l'altra la Zoanna. Ma la Zoanna non ebbe ben certo che la fussi insola.* When this was printed in the *Libretto de tutta la Navigation*, Venezia 1504, two islands discovered during the first voyage of Columbus, one of them called *Spagnola* and the other *Zoanna Mela*, are already spoken of. Thus *Zoana* might only be an Italian form of the name *Iuana* or *Johanna*, by which Columbus, after his return from the first voyage, designated one of the West-Indian islands. This explanation of Wieser is probably correct. Before I had seen it, I thought the name alluded to Cabot Senior, whose Christian name, JOHN or GIOVANNI, was also written ZOANNE. In a letter from the Venetian ambassador in London, RAYMONDO DI SONCINO, to the duke of Milano (HARRISSE, Cabot, p. 150 and 324) he is, for instance, called *Messer Zoanne*. As *in* and *m* are written in the same way in several places on the map, so *mela* possibly might have been erroneously written instead of *insula*. It was on the coast where this name is placed, that John Cabot landed in 1497.

It is uncertain whether this map was added to every copy of the Margarita Philosophica of 1515,[1] or not. It was certainly wanting or incomplete in most copies of the rare edition I examined, and to enable me to give a complete facsimile it became necessary to consult three copies of the map, one from the R. Library at Stockholm, one from the R. Library at Copenhagen, and one from the Imp. Library at Vienna. In both copies belonging to the Scandinavian libraries a small strip has been cut away in the same part of the map when the book was bound, which defect in the copy here communicated is supplied by a photograph of the Vienna copy. On the reverse of the map there is a geographical description divided in columns of the same size as the text of the book, but, as regards its contents, of scarcely any importance. It may finally be mentioned that the map of ROBERT THORNE of 1527 (N. T. XLI), is, with regard to the Old World, a minute copy of the map here under discussion.

If I except a few maps in the editions of Julius Cæsar and Macrobius, published in 1515—1519, and almost worthless in a geographical point of view, and the globe of Schöner of 1515 covered with a map printed in gores for that purpose, and of which an account will be given in the next chapter, the map in the Margarita of 1515 is the only printed map known to me between the years 1515 and 1519.

VII.

Terrestrial Globes from the 15th and the first part of the 16th century.

A. Globes from the 15th century.

1. *Behaim's* globe of 1492. It is generally assumed that the doctrine of the spherical form of the earth was established in the 6th century before our era, by PYTHAGORAS, or by some philosopher of his school, and that it was more generally adopted a couple of centuries later, in the times of PLATO. But this principle, so indispensable to scientific geography, was first fully proved in the fourth century, by ARISTOTLE (through the form of the shade of the earth during lunar eclipses), by DICAEARCHUS (through the different times of the setting and rising of the heavenly bodies in different latitudes) and others. ERATOSTHENES (276—195 B. C.), finally, made the first attempts to measure a degree of latitude for determining the circumference of the earth;[2] and HIPPARCHUS (160—125 B. C.) fixed the first geographical positions. Through these observations the most important scientific data, necessary for the construction of a globe of the earth, i. e. of a geographical representation of the lands and seas of the earth drawn on the surface of a globe, had been determined. Geographical globes probably existed from this time, although none of them are still extant. In the 22d and 23d chapters of his first book of the geography, Ptolemy also gives the necessary instructions for the delineation of the inhabited world (ἡ οἰκουμένη) on a sphere, but he does not mention that such a work had been actually executed. During the succeeding centuries, until the end of the Middle Ages, the doctrine of

antipodes and, as a corollary to this, the doctrine of the globular form of the earth, was most severely condemned by

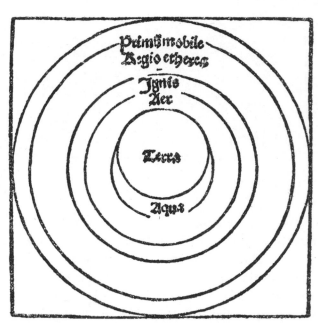

39. Section of the Cosmos from an edition of SACROBOSCO of the 15th century.

several of the most influential and distinguished men of the church. This condemnation was pronouced, in the first place, by LACTANTIUS, who, in the *Institutiones divinæ*, Lib. III cap.

[1] According to HARRISSE, *Bibl. Amer. Vetust.*, p. 341, the *Zoana Mela map* should also be found in the edition of 1535. It is, however, wanting in all the copies I have seen of this edition.

[2] Compare: FORBIGER, *Handbuch der alten Geographie aus den Quellen bearbeitet*, Leipzig 1842—1848. The numerous references to the classics make the voluminous treatise particularly valuable.

24, exclaims: *Est quisquam tam ineptus, qui credat esse homines quorum vestigia sint superiora, quam capita? aut ibi, quae apud nos jacent, inversa pendere? fruges et arbores deorsum versus crescere? pluvias et nives et grandinem sursum versus cadere in terram? ... Quid dicam de iis nescio, nisi quod eos interdum puto aut joci causa philosophari aut prudentes et scios mendacia defendenda suscipere, quasi ut ingenia sua in malis rebus excerceant.* Even AUGUSTINE adopted this opinion though, as appears in the *De civitate Dei*, Lib. XVI cap. IX, with some hesitation, while admitting that even if the doctrine of the existence of antipodes is regarded as absurd, the earth may yet be of a globular form. The figure on the preceding page from an edition of SACROBOSCO gives,

on their thrones, etc. It is rich in geographical details and in inscriptions of great importance to the history of geography. For these reasons, and owing to the prominent position occupied by Behaim with regard to the discoverers at the end of the 15th century, this globe has become not only the first, but also, without comparison, the most important document of this kind, of the period of the great geographical discoveries, that has been preserved. It has been the subject of a number of reproductions and monographs, of which the most important are inserted in the following works:

a. JOHAN GABRIEL DOPPELMAYER, *Historische Nachricht von den nürnbergischen Mathematicis und Künstlern*, Nürnberg 1730. On Tab. I of this work Doppelmayer gives

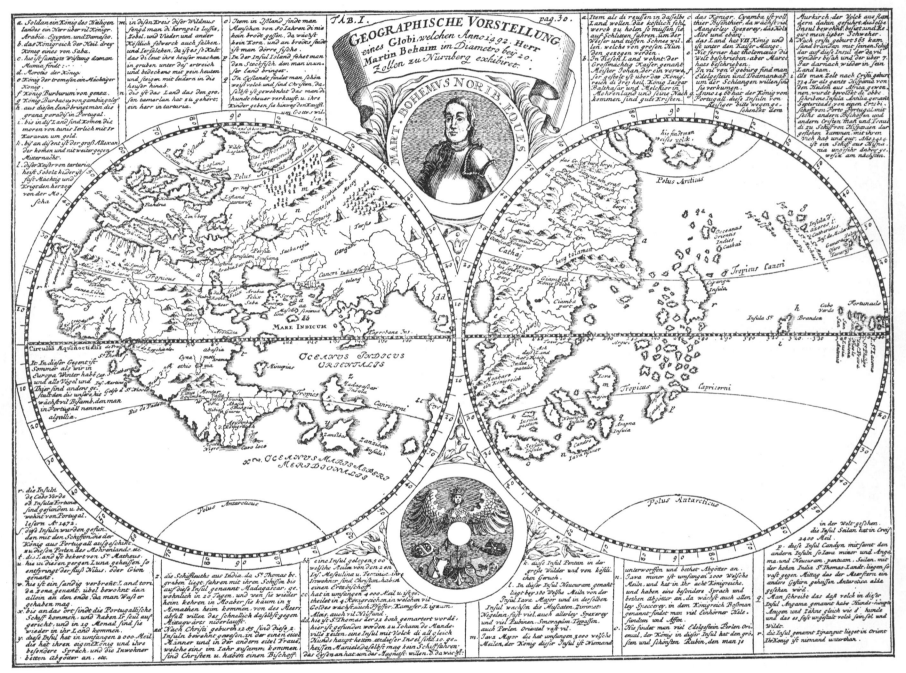

40. Martin Behaim's globe of 1492 from J. G. DOPPELMAYER. (Size of his drawing 414 × 302 m. m.)

without further explanation, a popular illustration of this hypothesis regarding the distribution of land and sea, according to which the theory of the impossibility of antipodes might be reconciled with the adoption of a globular form of the earth.

During the latter part of the Middle Ages, and especially after the circulation in the West of the Latin translations of Ptolemy's works, the doctrine of the globular form of the earth, and the possibility of antipodes, was again accepted by unprejudiced cosmographers. Yet no older globe, even of that time, than that which MARTIN BEHAIM presented to his native city of Nuremberg in 1492, has been preserved. This globe is thus the oldest at present known. It is drawn on parchment stretched on a sphere of a diameter of 1⅔ Paris-feet or 0ᵐ·541. In accordance with the custom of the period, the drawing is beautifully illuminated and ornamented with standards, Kings sitting

the first copy of the globe, although on a much reduced scale. As, on one hand, the size of the globe does not permit me to give a complete fac-simile, and, on the other, a reproduction of this important work, probably often copied in the 16th century, should not be wanting in an essay on the history of the oldest cartography, I here (fig. 40) give a reproduction of the drawing of Doppelmayer. It was published 158 years ago, when several inscriptions since effaced were still decipherable. It is, therefore, necessary that, in the study of this important geographical document, regard should always be paid to the versions on this first complete copy, which, for the rest, gives us a very good and comprehensive view of the principal features of the globe. The title-legend, within the south polar-circle, is left out by Doppelmayer. By Ghillany it is rendered thus:

Aus fürbitt und beger der fürsichtigen erbaren und weisen als der obesten haubtleut der loblichen Reichsstat Nürnberg die dan zu disen Zeiten regirt haben mit Nahmen hl. gabriel Nutzel hl. p. Volckamer und hl. Nicolaus Groland ist diese figur des apfels gepracticirt und gemacht worden aus gunst ausgebung vleys durch den gestrengen und erbar herrn Martin behaim Ritter, der sich dann in dieser kunst Cosmographia viel erfarhen hat und bey einen drittel der welt umfahren solches alles mit fleiss ausgezogen aus den büchern ptolom. plinii Strabonis und Marco Polo und also zuzam gefügt alles meer und erden jegliches nach seiner gestalt und form solches alles den erbarn Georgen hozschuer von rathswegen durch die gemelte hauptleuthe befohlen worden ist darzu er dan geholffen und gerathen hat mit möglichen fleiss solche kunst und apfel ist gepractisirt und gemacht worden nach Christi geb. 1492, der dan durch den gedachten herrn Martin peheim gemeiner Stadt Nürnberg zu ehren und letze hinter ihme gelassen hat sein zu allen Zeiten in gut zu gedencken nach dem er von hinen

c. F. W. Ghillany, *Geschichte des Seefahrers Ritter Martin Behaim . . . eingeleitet durch eine Abhandlung: Ueber die ältesten Karten des neuen Continents und den Namen America von* Alexander v. Humboldt, Nürnberg 1853. A drawing of the map on the globe is here reproduced in colours, on a stereographic projection, and with the legends complete.[1]

A full size drawing of this remarkable »Monument de géographie» is further given in Jomard's atlas. For various other memoires on the same subject, generally followed by drawings on a reduced scale, I may refer to the treatise by d'Avezac on the Laon globe cited below, and to Winsor's *Bibliography of Ptolemy's Geography*, p. 5.

The most complete of the works enumerated above, is that of Ghillany. But even his analysis is not exhaustive, and his handsome delineation scarcely comes up to the requirements of our time: nor are the inscriptions on it always rendered with perfect accuracy. Near the North-pole, for instance, we read in Ghillany: *Hier find man weisse völker*, words that

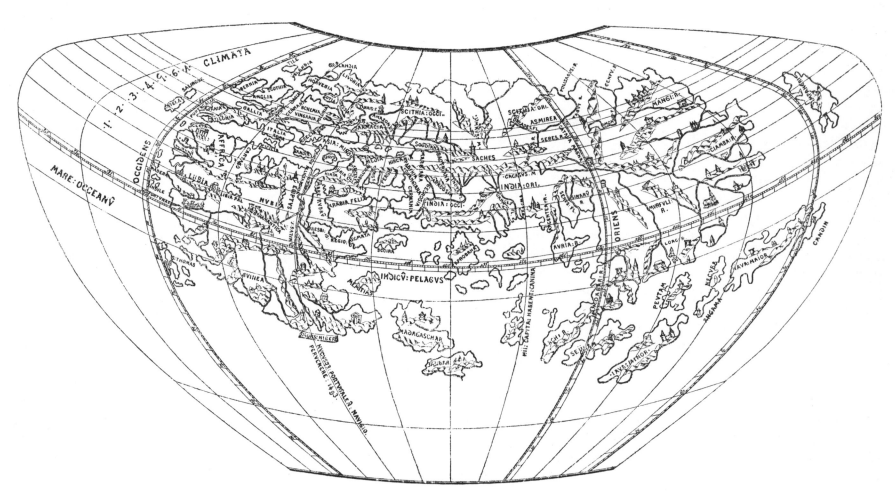

41. D'Avezac's reproduction of the map on the Laon globe. (Orig. size 320 × 169 m. m.).

wieder heim wendet zu seinem Gemahl das dann ob 700 mail von hinen ist da er hauss hält und sein tag in seiner Insel zu beschliessen da er daheimen ist.

b. Chr. Gottlieb v. Murr, *Diplomatische Geschichte des portugiesischen berühmten Ritters Martin Behaims*. This important contribution to the history of Behaim was first edited at Nuremberg, in 1778, then with considerable additions at Gotha, in 1801. A French translation by H. J. Jansen was published in the *Recueil de pièces intéressantes concernant les antiquités* etc., Paris 1787 (T. I. p. 317—363; T. II. p. 298—364), another in a French translation of Carlo Amoretti's *Pigafetta*, Paris 1801. A third French edition was prepared for the press by the author himself and published at Strasburg in 1802. Murr's description of Behaim's globe is minute and very meritorious for its time. But the reproduction he gives in his brochure, only embraces the portion between eastern Asia and the Azores.

have since become the subject of comment, whereas this passage in Murr is rendered by: *Hir fecht man weissen valken*, which no doubt is the correct reading. Ghillany says (p. 73) that the mechanician who restored Behaim's globe in 1823, declared it to be so decayed that before long it would have perished altogether. It is, therefore, desirable that no time should be lost in obtaining a new copy, which would be absolutely reliable, on a projection more fit for the reproduction of such a document in fac-simile than the stereographic. For this purpose, I should propose one of the globe-projections of Glareanus or Florianus (N. T. XXXVII and fig. 48), of course properly corrected according to our present knowledge of the length of lines drawn on the surface of a sphere.

On a closer examination of the drawings and legends on Behaim's globe we shall find it to be based 1st on Ptolemy's atlas; 2d on the narratives of the travels of Marco Polo and other medieval travellers in Asia; 3d on the Portuguese

[1] An incomplete representation of the western hemisphere was published by Ghillany already in the brochure: *Ueber Behaim und Schoner*, Nürnberg 1842 (Ghillany, *Gesch. Beh.*, p. III).

voyages of discovery; and 4th on the map of the northern countries of Europe in the Ptolemaeus Ulmæ 1482.

The delineation of the Mediterranean and Black Seas indicates ignorance of the Italian and Catalan portolanos, or rather, perhaps, that Behaim in the inland-town of Nuremberg had not access to these charts, exclusively intended for ship-owners and pilots. On the other hand the delineation of England, the Azores, the Canary Islands, the Cape Verde Islands, the western and southern coasts of Africa, and the long inscription at Iceland indicates personal observations or access to original documents now lost. To this it may be added, that the globe presents a faithful picture of the ideas regarding the distribution of land on the surface of the earth prevailing among the mariners of Europe, and especially among the mariners from the country of Henry the Navigator, at the period immediately before the first voyage of Columbus. All this makes the globe of Behaim one of the most important charters in the history of cartography. As regards its significance in this respect the reader is referred to the works of LELEWEL, HUMBOLDT, D'AVEZAC, KOHL, PESCHEL, WIESER, and others. Finally it may by mentioned that Behaim's globe, or the original documents on which it was based, had been used for the drawing of the maps of eastern Asia by RUYSCH and by the authors of the *Tabulæ Novæ Asiæ* in the Ptolemy of 1513.

According to Ghillany, Martin Behaim was born in about 1459. He belonged to a family which was originally Bohemian, had settled in Nuremberg, and had there been early included among the patrician families. After having in his youth been a disciple of REGIOMONTANUS,[1] he applied himself to commerce. He went to Antwerp in about 1475. Thence, in about 1480, he removed to Portugal, where, in 1486, he married a daughter of the hereditary governor of the islands of Fayal and Pico in the Azores. Owing to his mathematical insight, he seems soon to have acquired a high reputation in his new fatherland. He was made member of a commission charged to invent some practical method of determining a ship's position at sea by means of astronomical observations. He then, in the capacity of astronomer and cosmographer, accompanied the expedition of DIOGO CÃO in 1484 and 1485 along the western coast of Africa. In Portugal Behaim had, no doubt, had communication with Columbus. In the year 1491 he visited Nuremberg, probably on business, where he remained for two years and where he made his globe. In 1493 he returned to Portugal and died at Lisbon in 1506. Besides the globe here under discussion, Behaim (or his son, who was also called Martin) appears to have made another, which is mentioned in the accounts of Magellan's voyage, and which might have been similar to Schöner's globe of 1515. The report that Behaim had discovered America before Columbus, originated from an erroneous interpretation of a passage in Schedel's chronicle (Lat. edit. 1493 fol. ccxc), where it is said that *Jacobus Canus* and *Martinus Bohemus*, after having crossed the Equator, *in alterum orbem excepti sunt*. But here America is not meant by »Orbis Alter,« but southern Africa, a nomenclature fully justifiable, according to the older theories of the distribution of land on the earth. The same meaning belongs to »Novi et incogniti Orbes« in the *Oratio* by VALASCUS FERDINANDUS, cited by me at p. 62.

2. The *Laon globe of 1493* (fig. 41). Another globe, made at the end of the 15th century, was discovered in 1860 by M. LÉON LEROUX at an antiquarian's shop in Laon and described by D'AVEZAC in *Bulletin de la Société de Géographie*, Sér. 4: T. 20, Paris 1860. This globe consists of a gilded sphere of copper of 170 m. m. in diameter. I here give d'Avezac's representation of it on a slightly reduced scale. The projection chosen by d'Avezac is not very successful, and it is not possible to make out from his description whether the design reproduces the original completely and quite faithfully or not. The globe seems to a considerable extent to be based on that of Behaim and to have been made shortly after, the year 1493 occurring in a legend near the southern part of Africa. The discoveries of the Spaniards and Portuguese after 1497 were, according to d'Avezac, unknown to the engraver.

These two globes, the one drawn on a sphere covered with vellum, the other engraved on metal, are the only ones of the 15th century[2] at present known to exist. There are, of course, a greater number yet extant of the following century, but even of that period such works are scarce, although maps, in gores, intended for earth globes, had already been printed several times in the beginning of the 16th century.

The manner of drawing the twelve gores of which such globe-maps consist was described for the first time in HENRICI GLAREANI *Poetæ laureati de Geographia Liber unus*, of which

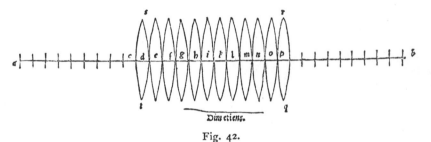

Fig. 42.

there are known editions printed at Basel in 1527, Friburg in 1529, Frankfort in 1532, Friburg in 1533, Venice in 1534, Friburg in 1535, (Brisgæ in 1536[3]), Venice in 1537, Venice in 1538, Friburg in 1539, Paris in 1542. I have had the opportunity of comparing the *editio princeps* and a number of the other older editions, with the Paris edition of 1542, and have found them to be quite identical. The work of Glareanus is a manual of geography of about 70 pages in small quarto. More than the half its contents is devoted to cosmography and spherical geography, of which, however, only two chapters are still of any interest, namely the last one *De regionibus extra Ptolemaeum*, in consequence of its being one of the many instances of the extremely small knowledge the learned world in the first part of the 16th century possessed of the great geographical discoveries, and of the total want of insight into their future importance, as regards the development of mankind; and chapter XIX, *De inducendo papyro in globum*. Here the method employed in delineating maps on the projection which I have designated by the name of Glareanus, is described in the following words (edit. Basileæ 1527): *Sit* (fig. 42) *linea a b, in triginta aequas divisa partes, quales singuli quadrantes æquinoctialis ternas habent. Ponito circini pedem alterum in b, alterum extende in o, sic enim dena*

---

[1] This is doubted by PESCHEL (*Geschichte der Erdkunde* 1865, p. 215), because Behaim, if the year of his birth was 1459, would have been only 16 years old, when Regiomontanus left Nuremberg. Peschel does not look upon Behaim as being very prominent as a cosmographer.

[2] In a letter from RAIMONDO DI SONCINO to the Duke of Milan, dated London, Dec. 18th 1497, a globe (*una sphera solida che lui a fatto*), made by *Messer Zoanne*, i. e. JOHN CABOT (HARRISSE, *Cabot*, p. 150) is mentioned. This globe is lost.

[3] This edition is not mentioned by Harrisse. I have seen it at the R. Library in Berlin. It contains a large map of the world on a double cordiform projection (added to the work at its binding?), which, however, is only a reprint of the heart-shaped map of the world by ORONTIUS FINAEUS (N. T. XLI). But the title-legend on the square field, below, in the middle of the map, is here altered to: *Christianus Wechelius lectori. S. Excudimus lector studiose veram et absolutam totius orbis descriptionem .... quae ad gemini humani cordis imitationem in plano ita exprimitur, ut in corpore sphærico vix possit absolutius.., Vale Ex scuto Basiliensi MDXXXV.*

transmittes spacia, duc arcum q r. Deinde ex b promove circinum uno puncto ita enim alter pes in n veniet. Tum rursus duc arcum, atque ita deinceps, donec in c deventum fuerit. Deinde in marginem alterum transfer circinum, ita ut in a posito uno pede, in d alterum extendas, atque illic duc arcum s t, et emerget duodecima pars superficiei quam quærimus c s d t. Deinde ex a promove uno puncto circinum ut antea in altero margine fecimus, ita enim in e pes alter veniet, ac deinceps promove donec ad p deveneris, ac habebis duodecim partes papyri, quam globo apte circumponere poteris, quanquam superne propter sphaerae coarctationem nonnihil superabit.

I have been unable to ascertain whether, as Myritius pretends, Glareanus was actually the inventor of this construction, which is practical and easy, though not perfectly correct. It does not appear to be mentioned by any other author of the beginning of the 16th century, though the invention must have been made at least fifteen years earlier, maps on such gores already existing in the third

lustre of the 16th century. HENRICUS LORITUS, called GLAREANUS after his birth-place, was a learned and distinguished humanist of the first half of the 16th century. He was born in 1488; in 1512 he was crowned a *poeta laureatus* at Cologne by the Emperor Maximilian, and in 1515 was summoned to the chair in mathematics at the university of Basel. Six years later he removed to Paris, where he professed humanities at the Collège de France. In 1524 he returned first to Basel and then to Friburg, where he died in 1563. I mention these dates, as they might possibly give a clue as to the age of mappemondes constructed in accordance with the instructions of Glareanus.

Owing to the circumstance that a mounted terrestrial globe is generally more easily destroyed than the sheets printed to cover its surface, especially if the latter happen to be inserted in some book or atlas; out of several globes of the 16th century, we only know the mappemonde generally printed on the projection of Glareanus. Fac-similes of such prints are here given T. XXXVII and XL.

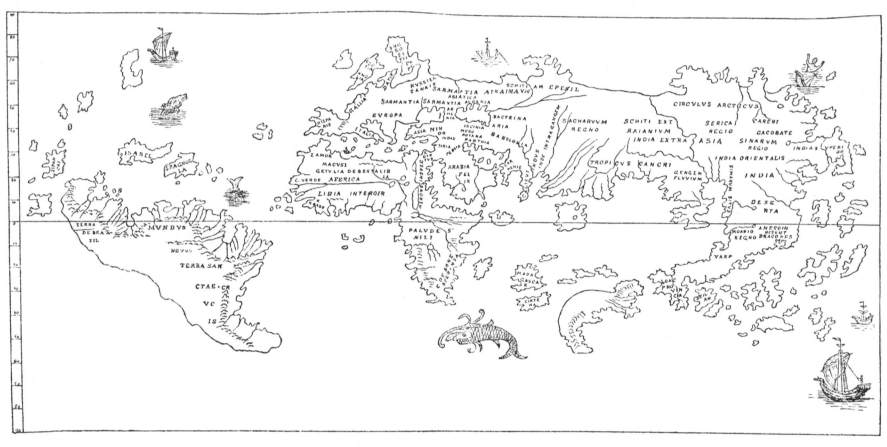

43. The globe of Lenox of the beginning of the 16th century, from B. F. DE COSTA'S drawing on an equidistant projection. (Orig. size 369 × 177 m. m.)

B. Globes and globe-prints from the beginning of the 16th century.[1]

3. *The globe of Lenox from c. 1510* (fig. 43). This small globe was found in 1855, at Paris, by Mr. RICHARD HUNT, who presented it to Mr. JAMES LENOX. It was described, by B. F. DE COSTA, in English in the *Magazine of American History*, Sept. 1879, and in French, with additions by GABRIEL GRAVIER, in the *Bulletin de la Société Normande de Géographie*, 1870. The globe forms a spherical box of copper of a diameter of only 0,m.127. There is no graduation on the map. Mr. de Costa assigns to it a date of 1508—1511, which seems to be confirmed by the general form of the continents and by several other peculiarities of the globe. The western coast of South America is here, as in other maps which were drawn before the news of Magellan's circum-

navigation had arrived in Europe, laid down not by direct observation but by estimation, and as may be concluded from the want of all inscriptions at Corte Real's land, the draughtsman has only had access to very vague reports of a continent or of larger islands to the north-west of the West Indies. The southern coasts of Asia are drawn less correctly than on the map of Ruysch and on the tabulæ novæ of Asia, inserted into the Ptolemy of 1513.

Notwithstanding all these defects and its small size, this globe, being the first post-Columbian globe at present known, is of considerable interest in the history of cartography. I, therefore, think it desirable to give here a slightly reduced fac-simile of DE COSTA'S reproduction.

---

[1] I have tried to follow the chronological order as far as possible in this catalogue, but, globes and globe-prints as a rule not being signed or dated, much uncertainty often prevails in this respect. This is for instance the case with Nos. 2, 3, and 4 in this catalogue. It is remarkable that no Italian globes are known of the beginning of the 16th century. That such have existed may be taken for granted. A detailed description of the construction of globes occurs in the beginning of the second part of Ruscelli's Ptolemaeus, Venetia 1561.

4. *A mappemonde in gores engraved in copper 1514 by* Ludovicus Boulenger (N. T. XXXVII). This map was found with two other copper-prints, of which one was signed *Artificio Ludovici Boulengier Allebie 1514*, in a copy of: *Cosmographiæ introductio cum quibusdam geometriæ ac astronomiæ principiis ad eamdem necessariis. Insuper quatuor Americi Vespucii navigationes . . . Impressa per Johannem de la Place*, s. l. et a., offered for sale by the antiquarian H. Tross at Paris in 1881. A fac-simile of it by Pilinski is given in his Cat. XIV. A copy of the book, but without the map, is described by d'Avezac in *Martin Hylacomylus Waltzemüller . . . . par un géographe bibliophile*, Paris 1867, p. 116. He supposes the book to have been printed in 1517—1518. This edition of the famous *Cosmographiæ Introductio* of Waldseemüller should not be confounded with the work of Apianus, published under the title of *Cosmographiae Introductio cum quibusdam Geometriæ ac Astronomiæ principiis ad eam rem necessariis*, and of which the first edition was printed at Ingolstadt in 1529. It is not quite certain either that the map of Boulenger originally belonged to the work in which it was found inserted, or that 1514 is really the year in which it was printed. At any rate, it is one of the earliest globe-maps, and one of the first prints, on which the name America has been applied to the New World. Otherwise this small mappemonde, with its rivers drawn in soft ornamental lines, and its oceans represented in the same manner as in the Ptolemæus Bononiæ 1462 is of no particular interest to the history of discovery.

44. The water-mark in the paper on which the globe No. 5 is printed.

5. *A mappemonde in gores, which I have found pasted on the reverse of the map of Switzerland* (No. 45) *in a copy of the Ptolemy of 1525* (N. T. XXXVII). This map was already cut up into its twelve segments destined to be mounted on a globe, whose radius was indicated by means of a broad double line on a slip of paper placed below the rest of the map. On the paper of the third segment, containing the map of the Red Sea etc., a water-mark can be discerned. I here give a fac-simile of it, as it might give some clue to the year of the printing of the map.

A fac-simile of the map and an analysis of it is given in the journal *Ymer* of 1884. Only two towns are laid down on the map, namely *S. Jacobus* (Santiago de Compostela) and *Ingolstad* (Ingolstadt). From this it may be inferred that the map was printed or drawn in the latter town, and that its author, in some way or other, had been in communication with Santiago; that, for instance, he had got his material for the map of the New World from that place. Concerning the author of the map, the name of Ingolstadt directs our thoughts to Petrus Apianus, who was professor of mathematics in the celebrated Academy at that place from 1527. Yet a comparison with other maps of the beginning of the 16th century, especially with the map of Apianus of 1520 (N. T. XXXVIII), shows that it cannot have been a work of this celebrated cosmographer. As to the outlines and inscriptions on the old hemisphere, this globe-map almost completely agrees with the map in the Ptolemy of 1511 of Bernardus Sylvanus (N. T. XXXIII), insomuch that, for its few inscriptions, a preference was given to the names printed in red by Bernardus Sylvanus. The form of South America

entirely differs from that on the map of Apianus of 1520, while it is identical with the form of South America on the globe of Schöner of 1515 (but not with that on his globe of 1520). For these and various other reasons, which space does not allow me here to explain, I have arrived at the following conclusions:

that the map under discussion, although not a work of Apianus, is printed or drawn at Ingolstadt;

that it is of a later date than 1511, the year when the original to the delineation of the Old World was printed;

that a common original has served for this map and for Schöner's globe of 1515;

that this map was probably printed before the time when the *Newe Zeytung auss Presillg Landt* (printed before 1515) and Schöner's *Luculentissima quædam terræ latius descriptio*, Nuremberg 1515, became generally known.

Further on I shall have occasion to return to these two insignificant pamphlets, so important, however, in the history of geography. Referring to the careful investigations of Humboldt, Varnhagen, Ruge, Wieser, and others, I shall here only remind the reader of the strange circumstance that, notwithstanding the *Newe Zeytung* as well as the *Luculentissima descriptio* were printed at least five years before Magellan's navigation round the southern point of America, yet it is expressly stated in these pamphlets, that the New World ends to the South with a cape surrounded by a strait, similar to that of Gibraltar. On Schöner's globe of 1515, in accordance with this, a large south-polar continent, *Brasilia regio*, is laid down to the south of the South American continent, the outlines of which are drawn with tolerable exactness. But on the globe-map here under discussion this ›terra australis‹ is wanting. The outlines of South America resemble those on Stobnicza's map of 1512. Perhaps a corrected and augmented copy of the map of Bernardus Sylvanus has served as a pattern to both. The globe-map found by me was probably drawn between 1511 and 1515.

6. *A globe from Hauslab's collection, before 1515.* This globe has been lately described by J. Luksch in *Zwei Denkmale alter Kartographie* (*Mittheil. der k. k. geograph. Gesellschaft,* Wien 1886, p. 364). It appears to have been drawn by hand on a wooden sphere covered with paste board. Its diameter is 0,ᵐ·368. So far as it is possible to judge from the imperfect copies of this globe given by Luksch, and from the globe of 1515 by Wieser, these works almost completely agree, except as regards the south polar continent and the large island drawn to the south of Java Minor on Schöner's globe, which are wanting on that of the Hauslab globe. Perhaps we here have one of the oldest of Schöner's works or, perhaps, the first prototype followed by him. F. A. de Varnhagen (Luksch, p. 370) considers the globe to have been drawn in 1513 and to have belonged to some ecclesiastical prince at Brixen.

7. *Leonardo da Vinci's globe.* A good representation of the geographical ideas prevailing in the period immediately preceding Magellan's circumnavigation of the earth, is further given by the globe-map, on a peculiar projection, found in a collection of drawings of Leonardo da Vinci and critically examined, in his usual masterly way, by R. H. Major (*Memoir on a Mappemonde by Leonardo da Vinci, being the earliest map hitherto known containing the name of America; now in the Royal collections at Windsor*, London 1865). Major, who conjectures the date of the map to be 1512—14 (Winsor considers it to be one or two years later) has tried to prove that it was actually drawn by the great artist among whose papers it was discovered. From this circumstance a certain interest is attached to this insignificant sketch, which is in no wise distinguished by such

accuracy and mastery in drawing, as might be expected from a map attributed to the great artist among whose papers it was found. It is, however, worthy of attention from a cartographical point of view, not merely on account of the remarkable projection, never before employed, but also because it is one of the first maps on which a south-polar-continent is laid down. It is likewise, if not the first, at least one of the first mappemondes with the name America.

That the da Vinci map is not an original drawing, but a careful copy of a globe, is obvious from the way the inscriptions on the northern coast of South America have been intersected without any reasonable cause, so that parts of the names are written on one, parts on the other segment. This circumstance seems likewise to make it probable, that the copy is not a work by Leonardo himself, but by some ignorant though trustworthy clerk or copyist. For the rest the map deviates considerably from all other maps of the beginning of the 16th century, with regard as well to the inscriptions

hos: cum Globis cosmographicis, sub mulcta quinquaginta florennorum Reneum et amissione omnium exemplarium.

The main interest of this work depends, as is shown by Prof. FRANZ WIESER, in his *Magalhães-Strasse und Austral-Continent auf den Globen des Johannes Schöner*, Innsbruck 1881, on the fact that it forms an explanatory text to a globe, ›*Globus noster Cosmographicus*,‹ of which three copies have lately been discovered, or rather identified, by Wieser, and partly on Schöner's mentioning in this brochure, printed seven years before the return of the »Victoria,» an otherwise unknown Portuguese exploring expedition, which reached Magellan's Strait previous to the voyage of Magellan. Schöner's words are (Tract. II cap. 11, fol. 61):

Brasiliæ regio.

A capite bonæ spei (quod Itali Capo de bona speranza vocitant) parum distat.

Circumnavigaverunt itaque Portugalienses eam regionem, et comperierunt illum transitum fere conformem nostræ

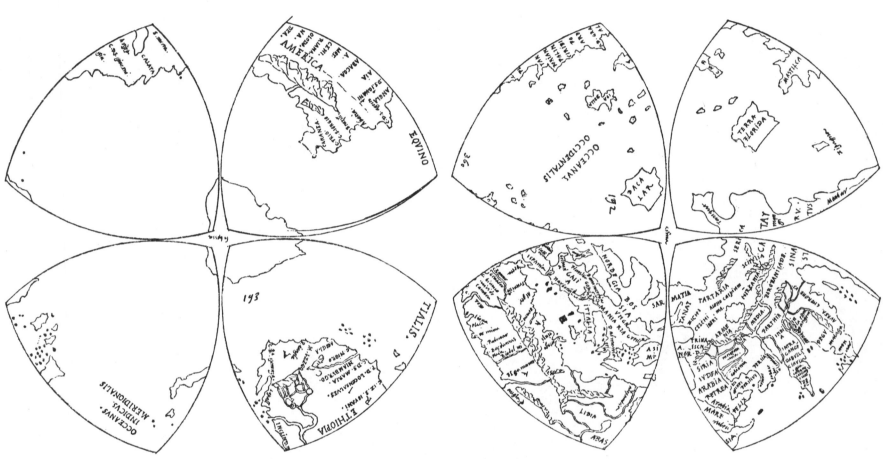

45. LEONARDO DA VINCI's mappemonde from c. 1514. (Distance from the Equator to the Pole on orig. 130 m. m.)

as to the outlines of the continents. That this map is based on Portuguese and not on Spanish originals appears to be deducible from the tolerably correct form of South Africa and from the outlines of the Indian peninsulas, which are here delineated more correctly than on the maps of RUYSCH, SYLVANUS, STOBNICZA, BORDONE, and in all editions of Ptolemy, before that of RUSCELLI of 1548.

8. *Schöner's globe of 1515* (fig. 46 and 47). In this year the celebrated mathematician and cosmographer JOHANNES SCHÖNER published a brochure of 81 quarto pages, with the long title:[1] *Luculentissima quædam terræ totius descriptio cum multis utilissimis cosmographiæ iniciis. Novaque et quam ante fuit verior Europæ nostræ formatio ... Multa etiamque diligens lector nova usuique futura inveniet.* (Colophon:) *Impressum Noribergæ in excusoria officina Ioannes Stuchssen. Anno domini 1515.* Immediately after the title an epigram is inserted, followed by: *Cum Privilegio Invictissimi Romanorum Imperatoris Maximiliani per Octo annos: ne quis imprimat aut imprimere procuret codices*

Europæ (quam nos incolimus) et lateraliter infra orientem et occidentem situm. Ex altero insuper latere etiam terra visa est, et penes caput hujus regionis circa miliaria 60, eo videlicet modo: ac si quis navigaret orientem versus, et transitum sive strictum Gibel terræ aut Sibiliæ navigaret, et Barbariam, hoc est Mauretaniam in Aphrica intueretur: ut ostendet Globus noster versus polum antarcticum.

Insuper modica est distantia ab hoc Brasiliæ regione ad Mallaquam, ubi Sanctus Thomas apostolus martyrio coronatus.

Sunt in hac regione loca montosa valde, et in quibusdam hisce locis nix toto anno nunquam dissolvitur. His in locis animalia comperiuntur plura et nobis incognita. Accolæ etiam eorum locorum pellibus animalium præciosis, nedum paratis (quia præparandi modum ignorant) se vestiunt. Ut sunt pelles Leonum, Leopardum, Castorum etc.

These lines prove that the delineation of the southern part of the New World on Schöner's globe of 1515 is founded on

[1] The title etc. of this rare work is given from HARRISSE and WIESER. I had not myself access to the original.

actual observations. Some further particulars about this voyage may be obtained from another rare German pamphlet, which, according to Sophus Ruge and Wieser, forms the source of Schöner's notices in the *Luculentissima descriptio*, and which consequently must have been printed before 1515, namely *Copia der Newen Zeytung aus Presillg Landt*. We here have the first print with the title *Zeitung*. Several editions of it are known, but all undated; two of them are printed in Augsburg. According to Wieser, a number of names and expressions show it to be a translation from a commercial

but that he had there been forced by contrary winds to return. The distance from the Straits to Malacca was said not to be very great. The day of the return of the expedition (the 12th of Oct.) is given, but unfortunately not the year, on the determination of which the Americanists have in vain exercised their learning and sagacity. Wieser supposes the expedition to have taken place before 1509.

Although these brochures were not provided with any maps, the *Terra Australis* was again introduced by them into geographical literature, as a large continent, surrounding

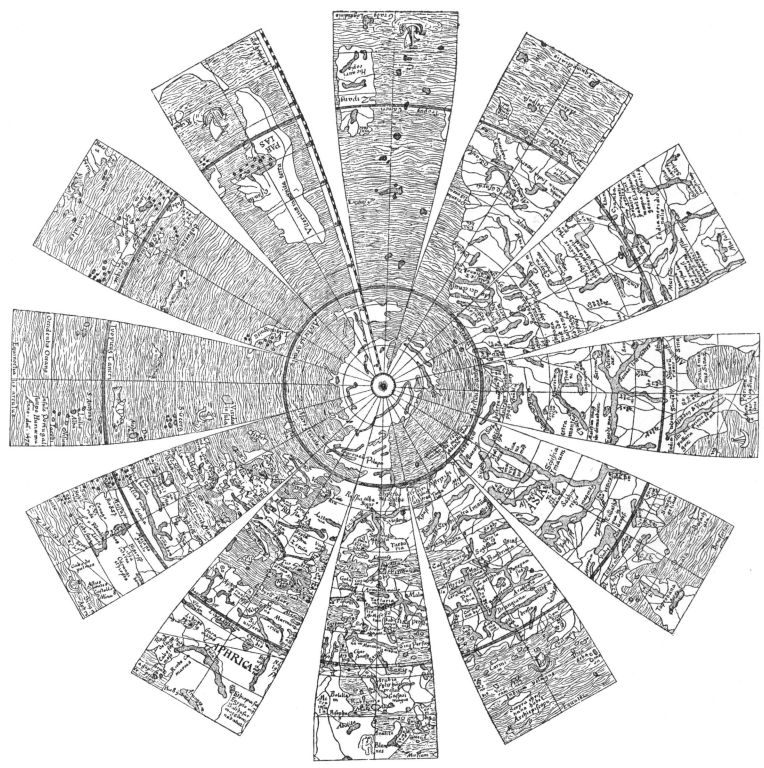

46. Schöner's globe of 1515. Northern Hemisphere. From Jomard. (Diameter of his drawing = 416 m. m.)

report, probably written by some Italian factor in Lisbon to the manager of the famous commercial-house of Welser at Augsburg. Among other news, some information is here given respecting a commercial voyage of discovery: *So dan Nono un Christoffel de Haro und andere gearmirt oder gerüst haben*. The expedition consisted of two Portuguese vessels, of which one returned during the stay of Welser's correspondent at Lisbon. He declares himself to have been a great friend of the captain *(piloto)*, who said that the expedition had sailed through a strait situated to the south of *Presill*,

the south pole and extending towards the equator far beyond the south polar circle, especially on the meridians of Australia and of the south point of America. From that time it was laid down on most mappemondes far into the 17th century, and is even met with during the 18th, if not on maps, at least in other geographical treatises, until it was definitively confined to the ice-covered regions of the South-Pole, in consequence of Cook's discoveries. Here the ancient inscription *Terra Australis Incognita* is still retained, and will for a long time to come continue to be convenient and adequate.

¹ *IV. und V. Jahresbericht des Vereins für Erdkunde zu Dresden*, 1868, p. 13—27, (with a reprint of *Die newe Zeytung*).

After the discovery of the New World, this continent was for the first time laid down on Leonardo da Vinci's globe and on Schöner's globe of 1515. Wieser has succeeded in identifying three copies still extant of this last mentioned globe, viz., one in the library at Frankfurt o. M., reproduced by JOMARD, and two others at the military library in Weimar. As for the numerous reproductions of this globe, and of the next one, I may refer the reader to WINSOR, *A Bibliography of Ptolemy's Geography*, p. 15. Unfortunately no exhaustive technical description of the globe

pressed himself as sure of success, because he had seen the straits laid down on a sea-chart by the Portuguese Martin de Bohemia, a native of the island of Fayal and a cosmographer of great reputation (GHILLANY, p. 62). The cosmographer generally designated in the history of geography by the name of Martin Behaim died in 1506. It is difficult to understand how Magellan, with reference to the voyage he wished to undertake round the New World to the Spice Islands, could have referred to a sea-chart or to a globe drawn so long before. Both must have become too antiquated

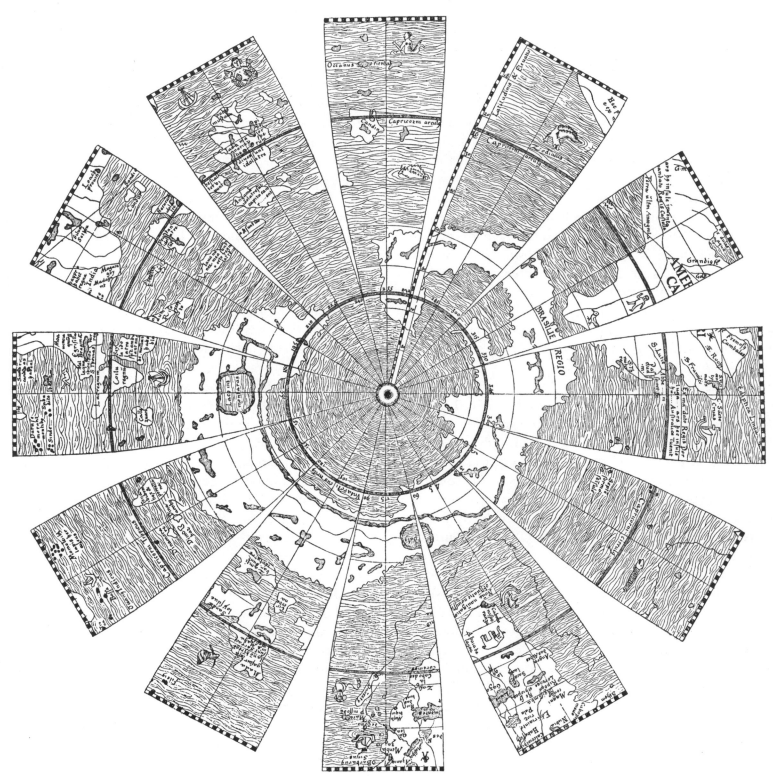

47. Schöner's globe of 1515. Southern Hemisphere. From JOMARD. (Diameter of his drawing = 419 m. m.)

is given by Wieser. He only mentions its diameter to be 0,^{m}.27 and that it is printed, not drawn by hand. I presume that it is printed on gores, similar to those of the globes on Pl. XXXVII.

In his account of the first circumnavigation of the earth PIGAFETTA says that Magellan, before his passage through the straits which now bear his name, had had access to a sea-chart by MARTIN DE BOHEMIA, and the Spanish historian HERRERA relates that Magellan had, in 1517, exhibited a globe to the Bishop of Burgos, on which the place where the straits were situated was left blank, but that he had ex-

in 1517. Neither was the separation of the Ocean into two parts by the New World likely to have been known in 1506, i. e. seven years previous to the discovery of the Pacific by Balboa, which discovery must be presupposed before a strait between those two parts of the Ocean could have been spoken of. This difficulty may perhaps be explained by assuming that MARTIN BEHAIM, the father, who was born in Nuremberg, who accompanied the expedition of Diogo Cão, and who constructed the globe at Nuremberg etc., has here been confounded with his son, who, according to Ghillany, was also called MARTIN. This Martin was actually born in Fayal. It is not at all improbable that he, following his father's example, had occupied himself with cosmographical labours and researches, and that he had registered the results of later voyages of discovery, of which nothing had been noted down on the pages of history, on his sea-charts, or on charts inherited from his father.

9. *A globe by Schöner, drawn by hand, dated 1520, and preserved at the town-library of Nuremberg.* This large globe, of which the diameter is 0,^m866, was described by MURR (*Dipl. Gesch. Behaims*, 2d edit., p. 47). It is critically examined and more or less completely copied by LELEWEL, GHILLANY, HUMBOLDT, KOHL (*Discovery of Maine*, Portland 1869, p. 158), and others. WIESER has also, in his often cited work, given a detailed account of it. Unfortunately the copies given, both by Ghillany and by Wieser, only embrace the New Hemisphere. However the discoveries in the Old Hemisphere, in the beginning of the 16th century, form an epoch of immense consequence to the development of that part of the world, which during several thousands of years previous to the discovery of the New World had been the centre of civilisation. It would be of no slight interest to get an easily accessible representation of the ideas respecting the distribution of land on the southern and eastern coasts of Asia which prevailed immediately before Magellan's circumnavigation. This globe has an advantage over those already mentioned, from its date being known. This may be deduced partly from the year 1520 being written in large golden letters at the South Polar-circle, and partly from an inscription in the vicinity of the South-pole, stating the globe to have been made at the expense of JOANNES SEYLER by Jo. SCHÖNER, *quando salutiferi partus numeravimus annos mille et quinquentos et quatuor addita lustra*.

Magellan left the port of San Lucar on the 10th of Sept. 1519. On the 11th of Oct. 1520 he arrived at the western entrance to the straits of Magellan, and after having circumnavigated the earth for the first time, his only remaining vessel anchored at San Lucar on the 27th of Aug. 1522. These dates show that Schöner's sketch of the straits to the south of continental America, on a globe made in 1520, cannot be based on discoveries during this voyage. It is evidently founded on a voyage along the eastern coast of South America, which would be perfectly unknown if it had not, as mentioned above, been occasionally referred to in some insignificant geographical pamphlets. For further details, the reader may consult WIESER's paper. He considers the delineation of the New World on the earlier of Schöner's globes to be founded on the *Tabula Terræ Novæ* in the Ptolemaeus 1513, on Stobnicza's map and on the map in the Margarita of Reisch of 1515 (N. T. XXXIV, XXXVI and XXXVIII). But this assumption seems to be inadmissible, because there are such essential differences between these documents and Schöner's globe, as that of North and South America being, on the maps, connected by an isthmus, while on the globes this isthmus is replaced by a strait. It seems to me more likely that the globe was based on the maps of Ruysch and Bernardus Sylvanus (N. T. XXXII and XXXIII). But, in the first instance, its authority was derived from some charts now lost, on which the delineation of the South American continent deviated from all other maps. The importance of this deviation to the commerce of the world may have been the immediate cause of the publication of the numerous globe prints in central Europe during the second decennary of the 16th century.

10. *A globe of Schöner in 1523*, mentioned in: *De nuper sub Castiliæ ac Portugaliæ Regibus Serenissimis repertis Insulis ac Regionibus, Ioannis Schöner Charolipolitani epistola et Globus Geographicus, seriem navigationum*

annotantibus. Clarissimo atque disertissimo viro Domino Reymero de Streytpergk, ecclesiæ Babenbergensis Canonico dicatæ. (Colophon:) *Timiripæ*[1] *Anno Incarnationis dominicae Millesimo quingentesimo vigesimotertio.* This brochure is now very rare, but it is reprinted in the above mentioned work of Wieser, and was published as a separate pamphlet by F. AD. DE VARNHAGEN at St. Petersburg in 1872. A fac-simile of the original, with an English translation, is inserted in the newly published elaborate monograph: *Johann Schöner Professor of Mathematics at Nuremberg. A reproduction of his Globe of 1523 long lost etc.* by HENRY STEVENS of Vermont, edited with an introduction and Bibliography by C. H. COOTE, London 1888. Schöner's brochure, the main part of which consists of an eulogy of the discoveries of Vasco da Gama, Columbus, and Magellan, often with incorrectly written names and erroneous data, is of importance to the history of cartography, because it forms a dedicatory letter to a globe presented by Schöner to the Bishop of Bamberg, and expressly constructed to illustrate the important letter of MAXIMILIAN TRANSYLVANUS, *De Moluccis insulis ad reverendissimum Cardinalem Salzburgensem.* A fac-simile and an English translation are given in the above mentioned work of Henry Stevens and Coote. Some information as to the source of Schöner's knowledge of modern geography is obtained, by the following words at the end of his letter: *Globum hunc in orbis modum effingere studui, exemplar haud fallibile aemulatus, quod Hispaniarum solertia cuidam viro honore conspicuo transmisit* (WIESER, p. 118). Schöner's globe of 1523 has been, but so far as I can see without sufficient reason, identified by Henry Stevens with the mappemonde in gores which was offered for sale in the catalogue XLII no. 136 of Rosenthal, and which I have reproduced on the pl. XL. It is obvious that the first circumnavigation of the earth was the immediate cause as well of the last mentioned globe-print, as of Schöner's letter and globe of 1523. But several circumstances militate against the identification of the mappemonde in gores sold by Rosenthal, with the globe made for the Bishop of Bamberg. Among the European towns engraved on the former we find the names of Nuremberg, Venice, and Constantinople, but not of Bamberg, which would have been little complimentary to the Bishop and contrary to the custom of the period, if the work had been intended for him. The outlines of the New World given on it differ completely from the geographical ideas of Schöner, and are copied from maps of Battista Agnese of the fourth decennary of the 16th century. From this and on other grounds mentioned below (p. 82), I conclude that this globe has nothing to do with Schöner's globe of 1523.

11. *Schöner's globe of 1533.* This globe is mentioned in *Iohannis Schoneri Carolostadii Opusculum Geographicum ex diversorum libris ac cartis summa cura et diligentia collectum, accomodatum ad recenter elaboratum ab eodem globum descriptionis terrenæ*, s. l. a. The dedication is dated *Ex urbe Norica 1533*. Besides a couple of worthless geographical drawings of the Old Hemisphere, the work contains the figure of a mounted globe reproduced in fig. 49. The globe alluded to on the title-page of this brochure was considered as lost, until a copy of it was lately discovered by Wieser in the military library at Weimar. Unfortunately it does not appear from Wieser's description, whether the map covering is, as I suppose, printed in gores on the projection of Glareanus, or drawn by hand, nor does he indicate the size of the globe. Wieser only gives a representation of the southern hemisphere on a reduced scale, and he points out the resemblance of the land-outlines on the globe to the heartshaped map

---

[1] Wieser supposed this name to be that of some place in the vicinity of Bamberg; Varnhagen that it signifies Erfurt (WIESER, p. 122). Coote gives good reasons for identifying it with Kirch-Ehrenbach, a place where Schöner performed the functions of a parochial vicar (above cited work, Introduction XLI).

48. Mappemonde of the middle of the 16th century in gores by Antonius Florianus, from »Lafreri's atlas.» (Orig. size 462 × 835 m. m.).

of Orontius Finæus of 1531 (N. T. XLI), on which North America forms a continuation of eastern Asia. In this respect this globe differs so entirely from the previous works of Schöner, that it would be doubtful whether it was really his, if various passages in the text did not prove that Schöner himself had altered his opinion. Chap. XX, Schöner says: *Post Ptolemæum vero ultra 180. gradum versus orientem multae regiones repertae per quendam Marcum Polum Venetum, ac alios, sed nunc a Columbo Genuensi & Americo Vesputio solum loca littoralia ex Hispaniis per Oceanum occidentalem illuc applicantes lustratae sunt, eam partem terrae insulam existimantes vocarunt Americam, quartam orbis partem.*

12. *An unsigned globe, probably made in Nuremberg about 1540.* The large mappemonde in gores given on pl. XL, and erroneously supposed to be the work of Schöner in 1523, may, from the general character of the map, with great probability be referred to this place and time. That it was printed in Nuremberg seems to be proved from the fact that *Nuremberga* is, with *Venetia* and *Constantin(opolis)*, the only name of a European town inscribed on it. The projection is that of Glareanus, although drawn so that the distance from the Equator to the Pole on the gores is exactly what it should be, i. e. 3 × the breadth of the segment at the Equator, and not as on the constructions of Glareanus 3,123 × that distance. There is, as above mentioned, no reason for ascribing this globe to Schöner and to the year 1523. The form of America, the Australian continent, and Asia altogether deviate, not only from the authentic works of this celebrated mathematician, but also from all other printed maps of the 2d and 3d decennaries of the 16th century. On the other hand, the outlines of America agree with the drawing on the earlier of the many portolanos published by Battista Agnese between 1536 and 1564. Such a map of Agnese is reproduced in Kohl's *Discovery of Maine*, p. 292, from two originals, of which one from Dresden is unsigned, and the other, belonging to the British Museum, signed *Bapt. Agnese Venetiis 1536*.[1] Moreover, the main form of the continents on this globe fairly corresponds with Mercator's double cordiform map of 1538 (N. T. XLIII) and with the *Novæ insulæ nova tabula* in Münster's Ptolemæus of 1540. The numerous inscriptions and drawings of monsters characteristic of the globes of Schöner, are wanting here. Instead of the name of *Madagascar* used by Schöner we here read that of *San Lorenzo*. But on the eastern coast of Africa, and on the western coast of India a greater number of names of towns etc. are given. All this seems to me to prove:

1st. That Schöner is not the author of this mappemonde in gores;

2d. That it is not a work of the 3d, but of the 4th or 5th decennary of the 16th century;

3d. That it was made in Nuremberg;

4th. That it may, in certain respects, be regarded as a copy on a sphere of the portolanos from Venice by Battista Agnese (about 1536);

5th. That it is, like most of the maps of the first part of the 16th century, based on information from Portugal, and not from Spain;

6th. That the globe seems to be the work of Georg Hartmann, a celebrated manufacturer of globes and cosmographical instruments in Nuremberg. He was born in 1489. In his youth he spent several years in Italy, probably in the Italian town of Venice, which is marked on the globe. In 1518 he settled in Nuremberg, where he died 1569. (Doppelmayer, *Hist. Nachr.*, p. 22 & 56.)

13. Mercator's *large globe of 1541.* This important »Monument de géographie» was also looked upon as lost, when the Royal Library at Brussels, at the sale of the effects of M. Benoni-Verelst in Ghent, succeeded in acquiring the engraved gores belonging both to it and to Mercator's celestial globe.[2] These have since been published in photolithographic fac-similes in: *Sphère terrestre et sphère céleste de Gérard Mercator, de Rupelmonde, éditées à Louvain en 1541 et 1551, édition nouvelle de 1875 d'après l'original appartenant à la Bibliothèque royale de Belgique*, Brussels 1875, preceded by a brief preface by M. Malou. A more detailed description and analysis of this important globe-print may be found in J. van Raemdonck: *Les sphères terrestre et céleste de Gerhard Mercator* (*Annales du Cercle Archéologique du Pays de Waes*, St. Nicolas 1875, V, p. 281). The map consists of twelve segments or gores intended to cover a globe 1,^{m}·29 in circumference. On the ninth segment we read: *Edebat Gerardus Mercator Rupelmundanus cum privilegio ces: Majestatis ad an. sex. Lovani an. 1541*, and on the seventh there is a dedication to Nicolaus Perrenotus Granvella. The celestial globe is dated 1551. The terrestrial globe is very rich in inscriptions and other geographical details, which are, in many respects, of great interest. The Scandinavian peninsula and the Baltic are, for instance, here for the first time tolerably correctly represented on a globe, through the guidance of the large map of Olaus Magnus published only two years previously in Venice, from which Mercator has also copied some of the marine monsters represented on his globe. The river-system of Russia is drawn from new sources. The outlines and rivers of Africa are drawn with less accuracy. Madagascar has received much too large an extension. Asia extends too far towards the east, but is separated from America by a strait, in which a walrus (from the drawing of Olaus Magnus) is swimming. America has almost the same form as on the largest of the globes reproduced on pl. XL. Round the South Pole an immense continent is laid down, for the existence of which Marco Polo is referred to. When Mercator's globe was published, it was, without comparison, the most complete work of its kind. Unfortunately its size does not permit me to give a satisfactory reproduction of it here.

14. *The Nancy globe of the middle of the 16th century.* This globe, presented by Charles V of Lorraine to a church in Nancy, where it was used as a wafer-box (ciboire), is now preserved in the city library of that place. It is a globe of silver gilt engraved with admirable skill and having a diameter of 0,^{m}·16. It was described for the first time by M. Blau in *Mémoires de la Société Royale de Nancy*, 1835, p. 97. For later publications of it, the reader is directed to Winsor, *Critical History*, III, p. 214. A comparison between the reproduction of it in Blau's memoire and Vopel's globe, preserved in Copenhagen (N. T. XL), will show that there exists a great resemblance between these two small globes.

15. *De Bure's globe* is nearly related to Schöner's globe of 1515. It is of copper gilt, formerly belonging to the brothers de Bure and now preserved in the Bibliothèque Nationale in Paris. As I have not seen either the original or a reproduction of it, I must refer the reader to the above mentioned paper of J. van Raemdonck, p. 281, and to Winsor, *Critical History*, III, p. 214.

---

[1] For further details regarding Agnese's maps the reader is referred to Harrisse, *Cabot*, p. 188, and Wieser, *Der Portulan des Infanten und nachmaligen Königs Philipp II von Spanien* (*Sitz.-ber. der k. Ak. d. Wiss. Philos.-hist. Cl.*, Vol. 82, Wien 1876, p. 541).

[2] According to a private communication by M. Ruelens, two more copies of these globes, already glued to their balls, have since been found. One of them was purchased by the Vienna Library, and the other by »la Société Archéologique du Pays de Waes à St. Nicolas» (Belgium). The Brussels edition of 1875 was only printed in 200 copies.

16. *Vopel's globe of 1543*, preserved at the *Oldnordiske Museum* in Copenhagen. According to Mr. KRISTIAN BAHNSON, this little globe is mounted as an armillary sphere surrounded by 11 brass-rings corresponding to the Equator, the Tropics, the Ecliptic, etc. On the brass-ring around the northern Tropic there is engraved: *Casper Vopell Medebach hanc sphaeram faciebat Colonie 1543*. On a paper fixed on the bottom of the case in which the globe is locked, there is written in a more modern hand:

Nicolaus Copernicus

1543

— ty — Brah (= Tycho Brahe?).

COPERNICUS died in May 1543; TYCHO BRAHE was born in 1546. The globe is supposed to have belonged to the last mentioned great astronomer. Its small size has, of course, not admitted of the insertion on it of any extensive geographical details. As to the general outlines of the continents, it makes North America the eastern part of Asia. This was, as is known, the opinion of Columbus, and the first printed map of the New World (Ruysch's map of 1508) is drawn in accordance with it. On the heartshaped map of Orontius Finæus of 1531 (N. T. XLI) we find this theory fully developed, and it is still adhered to on the map of the world in MYRITIUS, *Opusculum geographicum*, Ingolstadii, 1590. Yet it was never generally adopted. It was, for instance, severely censured by POSTELL in a letter of the 9th of April 1569 to ORTELIUS (ABRAHAMI ORTELII *Epistulae*, Cantabrigiæ 1884, p. 43).

Besides the globes here mentioned, there are various others of which I have neither seen the originals, copies nor satisfactory descriptions, and which I am consequently only able to enumerate. To this category belong:

A copper-globe made in Venice and signed EUPHROSYNUS ULPIUS 1542. This globe was found by BUCKINGHAM SMITH in Spain and now belongs to the New York Historical Society (HARRISSE, *Notes sur la Nouvelle France*, Paris 1872, p. 222; WINSOR, *Critical History*, III, p. 214).

A globe of 1524(?) (D'AVEZAC, *Bull. de la Société de Géographie*, 1860, p. 398; RAEMDONCK, *Les sphères de Mercator*, p. 28).

A globe made by Honter in the year 1542 (WINSOR, *Bibliogr. of Ptolemy's Geography*, p. 28).

As attention has been more generally drawn to the importance of these geographical documents, several other globes, as well printed as drawn by hand or engraved on metal, will doubtless soon be discovered. With regard to the latter class of globes it should, however, not be forgotten that they have often been erroneously looked upon as scientific documents worthy of critical examination by students of the history of geography. They are generally nothing but hand-specimens of goldsmiths' work, on which the map has merely been treated from an artistic or ornamental, and not from a geographical point of view. A nicely executed gilt globe, belonging to the Swedish regalia, and made in 1561 for the coronation of Erik XIV, the son of Gustavus Vasa, is an example of such work.

———

The terrestrial globes preserved from the 15th and the first half of the 16th century may, on the grounds indicated above, be divided into the following groups:

1. Globes made without any knowledge of the new world — Behaim's globe and the Laon globe.

2. Globes made from 1492 to 1515, i. e. after the discovery of the New World, but before the existence of a large South Polar-continent was admitted. To this group belong the globe of Lenox, that of Boulenger-Tross of 1514, the mappemonde in gores found by me, and the globe in Hauslab's collection mentioned under No. 6.

3. Globes made from 1515 to 1523, i. e. before the results of Magellan's voyage were known, but after the introduction or re-introduction of the South Polar-continent on the mappemondes. To these belong Schöner's globes of 1515 and 1520, and the globe attributed to Leonardo da Vinci.

4. Globes made after Magellan's circumnavigation of the earth, but whilst the northern portion of the New World was still believed to have an inconsiderable extension, or was

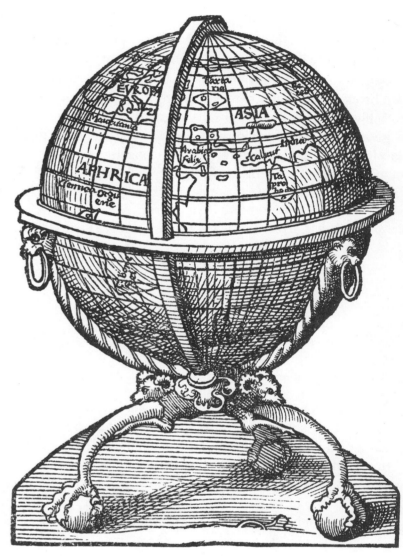

49. Drawing of a globe from SCHÖNER's *Opusculum geographicum*, Norimbergæ 1533.

considered as a complex mass of large and small islands — Schöner's globe of 1523 (?).

5. Globes on which North America is a continuation of Asia. The globe at Weimar, which has been identified by Wieser with Schöner's globe of 1533, Vopel's globe of 1543, and the Nancy globe, probably dating from about the same time.

6. Globes on which the Isthmus of Panama is laid down, and on which the northern part of the New World has a greater extension; North America is separated from Asia by a narrow strait drawn from the Gulf of California across Hudson Bay to Davis Strait. The globe No. 12 made at Nuremberg in about 1540, Mercator's globe of 1541, Demongenet's globe of 1552, and Floriani's mappemonde (fig. 48) belong to this group.

VIII.

Map-projections.

Before proceeding further it will be convenient to give a synopsis of the map-projections employed before the end of the 16th century. In doing this, I make use of the word projection in the same extensive sense as that which it has generally obtained. I signify by it not only delineations of maps obtained according to the principles of perspective or by strictly following certain mathematical rules, but generally every method of representing the earth's spherical surface on a plane for geographical purposes, even when the drawing is altogether dependent on conventional rules. It cannot, however, be expected that a detailed description of different kinds of projections or a mathematical development of the theme should be given here. On this subject I would direct the geographer who has no time or opportunity to consult the voluminous original literature on this subject, to: d'Avezac, *Coup d'oeil historique sur la projection de cartes de géographie* (*Bullet. de la Société de Géographie*, Paris 1863), A. Germain, *Traité des projections des cartes géographiques*, Paris 1868, H. Gret-

schel, *Lehrbuch der Karten-Projectionen*, Weimar 1873, and Matteo Fiorini, *Projezioni delle carte geografiche*, Bologna 1881. The last mentioned work, to which an atlas of the different methods of projection is added, is exhaustive, and is founded on an extensive knowledge of the old literature. I will here only, with constant references to the maps, try to give a chronological review of the projections which, until 1600, have been not only proposed, but actually used. Without such a review it would be difficult duly to understand and appreciate the work of many a distinguished early cartographer. In addition to this, the rich collection of old printed maps at my disposal has given me an opportunity of correcting several errors in modern literature on this subject.

If the period, when a projection was first employed for construction of a geographical map, is taken as a basis for classification, the projections of maps may be divided into the following groups:

A. Projections used before the beginning of the 15th century.

1. *Paratopical* maps. By this name, derived from παρά and τόπος, I indicate the first incomplete attempts to give a geographical picture of the earth or of some part of it, undertaken without any idea of its globular form, without any certain projection, any graduation or any system of loxodromes. Such were probably all the maps drawn before the time of Hipparchus and Eratosthenes, as well as all maps, based neither on the principles of Ptolemy, nor on sailing directions. Maps of this kind were only exceptionally published in print, as for instance the two maps in the *Rudimentum Novitiorum*, Lübeck 1475, of which the first reproduced by me on fig. 2, is the only paratopical map, which, as far as I know, has been published to serve as a general map of the world; the second (fig. 3), is a map of Palestine, scarcely more successful. A map of this kind is also inserted into the *Sanctarum peregrinationum in montem Syon ... opusculum*, by Bernardus de Breydenbach. It embraces (in the edition *per Petrum drach civem Spirensem impressum* MDCCCCXC) Palestine with surrounding lands, from Lebanon to Mecca and Alexandria. The same work also contains other geographical drawings, forming a mean between harbour sketches and views of towns.

2. *Sea-charts or portolanos of the Middle Ages.* Even in these maps the usual geographical coordinates are wanting. Yet the reciprocal positions of the harbour-places have, at least for the main part of the map, been determined with great care by means of the distances and the azimuths of the course-lines in sailing from one place to another. As I have previously (p. 43—51) given a more detailed account of these master-pieces in cartography, it is unnecessary here to return to this subject. The importance of the matter may, however, be an excuse, if I again point out that portolanos, and probably very excellent portolanos, had evidently been already drawn before the use of the compass became general among the navigators in the Mediterranean and Black Seas; that compass-

lines had consequently not constituted any original and characteristic feature of these charts, and that the custom of employing loxodrome-lines on maps had probably been introduced independently of the employment of the compass in navigation. It is even possible that these maps had originally been graduated, though the meridians and parallels were gradually exchanged for course-lines or loxodromes, which, at a time when the cosmographic ideas, even of learned men, were very confused and obscure, were not only of far more practical importance to merchants and shipowners than the meridians and parallel lines, but also greatly facilitated the copying of the maps. However, the portolano-draughtsman Grazioso Benincasa of Ancona, as early as the middle of the 15th century, began to introduce — or to restore (?) — geographical graduation (Fiorini, cited work, p. 353). But even if the portolanos had originally been founded on graduated maps, resembling those of Ptolemy, and constructed with due regard to the globular form of the earth, it is probable that the medieval *constructors* of such maps generally concurred with the most learned ecclesiastics in considering this doctrine as a dangerous heretical error. Nearly all the portolanos still extant are therefore nothing but plane maps (*cartes plates*, d'Avezac), only differing from the maps in the preceding group through the rich material at the disposal of the draughtsmen, and, above all, through the care with which this material was arranged and employed by generations of them.

When the navigators of Southern Europe, in the 15th century, had extended their voyages to the other side of the Equator and crossed the Atlantic, there was no longer any possibility, either for map-draughtsmen or navigators, to maintain the old idea of regarding the earth as a flat disk. In spite of this, and although far greater difficulties presented themselves than on the Mediterranean portolanos, in getting distant harbours and towns laid down in their proper places, the old medieval maps also served as models for the first charts of the whole Atlantic ocean. This could only be effected by sacrificing the correctness of the distances and azimuths, and of the longitudes, whose determination was very difficult before the invention of reliable chronometers. The first printed charts, »*Hydrographiæ sive chartæ Marinæ*,» thus form maps intermediate between true portolanos and maps having a rectangular net of graduation. How obscure the ideas even of very distinguished map-draughtsmen of this period were regarding the mathematical principles of their work, is shown from the circumstance that a common scale was given even for such maps, embracing a large extension in north and south of the earth's surface, e. g. on the Tabulæ novæ in Ptolemaeus 1513, here reproduced on T. XXXV and XXXVI, on the map of the world in Reisch's Margarita of 1515 (N. T. XXXVIII), and on the small map in Medina's *Arte de Navegar* of the middle of the 16th century.

3. *Zone-maps.* These maps form a transition from para-topical maps to maps intersected by a complete net of gra-duation. Here the meridians are altogether wanting, while the parallels, or at least the lines separating the climates, are indicated. The map of the earth, or rather of the old hemi-sphere, is placed within a circle, inside the circumference of which a space is generally left for the currents of the Ocean. Many an ancient map, now lost, might have belonged to this group, and likewise the majority of medieval maps, which neither directly nor indirectly are based on the works of Pto-lemy or of Marinus of Tyre. Generally these maps only form roughly designed drawings of the earth, void of details. They are found in geographical compendiums printed during the 15th century and in uncritical copies of them of a later period. A few such maps are here reproduced on pl. XXXI from the works of Macrobius, Escuidus, and Sacrobosco. To this category also belongs the map in Pierre d'Ailly's *Ymago Mundi* (fig. 19), drawn in about 1410, perhaps after an original by Roger Bacon of the 13th century, and published in about 1483. As stated on the legend at the top of the map, it was not finished, »*quia particularior distinctio majorem figuram requirit.*»

Fig. 20 shows a still ruder drawing, illustrating the rela-tive positions of Europe, Asia, and Africa often met with in medieval manuscripts and in prints of the 15th century.

I know of no zone-map on which any part of the new world was laid down. But the discoveries of the Portu-guese receive attention on the zone-maps in *Aristotelis Meteorologia*, Norimbergæ 1512 (N. T. XXXI), and in *La Salade nouvellement imprimée*, Paris 1522. The latter map (fig. 18) is evidently a copy of an original of the 15th century.

4. *Maps on the projection of Marinus* (Ptolemy's equi-distant-rectangular, or equidistant-cylindrical projection). For the sake of brevity I shall name this projection after Marinus of Tyre, who, according to Ptolemy, used it for his charts, but I suppose that it had already been employed by earlier unknown cartographers. The meridians and parallels are here equidistant straight lines, forming right angles to each other, and so drawn, that the proper ratio between the degrees of latitude and longitude are maintained on the map's mean or main parallel. When the Equator is selected for this purpose, the net of graduation becomes quadratic. The 26 spe-cial maps in all older manuscripts of Ptolemy are drawn on this projection, and likewise the Tabulæ Novæ added to the Latin translation of Jacobus Angelus, for instance the map of Claudius Clavus of 1427 (fig. 27), and the modern maps of the manuscript in Brussels (N. p. 56). When these Tabulæ

Novæ were printed, the projection of Marinus was sometimes maintained, even in editions in which Ptolemy's own maps had been reduced to the Donis projection (e. g. *Germania Nova* in edit. 1507, N. fig. 13). The projection of Marinus is also used for most of the maps in the first atlas of the New World, Wytfliet's *Descriptionis Ptolemaicæ Augmentum*, from the end of the 16th century, as well as for maps of minor territories of a much later period. Many charts, embracing extensive parts of the oceans, were drawn on this projection even in the last century, notwithstanding its unfitness for such a purpose, e. g. Waghe-naer's charts of the end of the 16th century; *De Lichtende Columne ofte Zee-Spiegel*, published by Jan Jansz. in 1653 and by Pieter Goos in 1658; most of the sea-charts in the famous marine-atlas of van Keulen of the end of the 17th and the beginning of the 18th century; and even many of the charts in Renard-Ottens' atlas of 1745. But in printed editions of Ptolemy this projection was only used for his own maps in Berlinghieri's versified Italian translation edited at Florence in about 1478, and for the map of Taprobane in Ptolem. 1478, 1490, 1507, and 1508.

5. *Maps on a conical projection.* Notwithstanding that Ptolemy, in his geography, mentions several different me-thods of delineating the earth's spherical surface on a plane, he only uses two projections for his maps, viz. the conical projection and the projection of Marinus; the former for the map of the world, the latter for the special maps. In his general map the northern hemisphere is projected on a cone, touching the earth's surface about the parallel of Rhodes, and with a height so calculated, that the proper ratio between the degrees of longitude at the Equator and at Thule (Lat. 63° N.) is maintained. In order to lay down on the same map the »oikumene» of the southern hemisphere, the projection is here modified in a manner explained in chap. XXIV of the first book (comp. p. 5) and which, without further description will be easily understood by a glance at the map (N. T. I). Also on Ruysch's map of the world (N. T. XXXII) the nor-thern hemisphere and the southern one as far as lat. 37° S. are laid down on a cone with the Equator as the base and its fourth part as the lateral height. The vertical height of the cone thus becomes 1,2113 × the radius of the base, and, no regard being given to the polar compression, it intersects the surface of the earth at the Equator and at Lat. 79° 5'. It is accord-ingly the first instance of an intersecting conical projection. But a glance at the map will suffice to show that the southern hemisphere is thrown considerably out of shape by such an attempt to project larger parts of both hemispheres on the surface of a single cone. This projection is evidently not at all adapted for such maps, and Ruysch's map is the single instance I know of employing an unmodified conical projection for a general map of the world. For special maps, on the contrary, the conical projection has often been used and still continues to be used in mapping more or less exten-sive parts either of the northern or of the southern hemi-sphere. This was already done in the edition »Bonnoniæ 1462» of Ptolemy's geography, without, however, the name of the cosmo-grapher being given, who introduced the remarkable modification in the method of projecting the earth's surface on a plane. But it is probable that the passage in the colophon about the share of the »accomplished astrologers» Hieronimus Manfredus and Petrus Bonus, in the redaction of the work, is to be referred to the revision of the maps. For as far as I know the co-nical projection was never used on the special maps of any of the Greek or older Latin manuscripts,[1] and if I except the map of the world in Ptolemy's geography, for which, as I

[1] Concerning a supposed codex to the Bologna edition, compare p. 12.

have before mentioned, a kind of conical projection was generally adopted, it was long before the example of the editors of the Bologna edition was followed, as may be seen by the following enumeration of the maps printed during the 15th and 16th centuries on a conical projection:

»1462» (1472): The maps in the above mentioned Bologna edition of Ptolemy's geography.

1478, 1490, 1507, 1508: The map of the world in editions of Ptolemy's geography for these years. The same projection was also used for the maps of the world in Pomponius Mela of 1482 (N. T. XXXI), and in the different editions of Schedel's *Liber chronicarum* 1493—1500 (N. p. 38).

1507 and 1508: Ruysch's map of the world (N. T. XXXII).

1558: The above (p. 57) mentioned Zeno map in the work published by Marcolini in Venice in that year (N. fig. 29). We have here a portolano of the northern countries graduated on the conical projection. The net, however, as a conical net of graduation, is not quite correct.

1561, 1562, 1564, 1574: The Zeno map slightly modified, printed from the same plate in the editions of Ptolemy of these years.

1596, 1598, 1621: Reprint of the last mentioned modification of the Zeno map in the editions of Ptolemy of these years.

1564 (?): »*Anglia, Scotia et Hibernia*,» »*Suecia et Norvegia cum confiniis*,» »*Russia cum confiniis*,» by Gerard Mercator, in Rumold Mercator's atlas of 1595. These maps are drawn on a slightly modified conical projection. The

cone, on which the map of *Anglia* is developed, intersects the surface of the earth at Lat. 50° and 60°. Corresponding numbers for the map of *Suecia et Norvegia* are 60° and 70°, and for *Russia* 50° and 65°. Mercator's net of graduation is here almost identical with that employed by De l'Isle for the map of Russia of 1745.

According to d'Avezac (*Coup d'oeil* etc., p. 61) the first printed map on an *intersecting conical* projection should be Mercator's map of Europe of 1554, but this is not correct, Ruysch's map having been, as mentioned above, already drawn on that projection. It is even possible that G. Mercator's celebrated map of Europe, which is now lost, was not delineated on the projection supposed by d'Avezac, but on the same projection as: *Europa ad Magnæ Europæ Gerardi Mercatoris P. imitationem edita*. Exact measurements on this map show that it is drawn on a modification of Werner's 2d projection (or a pseudo-conical projection) obtained by making the parallels describe circles with the pole as centre; the meridians straight lines, converging at a point beyond the pole, and preserving at 40° and 60° the proper ratio between the degrees of longitude at these parallels and the degrees of latitude. Such a modification, though less manifest, also occurs on the above enumerated maps of Mercator, who strictly speaking never constructed maps on a true conical projection. Gastaldi, as far as I know, never used conical projections: neither did Ortelius, at least not in his first edition of *Theatrum Orbis terrarum*.

B. Projections of maps introduced during the 15th century.

6. *The projection of Donis* (*Projection trapeziforme:* d'Avezac). This projection, characterized by equidistant, rectilinear parallels and rectilinear meridians converging towards the poles, was first employed for manuscript maps by Dominus Nicolaus Germanus, commonly, though erroneously, called Donis. For want of another name common to different languages, I have distinguished this projection by giving it the name by which its inventor (?) is generally designated. That it actually was »Donis» who first employed this important improvement of the original rectangular projection, seems to be proved by the above (p. 14) extract from his dedication to pope Paul II in the edition printed at Ulm in 1482.[1] Yet this projection had already been used in the Rome edition of 1478 for the special maps (with the exception of the last map of *Taprobane*). As the work of Donis, before being printed, existed in numerous manuscript copies, I conjecture that a manuscript of Donis was even used for the edition of Schweinheim-Buckinck. The projection was much liked, and it has since been used, not only for most maps in the subsequent editions of Ptolemy, but also in several other atlases, e. g. for many of the maps in the *Theatrum Orbis Terrarum* of Ortelius, and in Mercator's *Atlas*, as well as for quite modern maps of small districts. The Ptolemaic maps reproduced in fac-simile by me on pl. II—XXVI; the map of Scandinavia of which I have given a copy on pl. XXX; the Donis map of Scandinavia printed Ulmæ 1482 and 1486; the first map of Scandinavia engraved in copper (N. fig. 14); the map of South America (»Peru») by Forlani of 1566 (N. fig. 80); and the map of Africa by Ortelius of 1570 may be cited as examples of maps drawn on this projection. A special modification of the Donis projection was used on the

fine copper-printed maps of Africa IX, X, XI, and XII in Livio Sanuto's *Geographia*, Vinegia 1588, one of which is here reproduced on fig. 50.

7. *The homeother-projection of Ptolemy.* At the end of the first book, Ptolemy further develops the principles of a homeother projection, on which the true proportion between the areas is maintained. According to Ptolemy's rules for the construction of his net of graduation (N. p. 5), the meridians should be drawn so that the parts of the parallel-circles situated between two meridians always preserve the exact ratio to the scale of latitudes. Ptolemy, however, looked upon it as sufficient for practical purposes to maintain this ratio at the equator (e), and at the parallels through Thule (t), Syene (s) and Meroe (m). These quantities are estimated thus:

$$e: t: s: m: = 5: \quad 2\tfrac{1}{4} \quad : \quad 4\tfrac{1}{2}\tfrac{1}{12}: \quad 4\tfrac{1}{2}\tfrac{1}{3}$$
$$\text{or } 1: \quad 0,45 \quad : \quad 0,902: \quad 0,967$$
$$\text{should be } 1: \quad 0,454: \quad 0,915: \quad 0,959.$$

The agreement is here as complete as the system adopted in Ptolemy's geography for designating smaller parts of degrees admits of.

In the old codices this mode of constructing maps seems never to have been employed. But it is used for maps of the world in the editions for which Donis (N. T. XXIX), Münster, Porro, and Keschedt have constructed or engraved plates (comp. p. 7). It was also used for the map in Reisch's Margarita of 1503 (N. T. XXXI), for the maps in several later editions of this work, for the map in the *Cosmographia Pii Papæ*, Parrhisiis 1509 (see p. 40), and also, although in a modified form, for the maps mentioned below under Nos. 8—11.

---

[1] In the *Coup d'oeil sur la projection des cartes,* Paris 1863, p. 43, d'Avezac mentions a Ptolemy codex (No. 1401 in the *Bibliothèque Nationale* at Paris) »qui est réputé du XIV:me siècle,» and in which most maps are drawn on the projection of Donis. If there is no error either in the age or in the projection ascribed to these maps, the statement of Dominus Nicolaus Germanus, that he was the inventor of the new projection, cannot be correct.

C. Projections introduced during the first half of the 16th century.

8. *Stobnicza's homeother projection.* This is but a variety of the preceding one, characterized by a homeother-map of the New World being added to the old homeother map of Ptolemy's Oikumene. The general map of the earth's surface is here for the first time divided into two hemispheres, one of which embraces the Old World from 0° to 180° of longitude east of *Insulæ Fortunatæ,* and the other the newly discovered world between 0° and 180° west of the same departure. The new hemisphere comprises the greater part

formerly so much esteemed; even less than on maps of the world drawn on a stereographic projection, or on the cylindrical projections now generally used. Yet the example was not followed. The only early map constructed in this manner is the map in the edition of 1512 of Stobnicza's *Introductio in Ptholomei Cosmographiam* (N. T. XXXIV), of which I have given a detailed account above (p. 68).

9. *The cordiform projection of Sylvanus.* The only printed map of this class is the general map in the edition

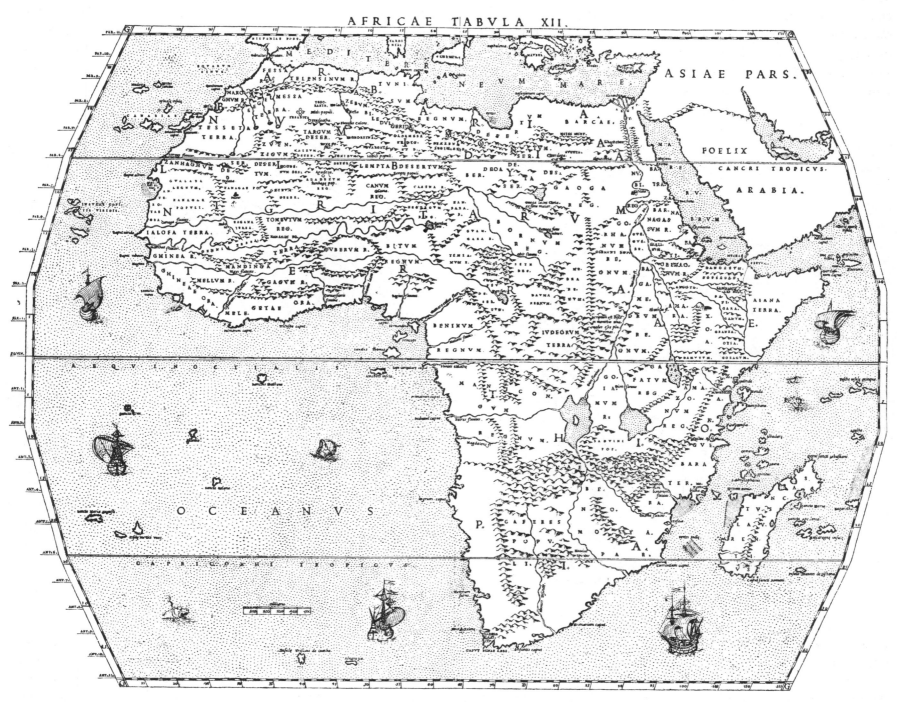

50. Map of Africa by LIVIO SANUTO, Vinegia 1588. (Orig. size 397 × 517 m. m.).

of the oceans, America and easternmost Asia, *Asia extra Ptolemaeum* or *Asia Marci Pauli Veneti.* The execution of Stobnicza's map is technically very rude, but his projection seems to me to be the best and most accurate of the map projections used during the first part of the 16th century for general maps of the world. With some modifications it might still be used with advantage for general maps on which the earth's surface is divided into two hemispheres. The general outlines of the continents on such a representation are less distorted than on the heart-shaped maps which were

of Ptolemy, Venetiis 1511, by Bernardus Sylvanus (N. T. XXXIII). This projection differs from Ptolemy's homeother-projection only in the common centre of the parallel-circles being placed 100° instead of 181° 8′ from the Equator, and in the parallel-circles being so extended, that they embrace not only the space between Long. 0° and 180°, but the whole surface of the earth, with the exception of the segment between Long. 250° and 290°, east of the *Insulæ Fortunatæ.* The map, in consequence of this extension, assumes a curvature at the pole which gives it a cordiform appearance. This would

---

[1] JOMARD gives a fac-simile of an elegant map engraved on the cover of a metal casket from the 16th century, signed »*Paulus Ageminius faciebat*». This map is a copy of that of Bernardus Sylvanus (compare: JOMARD's Atlas, Tab. XVIII; FIORINI, cited work, p. 592).

have been still more striking, had not the space between lat. 40° S. and the South pole, i. e. the very point of the heart, been left out.

10. *The cordiform projection of Apianus.* This projection was first used for the map of Apianus of 1520, remarkable as one of the first maps on which the New World is designated with the name of *»America.»* I shall have occasion to return to it in a following chapter. No description of it is given in the works in which it was inserted. But, as far as one can judge from measurements on the map, and on the assumption that the want of continuity in the meridians depends on technical difficulties in the execution, the net of graduation has been constructed according to the principles established by Ptolemy for his homeother-projection. The radius of the equatorial circle is here about 2 × the polar distance (the polar distance being 90° Ptolemy's projection will require 181° 50'). The parallels form concentric equidistant circles, and the proper ratio seems to have been re-

(N. T. XLV); a gigantic map of the world by VOPEL, which I have seen in the Hauslab collection, and which, its size excepted, seems to agree with the map in Girava's work.[2]

11. *Werner's cordiform projections.* Three years after the appearance of the edition of the map of Sylvanus, a work was published in Nuremberg by the celebrated mathematician JOHANNES WERNER, which, besides a critical revision of a part of Ptolemy's geography etc., contained: *Libellus de quatuor terrarum orbis in plano figurationibus ab eodem Joanne Vernero novissime compertis et enarratis.* Werner here[3] gives an account, with necessary numerical tables, of four methods, supposed by him to be new, of projecting the earth's surface on a plane. Three of these are modifications of the homeother-projections mentioned above. They differ from them, because WERNER always makes the pole a centre for the parallel-circles, and because he arbitrarily fixes the ratio between the length of the equatorial degree and that of the latitude, while the true proportion between the length of the degrees of the parallels is always maintained.

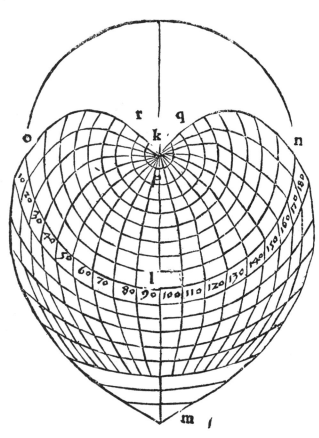

51. The first projection of WERNER, Nurenbergæ 1514.
(Orig. size 83 × 110 m. m.).

52. The second projection of WERNER, Nurenbergæ 1514.
(Orig. size 115 × 132 m. m.).

tained between the longitude of the Equator, of the Polar-circle, and of the mean-meridian.

It thus only differs from Ptolemy's projection through Apianus making the map embrace the whole earth, with the exception of the regions in the vicinity of the South pole, through which the map obtains its heart-shaped form. The following maps are constructed on this projection, or on modifications of it, regarding which the reader is referred to the accompanying fac-similes:

The map of the world by APIANUS of 1520 (N. T. XXXVIII); the map of GEMMA FRISIUS in the editions 1544, 1545, 1551 etc. of the cosmography of Apianus (N. T. XLIV);[1] the map of the world in HONTER's cosmography (N. T. XLIV); VOPEL's map of the world in GIRAVA's cosmography, Milano 1555

In Werner's first projection the semi-circle described, with the pole for centre and a radius of 90°, is divided into 180° equatorial degrees. Maintaining the proper proportion between the equatorial- and parallel-degrees we obtain by these means the net of graduation on fig. 51. As may easily be perceived, it is only possible to draw one of the hemispheres on each map constructed on these principles. To avoid such inconvenience Werner, in his second projection (fig. 52), reduces the length of the equatorial degree, assuming it to be the same as that of the degrees of latitude on the mean-meridian.

Werner finally proposes a third cordiform projection, for which, however, no net of graduation has been drawn. This only differs from the preceding one through the length of the equatorial degree being here assumed not to be

---

[1] The radius of the equatorial circle is here 4 × the distance from the Equator to the Pole. Judging from the rude drawing, the parallel-circles on this map are not equidistant. But as no description accompanies the map it is difficult to decide, whether this is made intentionally or whether it is caused by defective execution.

[2] On Honter's map the radius of the equatorial circle = 1,4, on Vopel's = 3 times the polar distance.

[3] According to BREUSING (*Gerhard Kremer gen. Mercator*, Duisburg 1869, p. 45), JOHANNES STABIUS (professor in mathematics at Ingolstadt and Vienna, † 1522) did first set forth the principles of Werner's *second* projection. The works of Stabius have not been at my disposal.

equal to the degrees on the mean-meridian, but $\frac{\pi}{3} \times$ the same.

If l is the distance from the equator to the pole, and q the length of 90° at the equator, the proportion between these quantities will be:

in Werner's first projection $\qquad q = \frac{\pi}{2} l.$

in Werner's second projection $\qquad q = l.$

and in Werner's third projection $\quad q = \frac{\pi}{3} l.$

Werner himself made no attempt to construct maps on the principles proposed by him. Nor do I know any maps

that the original map of Finæus also served as a model for the large Turkish cordiform map engraved on wood at Venice by HHÂGGY AHHMED from Tunis, and dated year 967 of the Mohammedan chronology, which corresponds with our year 1559. The blocks for this map which, for some reason or other, had been sequestrated, probably before the issue of the print, were discovered in 1795 in the depositories of the Venetian Council of Ten, and are now preserved at the Biblioteca Marciana. This discovery seems to have produced a certain sensation. It gave rise to a whole literature, intro-

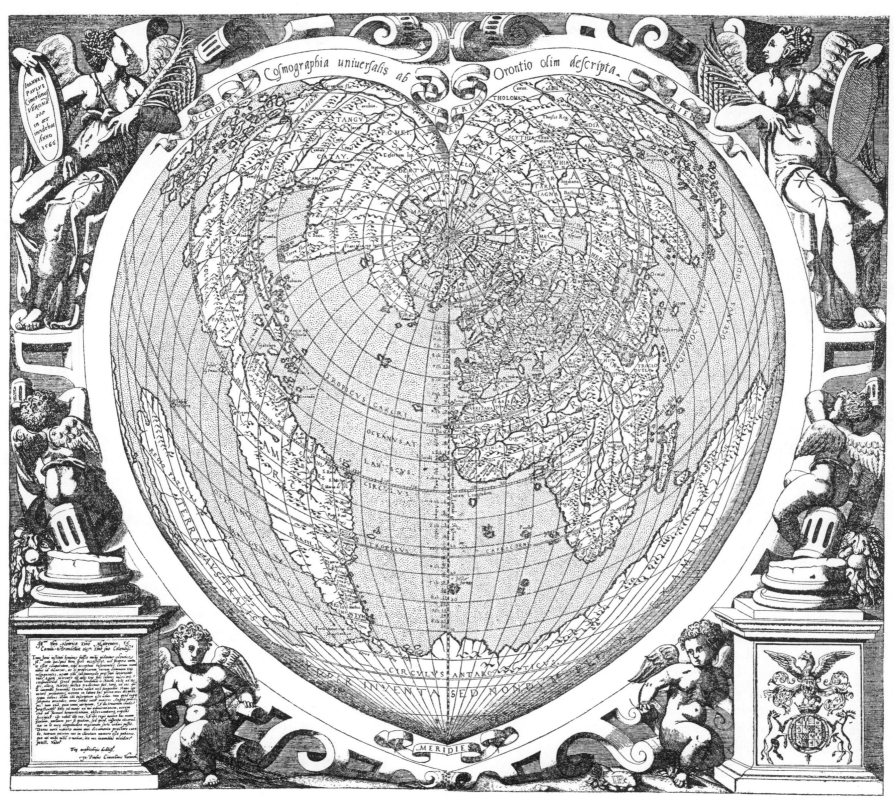

53. Cordiform map of the world by ORONTIUS FINÆUS. Copper-print by CIMERLINUS 1566. (Orig. size 211 × 580 m. m.).

drawn on his first and third projection. But his second projection is strictly applied to the handsome copper-engraving by JOH. PAULUS CIMERLINUS VERONENSIS, of which fig. 53 is a fac-simile. It is dated 1566 and inserted into my copy of Lafreri's atlas. The inscription »*Cosmographia universalis ab Orontio olim descripta*» indicates the map to be a copy from a work of ORONTIUS FINÆUS, the original of which probably was published in 1536[1] at Paris. I am not sure that it is still extant: at least no copies of it are to be found in the British Museum or in the Bibl. Nationale. It is evident

duced by a paper of Abbé SIMON ASSEMANI, and to various fables, concerning the manner in which the old blocks came into the possession of the Venetian government. Assemani got permission to draw 24 copies from them, but of these copies very few seem to be now extant. (Comp. D'AVEZAC, *Bulletin de la Société de Géographie*, Sér. 5: T. 10, 1885, p. 675). To judge from a much reduced copy of the Turkish map given by d'Avezac, it agrees, excepting some unimportant differences, with the map of Cimerlini.

[1] Comp. the biography of Orontius Finæus in HOEFER-DIDOT, *Nouvelle Biographie Générale*.

Several years before the publication of the cordiform map ORONTIUS FINÆUS had constructed another map also on Werner's second projection, but modified in such a manner, that the map of the world here is divided into two parts, the one embracing the northern hemisphere with the north-pole for a centre of the parallel-circles, and the other, the southern hemisphere, with the south-pole as a centre. It is is of this map that a facsimile is given on pl. XLI. It is dated 1531, but is generally found inserted in *Novus Orbis Regionum ac Insularum veteribus incognitarum*, Parisiis 1532. It was afterwards reprinted from the same block, but with a new title-legend from which the name of Orontius was omitted, in the edition of the Geography of GLAREANUS printed *Brisgae 1536* and in an edition of POMPONIUS MELA, *Parisiis apud Christianum Wechelum 1540*.

The map of Orontius Finæus finally had the honour of being copied, although with some modifications, by GERARD MERCATOR, for one of his first maps, of which I give a facsimile on the plate XLIII from a photo-lithograph of the only known copy, in the library of J. CARSON BREVOORT (WINSOR, *Bibliography of Ptolemy's Geography*, p. 22). Mercator afterwards constructed several special-maps of more or less extensive parts of the earth's surface on the same projection, e. g. his maps of *Africa*, of *Asia ex magna orbis terrae descriptione Gerardi Mercatoris desumpta studio et industria G. M. Junioris*, and probably also of *Taurica Chersonesus* (the southern part of European Russia) in RUMOLD MERCATOR's atlas, ed. 1595. This projection is also used for *Asia* in LORENZO ANANIA's *Universale Fabrica del Mondo*, 2d ed., Venetia 1582.

MERCATOR's double cordiform map was reproduced in copper by ANTON LAFRERI in Rome, probably in about 1560. The fine engraving, of which fig. 54 gives a fac-simile, faithfully follows the original. As to the title-legends the date (1538) is left out from the uppermost legend, and the key to the legends, placed in the original at the edge of *Terra Australis*, replaces the dedication of GERARDUS (MERCATOR) RUPELMUNDANUS to JOANNES DROSIUS on the Roman copy. Here the signature *Ant. Lafreri exc. Romae* is engraved underneath. This reproduction of Mercator's cordiform map is inserted in my copy of Lafreri's atlas, but appears to be rare. It is not to be found in the catalogue of printed maps in the British Museum, nor in Castellani's catalogue of the maps in »*Tavole Moderne di Geografia de la Maggior parte del Mondo*» (Lafreri's atlas) in the library of the Collegio Romano. A copy of this map, belonging to the city-archives of Turin, is described by FIORINI (cit. work p. 622).

12. *The oval projection of Bordone.* The map of the earth is here represented within an oval, of which the shorter axis, formed by the mean-meridian, is of half or about half the length of the longer axis, formed by the equator. The parallels form straight equidistant lines. The meridians are obtained by dividing the equator in parts of the same size, corresponding, for instance, to every 10th degree. The meridians are then drawn from these points of division to the pole in such a way that the parallels also become divided into parts of the same, or almost the same size. The projection is generally, though wrongly, designated by the name of APIANUS. Apianus has neither given any description of it nor left any map drawn according to it. In his cosmography we only find a few small wood-cuts with a net of graduation similar to that which belongs to the projection in question. But the figures are not accompanied by any explanation; the meridians are wanting, and the dimensions of the oval are far from correct.

In carefully examining the different oval maps of the world, printed in the 16th century, it will be found that,

although apparently much resembling each other, they present notable differences, as well in the form of the oval as in the way the meridians are drawn. On Bordone's map the oval appears to be formed by the line obtained, when an elastic ring is compressed between two parallel planes, until the longer axis becomes twice the length of the shorter one. In the Ptolemæus of GASTALDI, and on the general map of PORCACCHI, the oval is formed by two parallel lines and two circle-segments. On other maps arbitrary devices seem to have been applied, in drawing the meridians.

In the 16th century this projection was much used, e. g. on:

A map in *Libro di Benedetto Bordone, nel qual si ragiona del tutte l'Isole del Mondo* etc., Vinegia 1528. The map is supposed to have been finished before 1521 and was reproduced from the same block in several later editions of Bordone's *Isolario*. The Polar-axis (p): Equatorial axis (e) = 1 : 2 (N. T. XXXIX).

A map in *De Orbis Situ Epistola* by FRANCISCUS MONACHUS, Antverpiæ 1524. p : e = 1 : 2 (According to the much reduced reproduction by Lelewel).

A map in *Novus Orbis* by GRYNÆUS-HUTTICH, Basileæ 1532. p : e = 1 : 1,87. (N. T. XLII).

A map in the 2d ed. of *Isolario di* BARTOLOMEO DALLI SONETTI, Venetiis 1532 (There is no general-map in the editio princeps). p : e = 1 : 1,87.

A map in MÜNSTER's ed. of Ptolemy, Basileæ 1540. p : e = 1 : 2. (N. T. XLIV). Reproduced in several later editions, as well of Münster's Ptolemy as of his cosmography.

SEB. CABOT's large map of the world of 1544. p : e = 1 : 1,333 (according to the fac-simile published by JOMARD).

»*Universale Novo*» in GASTALDI's Ptolemy, Venetiis 1548. p : e = 1 : 1,77. (N. T. XLV).

A large map of the world by GASTALDI, of which two editions are described by FIORINI (cit. work, p. 601). On one of them is written: *Universale descriptione di tutta la terra conosciuta fin qui. In Venezia al segno del Pozzo*, and in the dedication of the copper-engraver, PAULO FORLANI, dated 1562, *Giaccomo Gastaldo cosmografo raro* is mentioned as the author of the map. The other edition bears the inscription: *Paulus de Forlanis Veronensis opus hoc Cosmographi Iacobi Gastaldi pedemontani instauravit et dicavit Paulo Michaeli Vicentino. Venetiis Ioan. Francisci Camotii aereis formis. Ad signum Pyramidis. Anno MDLXII.* A third edition, not dated, is cited by D'AVEZAC (*Coup d'oeil* etc. p. 72). These maps, like all maps printed on separate sheets, are now extremely scarce, though once much appreciated and often reproduced.

The map of the world in the *Theatrum Orbis terrarum* of ORTELIUS, Antverpiæ 1570. p : e = 1 : 2. (N. T. XLVI).

The map of the world in the *Opusculum geographicum* of MYRITIUS, Ingolstadii 1590. p : e = 1 : 1,8. (N. T. XLIX).

A map by HIERONYMUS PORRO, in the Ptolemæus Venetiis 1596 of MAGINUS. p : e = 1 : 2.

Maps of the world on this projection have further been constructed by BATTISTA AGNESE (in his portolanos), by BELLEFORESTE (*La Cosmographie Universelle*, Paris 1575), by GIOSEPPE ROSACCIO (*Il Mondo e sue parti*, Fiorenza 1595), and others.

13. *Stereographic projections.* The strictly perspective projections, the centrographic, the stereographic, the scenographic, and the orthographic, are considered to have been invented by THALES, ERATOSTHENES, and HIPPARCHUS. The most important among them, the stereographic and the orthographic, were under the name of *Planispherium* and *Analemma* described by Ptolemy in two treatises, of which that on the planisphere was first printed in Latin as an appendix to the Rome edition of 1507 of the geography (Comp. N.

54. G. MERCATOR's double cordiform map of the world of 1538. Copperprint, Rome about 1560. (Orig. size 324 × 519 m. m.).

p. 16). It was a translation from a Greek manuscript. But Ptolemy's description of the Analemma is only known from a very incomplete and faulty translation into Latin (probably from an Arabian translation of the original), which, revised and commented upon by FEDERIGO COMMANDINO, was published at Rome in 1552. Neither of these projections was used for cartographical purposes by Ptolemy, or by his successors during the Middle Ages. Even during the 16th century they were only occasionally employed for geographical purposes, in spite of the elaborate mathematical investigations of which they were often the object.

Only the following maps and map-nets on a stereographic projection were printed previous to the beginning of the 17th century.

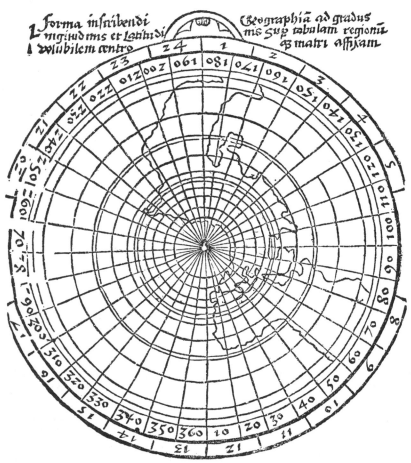

55. Stereographic net from REISCH, Margarita philosophica nova, 1512.
(Orig. diam. 131 m. m.).

REISCH, *Margarita Philosophica nova*, Argentine 1512. To this and several later editions of the famous Reisch encyclopedia was appended an *Appendix Matheseos in Margaritam Philosophicam* containing, among tracts on the Greek and Hebrew languages, on architecture and music etc. a *Tractatus de compositione astrolabii Messehalath*. We here find directions as to a polar-stereography, with two drawings of a stereographic net of graduation. On one of them, given here in fac-simile (fig. 55), the commencement of a geographical map can be discerned.

A net of graduation on a stereographic horizontal projection by JOHANNES WERNER 1514. (N. fig. 56.) The last projection of this eminent mathematician, in his above (p. 88) cited work, is such a projection on the horizon of Nuremberg. He praises several of its points, and invites the adoption of it in the following words: *Talis profecto terrarum orbis figuratio, plurimum honestatis atque ingens ornamentum viro adiiciet philosopho, si super ipsius mensæ plano depicta fuerit.* During the whole of the 16th century, however, no attempts were made to construct maps on this variety of the stereographic projection.

A map inserted into the different editions of *Cosmographicus Liber Petri Apiani Mathematici studiose collectus.* (Colophon in the first edition:) *Excusum Landshutæ . . . impensis Petri Apiani Anno . . . Millesimo quingentesimo vicesimo quarto.* This edition of 1524, as well as the numerous editions published afterwards, contain some diagrams cut out from stiff and strong paper, and invented by Apianus to explain the phenomena of astronomical geography. On the revolving diagram inserted at fol. 63 of the edit. 1524, or at fol. 32 of the edits. 1533 and 1534, a small map is delineated, embracing the whole of the northern hemisphere and the southern one to lat. 25° S. (N. fig. 57.) If we except the above mentioned very incomplete drawing in Reisch's Margarita (N. fig. 55), which can hardly be regarded as a map, this is the first printed map on a stereographic

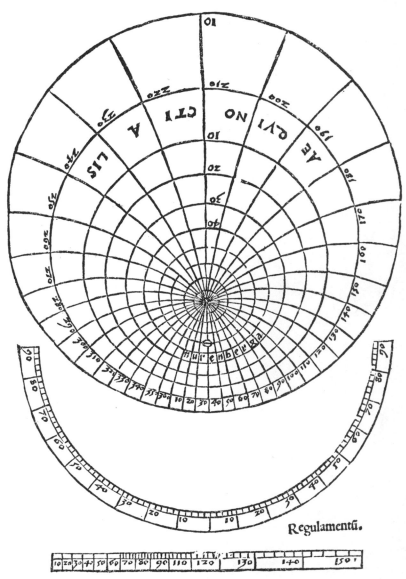

56. Net of a stereographic projection on Nuremberg's horizon, WERNER 1514.
(Orig. size 124 × 175 m. m.).

polar-projection. The table below shows that the drawing of the net, although rude, is very correct.

	Distance from the Equator.	
Latitude.	Stereographic polar-projection.	Map of Apianus (edit. 1533).
90°	1,000	1,000
80°	0,913	not determinable.
70°	0,824	
60°	0,732	0,742
50°	0,636	0,632
40°	0,534	0,532
30°	0,423	0,426
20°	0,300	0,292
10°	0,161	0,157
0°	0,000	0,000
—10°	0,192	0,210
—20°	0,428	0,458

Considering that the distance from the equator to the pole on the map of Apianus is only 34 m. m., the agreement between the numbers obtained by calculation and those obtained by measurement on the map is very satisfactory, and proves that we here actually have before us a map correctly drawn on the stereographic polar-projection.

A stereographic net constructed by Orontius Finæus. This celebrated mathematician described and delineated in his cosmography, published in *Orontii Finaei Delphinatis opus varium*, Parisiis 1532, a net of graduation on a stereographic meridian projection, and a few years later he, in his *Planisphærium geographicum*, Lutetiæ 1544, discussed the advantages of employing a stereographic polar projection for geographical maps. (Fiorini, cit. work, p. 127).

Rumoldus and Michael Mercator. In the first edition of Gerard Mercator's work *Atlantis Geographia Nova Totius Mundi. Authore Gerardo Mercatore Rupelmundano Illustriss. Ducis Iuliæ etc. Cosmographo*, Duysburgi Clivorum s. a. (the preface, dedicated by Rumoldus Mercator to Queen Elizabeth of England, is dated Duisburg first Apr. 1595), there are two maps on the stereographic projection, viz.

The map of the world by Rumoldus Mercator (N. T. XLVII). In the text printed beneath the map is written: *Sciet lector nos eam conplanandæ sphæræ rationem secutos esse,*

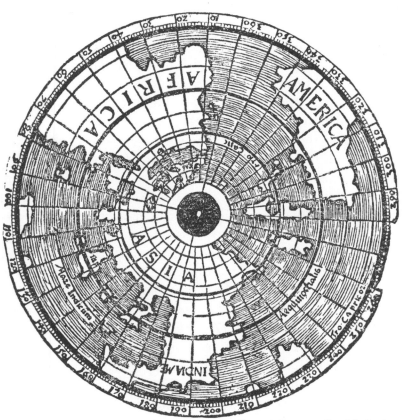

57. Map on stereographic polar-projection by Petrus Apianus, Landshut 1524. (Diam. of the orig. 112 m. m.).

quam Gema Frisius in suo planisphærio adinvenit, quæ omnium longe optima est.[1]

The map *America sive India nova ad magnae Gerardi Mercatoris avi universalis imitationem in compendium redacta per Michaelem Mercatorem Duysburgensem* from the same atlas. Fiorini pretends (cit. work, p. 129) that the map of Africa is also drawn on a stereographic meridian projection. But this is not correct, at least not as regards the handsome map of Africa inserted under Litt. C in my copy of edit. 1595, and constructed on Werner's 2d projection.

Rumold Mercator's map of the world was copied by Hieronimus Porro and published in the Ptolemy of Maginus of 1596, 1598, and 1621, and by Petrus Keschedt in the editions of 1597, 1608, and 1617.

No maps on the orthographic projection were, as far as I know, published before the 17th century, nor any on the scenographic, unless such woodcuts of globes as I have reproduced on pl. XLIV from the cosmography of Apianus, and

in fig. 49 from a work of Schöner, are regarded as scenographic representations of the earth.

14. *Bacon's meridian projection.* This projection is described by Roger Bacon († 1294) in the following words: *Sed in signatione civitatis in loco suo per longitudinem et latitudinem suam inventas ab auctoribus, superaddam artificium, quo locus civitatis habeatur per distantiam ejus a meridie et septentrione et oriente et occidente. Et hoc artificium consistit in concursu lineæ rectæ aequidistantis aequinoctiali signatæ in plano: secundum formam lineæ rectæ ductæ a numero graduum latitudinis regionis signato in quarta coluri ducta ab aequinoctiali ad polum mundi in concursum, inquam, cum arcu circuli magni qui transit per polos mundi et per numerum longitudinis civitatis signatum in aequinoctiali circulo. Hic autem modus*

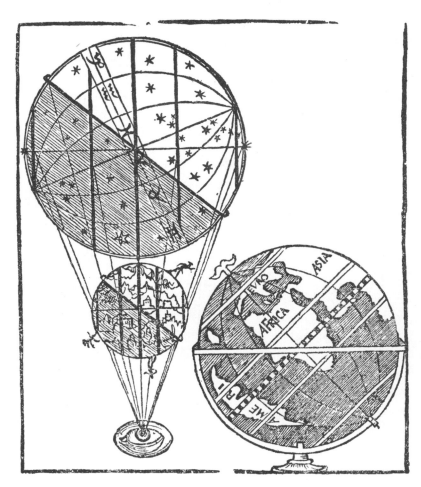

58. »In hoc sequenti typo totius Cosmographiæ Descriptio demonstratur». Petrus Apianus, Landshut 1524. (Orig. size 109 × 126 m. m.).

melior est et facilior, et sufficit considerationi locorum mundi in hujusmodi figuratione sensibili (*Opus Majus*, Ed. 1733, p. 186). In the beginning of the 15th century this remarkable passage was transcribed almost word for word by Pierre d'Ailly in the 17th chap. of his *Compendium cosmographicum*, which has caused d'Ailly to be quite undeservedly cited by Lelewel and Fiorini as the inventor of this projection. (Comp. Lelewel, II, p. 75; Fiorini, p. 604.) The projection is characterized by equidistant, rectilinear parallels and meridians formed by arcs of circles passing the poles and dividing the equator in equidistant parts. The maps of the whole surface of the earth on this projection were generally, when it became practically employed, divided into two hemispheres, and the great interest attached to it from a cartographical point of view, perhaps depends upon such a division of the general map of the earth being thus definitely introduced into cartography. But it should be remembered that in this respect these maps had a precedent in Stobnicza's

[1] It is evidently a mistake, when this map is ascribed to Rumold's celebrated father Gerard. The title expressly says: *Ex magna Universali Gerardi Mercatoris . . . Rumoldus Mercator fieri curabat.* But here by »Magna Universalis» is probably meant G. Mercator's chart on a cylindrical projection, as would appear from the enumeration of his cartographical works in the introduction to »Atlas» by Gualterus Ghymmius. At present no map drawn on the stereographic projection by G. Mercator sen. is known, but evidently it was he who superintended and inspired the works of his sons and grand-sons.

map of 1512 (N. T. XXXIV), which however seems to have been entirely overlooked by the cosmographers up to our time.

The first[1] map (N. fig. 58) printed on this projection is inserted in the text of the cosmography of APIANUS. It is an insignificant woodcut, which was subsequently reproduced on the title-page of *Dionysius Lybicus Poeta De situ habitabilis orbis a Simone Lemnio poeta Laureato nuper latinus factus*, Venetiis 1543. Bacon's projection was further employed on a map of the world, engraved in copper by GIULIO MUSI and published in Venetia by TRAMEZINI in 1554. On the upper borders of the map we read: *Cum priv. summi pont. et senat. veneti Michaelis Tramezini formis MDLIIII*, and beneath: *Julius de Musis Venet. in aes incidit MDLIII*. I have not seen this map. Fiorini describes it (cit. work, p. 605) from a copy preserved in the town-archives at Turin. What »mathematician» or cosmographer may have been its author, is not stated, but I suppose it to be a work of GASTALDI.

Bacon's projection has further been employed for:

The handsome map of the world, printed from the same plate in the editions of Ptolemy, Venetiis 1561, 1562, 1564,

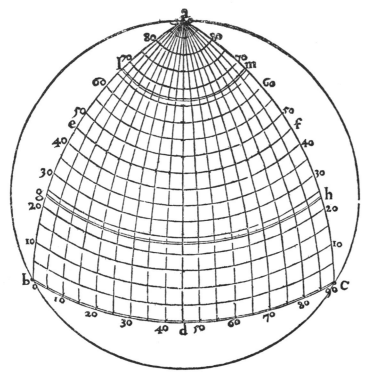

59. Net of graduation by ORONTIUS FINAEUS in 1551. (Orig. size).

and 1574, and in LORENZO D'ANANIA's *L'Universale Fabrica del Mondo*, 2d edit.,[2] Venetiis 1582 (N. T. XLV).

Maps of Africa, Asia, Europe, and America in ANDRÉ THEVET: *La Cosmographie universelle*, Paris 1575.

The map No. 2, *Americæ sive Novi Orbis nova descriptio*, in edit. 1570 of the *Theatrum Orbis Terrarum* of Ortelius.

The maps of the two hemispheres at the upper part of the first map in DE JUDAEIS' Atlas, Antverpiæ 1593.

15. *Da Vinci's projection.* The above mentioned geographical drawing, found among the papers of Leonardo da Vinci and described by R. H. Major, is drawn on a peculiar and well characterized projection, as may be seen by the fac-simile of Major's copy given above (fig. 45). I do not know of any other instance of such a construction having been employed for early maps, but a net of parallels and meridians is given in the *Sphæra Mundi* by ORONTIUS FINÆUS, Lutetiæ Parisiorum 1551 (N. fig. 59).

16 and 17. *The projections of Glareanus and Florianus.* The method of representing the spherical surface of the earth, employed in the globe-prints here given in fac-simile, on the plates XXXVII and XL, and also the method used for drawing the handsome map of which fig. 48 is a reduced copy, must further be enumerated among projections employed on maps of the 16th century. I have denominated the first of them after GLAREANUS, who according to Myritius *Opusculum Geographicum Rarum*, Ingolstadii 1590, p. 42, was its inventor. The last map, again, bears on the right upper corner, the medallion of ANTONIUS FLORIANUS UTINENSIS, which name I have adopted, for want of a better, to designate the method of projection.[3] I do not know any other map of this Italian artist from Udine, living in the second part of the 16th century, nor any other such map published in print. But a large (1440 × 790 m. m.) coloured map on this projection, drawn on vellum, is preserved at the Royal Library in Stockholm. It is signed: *Nova verior et integra totius orbis descriptio nunc primum in lucem edita per Alfonsum de Sancta Cruz[4] Cæsaris Charoli V. archicosmographum. A. D. MDXLII.* Floriani's map is bound in my copy of »Lafreri's Atlas.» It is a fine copper-print. As is the case with several other maps belonging to this atlas, the work of the engraver has not been finished, at least the medallions at the lower corners of the map and the two title-fields are left blank. The form of America and of the northern part of the Pacific indicates that the map had probably been engraved before 1566, the year when »Fretum Anian» was introduced into cartography.

D. Projections first employed between 1550 and 1600.

18. *Mercator-Postel's equidistant polar projection.* Although possessing many excellent qualities, the stereographic polar-projection suffers from the defect of having the degrees of latitude near the equator considerably larger than those in the vicinity of the pole. On the orthographic polar-projection, on the contrary, the degrees of latitude near the equator are too small. A construction, intermediate between these, is that with rectilinear meridians converging towards the

pole and circular equidistant parallels. This projection was generally supposed (for instance by D'AVEZAC, cit. work p. 63) to have been first used by POSTEL in 1581. But BREUSING, (*Gerhard Kræmer gen. Mercator*, Duisburg 1869, p. 51) claims priority for GERARD MERCATOR, who as early as 1569 employed it for the map of the polar regions, added to his celebrated chart on an isogonic cylindrical projection. I have, therefore, in the Swedish edition of this work desig-

---

[1] From the above quoted passage of Roger Bacon it may be concluded that he himself constructed a map on the newly-invented projection. This map is lost, but I suspect that the map-skeleton of d'Ailly on this projection (N. fig. 19) is copied from a manuscript of Bacon.

[2] The first edition of this work does not contain any maps, but the 2d edition contains, besides the map of the world, maps of Europe, Africa, and America on the Donis' projection and of Asia on Werner's 2d projection.

[3] This map is mentioned in Castellani's catalogue of the library of Collegio Romano (p. 239) by the following words: *Mappamondo in due emisferi formati a spicchi concentrici ai circoli e ai poli. Negli angoli, su in alto a sinistra è il ritratto di Tolomeo, alla diritta quello del Cosmografo »Antonius Florianus Vtin.»* Castellani may at least have known some biographical data regarding Florianus, as he is called a cosmographer.

[4] The University library of Upsala possesses a large map of the city of Mexico by the same imperial *archicosmographer* ALONZO DE SANTA CRUZ. It is not known how and when these two maps came to Sweden, but probably they belonged to the large collection of maps left by Santa Cruz at his death, 1572, and of which a list was deposited in Archivo de Indias at Seville. (*Relaciones geográficas de Indias publícalas el Ministerio de fomento. Perú.* T. 2, Madrid 1885, p. XXX).

nated it by the name of Mercator-Postel, a name which I shall also use here to avoid confusion. But exact measurements, which I have since had occasion to undertake on the special maps in the edition of Ptolemy of 1462, show that most of these maps are drawn on the equidistant polar-projection, which is of course only a variety of the conical development. The name which would most nearly express its origin would, therefore, perhaps be *the projection of Manfredus-Bonus*.

the large sea-chart, is in more than one respect of interest in the history of geography.[1] A fine copy of it had already been published in 1593 by DE JUDAEIS in *Speculum Orbis Terræ* (N. T. XLVIII). This projection was also used for the polar-map of MICHAEL LOK, *civis Londinensis*, inserted into HAKLUYT, *Divers Voyages*, London 1582, and for the celebrated map of Willem Barents' last voyage, published in LINSCHOTEN's *Navigatio ac Itinerarium*, Hagæ-Comitis 1599.

60. G. MERCATOR's map of the North-polar regions of 1569, from RUM. MERCATOR's atlas of 1595. (Orig. size 357 × 393 m. m.).

The polar-map of Mercator at first evidently created far more interest than the main chart, with which it was published. It was reproduced by Mercator himself in a copper-engraving inserted into the first edition (1595) of *Atlas*. The metallotype fig. 60 gives a fac-simile of this map, which, although by no means comparable, in importance, with

19. *Mercator's isogonic cylindrical projection (Mercator's projection).* The first map on this projection, which has exercised such powerful influence on the progress of navigation, was published in 1569 by GERARD MERCATOR. A long inscription on the map explains the principle of the new method of projection and its use for navigation. Mercator is

[1] To what I have said before regarding Mercator's remarkable delineation of the North, I may here add the following extract from the long legend attached to the polar-map on his chart: ... *Quod ad descriptionem attinet, eam nos accepimus ex itinerario Iacobi Cnoyen Buscoducensis, qui quidem ex rebus gestis Arturi Britanni citat, majorem autem partem et potiora a sacerdote quodam apud regem Norvegiae anno D. 1364 didicit* (Comp. LELEWEL, II: p. 231; JOMARD's fac-simile). From a passage in Purchas (*His pilgrimes*, III, London 1625, p. 518) we know that Iodocus Hondius possessed a copy of Ivar Baardson's description of Greenland of the middle of the 14th century. This copy probably once belonged to Gerard Mercator, whose copper-plates (and other geographical documents?) Hondius purchased in 1604. From this we may conclude that Ivar Baardson was the Norwegian priest of whom Mercator speaks in the above cited legend — and perhaps we here too have a clue to the origin of the Zamoiski-map (N. T. XXX).

thus incontestably its real inventor, notwithstanding that maps on such a projection, i. e. with rectilinear equidistant meridians and parallels so drawn that a proper ratio is always maintained between the longitudes and the latitudes, had, as I have already (p. 22) pointed out, been promised by Bilibaldus Pirckheimerus in the introduction to Ptolemy, printed at Strassburg in 1525, and although the mathematical principles on which it is based, and the tables necessary for its construction, were first published by Edw. Wright in his important work: *The correction of certain Errors in Navigation detected and corrected*, London 1599;[1] 2d edition 1610.

Of the large map of 1569 there is at present only one copy known, preserved in the Bibliothèque Nationale at Paris. A full-size fac-simile of it was published by Jomard, and a copy on a considerably reduced scale by Lelewel. The map scarcely appears to have been duly appreciated even by Mercator's nearest friends and admirers. It is not, like several other large maps of Mercator, reproduced on a reduced scale in the Atlas. Mercator's friend and biographer, Walter Ghymm, enumerates it among his works, but evidently without any idea of its real importance. Neither Waghenaer, nor Willem Barents employ it for the charts they published during the latter part of the 16th century. The length of time the reform introduced by *Magna Mercatoris* and Wright's *Errors of navigation*, needed for its general adoption is made evident from the circumstance, that all charts in *De Lichtende Columne ofte Zee-Spiegel*, published in Amsterdam by Jan Jansz in 1653 and by Pieter Goos in 1658, are still drawn on the rectangular projection of Marinus. This is also the case with most[2] charts in van Keulen's large atlas and even in the *Atlas van Zeevaert en Koophandel* by Renard-Ottens, Amsterdam 1745. That neither of these celebrated cartographers did fully appreciate the mathematical principles explained by Wright proceeds from the circumstance that the charts in their works, although drawn on the projection of Marinus, and often extending from 50° to 80° of latitude, are yet crossed by compass-bearings, in all directions.

The only printed maps of the 16th century known to me, which are drawn on Mercator's projection are:

1569: Mercator's large map: *Nova et aucta orbis terræ descriptio ad usum navigantium emendata, accomodata.... Aeditum autem est opus hoc Duysburgi an. D. 1569 mense Augusto.* Its dimensions (2,0 × 1,26 m.) prevent its reproduction here. A fullsize fac-simile is published by Jomard, but unfortunately with omission of several of the important inscriptions, for which Lelewel's *Géographie du Moyen âge*, II, p. 225, may be consulted.

1599: A map of Henricus Hondius in *Navigatio ac Itinerarium Johannis Hugonis Linscotani ...* Hagae-Comitis 1599 (N. fig. 61). Among the other maps in this work one (*Delineatio chartæ trium navigationum per Batavos ad Septentrionalem plagam*) is constructed on the equidistant polar-projection.

1599: The handsome map in Richard Hakluyt's *Principal Navigations*, 2d edition (N. T. L), which is supposed to be »the new map» of which Shakespeare speaks in »Twelfth Night» (Act. III, Sc. 2). Mr C. H. Coote suggests that Edward Wright is the true author of this map. It is one of the best general maps of the world of the 16th century (Comp. *The voyages and works of John Davis the Navigator.* By

Albert Hastings Markham. Works issued by the Hakluyt Society, London 1880, p. LXXXV).

The following table shows how nearly the constants of the nets of graduations in the oldest maps constructed on Mercator's projection, are calculated.

Distance to the equator in equatorial degrees:

Parallel at	Calculated[3] for Mercator's projection.	On Mercator's map of 1569 (Jomard's copy).	On Hakluyt's map of 1599.	On Hondius' map of 1599.
10°	10,05	10,1	10,1	10,1
20°	20,42	20,3	20,9	20,9
30°	31,47	31,0	31,3	32,0
40°	43,71	42,8	43,1	—
50°	57,91	56,5	57,2	—
60°	75,45	73,3	74,4	—
70°	99,43	96,3	99,0	—
80°	139,59	135,2	139,1	—

As may be perceived, but little remains to desire regarding the agreement between the numbers of the 2d column and the corresponding numbers on Hakluyt's map. On this map the equatorial degree is = 0,55 m. m., and the greatest difference between the calculated and the observed equatorial-distance only 1,05 × 0,55 = 0,6 m. m. At 80° the error amounts to 0,27 m. m. and, at 70°, to 0,24 m. m. On Mercator's map the differences at 10° and 20° are insignificant. At 30° the distance from the equator falls short of the calculated number by 0,47, at 40° by 0,91, at 50° by 1,41, at 60° by 2,15, at 70° by 3,15 and at 80° by 4,39 equatorial degrees. Such a degree has here a length of 1,73 m. m. An error occurs, gradually increasing towards the pole, and evidently arising from the imperfection of the mathematical resources of the map-constructors in the middle of the 16th century. Mercator seems to have calculated the length of the intervals between every tenth degree of the parallel by means of the approximate formula:

$$P_{\varphi+10} - P_\varphi = \frac{10}{\text{Cos}\,(\varphi + 5)}$$

The unity here is the length of the equatorial degree, and P_φ the equatorial distance on the map at the latitude φ.

By this formula the following numbers are obtained:

The parallel at	Distance from equator in equatorial degrees. Calculated	Mercator's map (Jomard's copy).
10°	10,04	10,1
20°	20,39	20,3
30°	31,42	31,0
40°	43,63	42,8
50°	57,77	56,5
60°	75,20	73,3
70°	98,86	96,3
80°	137,50	135,59

Even here the agreement is not so complete as might have been expected, but the differences can be explained by engraving-errors or by stretchings in the paper.

On printed maps which I have had an opportunity of examing, there have been employed down to A. D. 1600 nineteen different projections. This number might be further increased, if separate numbers were to be given to the pro-

---

[1] In his biography of Mercator Ghymmius says with regard to the new projection: ... *inventio nova et convenientissima ... quæ sic quadraturæ circuli respondet ut nihil deesse videatur, praeterquam quod demonstratione careat, ut ex illius* (Mercatoris) *ore aliquoties audivi.* It is not clear from this passage whether »*quod demonstratione careat*» concerns the projection or its connection with quadratura circuli. The latter appears to me more probable as fully harmonizing with the tendencies to speculations in the most heterogeneous branches of knowledge, which characterized the great cartographer.

[2] In the edition of 1683 of van Keulen's atlas only the first map (the map of the world) and the last one, of the north-eastern coasts of Asia, from Novaya Zemlya to Japan, are drawn on Mercator's projection. In the atlas of Renard-Ottens only the map of the world is drawn on Mercator's projection.

[3] Assuming the surface of the earth to be spherical.

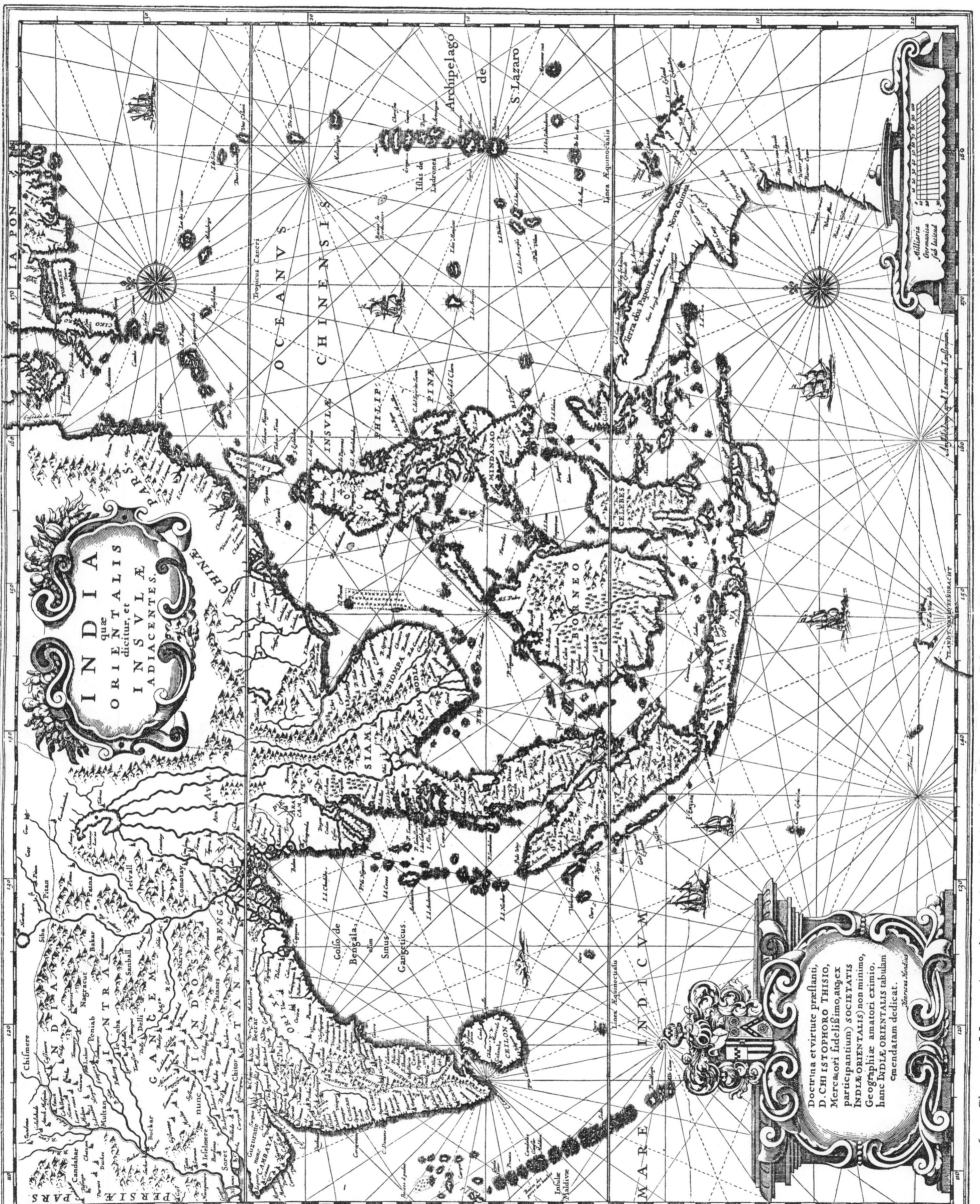

61. Chart on Mercator's projection in: *Navigatio ac Itinerarium Iohannis Hugonis Linscotani.* Hagae-Comitis 1599. (Orig. size 396 × 490 m. m.).

jections, apparently identical, but yet from a mathematical point of view very different, which have here been associated under No. 12 (Bordone's oval projection), and also to Ptolemy's tangent, and Ruysch's intersecting conical projections, and to the remarkable pseudo-conical or pseudo-Werner projection adopted by Mercator for his map of Europe. On the other hand it would, perhaps, be most correct, with d'Avezac, to unite the homeother projection of Ptolemy with that of Stobnicza, and to enter the cordiform projections (Nos. 9, 10, and 11) under a common number. This uncertainty arises from the difficulties connected with a strict limitation of the different kinds of early maps. It is an exception to find a mathematical description given of the net of graduation which has been employed, and the maps, especially when reproduced in wood-cut, are often of such inferior execution, that it is difficult to decide, even by careful measurements, what may have been the leading principles for the drawing. I hope, however, that the enumeration given above and illustrated by numerous facsimiles, will be sufficient for a general review of the early history of map-projections.

We learn from it that this chapter of mathematical geography had, at the end of the 16th century, already reached a very high development. It is true that no clear insight into the properties, advantages, and defects of the different projections had yet been obtained, nor was such knowledge then possible, owing to the deficiencies in mathematical resources. So that there could not be a critical inquiry into the conditions which it was possible to satisfy, in representing the spherical surface of the earth on a plane. But several of the most valuable methods of projection, for instance that of the development on an intersecting cone, the stereographic, the isogone-cylindrical, and the equidistant polar-projections, were then in use. Nor can it be said that any radical reform in cartography has been introduced owing to the complete mathematical analysis of the problem, on which the works of the modern cartographers can be based.

The merit of so early a development of the doctrine of projections must, in the first place, be ascribed to Ptolemy. It is true that he was unable to solve those analytical problems on which an exact theory of map-projections must be based. They were then insoluble, and are very difficult even in our days. Nevertheless he clearly understood the truth that the surface of a sphere cannot be exactly developed on a plane, and that consequently the problem must be solved by an approximation, for which he proposed not less than four different methods, two of them being practically applied by himself. At least two of these had already been employed or proposed by his predecessors, Hipparchus and Marinus. This important chapter of mathematical geography was further developed by Arabian writers, by Bacon and Nicolaus Germanus, by the authors of the maps in the Ptolemy edition Bononiæ 1462, by Ruysch, Bernardus Sylvanus, Bordone, Johannes Werner, Petrus Apianus, Glareanus, Orontius Finæus, Postel, and Mercator.

IX.

The end of the early period of cartography.

1520—1550.

Most of the printed maps, during these decennia, were still published as addenda to new editions of Ptolemy's geography. In eleven editions of this work, from 1520 to 1550, which were provided with maps, two hundred and sixty nine of the old maps and two hundred and forty four *tabulæ novæ* were printed, most of them in double folios, while the rest of the map-printing of the same period — if reprints in Münster's cosmography and the small wood-cuts in the works of Bordone and Apianus, are excepted — scarcely amount to one hundred. Thus the cartographical literature of these decennia is still very poor as well as regards its extent as with reference to the composition and execution. Yet the dawn of a new period might even then be discerned, partly from the appearance of new maps, founded on actual topographical investigations, in increasing numbers, partly from the attempts to employ improved methods of projection.

In another respect this period forms an epoch in the development of cartography. A couple of rough wood-cut maps had already been published at Lubeck in 1475, and two highly meritorious editions of Ptolemy, provided with large wood-cut maps, had been published at Ulm in 1482 and 1486, to which may be added some few other more or less important separate maps printed in Germany in the 15th or the first years of the 16th century. With these exceptions, almost all geographical maps had, until 1513, i. e. until the year when the large Strassburg edition of Ptolemy provided with 20 new maps was published, been printed in Italy, although often with the assistance of map-drawers and map-engravers from Gutenberg's fatherland. But from that year the principal seat of the industry of map-printing was transferred to the countries to the north of the Alps, although at first only for a short time. While only a few maps, generally of slight importance, were printed in Italy from 1513 to 1547, by far the greatest part of such works of the next period, 1548—1570, are of Italian origin. From 1570, i. e. from the year when the first edition of *Theatrum Orbis Terrarum* by Ortelius was published, the Netherlands became for a long time the principal seat of map-printing.

The first transfer of this industry to the countries north of the Alps was evidently effected at the expense of the finish of the execution. In this respect the maps of Mercator and Ortelius are the first that can be compared with the old copper-engraved maps from Rome and Venice. Hence the German maps were at first almost exclusively reproduced in a manner little adapted for large cartographical works, namely in wood-cut, whereas for the same purpose copper-print was almost exclusively used in Italy.[1]

The first half of the 16th century is thus characterized by an apparent retrogression in the cartographical art. But if the maps of this period are more closely examined, real progress will nevertheless be discovered under an almost grotesque exterior. This arises from the early attempts of the German, Dutch, and French geographers to emancipate themselves from the classical authors formerly so anxiously followed, and to base their maps, as well of the ancient hemisphere as of the New World, on modern geographical data. The difficulties encountered by geographers, partly due to the slowness with which the accounts of the exploring expeditions to the New World and to the eastern Asiatic Archipelago reached Europe, and partly from want of reliable data for longitudes and latitudes, will be conceived, if the maps enumerated on the list below are examined and compared with the contemporary history of geographical discoveries.

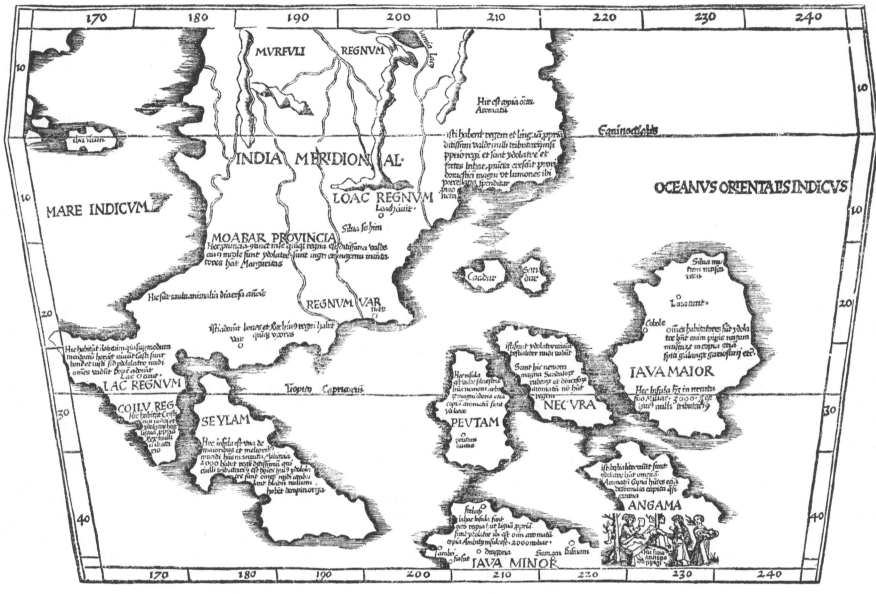

62. Tabula moderna Indiæ Orientalis, from Ptolem. Argent. 1522. (Orig. size 436 × 281 m. m.).

Maps printed from 1520—1550.

1. *Tipus orbis universalis iuxta Ptolomei Cosmographi traditionem et Americi Vespucii Aliorumque lustrationes a Petro Apiano Leysnico elucubratus. An. Do. MDXX.* (N. T. XXXVIII). This map had probably been originally printed separately in Vienna, though it was afterwards inserted in:

Ioannis Camertis Minoritani, Artium et sacrae theologiae doctoris, in C. Iulii Solini πολυΐστωρα enarrationes. . . . MDXX. Viennae Austriae, per Ioannem Singrenium; and in:

Pomponii Melae de Orbis Situ Libri Tres . . . . una cum commentariis Ioachimi Vadiani . . . Adiecta sunt praeterea loca aliquot ex Vadiani commentariis summatim repetita, et obiter explicata, in quibus aestimandis censendisque doctissimo viro Ioanni Camerti . . cum Ioachimo Vadiano, non admodum conuenit. Rursum, epistola Vadiani, ab eo pene

adulescente ad Rudolphum Agricolam iuniorem scripta . . . Basileæ Apud Andream Cratandrum Anno MDXXII.

These two works, printed in different towns and in different years, are often bound in one volume, to which even the map of Apianus is added, but without being mentioned in the prefaces, in the numerous dedicatorial poems, or in the detailed commentaries, by which the works of Solinus and Mela are explained. The same monogram by *Lucas Alantse civis et bibliopola Viennensis*, however, ornaments the left corner at the foot of the map, the title-page, and the colophon in Solinus. The map is a wood-cut, badly drawn and badly engraved. It appears to be based on the same sources as Schöner's globe of 1515 (fig. 46 and 47), and the globe of which I found the printed gores in a Ptolemy of 1525 (comp. p. 76). What renders this map remarkable

---

[1] From the forty years between the publication of Ruysch's map of the world and Gastaldi's maps in Ptolemæus Venetiis 1548 I know only the following copper-printed maps:

1514: Mappemonde in gores by Boulanger d'Albi (N. T. XXXVII).

1538: Map of the world by Mercator (N. T. XLIII).

1537—41: Mercator's maps of Palestine, and Flanders, and his large mappemonde in gores.

To these some of Gastaldi's first works may be added, which I have not seen. The art of reproducing maps by means of copper-print was thus almost forgotten, when Mercator and Gastaldi began to publish their works.

from a cartographical point of view, and causes it to be much sought after by collectors, is its peculiar projection and the inscription, *America prouincia*, on the South American continent.

If various interruptions in the continuity of the meridian lines may be considered to have originated from the woodcutter's want of skill, the whole globe, excepting the region in the vicinity of the South Pole, is here for the first time drawn on Ptolemy's homoeoter projection. As to the inscription, the opinion long prevailed that this map must be the first published in print, on which the New World was designated by the name of *America*, a circumstance which caused it to be often reproduced,[1] much sought after by collectors of early books on America, and an object of careful bibliographical research. At least three globe-prints with the name of America are at present known, older than this map of Apianus, namely:

1. The copper-printed map in gores for Boulanger's globe of 1514 (N. T. XXXVII);

2. The wood-cut map in gores, reproduced in fac-simile on pl. XXXVII; and

3. Schöner's wood-cut and similarly printed globe of 1515 (N. fig. 46 and 47).

The name was used in a printed book a few years earlier, viz.

1. *Deodate 1507*. (HYLACOMYLUS or WALDSEEMÜLLER) *Cosmographiæ introductio, cum quibusdam geometriæ ac astronomiæ principiis ad eam rem necessariis. Insuper quatuor Americi Vespucii navigationes*. Fol. 15 verso is written: *Nunc vero et hæ partes sunt latius lustratæ, et alia quarta pars per Americum Vesputium ... inventa est, quam non video cur quis jure vetet ab Americo inventore sagacis ingenii viro Amerigen quasi Americi terram, sive Americam dicendam: cum et Europa et Asia a mulieribus sua sortita sint nomina*. (HARRISSE, *Bibl. Americ. Vetust.*, p. 94).

2. *Argentinæ 1509*. (Anon.) *Globus Mundi Declaratio* etc. (HARRISSE, cit. work, p. 117). Even in this brochure it is occasionally spoken of: *Ipsa America Noviter reperta, quarta orbis pars*.

For a minute bibliographical description of these, in some respects, very remarkable brochures, I may refer to the above mentioned works of d'Avezac and Harrisse. According to d'Avezac six editions were published of Waldseemüller's *Cosmographiæ Introductio*: four in 1507 at Saint-Dié, one at Strassburg in 1509, and one, without date and place of printing, at Lyons. This work should not be confounded with: *Cosmographiæ introductio, cum quibusdam Geometriæ ac Astronomiæ principiis ad eam rem necessariis*, of which the first edition was printed in Ingolstadt 1529, and of which Petrus Apianus is regarded as the author. Several later editions are known.

3. *Cracoviæ 1512*. STOBNICZA, *Introductio in Ptholomei Cosmographiam*. Tab. VII verso, is written: *Non solum prædicte tres partes nunc sunt lacius lustrate verum et alia quarta pars ab Americo vesputio sagacis ingenii viro, inventa est, quam ab ipso Americo eius inventore amerigem quasi a americi terram sive americam appellari volunt*, and in the margin: *Quarta pars Orbis America*. In some foreign libraries I have seen a variety of this edition of which the colophon does not contain the year of printing. It was probably printed before 1512.

4. *Nurembergæ 1512. Aristotelis Meteorologia* (comp. p. 40).

That the name »America» was already generally adopted in 1512, may be seen by *Ioachimi Vadiani Helvetii poetæ laureati ad Rudolphum Agricolam Rhetum epistola*, dated *Viennæ Austriæ 1512*, but published in print a few years later. »America» is here used as the name of the New World without any further explanation. In an edition of Ptolemy, this name was used for the first time in 1522 on the map of Laurentius Frisius, reproduced on pl. XXXIX.

2. *Ptolemaeus Argentorati 1520*. All the maps of this edition, with the exception of *Tabula Nova Eremi Helvetiorum*, are reprints from the blocks used for the maps in the edition of 1513.

3. Schöner's globe of 1523 (comp. p. 80). Schöner's large globe of 1520 was drawn by hand; his globe of 1523 appears to have been printed in gores.

4. A map of the world in: (ANTOINE DE LA SALLE) *La salade nouvelle imprimee* etc.... (Colophon:) *Imprime en la Rue sainct iacques a lenseigne de la Rose blanche Et fut acheue le dixhuytiesme iour de januier*. s. l. et a. (Paris 1522 accord. to BRUNET). Fig. 18 is a reproduction of this map from the original edition in the Royal Library at Stockholm. The map, which was already antiquated when published in print, is evidently a copy from an original of the 15th century. La Salle was born in about 1398 and died after 1461. His *La Salade* was written in about 1440. Considering the time at which it was written, it contains (on fol. 49 edit. 1522) some very remarkable notices about *Greenland, Iceland*, and *Mare congelatum*. A comparison of the fac-simile published by Santarem with the map in the edition of 1522,[2] shows some discrepancies. Thus the lands *(frise, insequi* and *nasque?)* in the southern hemisphere on Santarem's copy are omitted on the original map, whereas an inscription, *mina mons*, on the mountains in the equatorial part of western Africa is found on the copy of 1522, but not on the reproduction of Santarem. Perhaps the original followed by Santarem belonged to the edition of 1527. If the map, as seems probable, had been constructed before the middle of the 15th century, it is of no slight interest, from the connection of the Indian and Atlantic oceans, and from the peninsular form of Africa here (if we except the maps of the Macrobius type) for the first time adopted.

Among the legends, the name *Patalie regio*, by which the continuation of Asia occupying the south-eastern corner of the map has been designated, may be mentioned. The name is found in Pliny (*Historiæ Mundi* Lib. II. cap. 73 and Lib. VI cap. 20 and 21). In Roger Bacon's *Opus Majus* (Londoni 1723, p. 192; the passage is copied by D'AILLY in the 2d chap. of *Ymago Mundi*) this region is described in the following remarkable words: *Sed, quod plus est, invenimus per eum* (Plinium) *habitationem fieri sub tropico Capricorni ultra. Nam regio Pathalis in India dicitur habens portum, ut dicit, celeberrimum, ubi umbræ solum in meridie cadunt; ergo habitatores ejus habent semper solem ad Aquilonem*.

Accordingly, Pliny and Bacon place *Regio Patalis in India*, with its celebrated harbour, to the south of the southern tropic. The only land found here, and to which *in India* would be applicable, is Australia. It is also in this part of the globe that the *Patalie regio* is placed on La Salle's map. On Behaim's globe (Ghillany's reproduction), a part of the East Indian peninsula is designated by that name.

5. The maps in Ptolemy's geography, edit. Argentorati 1522 and 1525, Lugduni 1535, and Lugduni-Viennæ 1541. I have already, pp. 20—24, given a brief account of these edi-

---

[1] Compare (D'AVEZAC) *Hylacomylus Waltzemüller*, Paris 1867; HUMBOLDT, *Kritische Untersuch.*, II: p. 318 and other passages; HARRISSE, *Bibl. Am. Vet.*; CARTER-BROWN, *Bibliogr. Notices*, Providence 1875. The map is completely reproduced in: A. E. NORDENSKIÖLD, *Om en märklig globkarta från början af sextonde seklet*, in *Ymer* 1884, p. 167.

[2] The map fig. 18 in the Swedish edition of this atlas is a fac-simile of Santarem's reproduction.

tions, whose maps are printed from the same blocks. Most of the maps are reduced copies of the maps in the edit. Argentinæ 1513. The reduction is made by Waldseemüller. Only three original maps, or maps not printed before, are added in these editions, namely,

> Tabula moderna Indiæ Orientalis (N. fig. 62).
>
> Tabula superioris Indiæ et Tartariæ majoris (N. fig. 63), and
>
> Orbis typus universalis iuxta hydrographorum traditionem exactissime depicta 1522. L. F. (N. T. XXXIX).

The last mentioned map, by LAURENTIUS FRISIUS, has obtained some notoriety, because it applies, for the first time in an edition of Ptolemy, the name of *America* to the southern part of the New World. For the rest it is, as well from an artistic as from a geographical point of view, one of the

auspicio praelo nuper demandari curavit. The lower part of the title-page is occupied by a small circular map of the old hemisphere. Colophon (above the printer's monogram:) *Impressum Landsshut per Ioannem Weyssenburger* (s. a.), four leaves in 4:o. After a »*Tetrastichon*» by *Joh. Aventinus*, printed on the reverse of the title-page, and an »*Elegidion*» by *Ioannes Dengkius*, follows a preface by *Petrus Apianus ex Leyssnigk*, where he speaks of a new map of the world in these words, *Terrestris convexitatis picturam nova quadam et vera magisque habitationi nostrae idonea imagine: quo Geographicæ picturæ usus intellectu facilior redderetur, elucubravi.* Then follow five pages of text: »*De diversis usibus hujus Mappae,*» divided into eleven »*propositiones.*» From this we may conclude that the brochure was destined to serve as an explanation of a new map of the world, which was probably

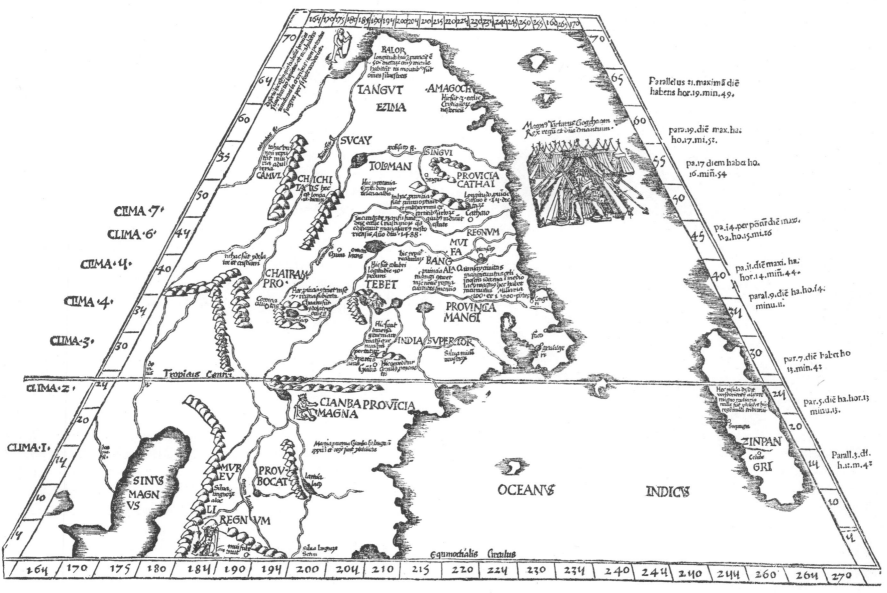

63. Tabula Superioris Indiæ et Tartariæ Majoris, from: Ptolemæus, Argentorati 1522. (Orig. size 461 × 293 m. m.).

rudest and most imperfectly drawn maps of the age. Even parts of the old world, which had been well known since the time of the Romans, are here so distorted as to become scarcely recognizable. The North American continent, already laid down by several previous cartographers, is here entirely omitted; Iceland is placed among the Britannic Islands; Greenland close to the western coast of Norway, etc. Also the two new maps of Asia (N. fig. 62 & 63) are coarse wood-cuts and, from a geographical point of view, by no means comparable to Ruysch's map of 1508 (N. T. XXXII) or to the modern maps of Africa in the Ptolemy of 1513 (N. fig. 8 and 9). They are almost exclusively based on Marco Polo and Martin Behaim, and convey, if regarded as an original production of Waldseemüller, no favorable idea of the skill and geographical learning of the author.

    6. *Apianus c. 1522.* About this year Apianus published: *Isagoge In typum Cosmographicum seu mappam mundi (ut vocant) quam Apianus sub Illustrissimi Saxoniæ Ducis*

identical with the above mentioned *Tipus Orbis Universalis* of 1520 by Apianus (N. T. XXXVIII), or with a new edition of it.

    Another small tract by Apianus, which is evidently only a slightly enlarged edition of that before mentioned, was offered for sale in 1885 by B. QUARITCH under No. 28131 of his catalogue 362. Its title is: *Declaratio et Usus Typi Cosmographici* (Landisutæ) 1522, eight leaves in 4:o. On the map of the title-page there is, to the east of Asia, an island indicated by the letters *A M*. I have not seen this edition, nor two others cited by Harrisse after Varnhagen, of which one was printed: *Ratisbonne per Paul Rhol 1522;* the other published without date or place of printing. These insignificant brochures are here mentioned on account of the information they give regarding earlier cartographical works of Apianus. They seem to indicate that his map of 1520, at present so rare, once had a wide circulation.

    7. The maps in the cosmography of PETRUS APIANUS. Of this work, which for a long time served as a cosmogra-

phical manual at the universities, a number of editions in different languages[1] were published during the 16th century, the first at Landshut, in 1524, with the title: *Cosmographicus liber Petri Apiani Mathematici studiose collectus.* All the editions contain several small maps or geographical diagrams serving to illustrate the text. Of these the following are of a certain interest in the history of cartography:

1. A drawing of a terrestrial globe, printed on the title-page. A similar globe from the edition *Parisiis 1551* is here reproduced on pl. XLIV.

2. A diagram (N. fig. 58) intended to illustrate the doctrine of antipodes, and a small map of one of the hemispheres on the equidistant meridian-projection of Bacon. America is here evidently recognized as the fourth part of the world.

3. Two sketches, the one a small map of the old hemisphere with a human head, and the other a castle with parts of the human face as a pendant. These figures are intended to explain Ptolemy's definition of the difference between geography and chorography.

64. Map to indicate the size of the earth. From: APIANUS, Antverpiæ 1545. (Orig. size).

4. A sketch illustrating the old theory of the architecture of the cosmos. The earth is here placed in the centre of the universe and surrounded by the celestial spheres: *Coelum Primum Mobile, Crystallinum; Firmamentum; Coelum Saturni, Jovis, Martis, Solis, Veneris, Mercurii,* and *Lunae.* Within the *Coelum Lunae* there are the spheres of fire and air, and finally, in the centre, the globe inhabited by man.

5. Various drawings illustrating the doctrines on climates, on latitude and longitude, and on map-projections.

6. The figure given above, illustrating the size of the earth.

7. The same wood-cut as that on the title of the above cited *Isagoge* by Apianus.

8. A small map of Greece.

9. A small map giving, quite correctly, the relative positions of various towns of central Europe.

10. The *Speculum cosmographicum,* a revolving geographical diagram of stiff paper, invented by Apianus to explain the fundamental doctrines of mathematical geography. Though long deemed insufficient for that purpose, this diagram merits a place in the history of cartography, because it is the first instance of a geographical map on the stereographic projection (N. fig. 57). Its net of graduation seems to have been copied from the drawing in Reisch's *Margarita philosophica* of 1512, of which I have given a reproduction on fig. 55.

11. A map of the world by Gemma Frisius (N. T. XLIV) which, after 1544 (?), was added to the different editions of Apianus' cosmography. Although very rude and void of details, this map was often reproduced during half a century. As to the outlines of the continents, it seems mainly to follow Münster's general map of 1540 (N. T. XLIV) and his map of America (N. fig. 71).

Although the above mentioned maps or geographical drawings of Apianus are of slight value, yet they introduced two important innovations into cartography. We here find two methods of constructing maps, proposed long before, for the first time practically employed, namely: *Bacon's equidistant meridian-projection* and the *stereographic polar-projection.* But on the other hand, I do not think it justifiable, on account of the insignificant diagram given by Apianus, to illustrate the delineations of meridians and parallels, that the priority in employing the oval projection of Bordone (comp. p. 90) should be ascribed to him.

8. Two maps in: *De orbis situ ac descriptione, ad reverendiss. D. archiepiscopum Panormitanum, Francisci, Monachi ordinis Franciscani, epistola... In qua Ptolemaei, caeterorumque superiorum geographorum hallucinatio refellitur, aliaque praeterea de recens inventis terris, mari, insulis. Deditio papae Ioannis De situ Paradisii, et dimensione miliarum ad proportionem graduum coeli, praeclara et memoratu digna recensentur.* (Colophon:) *Excudebat Martinus Cæsar, expensis honesti viri Rolandi Bollaert... Antverpiæ* (s. a. sed 1524), in 12:0.

I have only had access to a later edition of this work, (Antverpiæ 1565) containing no maps. According to HARRISSE (*Bibl. Am. Vetust.,* p. 243) there is, on the reverse of the title-page of the original edition, a map with the inscription: *Hoc orbis Hemisphærium cedit regi Lusitaniæ,* and, on the front-page of the following leaf, another: *Hoc orbis Hemisphærium cedit regi Hispaniæ.* Unfortunately no more detailed description is given of these maps, which I assume to be of slight importance. What Harrisse says regarding them hardly agrees with the map of *Franciscus Monachus ordinis Franciscanorum,* of 1526, of which a copy is given in LELEWEL's Atlas, pl. XLVI.

9. *Præclara Ferdinandi Cortesii de Nova Maris Oceani Hispaniæ Narratio Sacratissimo ac Invictissimo Carolo Romanorum Imperatori... Anno Domini MDXX transmissa.* (Colophon:) *...Impressa in Celebri Civitate Norimberga... MDXXIIII.*

This edition of the celebrated letters of Cortes contains, like the two editions of Venetiæ 1524, a plan of the city of Mexico which, to judge from the description of Harrisse, is not found in the Spanish editions of Sevilla 1522 and Saragossa 1523. This and analogous plans of the famous city have often been copied in the 16th century. It is of interest to compare them with an original drawing by ALONZO DE SANTA

---

[1] The following editions, printed before 1550, are cited by HARRISSE:
In Latin: Landshutae 1524*; Antverpiae 1529*; Antverpiae 1533*; Parisiis 1533; Antverpiae 1539; Antverpiae 1540; Norimbergae 1541; Antverpiae 1545*; and Antverpiae 1550. In French: Anvers 1544. In Spanish: Enveres 1548. Still, after 1550, editions were, according to SIEGMUND GÜNTHER (*Peter und Philip Apian,* Prague 1882) published at Antverp in 1553, 1564*, and two in 1584*. To these may further be added a French edition, Parisiis 1551, and one Dutch: *tot Amstelredam* 1598. The editions designated by an * exist in Swedish collections. This work of Apianus seems often to have been confounded with (Apianus) *Cosmographiae introductio cum quibusdam Geometriae ac Astronomiae principiis ad eam rem necessariis,* of which the first edition was published at Ingolstadt in 1529. This geographical compendium, which, in its turn, is not to be confounded with Waldseemüller's work with the same title, also contains, besides a number of cosmographical figures, of which one (in the 8th chapter) illustrates the variation of the compass, drawings of globes, and a small map of Greece.

Cruz of the Mexican capital and its environs, of which I have given a short account above and of which fig. 69 is a much reduced fac-simile.

10. (Laurentius Frisius:) *Yslegung der Mercarthen oder Cartha Marina, Darin man seken mag, wa einer in der welt sey, und wa ein ietlick Land, Wasser und Stat gelegen ist. Das als in den bucklin zefinden.* (Colophon:) ... *Strassburg, Johannes Grieninger 1525.* This work, of which new editions were published at Strassburg in 1527 and 1530, contains two maps. The first of these: *Tabula prima Navigationis Aloisii Cadamusti Mediram versus,* is a rudely drawn chart of the coasts of the southern part of the Pyrenean peninsula and northern Africa with the Atlantic ocean as far as Madeira; the second I have not seen. Nor have I had access to the work itself in which these maps were published, and regarding which the reader is referred to Harrisse, *Bibliotheca Americana Vetustissima,* p. 246. Laurentius Frisius or Laurens Fries was a physician at Metz, much interested in mathematical studies. Conrad Gessner characterizes one of his medical works as an *Opusculum ineptum, mille mendis refertum, ac eruditis auribus indignum.*

the map shows real progress. It is, also, specially interesting from the illustration here given of the controversies between the Spaniards and the Portuguese respecting the proper place of the famous Papal line of demarcation.

The map has previously been reproduced in the *Works issued by the Hakluyt Society,* London 1850. The fac-simile given on T. XLI is from a photograph of the original at the British Museum. Thorne was one of those wealthy, intelligent, and enterprising merchants who energetically contributed to the development of England's commercial predominance at sea. In order to make it possible for English tradesmen to compete with those of Spain and Portugal, he was eager for the discovery of a northern passage to China, Japan, and India, and to promote voyages of discovery to achieve such a passage he sent memorials to Edward Leigh, the English Ambassador in Madrid, and to King Henry VIII. It is these letters which Hakluyt has published in his above mentioned work, and which are of such importance to the history of geography. The other drawing in *Divers Voyages* is a map of the northern hemisphere on an equidistant polar-projection. It was constructed by Michael Lok, *civis Londinensis,* 1582, and dedicated *Illustri viro Domino Philippo Sidnaeo.*

65. Map of the world from: Pietro Coppo, *Portolano,* Venetia 1528. (Orig. size).

11. Robert Thorne's map of 1527 (N. T. XLI). This map is inserted into Richard Hakluyt's *Divers Voyages touching the discouerie of America and the Ilands adiacent vnto the same, made first of all by our Englishmen and afterwards by the Frenchmen and Britons ... with two mappes annexed heereunto for the plainer understanding of the whole matter...* London 1582; reprinted in 1850 with a valuable introduction and illustrative notes by John Winter Jones, in *Works issued by the Hakluyt Society.*

A long inscription on the right side of this map says that it was sent from Seville by the merchant Master Robert Thorne to *Doctor Ley, Embassadour for king Henry the 8. to Charles the Emperour.* Harrisse (*Cabot,* p. 176) supposes it to be based on the prototype of the planispheres in Weimar and on the map of Nuño Garcia de Torrena. This can only be the case as to the delineation of the New World, the work of Thorne being, as regards the Old World, so exact a copy of the map in Reisch's Margarita Philosophica of 1515 (N. T. XXXVIII), that no doubt is possible as to the principal source of Thorne's geographical knowledge of that part of the globe. But for the New World, he evidently had access to other sources, probably consisting of hand-drawn Spanish maps. As regards the delineation of South America and the Isthmus of Panama,

12. A small map in (Pietro Coppo) *Portolano,* (colophon:) *Stampata in Venetia per Augustino di Bindoni. 1528. Adi. 14. de Marzo;* very small quarto (Harrisse, *Bibl. Am. Vet.,* p. 264). The fac-simile given here (fig. 65) is a reproduction of the map in the copy at the British Museum. Its principal interest consists in its affording a proof of the erroneous notions of the New World still entertained, even in Venice, at the time when the map was printed. If any projection could be spoken of, with reference to a map so awkwardly drawn as this, we should here have the first map drawn on the oval equidistant projection of Bordone. The first edition of Bordone's work is dated a few months later than Coppo's Portolano. Besides the map of the world, this work is said to contain, on the reverse of the title-page, another probably still more insignificant map. Coppo's Portolano is also mentioned by Zurla in: *Di Marco Polo et degli altri viaggiatori Veneziani con Appendice sulle antiche mappe idrogeographice lavorate in Venezia,* Venezia 1818, II: p. 363. But the copy he describes contained seven wood-cut maps.

In Lafreri's Atlas there is a large map *Disegno dell' Istria di M. Pietro Copo,* engraved in copper by Ferrando Bertelli in 1569 and dedicated to Aldus Manutius.

13. The maps in: *Libro di Benedetto Bordone, Nel qual si ragiona de tutte l'Isole del mondo con li lor nomi*

*antichi & moderni, historie, fauole, & modi del loro
uiuere... Con il Breve di Papa Leone. Et gratia & pri-
uilegio della Illustrissima Signoria com' in quelli appare.
MDXXVIII.* As there has been some difference of opinion
about the real date of this work, it may here be remarked, that
the letter of privilege of the Pope, printed on the verso of the
title-page and dated 1521, does not specially refer to this book
of Bordone, but to all works which the printer, NICOLO D'ARI-
STOTELE, DETTO ZOPPINO, with the permission of the Pope, had
published or might hereafter publish. Another letter from the
Signoria of Venice, inserted immediately after the patent of the
Pope, and issued exclusively for *»Benedetto Bordone Miniator,»*
is dated 1526. But even then the work seems not to have been
finished, although the author declares himself to have been
employed on it for many years, night and day. Lelewel's
antedating of Bordone's maps to 1521 (comp. LELEWEL's Atlas,
pl. 46) does not, therefore, appear to me to be justifiable. Bordone's expression on fol. LXXIII: *Quando lo vescovo di Ra-
coscia scrive a Leone Summo pontifice, haver veduto, tutto
quello che io ho della Norbegia, ragionato,* is only a proof
that he had been occupied on the work previous to the year
of the death of Leo X, 1521.

Bordone's work begins with an unpaged introduction
containing three large maps printed on two folio-pages, viz.

A map of Europe and northern Africa. A rude, un-
graduated wood-engraving, on which only the towns *Lisbona* and *Venetia* are noted. It is, as far as I know, the first
printed special map of Europe, but in other respects it presents
nothing of interest to the cartographer.

A map in double folio of the Greek Archipelago and
surrounding lands. A portolano has evidently served as a
model for this very badly executed wood-cut.

A map of the world (N. pl. XXXIX). This map is also
a coarse wood-cut, poor in geographical details. But it is
of interest owing to the new, handsome projection on which
it is constructed.

This introduction is followed by the text of the *»isolario,»*
containing the following wood-cut maps, generally only occupying
⅓ of a page: *Islanda, Irlanda, Inghilterra secondo Moderni,
Inghilterra secondo tolemeo, Parte della bretagna, Parte de
hispagna, Norbegia* (from Ptolemaeus Ulmæ 1482), *Terra de
lavoratore, La gran citta di Temistitan,* the northern coast
of South America with adjacent islands, *Spagnola, Iamaiqua,
Cuba,* other West-Indian islands, *Guadalupe, Matinina;* five
maps of groups of islands on the western coast of Africa;
Gades; seven maps of islands in the western part of the Mediterranean; *Venetia* (large map in double folio); three other
views of cities; 64 maps of islands etc. in the Adriatic Sea
and the Archipelago; the Sea of Marmora with Constantinople;
the Crimea and Sea of Azof; Cyprus; 8 maps of islands off
the eastern coast of Asia, and in the Indian Ocean. This
enumeration shows that Bordone's work contains a considerable number of maps, among which there are 9 relating to
the New World and 8 of islands on the eastern coasts of
Asia and in the Indian Ocean. But unfortunately all these
maps, with a few exceptions (e. g. the map of *La gran citta
di Temistitan* and *Venetia*), are almost worthless.

Owing to the small number of modern geographical works
published in the beginning of the 16th century, Bordone's
book became very popular, as may be judged from the many
editions of it published under the title of *Isolario di Benedetto Bordone.* Harrisse cites the editions 1532, 1534, 1537,
1547. I have compared the maps of the editions Venetia
1534 and 1547 with the edition of 1528. They are all
printed from the same blocks. Even the text appears to be
unaltered, excepting an addition to the later editions: *Copia
delle lettere del Prefetto della India la nova Spagna detta,*
alla Cesarea Maesta rescritte, which is wanting in the original. The statement (HARRISSE, *Additions,* p. 113) that the
edition of 1534 was only a title-edition of Vinegia 1528 is
not correct.

I have not had an opportunity of comparing the maps of
Bordone's Isolario with those in the *Isolario di Bartolomeo da
li Sonetti,* Venetia 1477 (comp. p. 36), but I suppose that
several of Dali Sonetti's maps are copied by Bordone. Bordone was a renowned miniature-painter; which one would not
have suspected from the way his maps are drawn. A few
years after the issue of the last edition of the Isolario, an atlas
of the same kind was published by TOMASSO PORCACCHI under
the title of *l'Isole piu famose del Mondo,* of which several
editions were published in Italy from 1572 until late in the
17th century. I give a fac-simile of one of these maps on
pl. XLIX. They are engraved in copper by GIROLAMO PORRO,
and are incomparably better than the wood-cuts of Bordone.

14. A map of the region of Avignon in: *Il Petrarcha
con l'espositione d'Alessandro Vellutello... s. l. 1528. 4:0.*
The map occupies two pages. It is executed in wood-cut,
and renders in a clear and doubtless in a tolerably correct
manner the topography of the ancient Papal residence. On
account of the small number of special maps or *»chorogra-
phiae»* printed before 1540, or the year when Münster began
to publish his voluminous geographical works, it deserves to
be mentioned among prints of the early period of cartography.
This map probably occurs already in the edition of 1525 of
VELLUTELLO's Petrarca and probably also in other editions of
the great poet's work.

15. *Totius Galliae descriptio O. F. Delphinato* (Orontius Finæus) *autore.* An edition of 1525 is preserved in the
Bibliothèque Nationale at Paris (according to private communication by M. G. MARCEL). In the catalogue of printed
maps at the British Museum, editions of 1561 and 1563 are
mentioned. The map has probably been often reprinted, generally without mentioning the printer's name, e. g. by SEB.
MÜNSTER 1540. In Lafreri's atlas there is a fine reproduction,
which, according to a long legend in the right corner, is
printed *»ex aeneis formis Bolognini Zalterii 1566,»* and dedicated by PAULUS FURLANI to *Magnifico ac insigni viro
Marco Antonio Radici.*

16. *Apianus 1530.* This map was, in 1885, offered
for sale, for £ 40 by Quaritch (Catal. 362, No. 28142) under
the rubric: *»Apiani Universalior Cogniti Orbis tabula 21¾
× 15½ inches* (= 552 × 394 m. m.) *(Ingolstadii) 1530. Unique».*
It seems to be a reproduction on a larger scale of the map
of Apianus of 1520. The map is dedicated to LEONARDUS
AB ECK. At its upper part there are two small maps of the
world: *Observatio Ptolem.,* and *Observatio Vespu.*

17. *Figura e scrittura insomma di tutto lo abitato*
from *Isolario di Bartolomeo Dalli Sonetti,* Venezia 1532 (2d
edition, fol.) An interesting map of the world, on Bordone's
oval projection. The size of the oval is 295 × 155 m. m.
North America forms a continuation of Asia, being separated
from *»Terra S. Crucis sive Mundus Novus»* (South America)
by a broad strait, in which *»Zinpagu»* is placed. The draughtsman of the map has not taken any notice of Magellan's
voyage. I have seen this map in the Biblioteca Marciana.

18. EANDAVI's edition of Ptolemy, *Argentorati Apud
Petrum Opilionem 1532.* This edition, which I have never
seen and which I have in vain asked for in several of the
large libraries of Europe, is said to contain eight maps, of
which, to judge from the legends communicated by Harrisse
(*Terra Bacalaos, ulteriora incognita, Gronlandia, Hvit-
sargh Promont.),* at least one seems to be of a certain interest
from a geographical point of view. But I suppose these maps
to be identical with the eight maps in another work of Ziegler

cited below, which was published by the same printer, in the same year, and at the same town as the edition of Eandavi.

19. Maps in Jacobus Ziegler's *Quæ intus continentur Syria* etc., *Argentorati per Petrum Opilionem,* 1532. Fol. This book contains eight maps in double folio. I have before (p. 60, fig. 30) given an account of one of them (the map of Scandinavia). The remaining seven are maps of lands situated about the south-eastern part of the Mediterranean, from Marmarica, across Egypt, Suez, Syria with Palestine to Cilicia. These maps are also clumsy wood-cuts of no special interest from a cartographical point of view, if I except the fifth map, *Universalis Palestinæ,* which affords the first instance of the variation of the compass being indicated on a printed map or chart. On the same map the azimuths and the distances from Jerusalem to Rome, Venice, Reginospurgum, Carræ, Ekbatana, Ninus, Babylon, and Susa are indicated. We here

eodem interprete; Alberici Vesputij navigationum epitome; Petri Aliaris navigationis & epistolarum quorundam mercatorum opusculum; Iosephi Indi navigationes; Americi Vesputij navigationes IIII; Epistola Emanuelis regis Portugalliae ad Leonem X. Pont. Max. de victorijs habitis in India & Malacha etc.; Ludovici Rom. patritij navigationum Æthiopiæ, Ægypti, utriusque Arabiae, Persidis, Syriae, Indiae, intra & extra Gangem, libri VII. Archangelo Madrignano interprete; Locorum terrae sanctæ exactissima descriptio, autore F. Brocardo monacho; M. Pauli Veneti de regionibus Orientalibus libri III; Haithoni Armeni ordinis Praemonstrat. de Tartaris liber; Mathiæ a Michou de Sarmatia Asiana atque Europea lib. II; Pauli Iovij Novocomensis de Moschovitarum legatione liber; Petri Martyris de insulis nuper repertis liber; Erasmi Stellae de Borussiae antiquitatibus lib. II.

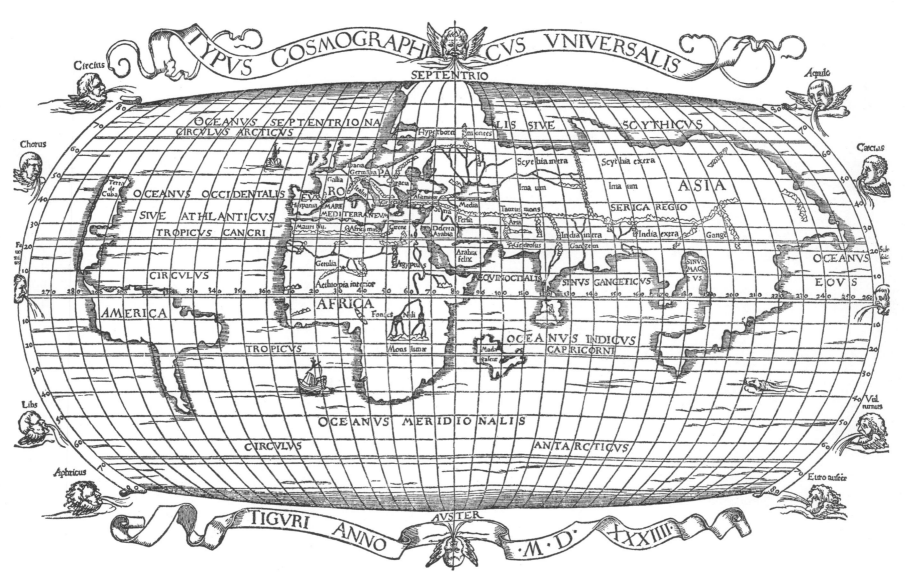

66. General map from: Joachimus Vadianus, *Epitome trium terræ partium,* Tiguri 1534. (Orig. size 375 × 235 m. m.).

find an illustration of the way the lands and seas had been mapped on the portolanos. Ziegler's work was reprinted Argentorati 1536. The new edition contains the same maps printed from the same blocks as those of the editio princeps.

20. Map in: (Simon Grynaeus and Ioan. Huttichius) *Novus Orbis Regionum ac Insularum veteribus incognitarum, una cum tabula cosmographica, et aliquot alijs consimilis argumenti libellis, quorum omnium catalogus sequenti patebit pagina... Basileae apud Io. Hervagium, Mense Martio, anno MDXXXII.* (N. T. XLII.) The contents of the work are given on the reverse of the title-page: *Catalogus eorum quae hoc volumine continentur. Praefatio Simonis Grynaei ad Collimitium; In tabulam cosmographiae introductio per Sebastianum Munsterum; Aloysij Cadamusti navigatio ad terras ignotas, Archangelo Madrignano interprete; Christophori Columbi navigatio ex iussu Hispaniæ regis, ad multas insulas hactenus incognitas, eodem Madrignano interprete; Petri Alonsi navigatio, eodem interprete; Pinzoni navigatio,*

The name of the compiler of this valuable collection of voyages, of which, however, the majority had been previously published, is neither recorded on the title-page nor in the prefaces. The work is usually cited under the name of Grynæus, on account of his preface written in the florid style of the age, and containing scarcely anything of interest to geography. It is addressed to *Georgius Collimitus Danstetterus Artis Medicae et disciplinarum Mathematicarum omnium facile princeps.* At the end of it Grynæus says that the editor, Hervagius, got the pamphlets here reprinted from *Ioan. Hutichius, vir doctus et antiquitatis mire studiosus.* The work is therefore sometimes cited under the name of *Collectio Huttichii-Grynaei-Hervagii.* Sebastian Münster has also contributed to this publication by a long introduction entitled: *Typi Cosmographici et declaratio et usus,* from which it has been concluded that he was the author of the map inserted in the editio princeps, Basileæ MDXXXII (N. T. XLII). But this seems scarcely admissible. For there is very little in common between this map and the general map in Münster's *Ptolemaeus*, Basileæ 1540. The inscriptions, the distribution of land, the method of drawing the meridians on the oval projection — all are different. The general character of the wood-cuts alone is the same, proving that these maps have issued from the same school of engravers.

This map of 1532 and Münster's *declaratio* were already quite antiquated when they were printed in *Novus Orbis*. The author of the map did not know e. g. of the first circumnavigation of the globe, as may be concluded from the omission of the south-polar continent, the discovery of which had been foretold by several cosmographers, while its existence was considered to have been confirmed by Magellan. Neither is there to be found, in Seb. Münster's introduction, any passage alluding to this memorable event, although Münster discusses the influence the voyages of the Portuguese were likely to exercise on the Indian trade of the Venetians via Suez, and although Münster praises the discoveries of Columbus and Vespucci in the following enthusiastic words: *Celebratur Alexander magnus toto orbe terrarum, quod primus fere penetrarit Orientem, non navigio, sed pedestri et tutiori itinere in illum deductus. Verum parva erit laus illius, si comparetur viris illis qui nostro aevo maria etiam incomperta sulcare tentarunt, et Occidentem sua exploratione aperuerunt.* This may be explained by assuming that the introduction was written in the year 1520, inscribed with the handsomely sculptured initial letter in the edition of 1532 of Münster's *Declaratio*, consequently before any accounts of the discovery of the straits of Magellan had reached Europe. The legends on the map are partly printed by movable types. Harrisse mentions two varieties, of which that given here on pl. XLII answers to his type B. From this the type A differs only by Asia being printed with somewhat larger types. Harrisse's third type, C, differs to a greater extent, and seems not to have been intended for *Novus Orbis*, but for Münster's own geographical works.

If the type B is compared with Schöner's globe of 1515 (N. fig. 46 and 47), there will be found so much affinity as regards the delineation of North America, the position of Iceland, and various other details, that it seems probable that a common original or prototype has served for both. The south-polar continent retained (or added) by Schöner is here omitted.

Among the legends on the map, the long inscription on the *Scytarum regio* deserves attention. North America is called *Terra Cuba*, and to the east of its most northern part a large island is drawn, called *Terra Cortesia*, instead of *Terra Corterealis*.

21. The map in: (Grynaeus-Huttichius) *Novus Orbis* etc. *Parisiis apud Joannem Parvum.* (Colophon:) *Impressum Parisiis apud Antonium Augerellum MDXXXII. VIII Calen. Novembris* (N. T. XLI). We have here an unmodified reprint in Paris of the edition of »Novus Orbis» published some months earlier at Basel. The original map is here replaced by a new one engraved in 1531 by Orontius Finæus, at the expense of Christian Wechel. I have above described the projection employed for it and also a later reprint on which the name of Orontius is suppressed and only Wechel mentioned on the title-legend, printed with moveable types. Another reprint, probably from the same block, but with a new modification of the title-legend, is employed for the: Pomponius Mela, *Parisiis apud Christianum Wechelum. MDXL.* (Comp. Harrisse, *Bibl. Am. Vetustissima, Addenda*, p. 133). In a geographical point of view this map far surpasses that of the Basel edition, as well in its greater richness of names and topographical details, as in due attention being here paid to the

latest geographical discoveries. The islands in the Polar-basin are here copied from Ruysch; Greenland is drawn as a large island; North America forms, in accordance with the conception of Columbus, a mere continuation of Asia, which, by a narrow isthmus, is connected with South »America.» In the south this part of the New World is separated by a narrow strait from a large south-polar continent: *Terra Australis recenter inventa sed nondum plene cognita*, on which the names *Regio Patalis* (comp. p. 100) and *Brasielie regio* are inscribed. No name is placed at the Straits of Magellan, but the sea to the west of it is called *Mare Magellanicum*, which as far as I know is the first time that the name of this discoverer occurs on a printed map of unquestioned date. On the isthmus connecting South America with *Parias* and *Teniscumatan*, is written *Furna* (Tierra) *Dariena*.

There is no reference to the map in the text. In the introduction Münster, on the contrary, expressly says: *Nam sub polo Antarctico compertum est nullam esse terram, saltem solidam*, which is altogether inconsistent with the leading geographical ideas as to the southern hemisphere, on the map of Orontius Finæus.

22. *Chorographia Franciae Orientalis. Das Franckenlandt von Seb. von Rotenhan* is cited with the date of 1520 (as still preserved?) in: *Beiträge zur Landeskunde Bayerns* (München 1884, p. 84). Another edition (?) is mentioned by Ortelius, under the year of 1533, in the *Catalogus Auctorum*. Later it was reproduced in his *Theatrum*, and in the Atlases of G. Mercator, Quad, and others.

23. A map of the world in: *Epitome trium terrae partium, Asiae, Africae et Europae compendiariam locorum descriptionem continens, praecipue autem quorum in Actis Lucas, passim autem Evangelistae et Apostoli meminere… per Joachimum Vadianum Medicum.* Of this work two editions, one in folio, the other in 12mo, were published at Zürich (»Tiguri») by Christophorus Frosch in 1534. The folio edition was provided with a general map, *Typus Cosmographicus Universalis, Tiguri anno MDXXXIIII* (N. fig. 66) drawn on the oval projection of Bordone and sometimes also bound up with the duodecimo edition (comp. Quaritch Catal. 352 No. 28144 & 28145). The map is a fine wood-cut, but from a geographical point of view it is only a poor copy of the map in *Novus Orbis*, Basileæ 1532. Europe as well as Asia and America have here got the same distorted shape as on the original. The West Indian Islands and Japan are altogether omitted. On the other hand *Madagascar*, whose place on the map printed at Basel is occupied by a long legend about India, has here got an excessive extension, with an incorrect and distorted form, and a too westerly position, all in conformity with the representation on Schöner's globe of 1515. The south-polar continent is omitted, a proof that no notices or at least no exact information about the circumnavigation of the world had, in 1534, penetrated to the learned humanist at St. Gallen. The work for which the map was drawn is a voluminous cosmography and geography on 274 pages in folio, or 564 pages in 12mo, ending with a long chapter on »*Insulae oceani praecipuae.*» But it only contains the following references to the New World (folio edition p. 267, 12:0 edition p. 551): *et in Africae parte quae ad occasum spectat, maxima insularum America cognominata obtenditur. Deinde longissimo ab occasu Continentis intervallo, Spagnolia, et ultra eam Isabella, dein Parias dicta, nuperrimis exploratoribus!*

24. A map in: Petrus Martyr & Oviedo, *Historia de l'Indie occidentali*, Vinegia MDXXXIIII.[1] Of this book only one copy containing the map is known. It belongs to the

[1] Concerning the complicated title of this work see Harrisse, *Bibliot.*, p. 313. I have given the dimension of the map from E. Uricoechea, *Mapoteca Colombiana*, Londres 1860, p. 2.

collection of Mr. JAMES LENOX. A reduced copy of the map was published by Mr. HENRY STEVENS, in his *Historical and geographical notes*, New Haven 1869. I have only had access to this reproduction, of which fig. 67 is a fac-simile on a still more reduced scale. As to the projection and the manner of drawing, the map much resembles the charts of Medina (N. fig. 75), with which it also agrees as to the outlines of the New World. On the map of 1534 the New World is designated by a long legend, *Terra Firma de le Indie occidentali*, running along the whole coast from the Straits of Magellan to Labrador. South America is called *Mondo Nuovo*. The West Indies are here laid down more correctly than on any map previously printed. It is also of

derable differences will be discovered. Finæus assumes North America to be a continuation of Asia; Mercator divides these parts of the world by a broad strait. The latter places a large continent in the vicinity of the North-pole; Finæus makes, in conformity with Ruysch's delineation, various large islands occupy this part of the earth's surface, a conception of the geography of the polar regions afterwards adhered to by Mercator himself. On Mercator's map we find one of the earliest indications of the river-system of La Plata. On the northern part of the American mainland, separated from a large polar-continent by a strait, *Fretum arcticum*, we read »*Hispania Major capta anno 1530*.» Its north-eastern portion is designated as *Baccalearum regio*, eastward of which a very

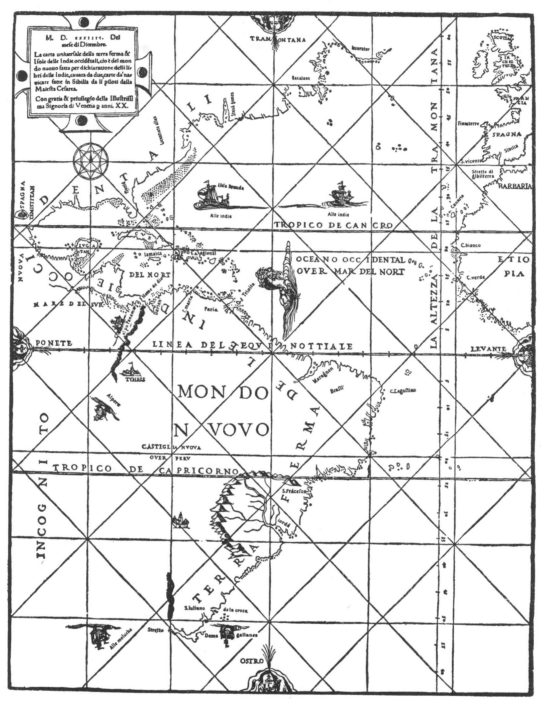

67. Map of America in PETRUS MARTYR, *Historia de l'Indie occidentali*, Vinegia 1534. (Orig. size 530 × 425 m. m.).

interest as being one of the few specimens preserved of the early cartographical works of the Spaniards.

25. A new edition of REISCH's *Margarita Philosophica* was published in 1535, which is sometimes said to contain the »Zoana Mela» map of the world (N. T. XXXVIII and p. 70) first inserted in the edition of 1515. But to judge from the many copies of the edition of 1535 which I had occasion to examine, it generally contains no other maps than some geographical diagrams printed in the text, and the map of the world originally drawn for the edition of 1503 (N. T. XXXI, p. 30).

26. MERCATOR's double cordiform map of 1538 (N. T. XLIII). This map is not enumerated by GHYMMIUS among the works of the great geographer. Superficially examined, it seems to be almost identical with the map of 1531 by Orontius Finæus (N. T. XLI), but on a closer examination consi-

large and several smaller islands, *Insulæ Corterealis*, are placed. Between Hispania and Terra Florida there is a large island, *Insula 7 civitatum*. Another mythical group of islands is laid down, under the name of *Losroccos insula*, between the south-polar continent and the Moluccas. On the south-polar continent we read: *Terras hic esse certum est, sed quantas quibusque limitibus finitas incertum*. This inscription reminds us of Mercator's design, in a large cosmographical work, of which the last book was to deal with geography, to divide the lands of the earth into three parts, viz. the *Old World* (Asia, Europe, and Africa), *India occidentalis* (America), and *Continens Australis*. Ghymmius thus sums up Mercator's opinion of the last named parts of the globe: *Tertiam etsi adhuc latentem et incognitam esse non ignoraverit, solidis tamen rationibus atque argumentis demonstrare ac evincere* se posse affirmabat, illam in sua proportione geometrica, magnitudine et pondere ac gravitate, ex duabus reliquis nulli cedere aut inferiorem vel minorem esse posse, alioquin mundi constitutionem in suo centro non posse consistere. Mercator here adopts or gives form to an old cosmographical theory, which, during centuries, exerted no slight influence on the voyages of discovery in the Pacific. It was founded on the supposition that the main part of the globe formed an hydrosphere, in which the earth-crust was freely floating, according to the laws of hydrostatics.

As mentioned above (p. 90), Mercator's cordiform map was reproduced with the utmost exactness and in full size by a fine copper-print (N. fig. 54), signed *Ant. Lafreri exc. Romae*, but not dated. In the British Museum there is a map of Malta by Lafreri dated 1551, and the last one (the map of Olaus Magnus), known to me, of this celebrated engraver is dated 1572. Some other important maps by him are dated about 1560, which I have adopted as the date for this remarkable reproduction of Mercator's map.

27. Some maps in C. Julius Solinus: *Rerum toto orbe memorabilium thesaurus locupletissimus, huic ob argumenti similitudinem Pomponii Melæ de Situ Orbis libros tres* ...

68. Moscovia from SOLINUS, Basileæ 1538. (Orig. size).

adjunximus. Accesserunt hic praeter nova scholia ... etiam tabulae geographicae permultae, regionum, locorum, marium, sinuumque diversorum situs pulchre deliniantes, Basileae 1538. This work contains the following maps in rude woodcut: 1. The ancient Rome; 2. Italy; 3. Morea; 4. *Typus Graeciae*; 5. *Figura Rhodi insulæ*; 6. The environs of *Fons Danubii*; 7. A map of Russia (N. fig. 68); 8. The environs of *Mare Hircanum seu Caspium*; 9. The Mediterranean and Black Seas; 10. *Helvetia*; 11. *Anglia*; 12. *Africa*; 13. *Palestina*; 14. *Asia Minor*; 15. Asia and parts of Europe, Africa, and America; 16. Europe with parts of Africa and Asia; 17. Europe; 18. *Palus Mæotica*; 19. Greece and a portion of the Balkan countries; 20. Sicily and southern Italy.

With the exception of 4, 12, 15, and 17, all these woodcuts are small and insignificant. The majority of them are remarkably poor, from a cartographical standpoint. Yet the maps 7 and 15 are of a certain interest, the former on account of the river-system of Russia being here for the first time represented with tolerable accuracy — even more correct than on the maps of ANT. WIED and HERBERSTEIN — the latter, on account of the delineation of the Pacific with a portion of the western coast of America. The map of Russia has already been reproduced in the Journal of the Swedish Geographical Society (*Ymer*, 1885, p. 262). It appears to be founded on commu-

nications from Herberstein and from the learned canon in Cracow, MATHIAS A MICHOU. This edition by Solinus-Mela was reprinted, with the maps from the old blocks, at Basel in 1543.

28. A map of La Maine by MATHEUS OGERIUS, printed *In urbe Cenomanorum* in 1539. A copy of the original map is preserved in the Bibliothèque Nationale at Paris (according to M. G. MARCEL). Reproduced by Ortelius, ed. 1595, pl. 22.

29. The map of Scandinavia by OLAUS MAGNUS (comp. p. 60) printed on 9 large folios at Venice in 1539 and reproduced by ANT. LAFRERI at Rome in 1572.

30. GERARD MERCATOR'S large map of Flanders of 1540. The copies of this excellent, and for a long time unsurpassed and unrivalled map, which once had a wide circulation, have been so destroyed and worn out, that the map until lately was considered as lost. But a copy having been newly discovered, it has been republished in fac-simile under the title: *De groote kaart van Vlaanderen vervaardigd in 1540 door G. Mercator, bij middel van lichtdruk weergeg. naar het ex. behoorende aan het Museum Plantin-Moretus ... en voorzien met eene verklarende inleiding door* J. VAN RAEMDONCK, Antwerp 1882, fol. obl.; text of 14 leaves in French and Dutch with 9 maps (according to P. A. TIELE, *Nederlandsche Bibliographie van Land- en Volkenkunde*, Amsterdam 1884, p. 168).

31. Printed gores to GERARD MERCATOR'S globe of 1541 (comp. p. 82).

32. *Palatinatus Bavariae Descriptio. Erhardo Reich Auctore* (1540). The original is cited (as still existing?) in *Beiträge zur Landeskunde Bayerns*, München 1884, p. 84. It was reproduced by ORTELIUS (edit. 1570, pl. 30), and by DE JUDAEIS (1593, II, pl. 26).

33. DEMONGENET c. 1541. I have not seen this map. In his Catalogue XLII ROSENTHAL cites it under No. 133 in the following words: *Planisphère. Carte de deux hémisphères. Dessinée par Franc. Demongenet et gravée par E. Vico. Venise c. 1541. Petit in fol. Haut. 145, Larg 263 m. m.* A globe-print on the projection of Glareanus by the same author, and dated 1552, is given in fac-simile by me on pl. XL. M. G. MARCEL, the Director of the collection of maps in the Bibliothèque Nationale at Paris, has found a globe by Demongenet, engraved in copper but not dated, and he is at present occupied with a memoir on this deserving but little known cartographer.

34. Maps in SEBASTIAN MÜNSTER'S editions of Ptolemy's geography of 1540, 1541, 1542, 1545, and 1552. I have already, p. 23, given an account of these different editions and of the maps in them. One of the maps is here reproduced in full size on pl. XLIV, and as further examples of Münster's cartographical productions I give on fig. 70 and 73 fac-similes of his map of Bohemia in the edit. 1545, and of the map of *Novæ Insulæ* in the edit. 1540. The former is remarkable as the first printed map illustrating the distribution of different religions, and may as such be regarded as the first statistical map. The latter seems chiefly to be based on the portolanos of Battista Agnese. It is of interest as being the first general map of the American continent, but it imparts no higher opinion of Münster's talent as a map-draughtsman, than his corresponding maps of *Europa*, *Ethiopia* (Africa), and *India Extrema* (Asia with the exception of its western-most part).

35. Maps in various editions of SEBASTIAN MÜNSTER'S cosmography, for the first time printed in German at Basel by HENRICUS PETRI in 1544, and often reprinted with considerable additions in several different languages. Of these editions I have examined those in German printed at Basel in 1544, 1567, 1578, 1592, 1628; those in Latin of 1550, 1552, 1559, also printed at Basel, and that in French published by FRANÇOIS DE BELLE-FOREST, Paris 1575. The first edition

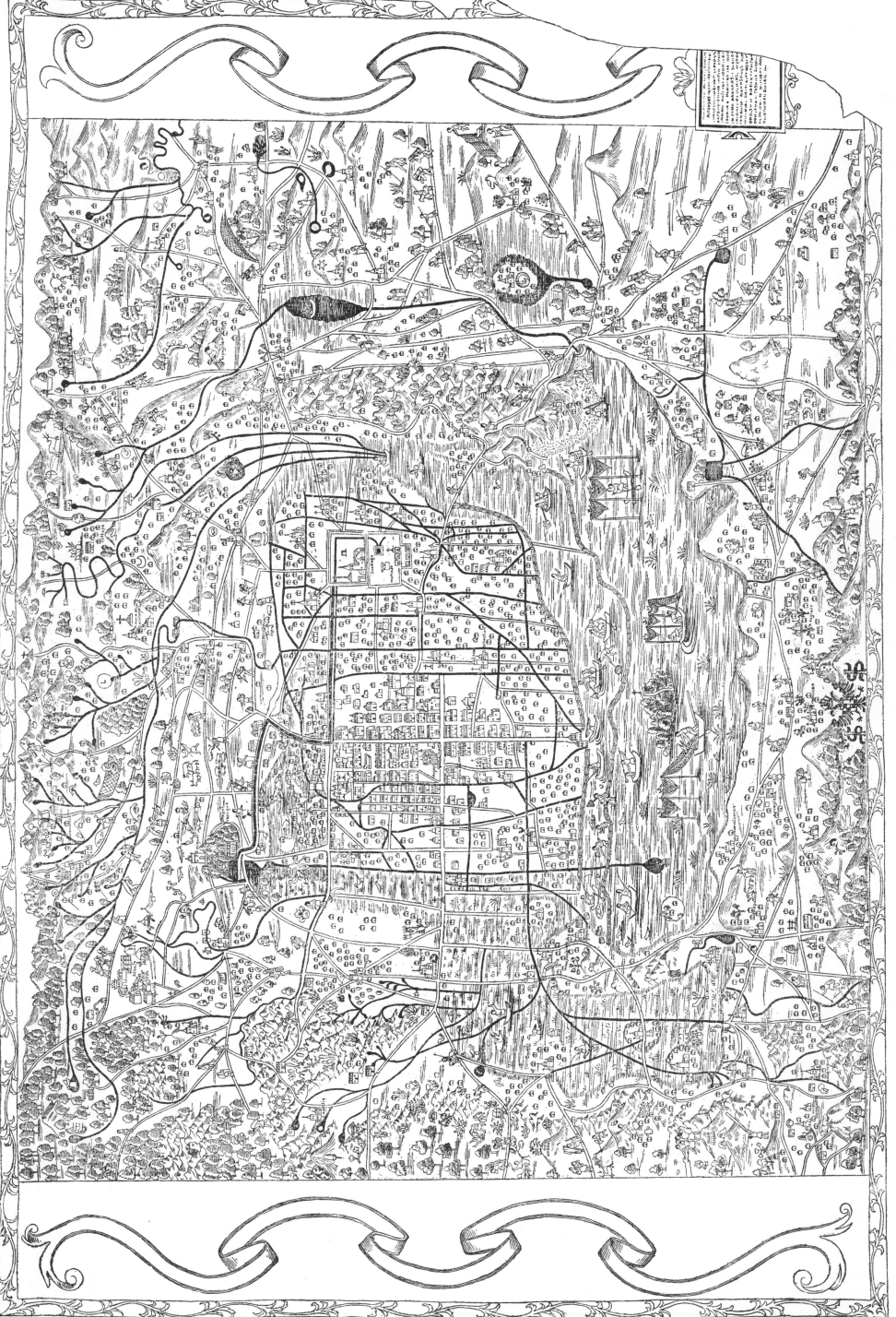

69. Map of the city of Mexico with environs, by ALONZO DE SANTA CRUZ, about 1550. (Orig. size 1140 × 780 m. m.).

only embraces 600 pages in folio with an introduction, and 24 large maps, printed from the same blocks as the maps in Münster's Ptolemy (edit. 1545). It contains numerous wood-cuts, often of infinite naivete,[1] representing cities, plans of towns, monstrous animals, coats of arms and other marvels, but, excepting the folio-plates, only one map, »*Moscoviter lands neue beschreibung*,» printed in the text (on p. 67). This map, of which fig. 74 is a fac-simile, was until a few years ago erroneously considered to be the first non-Ptolemaic map of Russia.

I have not seen the 2d edition, Basel 1545, which I suppose to be a very slightly altered reprint of the editio princeps. But the two editions printed in 1550 at the same place, one in Latin, the other in German, were considerably augmented. The Latin consists of 1164 folio-pages, the introduction and some large maps in the beginning of the work not included. These maps are in double folio and still printed from the same blocks as the maps in Münster's Ptolemy, but only 14 in number, some of the old maps being replaced by wood-cuts, printed in the text. These are very coarse and often quite worthless as geographical drawings, e. g. the maps of England and Italy, which may be quoted as examples of the worst special maps ever published in print. Others are better from a cartographical point of view, although equally imperfect artistically, and we find here the first printed map of more than one country, based on actual observations. I therefore suppose that the following catalogue of the maps in *Cosmographiæ Universalis Lib. VI. Autore Sebast. Munstero*, Basileae 1550, will be of interest for the student of early cartography.

A. Maps occupying two folios in the beginning of the work, printed from the same blocks as the maps in Münster's Ptolemy:

1. Universalis typus orbis terreni; 2. Terreni Orbis generalis et Ptolemaica descriptio; 3. Europae generalis descriptio; 4. Hispania; 5. Gallia; 6. Germania; 7. Helvetia; 8. Svevia et Bavaria; 9. Bohemici regnum; 10. Polonia et Ungaria; 11. Graecia; 12. Nova tabula Indiæ etc.; 13. Africa; 14. Novus Orbis.

B. Maps and more important plans of towns especially of the German Empire, printed in the text:

1. p. 40 Europe; 2. p. 42 Britain; 3. p. 56 Spain; 4. p. 75 France; 5. p. 82, 83 Trier 1548, signed *C. S.* and *D. K.*; 6. p. 88, 89 Paris 1543, signed *HR. MD.*; 7. p. 98, 99 Geneva 1548; 8. p. 115 Flanders; 9. p. 129 Amsterdam, signed *H. H.*; 10. p. 137 Italy; 11. p. 138 northern and central Italy; 12. p. 146 and 147 Rome, signed *H. H.*; 13. p. 150 and 151 View of the city of Rome, signed *C. S.*; 14. p. 158 and 159 Venice, signed *C. S.*; 15. p. 192, 193 Florence; 16. p. 233 Southern Italy; 17. p. 243 Sardinia; 18. p. 248 Cagliari; 19. p. 252 Sicily; 20. p. 255 Syracuse; 21. p. 261 Germany; 22. p. 331 Wallis; 23. p. 338 and 339 Sedunensis civitas (Sitten); 24. p. 351 The lake of Geneva; 25. p. 367 Luzern; 26. p. 369 Zurich; 27. p. 372 and 373 Solothurn; 28. p. 378 and 379 Bern 1549, signed *R. M. D.*; 29. p. 382 Wifelspurger Göw (the district between Biel, Thun, and Lausanne); 30. p. 390 and 391 Baden (in Switzerland), signed *D. K.*; 31. p. 402 and 403 Basel, signed *R. M. D.*; 32. p. 428 Alsace; 33. p. 433 The silver mines at Leberthal; 34. p. 442 and 443 Russach 1548, signed *C. S.* and *R. M. D.*; 35. p. 450 and 451 Colmar; 36. p. 454 and 455 Schletstadt, signed *R. M. D.* and *H. H.*; 37. p. 461 The Rhine between Strassburg and Bingen; 38. p. 466 and 467 Weissenburg, signed *C. S.*; 39. p. 470 and 471 Landau, signed *1547 W. S.*; 40. Speier,

signed *H. F.* on two unnumbered leaves, inserted between p. 474 and 475; 41. Worms, on two leaves inserted in the same manner between p. 480 and 481; 42. p. 495 Eifel; 43. p. 498 and 499 Coblenz 1549, signed *R. M. D.*; 44. p. 502 and 503 Cologne 1548, signed *H. R. M. D.* and *C. S.*; 45. p. 507 Brabant; 46. p. 513 Holland; 47. p. 522 and 523 Chur; 48. p. 525 Feldkirch; 49. p. 528 The Lake of Constance; 50. p. 532 and 533 Lindau; 51. p. 538 Regio Hegoiensis (Hegau); 52. p. 548 and 549 Friburg, signed *R. M. D. 1549*; 53. p. 556 Algoiensis regio (Algau); 54. p. 567 Suabia; 55. p. 576 and 577 Nördlingen, signed *R. M. D. 1549*; 56. p. 610 and 611 Augsburg; 57. p. 616 and 617 Heidelberg; 58. p. 627 Bavaria; 59. p. 644 and 645 Freising; 60. p. 647 Nordgau; 61. p. 650 Franconia; 62. p. 662 and 663 Würtzburg; 63. p. 674 and 675 Frankfort on the Maine, signed *M. H.*; 64. p. 678 Austria; 65. p. 682 and 683 Vienna by Wolfgang Lazius 1548, signed *H. H.* and *H. R. M. D.*; 66. p. 693 Istria; 67. p. 701 Hesse; 68. p. 702 Marburg; 69. p. 710 and 711 Erfurt and Fulda, signed *R. M. D.* and *C. S.*; 70. p. 713 Meissen; 71. p. 717 Saxonia vetus (North Germany); 72. p. 730 and 731 Lüneburg, signed *C. S.*; 73. p. 734 and 735 Lubeck, signed *C. S.*; 74. p. 752 Friesland; 75. p. 756 and 757 Frankfort on the Oder 1548, signed *R. M. D.*; 76. p. 768 and 769 Pomerania; 77. p. 776 Prussia; 78. p. 788 Riga; 79. p. 789 Bohemia; 80. p. 796 and 797 Eger, signed *H. K.*; 81. p. 813 Denmark and southern Sweden; 82. p. 830 Scandinavian peninsula; 83. p. 856 Hungaria; 84. p. 868 Ofen; 85. p. 869 Belgrad; 86. p. 886 Poland; 87. p. 887 Poland and western Russia; 88. p. 910 Moscovia; 89. p. 918 The Danube and Balkan countries; 90. p. 921 Greece; 91. p. 933 Crete; 92. p. 940 and 941 Constantinople and the Bosphorus, signed *C. S.* and *D. K.*; 93. p. 980 Asia Minor; 94. p. 994 The territory between the Wolga, Don, and Caucasus; 95. p. 997 Cyprus; 96. p. 1001 Syria and Palestine to the Euphrates; 97. p. 1002 The eastern coast of the Mediterranean, from Damiette to Cilicia; 98. p. 1016 and 1017 Jerusalem; signed *I. C.*; 99. p. 1113 Africa; 100. p. 1122 The town of Algiers; 101. p. 1127 Lower Egypt.

In the editions of 1552 and 1559 all the maps are unchanged and printed from the same blocks as those described above. Later some new wood-cuts were inserted and other modifications introduced in the work, but still in the last edition of 1628 the majority of the wood-cuts in the text are reprints of the old blocks, now considerably worn. In Belle Foreste's French edition most of Münster's maps are left out.

Münster occupies a peculiar position as a cartographer. With regard to the manner of drawing and the application of mathematical principles to cartography, he not only stands far below the better map-drawers of his time, but also below several cartographers of the latter part of the 15th and the beginning of the 16th century, e. g. Nicolaus Germanus, Schweinheim-Buckinck, Ruysch, Sylvanus, and others. On the other hand he far surpasses most of them in his exertions to get access to the latest information regarding the history, ethnology and geography of the countries he describes. His bulky cosmography will therefore always remain an important source for a history of civilisation of the period in which he lived.

36. A planisphere by SEBASTIAN CABOT, Anvers (?) 1544. Of this important map a single copy is extant. It is preserved at Paris in the »Bibliothèque Nationale.» It occupies an ellipse of 1,48 × 1,11 m., and is consequently too large to be repro-

---

[1] On the title-page of the earlier editions »Sultanus» is represented with a Turkish sword, but in the dress of a Roman warrior. In edit. 1559, p. 207 and 1120, Julius Cæsar is dressed in the armor of a medieval knight, and guns are seen to be used at the siege of Carthage.

[2] Münster has generally got the large town-views or town-panoramas, which are printed in his cosmography on two or more folio-leaves, from the Mayor and Aldermen of the respective towns, a circumstance he never neglects to mention in some complimentary strophes on the verso of the map. With their walls, towers, and pinnacles and with the gallows, often placed in the very foreground of the drawing, they manifestly give us a true, though perhaps, owing to a tendency of the artist to flatter local patriotism, a somewhat embellished representation of the cities in the middle of the 16th century.

duced here, especially as I have only had access to the very incomplete copy published by Jomard (*Monuments de la Géographie*, pl. 20,1—20,4.). Interesting notices of this map are found in Harrisse, *Jean et Sébastien Cabot*, Paris 1882.

37. A map of *Lacus Benacus* (the Garda-lake) in: Georgii Jodoci Bergani *Benacus*, Veronæ MDXLVI. 4:o. The wood-cut map occupies one folio-page.

38. A map in Pedro de Medina's *Arte de Navegar*, Sevilla 1545, afterwards translated into French by Nicolas de Nicolay 1554, into Italian by Fra Vicenze Paletino da Corsula 1554 (1555), into German by Michael Coignet 1576, and into English by J. Frampton 1581 (Harrisse,

respect, it has a predecessor in the drawing in Petrus Martyr, Venetia 1534 (compare above p. 106, fig. 67), which is also based on Spanish originals. Medina's map is further of interest as one of the few maps printed in Spain during the early period of cartography. The plate in the Italian edition is a faithful, somewhat reduced copy of the Spanish original, nicely cut in wood. Besides this map the *Arte de Navegar* (Italian edition) contains several geometrical figures and a small hemisphere surrounded by wind-heads. A map of Spain in Pedro de Medina's above cited *Cosas memorables de Espagna* is in Ortelius' *Catalogus auctorum* deservedly characterized as *valde rudis*. It is only of interest as a proof of the low

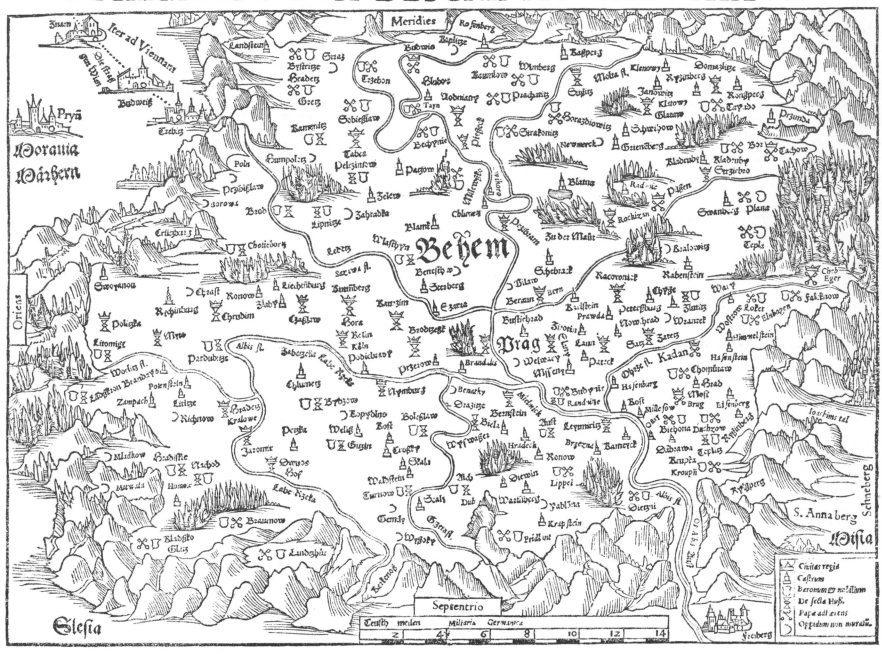

70. Map of Bohemia from: *Ptolemaeus*, Basileæ 1545. (Orig. size 357 × 254 m. m.).

Biblioth. Amer. Vetust., p. 413). The map in the Spanish edition is reprinted, probably from the original block, in P. de Medina's *Libro de grandezas y cosas memorables de Espagna*, s. l., first edit. 1548.[1] As may be perceived by the fac-simile given on fig. 75, this map, the only one to illustrate a manual of navigation which was once very popular, was very insignificant. The author commits on it the blunder, scarcely to be excused in a navigator of the middle of the 16th century, of marking a general scale on a map with equidistant rectilinear meridians stretching from pole to pole. This map, however, has the merit of rendering, with tolerable correctness, the outlines of the Isthmus of Panama and of the eastern coast of America. In this

stage of the art of map-printing in Spain in the middle of the 16th century.

39. Map in: (Joannis Honteri Coronensis) *Rudimenta Cosmographica*, Tiguri 1546, in-12:o, afterwards reprinted of the same size, under a somewhat different title, but with maps from the same blocks, Tiguri 1548[2] and 1549, Antverpiæ 1552,[2] s. l. 1583, Tiguri 1597. This small compendium, written in Latin hexameters, contains three geographical figures in the text intended to illustrate the »circles and zones of the sphere;» the distribution of the different celestial spheres around the earth, which is placed in the centre of the world; and (N. fig. 71) the names and positions of the winds. The

---

[1] Of these works I have only had access to the Italian edition of *Arte de Navegar* and an edition of *Cosas memorables de Espagna*, the title-page of which is dated 1548, but the last leaf: *Alcala de Henares, en casa de Pedro de Robles y Iuan de Villanueua, Año del Señor de 1566*. The map of Spain is here printed on the title-page, the sea-chart on the leaf lxiiij.

[2] Cited from Harrisse.

text is followed by a small atlas, which seems to have, in the middle of the 16th century, filled the same place in literature as the *Epitome Theatri Orteliani*, LANGENES' *Caert-Thresoor* etc. held, a few decennia later. It consists of a general map, here reproduced on pl. XLIV, and of 12 maps in the Ptolemaic style, though more or less modified, of Spain, France, Germany, Poland and a part of European Russia, Hungary, the Danube and Balkan countries, Greece, Italy, Syria, Asia Minor, Central Asia and India, Africa, as far as the country was known by Ptolemy, and Sicily. The general map is only a reduced copy of the map of APIANUS (N. T. XXXVIII). Of the others, those of central Europe are, due regard being taken to the time when they were printed and to their small size, tolerably good. None of the special maps embrace any portion of the New World, and, as far as I have been able to ascertain, not a single name from the New World is to be found among the thousand names of countries, towns, and peoples enumerated in the text. The following passage contains the only allusion to the newly discovered lands:

Est etiam ulterius non visa prioribus annis
Insula, dives opum cultuque immanis agresti,
Quam lucri studium, rerumque cupido novarum
Prima sub extremo conspexit sole cadente.

71. From: J. HONTERUS, *Rudimenta Cosmographica,* Tiguri 1546. (Orig. size).

Even in the description of Africa, which contains about 100 verses, I have in vain sought for a word referring to the Portuguese discoveries.

This pamphlet in hexameters, of which the editio princeps was printed in 1546, should not, as has often been the case, be confounded with:

Ioannis Honter Coronensis Rudimentorum Cosmographiae libri duo. Quorum prior Astronomiae, posterior Geographiae principia, brevissime complectitur. (Colophon:) *Cracoviae Mathias Scharfenbergius excudebat MDXXX.*

This work is not versified. It does not contain any collection of maps, but only an insignificant map of the Old World within a circle of 43 m. m. diameter, and the text only occupies 16 leaves in 12mo. In the chapter: *Nomina Insularum Oceani et Maris* we meet the following passage: *In Occiduo (oceano) Dorcades, Hesperides, Fortunatae, America, Parias, Isabella, Spagnolla et Gades.* Four unidentified names is all the student of this compendium will learn of the New World. This opusculum of Honter in prose, was again printed without any modifications, in: *Dionysii Aphri de totius Orbis situ ... Iohannis praeterea Coronensis de Cosmographiae rudimentis Libri duo.* (Colophon:) *Basileae ex aedibus Henrici Petri MDXXXIIII.*

Both works, the metrical, as well as the unrhymed, are published together in:

Procli de sphaera liber I; Cleomedis de Mundo ... libri II; Arati Solensis phaenomena ... Dionysii Afri descriptio orbis habitabilis ... una cum Io. Honteri Coronensis de Cosmographiae rudimentis duplici editione, ligata scilicet & soluta. (Colophon:) *Basilea Per Henricum Petri MDLXI.* Here the maps are printed from other blocks than those in edit. Tiguri 1546. Those of Poland and western Russia, of the Danube and Balkan countries and of Asia Minor are left out, whereas maps of Ireland, Majorca, England, Zealand, Euboea, Cyprus, »India intra Gangem,» »Iava Maior,» »India extra Gangem,» »Taprobana,» »Mædera,» South Africa, Malta, and Cuba are added. The general map, a fac-simile of which I give on fig. 76, is a copy of *Universale novo,* in Gastaldi's Ptolemy of 1548, redrawn on the cordiform projection of Apianus, and the copper-engravings in the same edition of Ptolemy have also served as types for several of the other maps, which, however, are very inferior

72. Seelandia from: J. HONTERUS, *De cosmographiæ rudimentis,* Basileæ 1561. (Orig. size).

to their Italian models. Some of them represent parts of the earth, of which no special maps seem to have been published before, for instance (p. 908) the map of *Seelandia,* of which a fac-simile is given on fig. 72. The blocks to the woodcuts in this small quasi-atlas were also used for other works published by the great Basel firm, with a rock blazing under the stroke of the hammer, as printer's mark, e. g. in an edition of Pomponius Mela and Solinus published by SEBASTIAN HENRICPETRI in 1595.

40. Sixty maps engraved in copper in MATTIOLO-GASTALDI's Ptolemy of 1548, of which two are reproduced on T. XLV. I have above (p. 25) given the full title and enumerated the maps of this handsome octavo-volume. Its 34 *tavole moderne,* and the new maps in the editions 1561, 1562, and 1564 copied from them on an enlarged scale, form the most complete collections of maps published in print between 1513 and 1570. On the title-page is written: *Con alcuni comenti et aggiunte fattevi da Sebastiano Munstero Alamanno,* but this evidently refers to the text, not to the maps, which, from a geographical point of view as well as in their technical execution, are immensely superior to those of the learned geographer and Hebrew professor at Basel. The work is dedicated by IACOBO GASTALDI to LEONE STROZZI, *dignissimo Signore di Capua*, whereupon follows an *A li lettori*, where the publisher, GIOVANBATISTA PEDREZANO, is highly complimented, for not having spared any expense in getting the maps as finely and well executed as possible, and for having reduced their size so as to allow of anybody, without difficulty, carrying the work »*nella manica.*» We thus have here the first atlas expressly said to be published in a pocket form.

41. Maps in SIGISMUND V. HERBERSTEIN's celebrated work on Moscovia or Russia, of which the editio princeps, *Rerum*

1. A map in the first latin edition of Herberstein's work.[2] On its upper left corner is written: *Moscovia Sigismundi Liberi Baronis in Herberstein Neiperg et Gutnhag. Anno MDXLIX*, and at the right border: *Hanc tabulam absolvit Aug. Hirsvogel Viennæ Austriæ Cum gra. et privilegio imp.* This copper-print (N. fig. 77) is not inserted into any other edition. It is remarkable as being one of the first maps reproduced by copper-print in Germany, as the original of Herberstein's map of Russia, and finally as perhaps the only map still extant engraved in copper and signed by Hirschvogel. The editio princeps further contains two copper-prints of Russian costumes and one of the Czar IVAN VASILJEVITSCH, signed *A. H. F. 1547.*

73. Map of America from *Ptolemaeus*, Basileæ 1540.   (Orig. size 257 × 348 m. m.).

Moscoviticarum comentarii, was published at Vienna in 1549. This latin edition was reprinted *Basileæ* 1551, 1556; *Antverpiæ* 1557 (without maps); *Francofurti* 1560; *Basileæ* 1567, 1571, 1573, 1574; *Francofurti* 1600. An Italian version was published in *Venetia* 1550 and reprinted in the collection of Voyages of Ramusio; a German: *Wien* 1557; *Basel* 1563 and 1567; a Russian and a Bohemian in the previous century.[1] I have examined the rare original-edition, of which a copy with marginal notes by GLAREANUS is preserved in the Royal Library at Stockholm, the Italian edition, the Latin editions of 1551, 1556, and 1571, and the German edition of 1567. They contain the following maps:

2. A wood-cut copy of the above mentioned map, inserted in the edition printed at Basel in 1551, and in several later ones. Here may also be read, at the upper left corner: *Moscovia Sigismundi Liberi Baronis in Herberstein, Neiperg et Gutenhag Anno MDXLIX.* That this date refers to the above mentioned original and not to the wood-cut copy of it, may be concluded from a letter of WOLFGANG LAZIUS to the editor IOANNES OPORINUS, printed on the reverse of the title-page of the edition *Basileæ* 1551.

3. A copy of the map No. 1, engraved in copper by GASTALDI for the edition *Venetia* 1550. Hirschvogel's original is not graduated; whereas the copy of Gastaldi is graduated

---

[1] A detailed description of these editions is given in FRIEDRICH ADELUNG, *Siegmund Freiherr von Herberstein, mit besonderer Rücksicht auf seine Reisen in Russland*, St. Petersburg 1818, containing an excellent lithographic copy of the map in *editio princeps*.

[2] According to the Catalogue of printed maps in the British Museum, p. 1820, this map of Hirschvogel had already been printed separately in 1546.

on the Donis projection, but under the evidently incorrect supposition that the original had been drawn on the equidistant cylindrical projection of Marinus. There are, besides, so many additions and essential improvements made in Gastaldi's map, that it may be considered as an almost independent work. It is reproduced in the 2d part of *The Voyage of the Vega round Asia and Europe*, London 1881.

4. A large map of the forests of Russia, first inserted into the edition of 1556.

5. A plan of Moscow first inserted into the same edition.

The principal map (No. 1) of Herberstein embraces the major part of Poland and European Russia, with a portion of western Siberia. That this map was the first by which an approximate geographical knowledge of European Russia was spread in western Europe, is shown by the following

Catalogue of maps of European Russia down to 1550.

1. 2d century. Ptolemy's *Tabula quarta et octava Europæ* and *secunda Asiæ*. These maps serve to illustrate the scarcity and insufficiency of the notices which had in Ptolemy's time penetrated to the Greeks and Romans, of the lands situated to the north of the Black and Caspian Seas.

2. 14th century. Portolanos of these regions. They are generally copies of the previous ones, with inconsiderable additions and corrections, founded on modern information collected from travellers and merchant-adventurers.

3. 1493. A wood-cut map of central Europe in SCHEDEL's Chronica Nurembergensis 1493 (N. fig. 5). This map also embraces Poland and the most western part of European Russia.

4. 1507. A map (N. fig. 13) engraved on copper for the edition of Ptolemy's geography published at Rome in 1507 and 1508. This map agrees, in its main features, with the preceding one, but is of better execution and more rich in detail. We here for the first time, on a printed map, meet the names of *Moskva, Smolensko, Kowno, Grodno, Wylno, Neper, Kyow, Cerkaszy, Bratslav, Chmelnik* etc. This and the preceding one are evidently founded on the same original, but in the works in which they were published, we look in vain for any information as to its author. As will be shown in a subsequent chapter, giving an analysis of the *Catalogus auctorum* of Ortelius, this map is probably a reproduction of the map of NICOLAUS A CUSA of the middle of the 15th century (comp. Addenda).

5. 1513. *Tabula moderna Sarmatie Eur. sive Hungarie, Polonie, Russie, Prussie et Walachie*, first published in Ptolemaeus 1513, then reproduced from the same block in the edition of 1520. A rough and but little modified copy of the map in Schedel's chronicle. A reduced, poor and defective copy of this map was also published in the Ptolemy editions of 1522, 1525, 1535 and Viennæ 1541.

6. 1525. A map, drawn by PAULUS JOVIUS (or PAULO GIOVIO) in Rome from communications by the ambassador, DMITRI GERASSIMOW, sent by the Grand-duke of Russia to the Pope. Of this map, mentioned in PETRUS IOVIUS, *Libellus de legatione Basilii magni Principis Moschoviae ad Clementem VII*, Romæ 1525, no printed copy is known, and it seems doubtful if the original was ever published in print. But an early hand-drawn copy by BATTISTA AGNESE still exists, which has recently been reproduced by THEOB. FISCHER and H. MICHOW in their works.

7. 1528. A map mentioned in the *Catalogus auctorum* of Ortelius (edit. 1584) in these words: *Florianus tabulam* (edidit) *Sarmatiae, Regni Poloniae et Hungariae, utriusque Valachiae, necnon Turciae, Tartariae, Moscoviae et Lithuaniae partem comprehendentem, Cracoviae 1525*. This map

appears to be lost; I have not found it mentioned by any one but Ortelius. Its author is probably identical with FLORIANUS UNGLERIUS, who, in 1512, at Cracow printed Stobnicza's *Introductio in Ptholomei Cosmographiam* (compare p. 68), and should not be confounded with the artist *Antonius Florianus*, who constructed the map reproduced on fig. 48.

8. 1532. The map in ZIEGLER's *Schondia* (comp. p. 60 and fig. 31) is extended over a part of Russia. But this part of the map is extremely incorrect. Moscow, for instance, is placed due south of Hangö; Boristhenes and Tanais commence from *Lacus albus*, which communicates with the Baltic by another river. As this map is founded on information from four learned bishops, who had taken a leading part in the public affairs of the Scandinavian countries, it may serve as a proof of the confused ideas respecting the large dominion of the Czar still prevailing among their countrymen.

9. 1538. The rough, but correct map based on actual observations, in SOLINUS, Basileæ 1538 (N. fig. 68).

10. 1540—44. Various maps in Münster's editions of Ptolemy and in his cosmography. The most remarkable are Tab. XV in Ptolemaeus 1540 called *»Polonia et Ungaria, nova tabula,»* but embracing a great portion of European Russia, and the map of Russia reproduced on fig. 74 and inserted between the pages 656 and 657 of the first edition of the cosmography of 1544.

11. 1545. A map, very rich in details, of *Moscovia* by ANTONIUS WIED. Of this map there is only extant a copperengraving of 1570 by FRANCISCUS HOGENBERG, of which only two copies are known, one belonging to the British Museum and the other to Dr. MICHOW in Hamburg, who has given a fac-simile of the map and described it in a memoir: *Die ältesten Karten von Russland*, Hamburg 1884. Hogenberg's copper-engraving was probably originally intended for the *Theatrum Orbis terrarum* of Ortelius. High up, in the left corner, we read: *Franciscus Hogenb. ex vero sculpsit 1570*. Two long legends beneath are dated 1555. By means of an analysis of various details, and a comparison with Münster's map of 1544, Dr. Michow finds it probable that Ant. Wied's map was constructed between 1537 and 1544.

12. The above mentioned map of Herberstein engraved in copper by Hirschvogel, first printed separately in 1546, and then inserted in the editio princeps of Herberstein's Moscovia.

42. Maps by GASTALDI. Besides the above cited maps in Gastaldi-Mattiolo's edition of Ptolemy of 1548, this distinguished cosmographer, to whose works I shall return in the succeeding chapter, had already, before 1550, published various other maps generally printed separately, and, like all early cartographical productions of that kind, they are at present extremely rare or entirely lost.

Of these maps I have seen, or found mentioned in literature, the following:

1544. *La vera descrittione de tutta la Spagna (Br. Mus. Cat. of printed maps*, I: 1497).

1546. *»Universale,»* signed: *Giacomo Cosmographo in Venetia MDXXXXVI*. 380 × 640 m.m. (CASTELLANI's catalogue, p. 248. *Br. Mus. Cat. of printed maps*, II: 4544). A general map on Bordone's oval projection, probably agreeing, except in size, with Gastaldi's *»Universale Novo,»* of 1548 (N. T. XLV).

1545. *La Sicilia per Giacomo Gastaldo Piemontese Cosmographo in Venetia 1545*. 370 × 530 m.m. (CASTELLANI's catalogue, p. 249). Reproduced by Ortelius in 1570, pl. 38, and probably several times in Italy.

1550. *Descriptione de la Moscovia per Giacomo Gastaldo piamontese Cosmographo in Venetia MDL*, mentioned above.

c. 1550. Some maps of Gastaldi, which were first inserted in the third volume of RAMUSIO's known collection of voyages printed in 1556. The maps are considered to have been finished before 1550.[1] In the *Discorso di M. Gio. Battista Ramusio sopra il terzo volume delle Navigationi & Viaggi nella parte del Mondo Nuovo all' eccellente M. Hieronymo Fracastoro*, dated the 20th of June 1553 and forming the introduction to the third volume of Ramusio's work, he says that, on Fracastero's repeated request to have four or five

produced by wood-cut. The first mentioned four are so imperfect, and so different from all other maps of Gastaldi which I have seen, that they can scarcely be considered as original works of this distinguished cosmographer. They seem rather to be copies ornamented in Münster's style by some ignorant wood-cutter, from originals of Gastaldi. The map of Nova Francia is reproduced in several modern works on America, for which I may refer to HARRISSE, *Cabot*, p. 236; KOHL, *Discovery of Maine*, p. 226; and WINSOR, *Critical History*.

As most of the maps of Gastaldi were published after 1550, I shall have occasion to return to them in the next chapter.

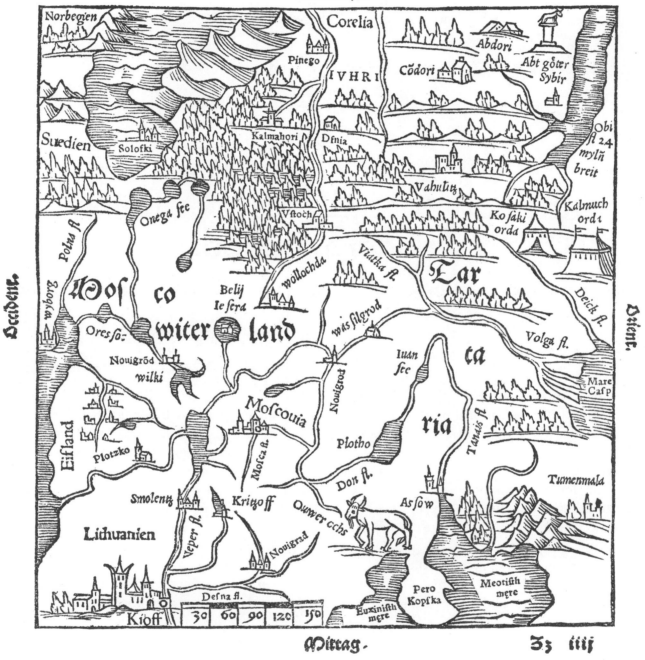

74. Map of Russia in SEB. MÜNSTER's cosmography, Basel 1544. (Orig. size).

maps of the newly discovered lands made in the Ptolemaic style, he had asked the distinguished cosmographer GIACOMO GASTALDI to construct a general and four special maps of those parts of the earth. In Vol. III there are five maps, corresponding to the above communication, viz. *Nuova Francia* (a part of Labrador, New Foundland and Canada); *Brasil; Parte de l'Africa; Taprobana* and *Universale della parte del Mondo nuovamente ritrovata*. The maps are re-

43. A map of Britain printed by GOURMONT, Paris 1548. One copy of this map is, according to a private communication by M. G. MARCEL, preserved in the Bibliothèque Nationale at Paris.

I have excluded from the above catalogue of maps printed between 1520 and 1550, several maps drawn in the

[1] In a note at the end of *Nomi de gli Autori* in the first volume Ramusio says: »To the year MDL, *when the first volume of this work was laid under the press*, answers Hegira's year of DCCCCLVII.» The third volume was printed in 1556, before the second, which was laid under the press in 1559, or two years after the death of Ramusio (comp. THOMASSO GIUNTI's preface to the 2d volume). All the volumes have since been reprinted several times, partly with addenda.

Ptolemaic style, inserted in editions of Julius Cæsar, Justinus' Trogus Pompeius, and other classical authors; maps printed in theological works or books of devotion, of Palestine, Egypt, Sinai etc., founded on older prototypes or only invented at the writing-table; general maps similar to those reproduced on T. XXXII from MACROBIUS and SACROBOSCO and still employed without any modifications in editions of these authors during the whole of the 16th century; small schematical maps in such geographical compendia as *De geographia liber unus* by HENRICUS GLAREANUS, of which a number of editions were published in Basel, Friburg, Venice, and Paris, after 1527, or *Cosmographiae introductio* (by APIANUS), the first edition of which was printed at Ingolstadt in 1529 etc. All these appear to be worthless from a geographical point of view. As regards the maps in Macrobius it may be remarked that, in the Basel edition of 1535, the typical form has been exchanged for a circular planisphere of 89 m. m. diameter, on which the outlines of Africa are laid down with tolerable correctness.

No copies are at present known of a number of maps belonging to this period, especially of those published separately. But it is to be hoped that several of them, as lately happened with the important map of Olaus Magnus, may hereafter be exhumed from the dust of the libraries, and I should feel it to be a great success, if this fac-simile atlas should be the cause of at least some discoveries of that kind. As examples of lost maps mentioned in literature may be cited: the Sarmatia by FLORIANUS, the map of Hungaria by LAZARUS, CRATANDER's reproduction printed at Basel in 1530 of NICOLAUS A CUSA's Germania, GERARD MERCATOR's map of Palestine printed in 1537, and a number of other maps mentioned in the *Catalogus auctorum* by Ortelius.

I hope, however, that comparatively few of the maps at present known have escaped my attention, and that the catalogue given above may suffice for an objective appreciation of the merits and defects of the period (1520—1550) under discussion.

X.

The transition to and the beginning of the modern period.

Jacopo Gastaldi. Philip Apianus. Abraham Ortelius. Gerard Mercator.

A decided change in the development of cartography occurred in the middle of the 16th century. The geographer had, until then, been satisfied with general maps based on the geographical data enumerated and commented upon in Ptolemy's cosmography, namely itineraries, valuations of the distances between different places and the bearings between them, and finally a few astro-geographical observations which, however, were almost always incomplete, no means existing before the discovery of chronometers, for determining the longitudes with even approximate exactness. The cartography of the New World was still almost exclusively limited to a general outline of the coast, and geographers scarcely attached as much importance to it as they now give to the mapping of the uninhabited north and south polar regions. As regards the Old World, faith in the infallibility of Ptolemy was yet almost undisturbed, and it was regarded as the greatest merit for a cartographer, to reconcile the newly collected data with the classical types of the 2d century. However, a few chorographical or topographical maps, perhaps directly called forth by the first chapter of Ptolemy's geography, were already published in print, and from the middle of the 16th century such special maps, founded upon actual surveys, became more and more common. They then reacted on the general maps, and communicated to these, even when their technical execution was defective, a completeness which is wanting in the productions of the time which may be called the period of incunabula of cartography. It was this breaking with classical authorities that formed the real source and cause of the modern period introduced by the works of Gastaldi, Philip Apianus, Ortelius, and Mercator.

The following chorographies or detailed maps of smaller districts were already published in print before the middle of the 16th century:

1. A map of *Lotharingia*, commenced in 1507 »Vosagi rupibus,» and published in print in 1513. As the first printed topographical map founded on actual measurements, it is of great interest, and really deserves the epithet *nobile opus*, by which it seems to be designated in the author's preface (comp. above p. 69). The map, however, is far from exact. The skill of King René's cosmographer in making and calculating astronomical observations appears to have been very defective. For, even deducing a constant error of a whole degree, there still remain on the first modern topographical map greater errors of latitudes than those found on the better maps of Ptolemy, e. g. on his map of Egypt.

2. *Chorographia Eremi Elvetiorum*, *Chorographia Rheni* and *Chorographia Cretae*, printed for the first time in Ptolemaeus 1513, then reproduced from the same block in the edition of 1520, and on a reduced scale in the editions 1522, 1525, 1535 and Viennæ 1541. On these maps the geographical co-ordinates are also inexact, and the drawing is so rough, and so different from the style as well on modern maps as on the first printed maps of Ptolemy, that it requires some time to become conversant with them. But they are rich in details and have the unmistakable character of having been drawn by cartographers well acquainted with the countries. The map of Crete seems to be copied from a portolano, or at least to be founded on a Venetian original. I have not succeeded in obtaining any information as to the author of *Chorographia Provinciæ Rheni*. The third map, *Chorographia Eremi Elvetiorum*, is a tolerably accurate copy of the map of Switzerland constructed in 1496 by CONRAD TYRST. Some of the legends of the Tyrst map, however, are here omitted, evidently on account of the technical difficulty of rendering all the inscriptions on a wood-cut. A manuscript copy of the original on vellum is still preserved in the Imperial

library at Vienna. It has lately been reproduced in the 6th volume of *Quellen zur Schweizer Geschichte*, Basel 1884, but the publisher was unaware that the greatest portion of Tyrst's map had already been engraved in wood and printed in 1513.

3. A map of Avignon in: *Il Petrarca con l'espositione d'Alessandro Vellutello*, s. l. 1528. This insignificant wood-cut is only mentioned here as one of the few early topographical maps published in print.

4. Several of the new maps in the different editions of Ptolemy's geography by MÜNSTER, and in his cosmography (comp. p. 108—110).

5. Some of the modern maps of smaller districts in MATTIOLO-GASTALDIS *Ptolemeo*, Venetia 1548 (comp. p. 25).

To these may be added maps of islands etc. in DALLI SONETTI's and BORDONE's works, generally consisting of rudi-

1550 and 1570 by Gastaldi and other Italian geographers or artists, in Germany by Philip Apianus (1568), in the Netherlands by Ortelius (from 1562), and above all by Mercator (from 1537). It is by the maps of these eminent reformers that the period of incunabula in cartography was closed and the modern period introduced. The last chapter of this essay on early cartography may, therefore, be devoted to an enumeration of the most important of their works, and a short review of their influence on the development of the science.

———

When, as is generally the case, the new era in cartography is counted from the publication of the *Theatrum Orbis terrarum* and the merit of the reform is exclusively attributed to ORTELIUS, and to GERARD MERCATOR, great injustice is done to the draughtsmen of the excellent maps

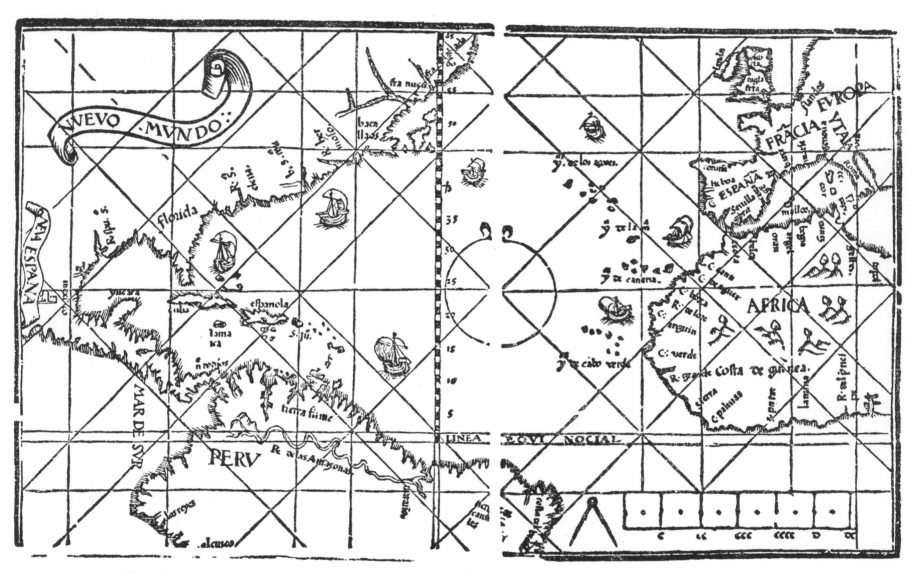

75. From: PEDRO DE MEDINA, *Libro de grandezas e cosas memorables de España*, Alcala de Henares 1548—66. (Orig. size 263 × 150 m. m.).

mentary sketches, hardly deserving the name of maps; maps of the Holy Land based on biblical traditions, on portolanos and on observations by pilgrims; plans and views of ports and cities, often merely fictitious, yet sometimes with internal evidence that the draughtsman had been acquainted with the place represented; Ziegler's *Schondia* of 1532; Olaus Magnus' map of Scandinavia of 1539; Mathias a Michou's *Sarmatia Europea* of 1538; Herberstein's maps of *Moscovia* etc. Several of these maps may certainly be regarded as precursors of modern cartography, on account of their character of special maps, founded, not on a study of the old authors or on works of the Middle Ages, but on actual, more or less correct observations, and on modern itineraries, or surveys. Nevertheless very few of them are, as regards their execution, comparable to the maps added to some of the oldest editions of Ptolemy. From a cartographical point of view again, they are much inferior to the maps published at Rome and Venice between

printed in Italy during the 6th and 7th decennia of the 16th century. These maps often served as models to Ortelius, and many of them may, without disadvantage, be compared with works of Mercator. Their neglect, in the history of cartography, is evidently due to the circumstance that the Italian maps were published on separate sheets and never united into such a complete and systematic work as that of Ortelius' Theatrum or Mercator's Atlas. Many of them are, therefore, as are several of the most important maps of the same period published north of the Alps in separate sheets, entirely lost. Others have become so rare, even in Italy, that they have generally escaped the attention of investigators into the early history of cartography. Fortunately the most important maps printed separately in different towns of Italy had, in the 16th century, already been collected in one or more folio volumes, and provided with a common title, forming a magnificent copper-engraving, on whose inner field is written:

Geografia.
Tavole moderne di geografia
de la maggior parte del mondo
di diversi autori
raccolte et messe secondo l'ordine
di Tolomeo
con idisegni di molte citta et
fortezze di diverse provintie
Stampate in rame con studio
et diligenza
in Roma.

The rich border, surrounding this title, of which a fac-simile is given on the title of the present work, is interesting, because it is the first instance of *Atlas* supporting the earth globe being used in print, as a symbol for a collection of maps. Neither the year of printing nor the name of the publisher are mentioned. But the engraved title is probably the work of ANTONIO or ANTOINE LAFRERI, a French artist who, together with his uncle DUCHET, founded a celebrated *atelier* for copper-engraving at Rome in 1540. This atlas is therefore generally cited under the name of Lafreri's Atlas or the Roman Atlas.

Many of the maps were engraved in Rome by Lafreri or Duchet, but notwithstanding the express statement on the title »*Stampate con studio et diligenza in Roma,*» the majority of them were printed in Venice or elsewhere in Italy. The unfinished state of several of the maps, their curious mounting, difference in size etc., seem to indicate that we here only have to deal with a collection of the necessary material for a work resembling that of Ortelius, which Lafreri or some other geographer or editor intended to publish at Rome, and for which he perhaps had bought convenient plates from different Italian engravers.

Such an intention, if it ever existed, was never realised. But this collection was the cause that a number of maps were saved from destruction, of which probably no traces would otherwise have been found in literature. We here get an insight into the high development of cartography and of the industry of map-printing in Italy in the middle of the 16th century.

This collection is seldom mentioned in geographical literature. For example it is never alluded to by Lelewel, Humboldt, Peschel, Vivien de St. Martin, Ruge, Breusing. Probably only a very limited number of copies were originally issued, and at present this atlas is one of the greatest rarities in cartographical literature. Hence a notable difference exists as to the contents of different copies. The following catalogue of the maps in this collection may, therefore, be of use to students of the history of cartography. It is founded on two copies, of which one belongs to the library of Collegio Romano, the other to my private collection. Of the former a detailed description is given in CARLO CASTELLANI's *Catalogo ragionato delle opere geografiche a stampa che si conservano nella biblioteca del Collegio Romano*, Roma 1876.

Maps in Lafreri's Atlas.[1]

1. N. 1572. 0,805 × 0,533.[2] — A faithful though much reduced copy, engraved in copper at Rome in 1572 by ANTONIUS LAFRERI, of the large map of OLAUS MAGNUS, first published in Venice 1539 (N. fig. 32).

2. CR, I: 1. N. s. a. 0,835 × 0,462. — General map by *Antonius Florianus Utin.* (comp. p. 94). There are no inscriptions or drawings on the title-fields or on the two medallions at the lower corner of the maps, which seems to indicate that the engraving never was finished (N. fig. 48).

3. CR, I: 2. 1565. 0,50 × 0,30. — *Universale descrittione di tutta la terra conosciuta fin qui. Paulo Forlani Veronense fecit. Ferando berteli exc. 1565.*

4. CR, I: 3. N. 1566. 0,579 × 0,518. — *Cosmographia universalis ab Orontio olim descripta.* JOANNES PAULUS CIMERLINUS *Veronensis in æs incidebat Anno 1566.* Dedicated to *Ill:mo viro Henrico Domino Matrevors, Comiti Arandelliae etc.* (N. fig. 53).

5. CR, I: 4. 1562. 0,52 × 0,30. — Planisphere. In the left corner there is engraved: *Paulus de furlanis Veronensis opus hoc ex:mi Cosmographi Domini Iacobi Gastaldi Pedemontani instaurcvit et dicavit ex:mo I. U. D. et aurato Aequiti Domino Paulo Michaeli Vincentino. Venetiis Ioan. Francisci Camotii aereis formis ad signum Pyramidis Anno MDLXII.*

6. CR, I: 5. s. a. 0,50 × 0,32. — General map. »*Ant. Lafreri, exc. Romæ*».

7. CR, I: 6. s. a. 1: 0,26 × 0,19. 2: 0,24 × 0,18. — General map in two parts, 1: »*Ptolomaei typus*», and 2: »*Septentrionalium partium nova tabula*».

8. CR, I: 7. 1569. 0,41 × 0,82. — Map of Europe with the countries to the east and south of the Mediterranean, dedicated by PAULO FURLANI to GIACOMO MURARI MDLXIX.

9. N. s. a. 0,507 × 0,323. — Double cordiform map of the world. »*Ant. Lafreri exc. Roma*». An elegant copy of G. Mercator's map of 1538, as copper-engraving far surpassing the original (comp. p. 90, N. Pl. XLIII and fig. 54).

10. N. s. a. 0,247 × 0,182 and

11. N. s. a. 0,242 × 0,182. — Two copper-engraved maps of the islands *Frisland* and *Estland* mentioned in the voyage of the brothers Zeno, published in Venice 1558. They nearly agree with the delineation of these islands on Zeno's map (comp. above p. 58). Even here, as in many other maps of Lafreri's atlas, the title-fields are left unfinished. These maps were evidently engraved in Venice by BERTELI.

12. CR, I: 8. s. a. 0,35 × 0,24. — Chart (portolano) of the South Atlantic Ocean with eight wind-roses. Dedicated by BERTELI to MARCO DEL SOLE.

13. N. s. a. 0,285 × 0,160. — *Hibernia sive Irlanda insula maxima inter Brittanniam & Hispaniam sita, longitudine mill. 260, in regiones quatuor dividitur, habet miram coeli temperiem, episcopatus 50, nihil venenatum gignit, gens moribus incultior, bello, latrociniis et musica gaudent.* The orientation of the map is South upwards.

14. N. 1562. 0,468 × 0,340. — *Britania Insula quae duo regna continet Angliam et Scotiam cum Hibernia adiacente.* Two long title-legends. Beneath the right is written: *Venetiis Anno MDLXII*, beneath the left: *Ferando de Berteli exc. 1561.*

15. CR, I: 12. N. 1558. 0,541 × 0,398. — *Britanniae insulae quae nunc Angliae et Scotiae Regna continet cum Hibernia adiacente nova descriptio.* Two long title-legends; beneath that in the right corner is engraved: *Romæ Anglorum studio et diligentia MDLVIII*, and »*Sebastianus a Regibus Clodiensis in aes incidebat*». This map entirely agrees with the map of Britain in *Carte Nautiche di Battista Agnese dell' anno 1554*, preserved at *Biblioteca Marciana* in Venice and reproduced in fac-simile by THEOBALD FISCHER, Venetia 1881 (N. fig. 78).

---

[1] With regard to some of these maps I may refer to G. MARINELLI: *Saggio di cartografia della Regione Veneta*, Venezia 1881, where we find a few meagre notices regarding Gastaldi, Furlani, Camocio, and other Italian cartographers or engravers.

During the middle of the 16th century map printing reached a high development in Italy, especially in Venice. Lafreri's atlas is not the only instance of maps first published on detached leaves and afterwards collected and brought into the market with a common title-page. Such a collection, in oblong folio, was published about 1572 by Camocio with the title: *Isole famose, porti, fortezze, et terre maritime sottoposte alla Ser:ma Sig:ria di Venetia, ad altri Principi christiani, et al Sig:or Turco, nuovamente poste in luce, In Venetia alla libraria del Segno di S. Marco.* This collection contains 88 maps generally insignificant in a geographical respect, though often excellent as copper-engravings. An almost complete copy is preserved in the Royal Library at Stockholm. Copper-engraved maps were also inserted in the text of printed works, e. g. in *L'isole piu famose del Mondo descritte da Thomaso Porcacchi da Castiglione Arretino e intagliate da Girolamo Porro Padovano*, of which several editions were published, the first *In Venetia MDLXXVI*. The work contains a number of small well executed maps from different parts of the earth. I have reproduced one of them on pl. XLIX. Another collection of small maps engraved in copper by ANGELO MARELLI and representing islands in the Mediterranean (and Britain) was published in *Francesco Ferretti's Diporti Notturni* etc., Ancona 1580.

[2] CR = Biblioteca del Collegio Romano, according to Castellani's catalogue. N = Nordenskiöld's copy of Lafreri's atlas. The numbers following indicate the year of printing (s. a. = no date) and the size of the map in metres.

[N. s. a. — On this map follows, in my copy of the atlas, a printed leaf: »*Nova et antiqua locorum nomina in Anglia*» and »*Nova et antiqua locorum nomina in Scotia*».

16. CR, I: 13. N. 1554. 0,475 × 0,375. — *La vera descrittione di tutta la Francia, et la Spagna, et la Fiandra . . . MDLIIII.*

17. CR, I: 14. N. 1560. 0,559 × 0,437. — *Hispaniae Descriptio. Dominicus Zenoi* (not Zendi) *Venetus restituit. Venetiis MDLX.*

18. N. s. a. 0,240 × 0,180. — *Minorca.*

19. CR, I: 17. N. 1558. 0,51 × 0,38. — *Totius Galliae descriptio cum parte Angliæ, Germaniæ Flandriæ, Brabantiæ, Italiæ, Romam usque, Pyrrho Ligorio Neap. auctore. Romae MDLVIII. Michaelis Tramezini Formis. Cum Pontificis Maximi ac Veneti Senatus privilegio ad decennium. Sebastianus a Regibus Clodiensis incidebat.*

20. N. 1571. 0,498 × 0,372. — The same map, but the end of title-legend changed into: *Pyrrho Ligorio Neap. auctore. Claudii Ducheti formis 1571.*

21. CR, II: 16. N. 1556. 0,628 × 0,456. — *Totius Galliae exactissima Descriptio.* According to a long dedication at the upper corner to Marcus Antonius Radici, this map was engraved by Paulus Forlani Veronensis S. D. from an original by the distinguished mathematician Orontius, and printed at Venice, *ex aeneis formis Bolognini Zalterii MDLXVI.*

31. CR, I: 25. 1566. 0,47 × 0,38. — *Frisia antiquissima trans Rhenum provincia etc. a Jacobo Darent. Belga descripta. Romae MDLXVI.*

32. CR, II: 62. N. 1572. 0,520 × 0,384. — Chart of the Baltic and North Sea, *Venetiis MDLXII apud Ioannem Franciscum Camocium* (N. fig. 25).

33. CR, I: 26. s. a. 0,52 × 0,36. — *Europa settentrionale. Dalla Svezia e dal ducato de Moscovia alla Lapponia e al Mare Sitichum.*

34. N. 1568. 0,517 × 0,373. — *Di M. Iacomo Castaldo vi si rapresenta la prima parte della descrittione del regno di Polonia, con la sua scala di miglia, intagliata da Paolo furlani veronese al segno della Colonna. Venetia l'anno 1568.* (N. fig. 79).

35. N. 1568. 0,497 × 0,382. — *Il vero disegno della seconda parte dil Regno di Polonia, dell' ecc:mo m. Giacomo Gastaldo Piamontese. In Venetia l'anno MDLVIII. Intagliata da Paolo Forlani Veronese al segno della Colonna immerzaria.* Embracing not only Poland but also the main part of the European Russia.

Instead of the above mentioned two maps, the Collegio Romano collection contains two others, viz.

36. CR, I: 27. 1562. 0,52 × 0,38. — *Il disegno . . . del regno di Polonia, e parte del ducato di Moscovia . . . Scandia, Svetia . . . Ustinga . . Severa in sino al Mare Maggiore. Giac:o de Castaldi piamontese cosmogr. MDLXII f. in Venecia.*

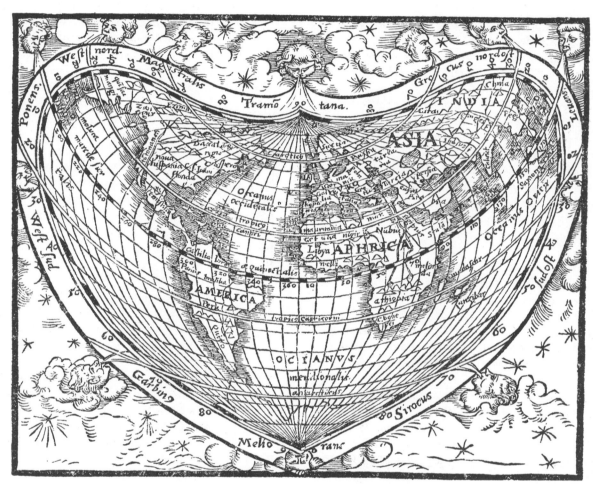

76. General map from: Johannes Honterus, *De Cosmographiæ rudimentis*, Basileæ 1561. (Orig. size).

22. CR, II: 19. N. 1562. 0,438 × 0,328. — *Descrittione del Ducato di Savoia.* Dedicated by Paulo Forlani to Luigi Balbi, Venetia MDLXII.

23. CR, I: 18. N. 1555. 0,610 × 0,444. — Switzerland. Dedicated by Antonio Salamanca to Jodocus a Meggen, *Praetorianorum Praefectus. Romae MDLV.* In the left corner: *Iacobus Bossius Belga in æs incidebat.*

24. CR, I: 19. 1558. 0,47 × 0,35. — *Gallia Belgica. Romae MDLVIII.*

25. CR, I: 20. N. s. a. 0,505 × 0,376. — *Flandriae recens exactaque descriptio. Claudio Ducheto formis.*

26. CR, I: 21. N. s. a. 0,505 × 0,397. — *Brabantiae Belgarum provinciae recens exactaque descriptio. Venetiis. Bolognini Zalterii formis.*

27. N. 1563. 0,490 × 0,385. — *Gelriae, Cliviae, Juliae, nec non aliarum regionum adjacentium nova descriptio. Anno MDLXIII. Venetiis. Apud Ioannem Franciscum Camocium. Pauli Forlani Veronensis incidente.*

28. CR, I: 23. 1566. 0,49 × 0,37. — The same map, but »*Per Iacobum Darent. Belgam, Romae MDLXVI.*»

29. CR, I: 24. N. s. a. 0,495 × 0,378. — *Hollandiae Batavorum veteris insulae et locorum adjacentium exacta descriptio. Bolognini Zalterii formis.*

30. N. 1566. 0,475 × 0,385. — *Frisiae antiquissimae trans Rhenum Provinc. et adjacentium regionum nova et exacta descriptio. Venetiis. Io. Francisci Camotii formis ad signum Piramidis MDLXVI.*

37. CR, I: 28. 1562. 0,36 × 0,25. — *Nova descripcione de la Moscovia per lecce:te M. Giacomo Castaldo piamontese cosmographo. In Venetia MDLXII. Ferando berteli exc.*

38. N. s. a. 0,248 × 0,183. — *Gotlandia.* No inscription on the title-field of the map, bound in the atlas, but on an almost identical map in my collection, printed separately from another engraving, there is written, at the base of the otherwise blank title-field: *Ferando Bertelli.*

39. N. 1562. 0,503 × 0,374. — *Germaniae omniumque ejus provinciarum, atque Austriae, Boemiae, Ungariae . . . descriptio. Ferando Berteli exc. 1562.* An almost exact copy of the map (by Nicolaus a Cusa) of central Europe printed at Rome in 1507 (N. fig. 13).

40. CR, I: 31. 1564. 0,35 × 0,24. — *Germania del Gastaldo. Paulo Forl. Veronese f. 1564. Ferando Berteli exc.*

41. CR, II: 30. N. s. a. 0,642 × 0,472. — *Bohemiae nova et exacta descriptio . . . Bolognini Zalterii formis.* The map is drawn in a manner deviating considerably from the style of drawing of other maps in Lafreri's collection.

42. CR, I: 32. N. 1570. 0,355 × 0,280. — *Descrittione del Ducato di Baviera . . . MDLXX. Paolo Forlani Veronese f.*

43. CR, I: 33. N. 1559. 0,446 × 0,378. — *Nova descriptio totius Ungariae. Romae MDLVIII.*

44. CR, I: 34. N. s. a. 0,413 × 0,283. — *Austria et Ungaria.*

45. CR, I: 35. 1566. 0,48 × 0,34. — *Vera et ultima discrittione di tutta l'Austria, Ungheria, Transilvania, Dalmatia . . . In Venetia MDLXVI. Intagliato da Paolo Furlani Veronese al segno della Colonna in Merzaria.*

46. CR, I: 36. s. a. 0,50 × 0,35. — Hungary with Servia, Transylvania, Austria, Friaul, northern portion of Adria, and Dalmatia.

47. CR, I: 37. N. 1560. 0,510 × 0,350. — The country about the lower part of Danube, between Lat. 43° and 48° 20′ from Belgrad to the Black Sea. *In Roma per Ant. Lafreri. Fabius Licinius fec.*

48. N. 1560? 0,510 × 0,350. — A continuation of the last map of the Donau countries, between Lat. 43° and 48° ½, extending westward as far as to Bavaria and Venice. As the want of border seems to indicate, these fine maps seem not to have been finished.

49. CR, I: 38. N. 1564. 0,380 × 0,272. — *Nova descrittione del Friuli. Anno MDLXIV. Paulo Forlano Veronese f.*

50. CR, I: 39 and II: 32. 1564. 0,42 × 0,32. — Illyria (Carinthia, Croatia and Dalmatia). *Bologninus Zalterius. Venetiis MDLXIIII.*

51. CR, I: 40. N. 1569. 0,496 × 0,317. — *Istria di M. Pietro Copo.* Engraved by FERRANDO BERTELLI and dedicated to ALDUS MANUTIUS. This COPO, whose old map had been engraved by the excellent and industrious engraver, but weak cartographer, Bertelli, is probably the same COPPO of whose general map of the world I have given a fac-simile above (fig. 65 p. 103).

52. CR, II: 38. N. 1570. 0,519 × 0,353. — *Il vero ritratto di Zarra et di Sebenico . . . MDLXX da Martino Rota Sebenzan.*

53. CR, I: 41. 1570. 0,40 × 0,29. — *La vera et fidele descrittione di tutto il Contado di Zara e Sebenico . . . intagliato da Paolo Forlani Veronese. Venetia 1570.*

54. CR, I: 42. N. 1565. 0,405 × 0,284. — *Nova descrittione dela Dalmatia et Crovatia. MDLXV. Ferando Berteli exc. in Venetia.*

55. CR, I: 44. N. 1561. 0,770 × 0,536. — *Il disegno della geografia moderna de tutta la provincia de la Italia . . . . all' Ill:mo Sig:r il:sr Alfonso secondo da Este, duca di Ferrara quinto Giacopo di Castaldi Piamontese cosmografo in Venetia . . . . Fabio Licinio exc.* Copied on a reduced scale by Ortelius 1570.

56. CR, I: 45. 1566. 0,50 × 0,36. — *Descrittione del Piamonte, Monferra, et la maggior parte della riviera di Genova . . . . Opera dell' Ecc:te M. Jacomo Gastaldo . . . Venetiis MDLXVI.*

57. CR, I: 46. N. s. a. 0,510 × 0,400. — *Regionis subalpinae vulgo Piemonte appellatæ descriptio, aeneis nostris formis excussa.*

58. CR, I: 47. 1558. 0,40 × 0,30. — Upper Italy; *Romae Vicentii Luchini aereis formis ad Peregrinum 1558.*

59. CR, I: 48. 1570. 0,74 × 0,48. — *La Nuova Descrittione della Lombardia.* Dedicated by the cosmographer GIORGIO TILMAN to CRISTOFORO MADRUTIO; *stampata in Roma appresso Laffreri. L'A. 1570.*

60. CR, I: 49. 1567. 0,45 × 0,29. — *Nova descrittione di tutto il ducado di Milano . . . Venetia appresso Ferrando Bertelli Nel MDLXVII.*

61. CR, I: 49. N. 1564. 0,485 × 0,390. — *Marchia Anconitana, Picaenum olim dicta . . . Romae apud Vincentium Luchinum 1564.*

62. CR, I: 51. 1559. 0,50 × 0,38. — Tuscany, *Romae Anno MDLVIIII.*

63. CR, I: 52. 1564. 0,47 × 0,32. — *La descrittione della Campagna di Roma . . . Roma l'Anno MDLXIIII.*

64. N. 1560. 0,445 × 0,318. — *Paese di Roma.* Neither engraver nor place of printing are stated, only the year 1560, at the right corner below a papal coat of arms.

65. CR, I: 53. N. s. a. 0,477 × 0,301. — *Regno di Napoli.*

66. CR, I: 54. 1567. 0,36 × 0,21. — *La descriptione dela Puglia. Opera di Giacomo Gastaldo Cosmographo in Venetia. Ferando Berteli 1567.*

67. CR, I: 55. N. 1560? 0,520 × 0,483. — *Geographia particolare d'una gran parte dell' Europa . . . Opera nuova di Giacopo di Castaldi Piamontese.* Dedicated to GIO. GIACOMO FUCCARI CONTE DI KIRCHBERG E DI WEISEMHOM. This map is a continuation of the maps 47, 48 and 70.

68. CR, I: 56. N. s. a. 0,493 × 0,370. — *Sicilia insularum omnium (ut inquit Diodorus) optima.*

69. CR, I: 57. N. s. a. 0,304 × 0,202. — *Sardinia insula.*

70. CR, I: 57. N. s. a. 0,302 × 0,198. — *Cirnus sive Corsica.*

71. CR, I: 58. s. a. 0,28 × 0,20. — Elba.
    [N. 1564. — Two printed folios: *I nomi antichi et moderni della Italia . . . di Giacobo de Gastaldi Piamontese, Cosmografo.* (Colophon:) *In Venetia MDLXIIII Con privilegio.*]

72. CR, I: 59. N. 1565. 0,447 × 0,375. — *Nuovo disegno dell' Isola di Malta . . . Ant. Lafreri Romae Anno 1565.* Another impression probably from the same plate, but dated 1569, occurs in CR, II: 47.

73. CR, I: 60. N. s. a. 0,425 × 0,290. — *La dimostratione del luogo dove al presente si trova l'armat di Barbarossa et de Christiani, detto il golfo dell' Artha . . .*

74. CR, I: 61. N. 1564. 0,369 × 0,270. — *Corfu. Ferando Berteli exc. 1564.*

75. CR, I: 62. s. a. 0,28 × 0,21. — Another map of Corfu.

76. CR, I: 63. s. a. 0,47 × 0,34. — *Fortezza di Soppolto.*

77. CR, I: 64. 1569. 0,45 × 0,32. — Morea. *Appresso Gio. Francesco Camocio 1569 con privilegio.*

78. CR, II: 54. N. 1570. 0,328 × 0,245. — *Peloponnesus Nunc Morea l'Anno 1570. Per Claudio Ducheto.*

79. CR, I: 66. 1564. 0,38 × 0,27. — *El vero et nuovo disegno di tutta la isola di Candia. Venetiis, Io. Francisci Camotii aereis formis ad signum Pyramidis 1564.*

80. N. s. a. 0,281 × 0,205. — *Creta Insula, hodie Candia.*

81. CR, I: 67. N. s. a. 0,267 × 0,203. — *Rhodus.*

82. CR, I: 68. s. a. 0,56 × 0,47. — The Archipelago.

83. N. s. a. 0,530 × 0,486. — Greece and the Archipelago. Beneath in the right corner: *fabius Licinius fecit Venetiis.* This map and the maps 47, 48 and 67, form together Gastaldi's »*Geographia particolare d'una gran parte dell' Europa*» printed on four sheets, of which the first is not signed, the second is signed: »*In Roma per Ant. Lafreri Fabius Licinius fecit,*» on the third we only meet the cosmographer's name, and the fourth is signed: »*fabius Licinius fecit Venetiis.*»
    [N. 1570. — A printed folio leaf: *Nomina antiqua et recentia urbium Graeciae descriptionis a N. Sophiano jam aeditae;* (colophon:) *Romae sub anno Domini MDLXX Typis Antonii Lafreri.*]

84. CR, I: 69. 1558. 0,61 × 0,41. — *Totius Graeciae descriptio. Romæ Vincentii Luchini aereis formis ad Peregrinum 1558.*

85. CR, I: 70. s. a. 0,61 × 0,41. — *Graeciae Chorographia* by FRANCESCO SALAMANCA. »*Sebastianus Clodiensis incidebat*».

86. CR, I: 71. N. 1562. 0,596 × 0,437. — Africa. The dedication to *Ecc:mo Philosopho, Mathematico . . . Guardiano grande della Scola de S. Marco il Sig:or Thomaso Ravenna*, is signed: PAULO FORLANI VERONESE and dated *Venetia MDLXII.*

87. CR, I: 72. N. s. a. 0,437 × 0,300. — *Disigno dell' Isola de Gerbi.*

88. N. s. a. 0,400 × 0,245. — *Fortezza di Gerbi.*

89. CR, I: 73. N. 1570. 0,338 × 0,260. — *La nuova et copiosa descrittione di tutto l'Egitto l'Anno MDLXX.*

90. CR, I: 74. s. a. 0,25 × 0,19. — *S. Lorenzo* (Madagascar).

91. CR, I: 75. N. 1566. 0,584 × 0,421. — Asia Minor. *Opera dell' Ecc:mo M. Giac:o Castaldo Piamontese. In Venetia Apresso Gio. Franc. Camotio 1566.*

92. CR, I: 76. 1570. 0,44 × 0,33. — *Il vero disegno della Natolia e Caramania . . . di Giacomo Gastaldo Cosmographo. Venetiis MDLXX. Bolognini Zalterii formis.* Probably another edition of the last map.

93. CR, III: 78. N. 1561. 0,735 × 0,428. — *La descrittione della prima parte dell' Asia, con i nomi antichi e moderni di Jacopo Gastaldi Piemontese Cosmografo . . . Restituta da Antonio Lafreri. L'anno MDLXI. Jacobus Bossius Belga incidebat.* Two years previously the same map was engraved by FABIUS LICINIUS (comp. CR, II: 66).

94. CR, I: 78. 1566. 0,33 × 0,26. — Syria and Palestine; *In Venetia l'Anno MDLXVI.*

95. CR, I: 79. N. s. a. 0,504 × 0,256. — *Tabula Moderna Terrae Sanctae.*

96. CR, I: 80. N. s. a. 0,535 × 0,382. — *Palestinae sive Terrae Sanctae descriptio. Romae apud Joannem Franciscum vulgo Della Gatta.*
    [N. 1564. — A printed folio leaf: *Nomi antichi e moderni della prima parte dell' Asia . . di Giacobo de Gastaldi, Piamontese Cosmografo.* (Colophon:) *In Venetia 1564.*]

97. CR, I: 81. N. 1561. 0,748 × 0,473. — *Il disegno della seconda parte dell' Asia.* Dedicated *all' ill:mo sig:r Marcho fucharo, Barone di Kirchberg e d'Waissenhoven* by *Giac. di Castaldi in Venetia 1561.*
    [N. 1564. — A folio leaf: *I nomi antichi e moderni della seconda parte dell' Asia . . di Giacomo di Castaldi . . . in Venetia 1564.*]

98. CR, I: 82. N. 1561. 0,733 × 0,632. — *Il disegno della terza parte dell' Asia.* On the right side a larger field with the superscription: *I nomi antichi e moderni della terza parte dell' Asia per me Giacomo di Castaldi . . in Venetia 1561;* farther down: *Fabius Licinius excudebat.* In the north-eastern corner of the map is written *Ania Pro.* As far as I know, it is the first time this name, which was afterwards transferred to the strait between Asia and America (*Fretum Anian*, comp. fig. 81), occurs in cartographical literature.

99. CR, I: 83. N. s. a. 0,263 × 0,202. — *Taprobana.*

100. N. 1566. 0,243 × 0,175 and

101. N. 1566. 0,245 × 0,170. — Two maps engraved on the same plate; the former, representing *l'Isola Cuba*, is signed *F. B.*; the latter *l'Isola Spagnola . . . in Venecia l'ano 1566. Ferando Berteli exc.*

102. CR, I: 84. N. s. a. 0,505 × 0,365. — *La descrittione di tutto il Peru* (South America). Dedicated to Gio. Pietro Contarini by Paulo di Forlani. A fac-simile on a reduced scale is given by me, fig. 80.

103. CR, I: 86. N. 1566. — *Il disegno del discoperto della Nova Franza ... Venetiis æneis formis Bolognini Zalterii. Anno MDLXVI.* An interesting map of North America. (N. fig. 81).

104. CR, I: 86. N. s. a. 0,417 × 0,375. — *Victoria di Chatolici contra Hugonoti.* Plan of a battle delivered in the vicinity of La Rochelle between Papists and Hugenots, »il settimo giorno di quest' anno 1569».

105. CR, I: 87. 1567. 0,26 × 0,19. — Paris. *In Venetia l' Anno MDLXVII.*

106. CR, I: 90. s. a. 0,38 × 0,18. — Jerusalem.

107. CR, I: 101. s. a. 0,50 × 0,36. — Augsburg.

108. CR, I: 102. s. a. 0,19 × 0,25. — Mirandola.

109. CR, I: 102 A. 1567. 0,18 × 0,25. — Parma. *Nell' anno del Signore MDLXVII.*

117. CR, I: 112. 1567. 0,54 × 0,41. — Messina. The dedication signed: *1567 per Antonio Lafreri.*

118. CR, I: 116. s. a. 0,41 × 0,28. — *Disegno de' porti e forti del Isola de Malta.*

119. CR, II: 100. N. 1565. 0,514 × 0,372. — *Disegno dell Isola di Malta. In Roma per Antonio Lafreri nel anno 1565.*

120. CR, II: 98. N. 0,383 × 0,278. — *Il vero ritratto di Nettuno.*

121. N. 0,463 × 0,245. — *El Pignon ... Apresso Gio. Francesco Camotio Domenico Zenoi.*

Nos. 1—121 comprise all the maps as well in the first volume of Collegio Romano's as in my copy of Lafreri's Atlas. In the library of the Collegio Romano there are further two other volumes belonging to the same collection. These consist, according to the catalogue of Castellani, partly of some of the above recorded maps, here bound not in the first but in the second or third volume, partly of new editions of the

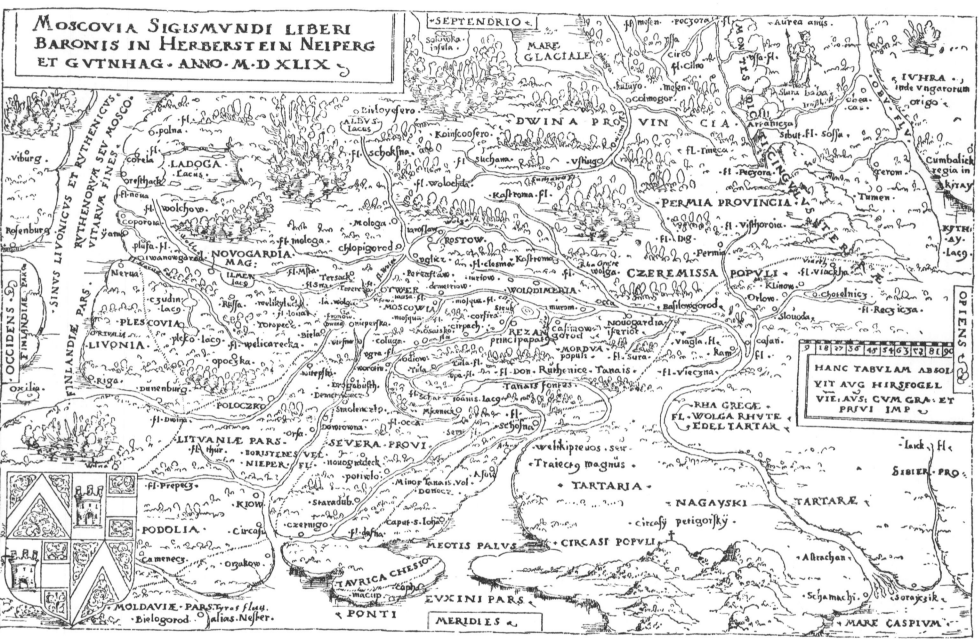

77. Moscovia by Herberstein, copper-engraving by Hirschvogel 1549. (Orig. size 261 × 164 m. m.).

110. CR, I: 103. s. a. 0,72 × 0,44. — Venice. *Ex aeneis formis Bolognini Zalterii.*

111. CR, I: 104. 1569. 0,70 × 0,42. — Ancona. Dedicated by Giacomo Fontana to the duke of Urbino; *Ancona a di 3 di Marzo 1569.*

112. CR, I: 105. 1561. 0,54 × 0,40. — Rome. The dedication signed: Bartolomeus Phaletius, *Romae Calendis Ianuarii MDLXI. Sebastianus a Regibus Clodiensis in aere incidebat.*

113. CR, I: 106. N. 1557. 0,465 × 0,346. — *Urb. Romae topographia ... public. impen. fieri curavit Paul IIII pont. max. dum bello parthenop. premeretur An. MDLXII. Sebastianus a Regibus Clodiensis in aes incidebat.*

114. CR, I: 107. N. 1557. 0,388 × 0,255. — *Il vero disegno del sito di Hostia e di Porto ... MDLVII.*

115. CR, I: 111. 1566. 0,55 × 0,38. — Naples; *Ant. Lafrerii formis Romae MDLXVI.*

116. N. s. a. 0,430 × 0,295. — *Il vero disegno in sui propio luogho ritratto del infelice paese di Posuolo.* The environs of Naples with Monte Nuovo.

maps in the first volume, and of new impressions from the old plates, but with altered dates and sometimes also with altered names of the editor and engraver, and finally of the following maps not before mentioned:

122. CR, II: 42. 1567. 0,39 × 0,34. — Piedmont; *per Paolo Furlani, MDLXVII.*

123. CR, II: 65. s. a. 0,55 × 0,44. — Cyprus with the opposite shores of Egypt and Asia minor.

124. CR, II: 104. s. a. 0,45 × 0,30. — Constantinople.

125. CR, III: 3. 1546. 0,64 × 0,38. — *Universale.* A planisphere by *Giacomo Cosmographo in Venetia MDXXXXVI.* Probably an older edition of No. 4, and one of Gastaldi's first works.

126. CR, III: 14. 1559. 0,53 × 0,39. — *Nova totius Hispaniae descriptio. Pyrrho Ligorio Neap. Auctore Romae MDLVIIII. Michaelis Tramezzini formis ... Sebastianus de Regibus Clodiensis in aere excidebat.*

127. CR, III: 16. 1561. 0,66 × 0,35. — Map of Portugal by ACHILLE STAZIO, dedicated to the cardinal GUIDO ASCANIO SFORZA; *Romae... MDLXI. Michaelis Tramezini formis. Sebastianus a Regibus Clodiensis in aere incidebat.*

128. CR, III: 22. 1563. 0,59 × 0,43. — Switzerland; *Venetiis Anno MDLXIII. Paulus de Furlanis Veronensis fecit.* Dedicated to JODOCUS A MEGGEN. Probably a new edition of the map No. 23.

129. CR, III: 27. 1559. 0,49 × 0,41. — Flanders; *Venetiis MDLVIIII.*

130. CR, III: 32. 1558. 0,52 × 0,38. — Chart of the seas surrounding Scandinavia; *Michaelis Tramezini formis MDLVIII. Jac. Bossius Belga in aes incidebat.* Probably a new edition of No. 32.

131. CR, III: 33. 1561. 0,48 × 0,37. — Belgium; *MDLXI.*

132. CR, III: 35. 1552. 0,35 × 0,25. — Germania; *Opera di Jacopo di Gastaldi. In Venetia 1552. Appresso Gabriel Giolito al segno della Fenice.* An edition of Gastaldi's Germania older than No. 40.

133. CR, III: 37. s. a. 0,50 × 0,38. — *Tabula Moderna Poloniae, Ungariae, Boemiae, Germaniae, Russiae, Lithuaniae. Ant. Sa(lomon) exc.*

134. CR, III: 45. 1563. 0,66 × 0,40. — Friaul; *Per Pyrrho Ligorio Napolitano... In Roma del MDLXIII, con le forme di M. Michele Tramezzino. Sebastiano di re da Chioggia intagliava in rame.*

135. CR, III: 47. 1567. 0,48 × 0,34. — Piedmont. Dedicated by PAOLO FURLANI to ANDREA DEGLI OREFICI, *di Venetia, l'Anno MDLXVII.*

136. CR, III: 54. s. a. 0,55 × 0,37. — *Sacra Tuscia.*

137. CR, III: 55. s. a. 0,24 × 0,32. — The Papal states.

138. CR, III: 57. 1558. 0,69 × 0,44. — *Nova regni Neapolit. Descriptio ... Pyrrho Ligorio Neap. auctore. Romae MDLVIII. Michaelis Tramezini formis, Sebastianus a Regibus Clodiensis in aes incidebat.*

139. CR, III: 63. 1545. 0,53 × 0,37. — Sicily; *Per Giacomo Gastaldo Piemontese Cosmographo in Venetia 1545.*

140. CR, III: 72. 1566. 0,40 × 0,26. — Cyprus; *Expensis Io. Fr. Camotii ... Venetiis ad Signum Pyramidis MDLXVI.*

141. CR, III: 89. s. a. 0,35 × 0,24. — Chart of the Atlantic Ocean from the western coast of Africa and the British islands to Brasil and Canada. Dedicated by FER. BERTELI to MARCO DEL SOLE.

142. CR, III: 114. 1555. 0,94 × 0,52. — Roma; *Ex typis et diligentia Ant. Lafreri. Jac. Bossius Belga in aere incidebat.*[1]

The total number of maps in the above catalogue is 142. 87 of them are dated, viz.

Printed before 1556 .... 3.
» 1556—1560 .... 21.
» 1561—1565 .... 27.
» 1566—1572 .... 36.

The maps in Lafreri's atlas are consequently, with few exceptions, printed or dated between 1556 and 1572. The most important of them are of GASTALDI. Others are drawn by PYRRHUS LIGORIUS, ANTONIUS SALAMANCA, JACOBUS DAVENTERIUS, PIETRO COPPO, MARCUS ROTA SEBENZAN, GEORGIUS TILMAN, GIACOMO FONTANA, ACHILLES STAZIO, BARTOLOMEUS PHALETIUS, JOHANNES FRANCISCUS, ANTONIUS FLORIANUS, and DOMINICUS ZENOI. The most prolific engravers have been FERRANDO BERTELLI, PAULO FORLANI, ANTONIO LAFRERI, SEBASTIANUS A REGIBUS CLODIENSIS, JACOBUS BOSSIUS, FABIUS LICINIUS, and FABIA LICINIA. Forlani and Lafreri were also publishers and dealers of maps, and judging from the expressions »apud» or »ex aereis formis», we may even be permitted to consider as such JOANNES FRANCISCUS CAMOTIUS, MICHAEL TRAMEZINI, CLAUDIUS DUCHETUS, VINCENTIUS LUCHINI, and BOLOGNINUS ZALTERIUS.

The maps are all printed from copper-engravings, and many, from this point of view, are real masterpieces. Several

of them, especially maps of the Mediterranean countries, were copied by Ortelius. Of the maps in Lafreri's atlas the following have been here reproduced, generally on a reduced scale: No. 1 (fig. 32); No. 2 (fig. 48); No. 4 (fig. 53); No. 9 (fig. 54); No. 15 (fig. 78); No. 32 (fig. 25); No. 34 (fig. 79); No. 102 (fig. 80) and No. 103 (fig. 81).

The reason why I have occupied myself so fully with Lafreri's collection of maps is, that Italy, through the maps from different parts of that country, which were saved from destruction by this collection, contributed to the development of cartography in a degree as yet almost entirely overlooked. This entitled Italian cartographers, for a short time, again to occupy the foremost place in that science or art, the aim of which is to give us a cartographical representation of the earth's surface.[2] The most prominent cartographer or, as he was generally styled, cosmographer of this period in Italy, and indeed in the whole world, was no doubt JACOPO GASTALDI or CASTALDI. Unfortunately I have not been able, in the literature accessible to me, to find any other date relating to his biography than that he was born at Villa Franca in Piedmont, and that he lived in Venice, where the main part of his cartographical work was published. A memoir: *Notizie di Jacopo Gastaldi, cartografo Piemontese del secolo XVI*, by BARONE MANNO and CAV. VINCENZO PROMIS, inserted in *Atti della R. Accademia delle Scienze di Torino*, Vol. 16 (1881), contains only an incomplete enumeration of his works, without any other contribution to his biography than the mentioning of his birthplace. It is to be hoped that further investigations among the archives and libraries of Piedmont and Venice may bring to light at least a few data concerning the life of this geographer, one of the most productive and prominent of the sixteenth century and fully deserving the epithet: *eccellentissimo cosmografo Piamontese*, by which he was generally designated by contemporary authors. His later works are not only equal to the maps in the first edition of *Theatrum Orbis terrarum* but often superior, as well as regards originality as execution. It would be unjust not to cite his name in the history of cartography, together with those of Ortelius and Gerard Mercator, among the promotors of the great reform in cartography accomplished in the latter part of the 16th century.

Among the reformers of cartography PHILIP APIANUS, a son of PETRUS APIANUS or BIENEWITZ, also occupies a prominent place.[3] He was born in 1531 and died in 1580. In 1552 he had already been nominated successor to his illustrious father in the mathematical professor's chair, at the university of Ingolstadt. In 1554 the construction of a modern map of Bavaria was entrusted to him, and he applied himself to this work with such energy that in 1561 he had already finished the measurements and triangulation on which the new map was to be based. Its first edition was published at Munich in 1566; the second at Ingolstadt in 1568. The title (of the edit. 1568) is: *Bairische Landtafeln XXIIII Darinnen das Hochlöblich Fürstenthumb Obern unnd Nidern Bayrn, sambt der Obern Pfaltz, Ertz und Stifft Saltzburg, Eichstet unnd andern mehrern anstossenden Herschafften mit vleiss beschriben und in druck gegeben Durch Ph. Apianum.* My copy contains, besides 24 special maps printed on 22 double folio-leaves and mentioned on the title-page, a general map: *»Ein klaine*

---

[1] One map in the copy of Collegio Romano (II: 1) Viterbo, »*Tarquinio Ligustri Viterb. fece 1596*», is here omitted, because it evidently has been inserted in the collection without originally having belonged to it.

[2] What a prominent place the art of map-drawing and the map industry occupied in Italy in the middle of the 16th century is evident e. g. from the fact that there is in the *Catalogue of Printed Maps* of the British Museum, under the subsection *General Maps*, to be found from the period between 1550 and 1570, six maps of France, six of Italy, six of Britain, seven of Germany, four of Spain, four of Poland, six of Greece, three of Ireland, all engraved in copper and printed in Italy (Rome and Venice); whereas this catalogue does not, from the same time, contain one single general map of these countries printed to the north of the Alps.

[3] Comp. *Peter und Philip Apian, zwei deutsche Mathematiker u. Kartographen* von Dr. SIEGMUND GÜNTHER (*Abhandl. der Königl. Böhm. Gesellschaft der Wissenschaften*, VI Folge, Bd. 11) Prag 1882.

Landtafel des Lands Obern und Nidern Bayrn, surrounded by a border formed of the coats of arms of the most important towns etc. of Bavaria, and divided into 24 rectangles or squares each corresponding to one of the special maps. According to GÜNTHER a reprint of the Bairische Landtafeln was published in the 17th century, and a new edition with maps engraved in copper was published at Berlin in 1766. In 1802 and 1881, some new copies were drawn from the old blocks, which are yet preserved at Munich, and are remarkable as being provided with stereotyped legends.[1] The Bairische Landtafeln are of coarse execution, and are disfigured by ornaments (coats of arms etc.) which are out of proportion and are foreign to cartography. They are, in technical respects, by no means to be compared with the above mentioned Italian copper-

	Ph. Apianus.		The true position.[2]	
	Lat.	Long.	Lat.	Long.
Nuremberg	49° 27'	31° 41'	49° 28'	31° 45'
Regensburg	48° 57'	32° 45'	49° 1'	32° 47'
Ingolstadt	48° 42'	32° 6'	48° 45'	32° 7'
Augsburg	48° 18'	31° 36'	48° 21'	31° 35'
Munich	48° 1'	32° 16'	48° 8'	32° 16'
Salzburg	47° 42'	33° 47'	47° 48'	33° 43'
Passau	48° 28'	34° 10'	48° 34'	34° 7'

This table shows that the latitudes of APIANUS are affected by a constant error of about 4'. The corresponding error in the longitudes is included in the difference of 21° 41' which I have adopted between the first meridian of Apianus and Greenwich. With due regard to this correction and reduction, the agreement between the true positions and the positions given by Apianus is as complete as possible, without

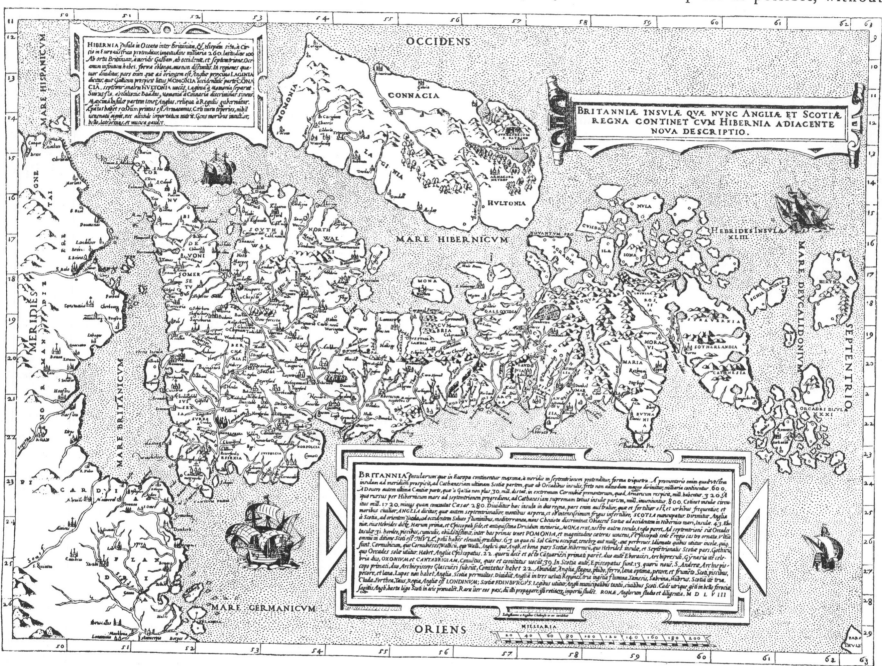

78. Map of Britain 1558. From LAFRERI'S atlas. (Orig. size 541 × 398 m. m.).

engravings. But as to the correctness of the relative positions of the towns, rivers, mountains etc., the maps of Philip Apianus, based as they are on real triangulation, cannot be too highly appreciated. In this respect they are far more accurate than any similar previous or contemporary work, perhaps with the exception of Mercator's map of Flanders, which I have not had an opportunity of examining.

The following comparison between the astronomical positions of the places on his *Bairische Landtafeln* and their true positions, will show that I do not exaggerate the merit of PHILIP APIANUS in this respect.

stating the exact point in the town, to which the geographical coordinates are referable.

In order to show the enormous progress introduced by the map of Apianus I here give a table of latitudes and longitudes adopted for Italy by Gastaldi in his above (p. 120) mentioned: *Nomi antichi et moderni della Italia*, inserted in Lafreri's atlas.

	Gastaldi.		The true position.[3]	
	Lat.	Long.	Lat.	Long.
Milan	45° 58'	30° 35'	45° 27'	32° 0'
Trieste	45° 37'	35° 50'	45° 38'	36° 37'
Venice	45° 0'	34° 0'	45° 30'	35° 13'

---

[1] According to GÜNTHER, who considers this to be the first instance of the employment of stereotype.

[2] According to *Philips' Imperial Library atlas*. In order to facilitate a comparison between the true longitudes and the longitudes of Apianus, the former are increased by 21° 41', corresponding to the difference between Greenwich and the point of departure of Ph. Apianus. Ferro is situated 18° 10' W. from Greenwich.

[3] In order to refer the longitudes to the same point of departure I have here increased the numbers given in Philips' Atlas by 22° 49'.

	Gastaldi		The true position.	
	Lat.	Long.	Lat.	Long.
Bologna	43° 34′	33° 36′	44° 30′	34° 40′
Ancona	43° 12′	37° 0′	43° 37′	36° 20′
Genoa	42° 40′	30° 56′	44° 25′	31° 44′
Nice	42° 18′	28° 25′	43° 41′	30° 4′
Rome	41° 12′	36° 30′	41° 54′	35° 22′
Naples	40° 37′	38° 5′	40° 50′	37° 4′
Messina	37° 48′	40° 12′	38° 11′	38° 23′
Palermo	37° 18′	37° 20′	38° 6′	36° 10′
Syracuse	36° 19′	40° 4′	37° 3′	37° 55′

These tables prove that the geographical coordinates of Apianus are far more exact than those of Gastaldi, which no doubt principally depends on the fact that whilst the map of Apianus is based on trigonometrical surveys, the map of the sea-encircled Italy is founded on portolanos and on compass-bearings taken, perhaps, during the 13th or 14th century, before the variation of the compass was known. Furthermore, a comparison between the maps of Ph. Apianus, the best of the early German cartographers, and of Gastaldi, the most prominent Italian cosmographer from the same time, shows, that if the Italians, in the middle of the 16th century, were still foremost in the art of drawing and engraving maps, the art of constructing them on mathematical principles and by the aid of astronomical observations had already been transferred to the countries to the north of the Alps. It was not long, either, before the cosmographers from these countries were able, as regards technical execution, to compete with copper-engravers in the native country of the art. From that time the most prominent and productive cartographers are no longer found in Venice, or Rome, but in the Netherlands, France, Germany, and England, to which countries the whole industry of cartography was soon transferred. This change was connected with a complete reform in the science. The period of incunabula, characterized by a slavish following of the old doctrines and types of Ptolemy, was closed, and a new period succeeded, which was characterized by the effort to found the knowledge of the lands and seas of the earth, not on commentaries of the writings of more or less classical authors, but on new and careful investigations, based if possible on topographical surveys and astronomical observations. The reform had been prepared by Gastaldi and Ph. Apianus, but it was first accomplished by ORTELIUS and GERARD MERCATOR. As the history of the early period of cartography would be incomplete and difficult to understand without a comparison with the time immediately succeeding it, I shall yet here give a brief sketch of the significance of the work of the last mentioned eminent cartographers, and of their relation to the preceding period, from which the main parts of the material they employed were borrowed.

———————

ABRAHAM ORTELIUS or ORTEL (ORTELS), was born at Antwerp in 1527, and he died at that place in 1598.[1] In his youth he seems to have established himself as a map-dealer in his native city. In 1547 he became a member of a guild in Antwerp, as a colourist of maps, and several years later he still, according to a very remarkable letter from JOHANNES RADERMACHER to JACOB COOL (HESSELS, p. 772), contributed to the support of his family by purchasing the best maps he could get hold of, pasting them on canvas, colouring them, and then selling them either in his own country or abroad. In connection with this business he seems to have undertaken

repeated voyages to different countries, and by it he acquired an extensive knowledge of the existing map-literature. Whole map-collections were united occasionally into a volume, and new editions of rarer maps were published. Finally the results of his labours were embodied in his great systematic collection of maps under the title of *Theatrum orbis terrarum*.

The first edition was published at Antwerp in 1570, with 53 plates in double folio. A second edition left the press in the same year, and then edition followed rapidly on edition until 1612, when the last one was published. It contained 128 modern maps + 38 ancient or historical maps collected in a separate appendix: *Parergon sive veteris Geographiae aliquot tabulae.* For a more particular description of the different editions, and the many additions, at first printed separately, and afterwards incorporated in the main work, I may refer to: P. A. TIELE, *Bibliographische Adversaria*, Haag 1876, to the *Nederlandsche Bibliographie van Land- en Volkenkunde*, as well as to HESSELS' above cited work.

In 1573 Ortelius was nominated Geographus Regius by Philip II. Before 1570 he had already published maps, e. g. *Typus Orbis Terrarum*, sold separately already about 1560 and later inserted as the first map into the *Theatrum;* a map of Egypt no longer extant; a map of Asia published in 1567 »in majori tabula», according to the text on the first page of *Asiae nova descriptio* in the *Theatrum* of 1570. Among maps now extant the *Typus Orbis terrarum* seems to be his first work. I have given a full size fac-simile of it on tab. XLVI. Most of the maps in the *Parergon* were, according to the title-vignettes, constructed by Ortelius himself, whereas the modern maps in the Theatrum are almost always copies of the works of other cartographers whose names are given.

Ortelius was thus an enterprizing dealer in maps, a zealous collector, and an intelligent publisher, but with the exception of the above named map of the world and of a few other works, he was neither an author of modern maps nor a map-draughtsman or map-engraver. He never seems to have executed any topographical survey; nor did he contribute to the development of science by the introduction of any new projections, or by setting his own hands to the engraving of maps. The great influence he exercised and the great fame he acquired depend on the fact that he was the first who collected all the map-material accessible, and employed it with great discernment and skill for a systematical collection of maps, by which Ptolemy's classical work was definitively supplanted in modern literature.

Ortelius is, moreover, entitled to great praise for giving the author's name on almost every map he copied. His *Theatrum* is thus, to this very day, of immense importance in the history of geography, as a collection of geographical documents, by which many a map, otherwise entirely lost, has been saved. In the first edition of the *Theatrum* he gives a valuable contribution of his own to the history of cartographical literature by inserting in the introduction a catalogue of all the maps that the indefatigable collector had been able to accumulate. As this catalogue gives a faithful though by no means complete, idea of the maps most extensively in circulation in central Europe before 1570, I shall here reproduce it together with a few illustrative remarks, which, however, the want of space and of time for necessary researches have prevented me from making so complete as the importance of the subject merits.

———————

[1] For the biography and bibliography of Ortelius we have, besides various older sources, a recently published work of great value to the whole history of the geography during the 16th century, viz: *Ecclesiae Londinae-Batavae Archivum. Tomus primus: Abrahami Ortelii (Geographi Antverpensis) et virorum eruditorum ad eundem et ad Jacobum Colium Ortelianum (Abrahami Ortelii sororis filium) epistulae, cum aliquot aliis epistulis et tractatibus quibusdam ab utroque collectis (1524—1628). Ex autographis mandante Ecclesia Londina-Batava edidit* JOANNES HENRICUS HESSELS, Cantabrigiae 1887. The biographical data which I have given here are borrowed from this work.

— 125 —

Catalogus Auctorum

tabularum geographicarum, quotquot ad nostram cognitionem hactenus pervenere; quibus addidimus, ubi locorum, quando, et a quibus excusi sunt.[1]

Ægidius Bulionius Belga, *Galliam Belgicam descripsit; quam edidit Antverpiæ Joannes Liefrinck; et Sabaudiam cum Burgundiæ Comitatu, evulgatam apud Hieronymum Cock, Antverpiæ.* — Of the last mentioned map two editions s. l. et a. are preserved at the Bibliothèque nationale in Paris. It is reproduced by Ortelius in *Theatrum Orbis Terrarum,* edit. 1570, Tab. 12.

Ægidius Tschudus, *Rhetiam, Helvetiamque; Basileæ apud Isingrinum.* — Reproduced by Seb. Münster in several of his geographical works;

As cartographer of Switzerland Tschudi had a predecessor in Conrad Türst of Zürich, who, as early as in 1496, delineated a map of this country, which for its time was very meritorious, and which was printed as woodcut in the editions of Ptolemy of 1513 and 1520. Tschudi was born in 1505 and died in 1571 (Jöcher).

[**Andreas Pagradus Pilsniensis,** *Sarmatiæ Europææ partem, quæ subiacet Sigismundo Poloniæ Regi; Venetiis 1569.*] — In a letter to Ortelius

79. The north-eastern Europe by Jacopo Gastaldi, Venetia 1568. From Lafreri's atlas. (Orig. size 517 × 373 m. m.).

by Ortelius 1570, 31, and by Quad in *Geogr. Handtbuch,* 1600, 12. Gesner (*Bibliotheca Universalis,* Tiguri 1545, fol. 5) mentions with much praise Tschudi's description and map of Switzerland. The description was, according to Gesner, originally written in German. Later it was translated into Latin by Seb. Münster, and published at Basel in 1538 with the map on nine leaves. An edition of this map, published at Basel MDLX by Michael Isingrinius, was reproduced by photolithography at Zürich in 1883, by Hoefer & Burger.

by Nicolaus Secovius, *eques Polonus* (Hessels, *Ortelii Epistulæ,* p. 217) the name is corrected to Pograbius. This important map of the region between the Oder and Dnieper, from 47° 50′ to 56° latit., was published at Venice in 1570, *Nicolai Nelli aereis formis.* In the upper corner of it is engraved: *Partis Sarmatiae Europeae, quae Sigismundo Augusto Regi Poloniae potentissimo subiacet, nova descriptio.* In the lower corner is a long dedication: *Generoso domino Nicolao Tomicio ... Domini Joannis de Tomice, Castellani Gnesnen. filio Andrea Pograbius Pilsnensis S. P. D. In*

---

[1] According to the edition Antverpiae 1570. I have given, in brackets, the names of a few authors of maps previous to 1570, added to the *Catalogus* in later editions. The success of the *Theatrum* led to several other similar publications at the end of the 16th century, such as Mercator's *Atlas,* the publication of which had been planned almost simultaneously with the work of Ortelius, though its first edition did not appear before 1595, and the *Speculum Orbis Terrarum,* Antverpiae 1578, of Gerard de Judaeis, containing 38 maps in folio, drawn (or copied) by Ger. de Judaeis and engraved partly by him, partly by Joa. and Luc. a Deutecum. I have only seen the second edition of this rare work, *Speculum Orbis terrae, Antverpiae, Sumptibus viduae et heredum Gerardi de Judaeis 1593,* published after Gerard's death by his son Cornelius de Judaeis. The number of maps is here augmented to 84, of which 33 are in the first and 51 in the second volume. The maps resemble those in the Theatrum. De Judaeis generally gives the names of the authors of the maps, and owing to this circumstance his *Speculum Orbis terrae* has obtained a lasting place in cartographical literature. As an example of the maps of de Judaeis a fac-simile of pl. 2 in his work is given on T. XLVIII.

Among the smaller atlases Matthias Quad's *Geographisch Handtbuch,* Cöln 1600 (a Latin edition: *Fasciculus Geographicus,* Coln am Rein 1608) deserves to be mentioned, because even here the names of the authors of the maps are given, and because it contains reproductions of some few maps, the originals of which appear to be lost. On pl. XLIX one of Quad's maps is given in fac-simile.

chorographiis, quas Venetiis editas vidi, Generose Tomici, non rectam Poloniae, nostrae patriae, descriptionem animadverti etc... *Patavii 1569.* The map is remarkable for its richness in topographical details and for accuracy in the spelling of the names. Its size is 0,693 × 0,476. The map 98 of the edition of the *Theatrum* of 1595 is partly based on the work of Pograbius.

Andreas Thevetus, *Galliam [Parisiis 1578. Ibidem idem quoque Orbem terrarum, sub lilii forma.]* — In *Cosmographie universelle* d'ANDRÉ THEVET, *Cosmographe du Roy,* Paris 1575, there are maps of Africa, Asia, Europe, and America, but none of France. Nor is there any map of France by Thevet cited in the *British Museum Catalogue of Printed Maps.* He was born in 1502 and died in 1590.

Antonius Ienkinsonus, *Russiam; Londini 1562.* — ORTELIUS 1570, 46; DE JUDAEIS, *Speculum Orbis Terrae,* Antverpiæ 1593, II: 9.

Antonius Wied, *Moscouiam; Antverpiæ.* — Anton Wied's Moscovia, dated: *ex Wilda Lithauiæ 1555,* but engraved in 1570, is reproduced in fac-simile by Dr. H. MICHOW, *Die ältesten Karten von Russland,* Hamburg 1884. Also in the *Catalogue of Brit. Mus.*

Augustinus Hirsvogel, *Regionum hactenus non visarum (uti titulus habet) Tabulam edidit; continet vero Slauoniam, Carinthiam, Styriam, Goritziam, etc. vicinasque regiones; Nurenbergæ apud Joannem Weygel.* — Slavonia, Carinthia etc. reproduced by ORTELIUS 1570, 41; Oesterreich ober Enns: DE JUDAEIS II, 16; Illyricum seu Sclavonia: DE JUDAEIS II, 17. Hirschvogel was a celebrated copper-engraver and painter in glass and enamel at Nuremberg. He also occupied himself with mathematical and cartographical works. Among others he contributed several engravings to Herberstein's Moscovia (comp. p. 113 and fig. 77). † 1560. (DOPPELMAYER, *Nachricht von den nürnbergischen Mathematicis und Künstlern,* p. 156 and 199.)

Augustinus Iustinianus, *Nebiæ Espisc. Corsicæ descriptionem in Tabulam redigit, ut ipsemet inquit in sua Historia Genuensi.* — A map based on the description of Justinianus was published in the work cited below of LEANDRO ALBERTI and in QUAD's atlas (the map 64 is signed: *Leander Albertus ex commentariis Augustini Justiniani*). Aug. Justinianus published: *Psalterium Hebræum, Græcum, Arabicum et Chaldeum,* Genua 1516, celebrated for an interesting annotation about Columbus and his family, inserted at the Psalm XIX. Justinianus was born at Genoa in 1470 and died in 1536 (JÖCHER).

Bartholemæus Scultetus, *Misniæ et Lusatiæ Corographiam; Gorlitzii, anno 1569.* — Misnia: ORTELIUS 1573, 29; DE JUDAEIS, II: 29; QUAD 20; Lusatia: BLAEU, *Geographia Blaviana,* Amsterdam 1662, III s. 121. Four maps of Bartholomæus Scultetus are mentioned in the *Brit. Mus. Cat. of Printed Maps,* p. 3739. Scultetus died in 1614 (JÖCHER).

Benedictus Bordonius, *Italiæ Tabulam; uti habet Leander in sua Italiæ descriptione.* — I have above (p. 104) mentioned the maps in Bordone's *Isolario,* but I have not seen any other cartographical works by him.

[**Bernardus Brognolus,** *Veronens. Territorium evulgavit; Venetiis 1564.* — ORTELIUS 1584, 66.]

Bernardus Syluanus, *cuius Galliam, et Italiam, citat Robertus Cænalis in Gallica sua historia.* — The maps of Gallia and Italia cited by Caenalis are probably nothing but the slightly modernized classical maps which Sylvanus published in his edition of Ptolemy, Venetiis 1511. A general map of the world by Sylvanus is reproduced N. T. XXXIII.

Bonauentura Brochardus, *Palæstinam; Parisiis, apud Poncetum le Preux.* — Bonaventura Brochard was a French monk of the beginning of the 16th century. He wrote an account of a journey to Sinai and Jerusalem.

Bonauentura Castilioneus, *Longobardiam; auctor Joannes Antonius Castilioneus, in libello qui de Insubrum antiquis sedibus inscribitur.* — Bonaventura Castilioneus was a canon at Milano living near the middle of the 16th century. The work here cited was published, contrary to his wish, by his kinsman ANTONIUS. I have not had access to the original edition, which was printed in 1541 in quarto under the title of: *De Gallorum Insubrium antiquis sedibus, Mediolani apud Io. Antonium Castilioneum.* There is a later edition at the Royal Library of Stockholm, Bergami 1593, but without map.

Carolus Heydanus, *Germaniæ Typum; Antverpiæ apud Hieronymum Cock.*

Carolus Clusius A., *Hispaniam; antiquis ac recentibus locorum in ea nominibus inscriptam; quam nos propediem edituri sumus.* — Carolus Clusius or Charles Lescluse was a celebrated botanist, born in 1525, † 1609. His map of Spain is not inserted in the *Theatrum,* but on T. 12, ed. 1570, Ortelius reproduces his map of *Gallia Narbonensis.*

Caspar Vopellius Medebach, *Descriptionem Orbis terrarum; Item Europæ totius; ac Rheni tractum; omnia Coloniæ.* — I have reproduced a globe of Vopel on T. XL (comp. p. 82). An enormous cordiform map of the same geographer is preserved in the Hauslab Collection at Vienna. In a letter to Ortelius (HESSELS, p. 43) Vopel is blamed by POSTELL, because he makes North America communicate with Asia. A copy of his large map of Europe on 10 sheets is preserved in the Bibliothèque Nationale at Paris; it is printed at Antwerp by *Bernart van den Putte Figuersnyder 1566.*

Caspar Bruschius Egranus, *Montis Piniferi (quem Fiechtelberg vulgo nuncupant) Tabulam; Ulmæ apud Sebastianum Francum.* — Caspar Brusch, *Comes Palatinus* and *Poeta laureatus,* was born at Schlackenwald in Bohemia in 1518, † 1559. The map mentioned here was probably intended for his description of Fichtelgebirge. I have not seen the original edition. In an edition of 1663, in the Royal Library at Stockholm, the map is wanting.

Christianus Schrot Sonsbekensis, *Gelriam cum Cliuia, vicinasque Regiones, Antuerpiæ apud Bernardum Puteanum; Eandem tabulam idem recognouit, edique curauit per Hieronymum Cock, Antuerpiæ; Descripsit quoque universam Germaniam, quam idem Cock prælo excudit.* — Geldria et Clivia etc.: ORTELIUS 1570, 15; Terra Sancta: ORTELIUS 1584, 97; Saxonia: DE JUDAEIS, II: 8; Tractus Danubii: DE JUDAEIS, II: 27 et 28; Westphalia: ORTELIUS 1584, 46.

Christophorus Zellius, *Europæ Typum; Nurenbergæ.* — According to Doppelmayer (p. 143 and 207) »ein Formschneider aus Nürnberg». He published maps of Europe, Prussia etc., † c. 1590.

Christophorus Pyramius, *Germaniæ Tabulam; Bruxellis Brabantiæ.* — In the edition of 1595 is written that this map was printed *Bruxellis Brabantiæ 1548.*

Cornelius Antonij, *Regionum Orientalium Tabulam (uti titulus habet) Continet autem Daniæ Regnum, et circumiacentes Regiones, Excusam Amstelrodami. Idem descripsit Europam, editam Francofurti ad Mœnum.* — Daniæ regni typus: ORTELIUS 1584, 44. Cornelis Antoniszoon published in 1544 a large wood-cut map of Amsterdam. Compare *Geschiedenis van Amsterdam door* J. TER GOUW, V, Amst. 1886, and C. G. BRUUN, *Cornelius Antoniades Kaart over Danmark* (Geogr. Tidskr. Khvn 1888, p. 148).

Diegus Gutierus, *Americam; Antuerpiæ apud Hieronymum Cock.* — British Mus. Catal. of printed maps, p. 1704. The map is dated 1562. The name is here written DIEGO GUITEREZ. Concerning the two Spanish cosmographers of this name, see: HARRISSE, *Cabot,* p. 231.

Dominicus Machaneus, *Benaci lacus Corographiam; a Leandro Alberto citatam legimus.* — Domin. Macaneus' *Verbani Lacus Chorographica descriptio,* inserted in *Thesaurus antiquitatum et historiarum Italiæ,* IX: 7, Lugduni Batavorum 1723, does not contain any maps. Machaneus died in 1530 (JÖCHER). A map of the Garda lake (Lacus Benacus), possibly a reproduction of the map of Machaneus, is inserted in *Georgii Jodoci Bergani Benacus...* Verona apud Antonium Puteolum 1546.

Erhardus Reych Tyrolensis, *Palatinatus Bauariæ tractum; Nurenbergæ 1540.* — The original map, which I have not seen, is mentioned in the *Beiträge zur Landskunde Bayerns,* München 1884, p. 84. It is reproduced by ORTELIUS 1570, 30, and by DE JUDAEIS, II: 26.

Ferdinandus à Lannoy, *Burgundiæ Comitatus Tabulam; apud Hieronymum Cock, Antuerpiæ. Sed nondum edita est,* — Lannoy's map is published by ORTELIUS (1579, 23) and according to information in the edit. 1595 of his Theatrum, it seems not to have been published separately.

Fernandus Aluares Zeccus, *Lusitaniam; Romæ, apud Michaëlem Tramezzinum 1560.* — ORTELIUS 1570, 8.; the map is dated: Roma 1560. Portugallia: QUAD 56; BLAEU, IX. p. 92.

[**Florianus,** *Tabulam Sarmatiæ, Regna Poloniæ et Hungariæ utrusque Valachiæ; nec non Turciæ, Tartariæ; Moscouiæ, et Lithuaniæ partem comprehendentem, Cracouiæ 1528].* — The publisher of this map was probably the same Florianus Unglerius, who in 1512, at Cracow, published Stobnicza's *Introductio in Ptholomei Cosmographiam* (comp. p. 68). He should not be confounded with the Florianus who published the general map which I have reproduced on p. 81, fig. 48.

Franciscus Monachi Mechliniensis, *Regiones Septentrionales; Antuerpiæ, apud Syluestrum à Parisiis.* — Comp. above p. 102.

Gabriel Symeoneus, *Almaniæ tabulam; in libello inscripto Dialogus Pius et Speculativus [Lugduni, apud Guilielmum Rovilium 1560].* — Limaniæ topographia: ORTELIUS 1570, 10; BLAEU VII, p. 131. Gabriel Simeon was born at Florence and lived in the middle of the 16th century (JÖCHER). A copy of *Description de la Limagne d'Auvergne... Traduite du livre Italien de Gabriel Simeon en langue Françoyse par Antoine Chappuys du Dauphiné.... à Lyon par Guillaume Roville 1561,* in the Royal Library of Stockholm, does not contain any map.

Gemma Frisius, *Universi Orbis Tabulam, Antuerpiæ.* — N. T. XLIV; p. 102. GESNER (fol. 267) says respecting the map of Gemma Frisius: *Impressa anno 1540 Lovanii, ut videtur.* It thus seems to have been originally published independently of the cosmography of Apianus. Reinerus Gemma Frisius was born in 1508 at Dockum in Frisland. He devoted himself to the study of medicine and mathematics; † 1555 or 1558 (JÖCHER).

Georgius Collimitius, *Hungariæ Tabulam [Lazari] (quam Cuspinianus edidit) recognouit.* — Collimitius was a physician and mathematician of the court at Vienna. He has written a good deal about medicine and astrology (JÖCHER). In the edition of 1595 of Ortelius it is said, that it was the map of Hungaria of Lazarus which was published by Collimitius.

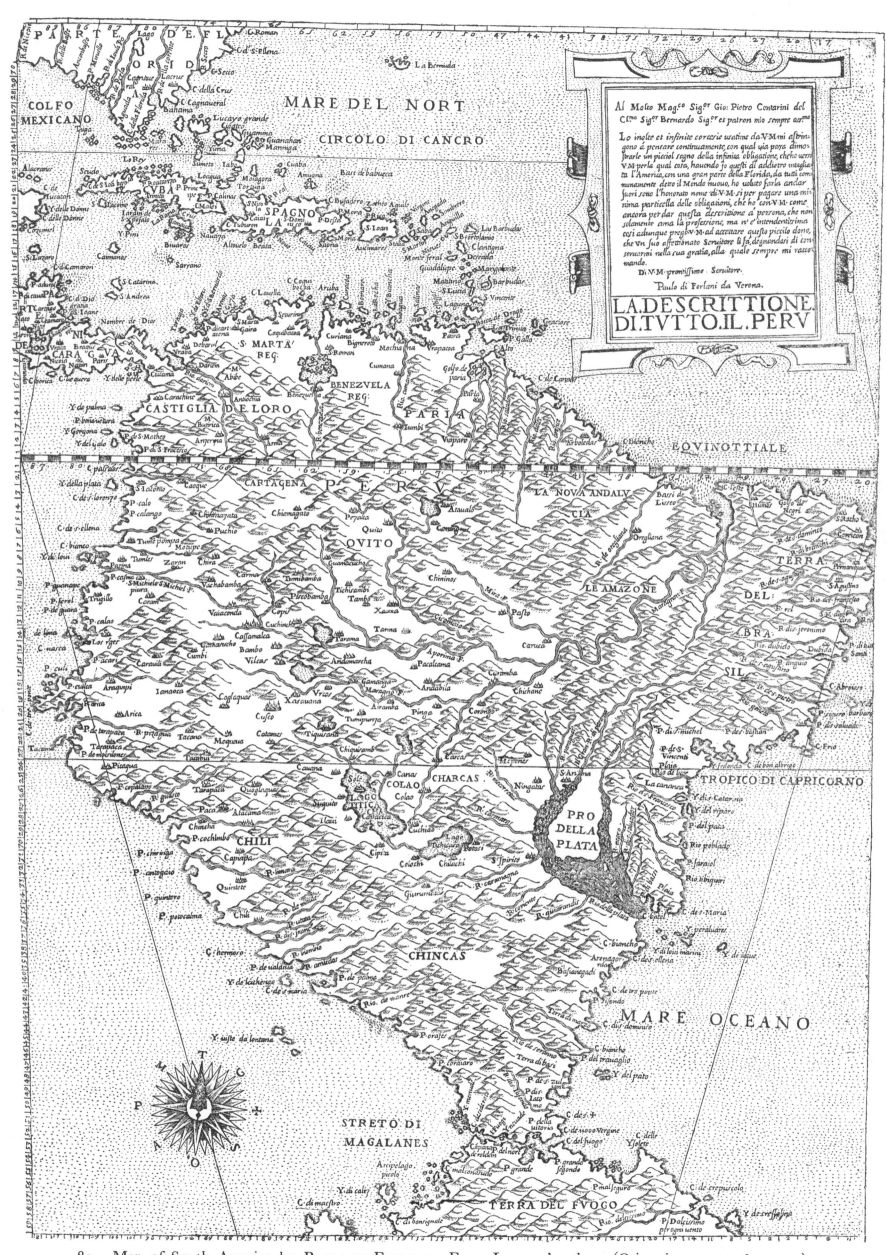

80. Map of South America by PAULO DI FORLANI. From LAFRERI's atlas. (Orig. size 505 × 365 m. m.).

Gerardus Mercator Rupelmundanus, *Palæstinæ, siue Terræ Sanctæ; Item Flandriæ, Louanii. Postea Europæ; Deinde Orbis Uniuersi ad usum nauigantium accommodati Tabulam, Duisburgi ædidit. Excudit quoque Britannicæ Insulæ Tabulam, ab alio quopiam descriptam.* — I shall finish this chapter with a short review of Mercator's well known merits as a cosmographer.

Godefridus Mascopius Embricensis, *Diæcesis Monasteriensis et Osnaburgensis Typum; Embricæ edidit per Remigium Hogenbergum 1558.* — ORTELIUS 1570, 24.

Gregorius Amaseus, *Fori Iulij Tabulam descripsit; quam ab Auctore se habuisse, inquit Leander in sua Italia.* — Gregorio Amaseo was an Italian author, † 1541. The map has probably belonged to his *Descriptio geographica Italiæ et Forojuliensis ad Leandrum Bonnoniensem* of which a manuscript exists at the Abbey of St. Germain (HOEFER-DIDOT). It seems uncertain whether the map was ever published in print.

Henricus Zellius, *Europam; Nurembergæ.* — Prussiæ descriptio: ORTELIUS 1570, 22; QUAD 17. His name is mentioned by DOPPELMAYER with that of his kinsman Christophorus.

Hieronymus Chiauez, *Americam descripsit, quæ nondum in lucem prodiit.* — La Florida: ORTELIUS 1584, 8; Hispalensis conventus: ORTELIUS 1579, 15. It is possible that the map 6 in the edition of 1595 (*Mare Pacificum*) is also based on a work by the same cosmographer. The »*Victoria*» is here represented under square sails, with the inscription:

Prima ego velivolis ambivi cursibus Orbem
Magellane novo te duce ducta freto,
Ambivi, meritoque vocor Victoria. Sunt mi
Vela, alae; precium, gloria; pugna, mare.

Hieronymus Bellarmatus, *Tusciam; Romæ.* — ORTELIUS 1570, 36; DE JUDAEIS, I: 21.

Humfredus Lhuyd Denbygiensis, *Angliæ Regni Tabulam; Item Cambriæ Corographiam, quas nos Deo fauente aliquando publicabimus.* — Anglia: ORTELIUS 1573, 8; Cambria & Wallia: ORTELIUS 1573, 9; Anglia: QUAD 52. Lhuyd or Llwyd was an English archæologist, † about 1570.

Iacobus à Dauentria, *Brabantiæ; Hollandiæ; Gelriæ; Frisiæ; Zelandiæ Tabulas descripsit et edidit, Mechliniæ.* — Brabantia: ORTELIUS 1570, 16; Zelandicarum Insularum descriptio: ORTELIUS 1570, 18; Hollandia: ORTELIUS 1570, 19; Zelandicarum Insulæ: QUAD, 45.

Iacobus Castaldus Pedemontanus, *Orbis Universalis typum, magna forma; eandem minori forma; Item Asiæ; Africæ; Hispaniæ; Italiæ; Siciliæ; Hungariæ; et Pedemontanæ Tabulas; Venetiis, omnia per Matthæum Paganum.* — I have before given a short account of the works of this distinguished cosmographer. To what extent these were used by the Dutch cosmographers appears from the following list of maps by Gastaldi copied by Ortelius and de Judaeis. *Italia:* ORTELIUS 1570, 32; DE JUDAEIS, I: 18; *Pedemontanæ, vicinorumque regionum descriptio:* ORTELIUS 1570, 34; *Patavini territorii chorographia:* ORTELIUS 1573, 45; *Apulia:* ORTELIUS 1573, 45; *Grecia:* ORTELIUS 1570, 40, QUAD, 66; *Sicilia:* ORTELIUS 1570, 38; *Romania:* ORTELIUS 1584, 89; *Asia* (three large maps) DE JUDAEIS, I: 8, 9, 10.

[**Iacobus Homen Lusitanus**, *Europam navigatoriam descripsit, quæ edita fuit Venetiis 1569*]. — In the British Museum are kept: HOMEM DIEGO; *La vera descrittione della navigatione de tutta Europa, et parte dell' Africa et dell' Asia … Fatta dell' excellente cosmografo G. Home Portugense, 1572.* Some other maps of this cosmographer are also extant, partly printed, partly handdrawn (comp. CASTELLANI, *Catalogo*, p. 250; HARRISSE, *Cabot*, p. 243).

Iacobus Zieglerus, *Palæstinæ, Scondiæ, Ægyti et Arabiæ, libri forma, et in iis Commentaria.* Argentorati, apud Petrum Opilionem 1532. — I have above (p. 60 and 104) given an account of Ziegler's maps and a fac-simile (fig. 30) of one of them. Ziegler was born at Landshut and died in 1549 (JÖCHER).

Ioannes Auentinus, *Bauariæ Tabulam; Landshuti Anno 1533.* — Reproduced by SEB. MÜNSTER from 1540; ORTELIUS 1570, 29. Aventinus was born in 1466 at Abensberg in Bavaria, † at Regensburg in 1534 (JÖCHER). The map of Aventinus was probably constructed for his great work *Annales Boiorum.* Of this work the Royal Library at Stockholm has the edition, Ingolstadii 1554. But it does not contain any map.

Ioannes Baptista Guicciardinus, *Universi terrarum Orbis imaginem, maxima forma; quam Aquila compræhendit, Antuerpiae 1549.*

Ioannes Bucius Aenicola, *Europam; sub forma puellæ, Parisiis apud Christianum Wechelum.* — In GESNER'S *Bibliotheca*, Tiguri 1545, fol. 393, this map is mentioned, with the addition that it is printed in Paris on two leaves, *altera tantum facie impressa ut liceat affigi ad parietem.*

Ioannes Calameus, *Biturigum Regionem; Lugduni apud Gryphium.* — ORTELIUS 1570, 10.

Ioannes Crigingerus, *Bohemiæ, Misniæ, Turingiæ, et collateralium Regionum Tabulam; Pragæ 1568.* — Bohemia: DE JUDAEIS, II: 11; Saxonia: ORTELIUS 1570, 23. The name is also written CRIGINGER, GRIGVIGERUS

or CRUGINGERUS. He was a dean at Marienburg in Bohemia and published a map of Meissen, Bohemia, and Thuringia at Prague in 1568 (ADELUNG's continuation of JÖCHER).

Ioannes Cuspinianus, *Hungariam; quam Petrus Apianus edidit; uti Auctor est Wolfg. Lazius in sua Hungariæ Tabula.* — Johannes Cuspinianus, or Spieshammer, was physician, ambassador (*Orator*) and librarian of the emperor Maximilian I; he was born in 1473, † 1529. A detailed biography of him is found in *Vitae Clarissimorum Historicorum*, Jena 1740.

Ioannes Dominicus Methoneus, *Europam; Venetiis, apud Matthæum Paganum.*

Ioannes Georgius Septala Mediol. *Ducatum Mediolanensem, et Regiones vicinas, Antuerp. apud Hieronymum Cock.* — ORTELIUS 1570, 33.

Ioannes a Horn, *Germaniae Inferioris Tabulam; Antuerpiæ.*

Ioannes Honterus, *Tabulas Geographicas edidit libelli forma; sub titulo, Rudimentorum Cosmographicorum, Tiguri apud Christoph. Froschouerum.* — I have at p. 112 given an account of Honter's cosmographical works. Some of his maps, or some maps printed in his cosmography, are reproduced on N. T. XLIV, and fig. 71, 72, 76. Honter was an evangelical theologian from Kronstadt in Transylvania, † 1549 (JÖCHER).

Ioannes Ioliuetus, *Galliam; Parisiis, aut Lugduni [apud Oliverium Truchetum 1560].* — ORTELIUS 1570, 9. He was a native of Limoges and geographer to Francis I. A map of Berry by him was printed in 1545. Several editions were published of his map of France (HOEFER-DIDOT). In the Bibliothèque Nationale at Paris there is a large wood-cut map (size 0,853 × 0,552 m.) *Descriptions des Gaules, avec les confins d'Allemagne, et d'Italie.* JOANNES JOLIVET *inventor. A Paris par Marc du Chesne 1570.*

Ioannes Mellinger Halens. *Turingiæ Tabulam, Wimariæ.* — ORTELIUS 1573, 29; DE JUDAEIS, II: 29. Luneburgensis Ducatus: QUAD 1608, 34; BLAEU III, p. 189.

Ioannes Stumpfius, *Heluetiæ Tabulas; in historia Heluetica volumine, Tiguri excuso, apud Christophorum Froschouerum.* — A narrative of a journey in Switzerland in 1544 by Johannes Stumpf is entered into *Quellen zur Schweizer Geschichte*, Vol. 6. GESNER relates (fol. 456), that Stumpf in 1545 was occupied with a description of Switzerland and neighbouring countries, »*cum tabulis locorum exactissime depictis*». I suppose that Stumpf was the author of an Atlas published at Zürich 1562 by »*Cristoffel Froschower*» under the title: *Hierum findst du lieber Läser schoner recht und wolgemachter Landtaflen XII* etc. (Bibl. Nationale in Paris). Stumpf was born in 1500 at Bruchsal (Speyer), † 1566 at Zürich (JÖCHER).

Ioannes Sambucus Pannonius, *Transsyluaniam; Viennæ Austriæ. 1566. Item Hungariam, ibidem; 1570.* — Fori Julii descriptio: ORTELIUS 1573, 42; Transsilvania: ORTELIUS 1570, 43; Hungaria: ORTELIUS 1579, 77, QUAD, 69; Illyricum: ORTELIUS 1573, 54. Sambucus was born in 1531 at Turnow in Hungaria, † 1584 at Vienna (JÖCHER).

Ioannes Surhonius, *Veromanduorum Regionem; Antuerp. apud Christophorum Plantinum 1558.* — Veromandois: ORTELIUS 1570, 11. Several other maps (Lutzenburgum ducatus, Namurcum, Veromanduum, Artesia, Picardia) by Ioannes Surhonius are from 1579 reproduced in different editions of Ortelius' Theatrum. The *Artesia* of Surhonius is reproduced by DE JUDAEIS, II: 50.

Laurentius Frisius, *Cartam uniuersalem Marinam (ut vocant) alicubi in Germania.* — Occurs for the first time in Ptolemaeus 1522 (N. T. XXXIX).

Lazarus Secretarius Cardinalis Strigon. *Hungariæ typum primus descripsit, qui editus est Ingolstadii per Apianum, Anno 1528.* — According to information in the edition of 1595 of Ortelius, this map was published by Georgius Collimitius.

[**Leander Albertus**, *Corsicam, Siciliam, Sardiniam; in libro cui titulus est de Insulis Italicis, Veneiis impresso, 1568.*] — In the *Isole appartenenti alla Italia di F. Leandro Alberti Bolognese*, Venetia 1567, there are copperprinted maps of Venetia, Corsica, Sardinia, Sicily, and some small islands (Is. Tremiti) near the eastern coast of Italy.

Leuinus Algoet, *Regionum Septentrionalium Typum; apud Girardum Iudæum, Antuerp.* — DE JUDAEIS, II: 4. Remarkable map of the Scandinavian peninsula, dated 1570. Algoet was, according to JÖCHER, a mathematician from Ghent.

[**Macæus Ogerius**, *descripsit regionem et Comitatum de La Maine Galliae provinciam, impressam in urbe Cenomanorum ibidem, 1539.*] — A copy of this map is, according to M. G. MARCEL, preserved in the Bibliothèque nationale at Paris. Reproduced in the later editions of the Theatrum of Ortelius.

Marcus Ambrosius Nissensis, *Liuoniam, Vicinasque Regiones, Antverpiæ, sed nondum edita.* — According to Ortelius the map was not yet published in 1595.

Marcus Iordanus, *Daniæ Regni Typum; Hafniæ, apud Ioannem Vinitorem 1552. [Idem Holsatiæ, Sleswig etc. Hamburgi, apud Joachimum Leoninum, 1559. Et typum Corographicum Itinerum D. Pauli, necnon Abrahami Patriarchæ etc. Wittembergæ, apud Joannem Cratonem, 1562.*

Idem Iutiæ peninsulæ tabulam conscripsit, cujus autographum apud me est.]
— Dania: ORTELIUS 1595, 48 b; Holsatia: ORTELIUS 1579, 43. Excellent data for the biography of Jordanus are given in EDV. ERSLEV'S *Jylland*, Copenhagen 1886, p. 144. Jordanus was, for some time, professor of mathematics at the University of Copenhagen, † in Holstein in 1595.

Marcus Zecsnagel Salisburg. *Ditionem Salisburgensem, Salisburgi.* — Salisburgensis jurisdictionis... vera descriptio: ORTELIUS 1570, 28; Salzburgensis Episcopatus: DE JUDAEIS, II: 21.

Martinus de Brion, *Palæstinam; Parisiis, apud Hieronymum Gormontium.* — He published at Paris in 1540: Totius terræ sanctæ descriptio; without map (TOBLER).

Martinus Helwig Neissensis, *Silesiæ Tabulam, quæ Nissæ excusa est. 1561.* — ORTELIUS 1570, 26; BLAEU, III: p. 43; MÜNSTER, edition 1628. Rector of a gymnasium in Breslau, born in 1516, † 1574.

Martinus Ilacomilus Friburgensis, *Europam; eam alicubi in Germania impressam habemus.*

Nicolaus a Cusa, *Huius Chartam Germaniæ, citat Althamerus.* — Nicolaus a Cusa's map of central Europe was published at Eystat in 1491. Of this edition I have only seen a photograph of a copy preserved at the British Museum. It is a copper-engraving on Donis' projection, very remarkable considering the early time at which it was executed. This map seems also to have been published, probably in wood-cut, about 1530 at Basel by ANDREAS CRATANDER, as proceeds from: *Germaniæ atque aliarum regionum; quæ ad imperium usque Constantinopolitanum protenduntur, descriptio, per Sebastianum Munsterum ex Historicis atque Cosmographis, pro tabula Nicolai Cusæ intelligenda excerpta. Item eiusdem tabulæ Canon. s. l.* (The preface dated: *Basileae Mense Augusto anno MDXXX.*) I do not know whether the map described by Münster is still extant. The rich and masterly engraved map of central Europe in the editions of 1507 and 1508 of Ptolemy's geography, is also a copy of Nicolaus a Cusa's map, as may be concluded from a comparison with the above mentioned map of 1491 and with Münster's description. On the map of 1507 the insignificant birth place, Cusa, of the author is

81. Map of North America, Venice 1566. From LAFRERI's atlas. (Orig. size 393 × 269 m. m.).

Martinus Waldseemuller, *Universalem navigatoriam (quam Marinam vulgo appellant) in Germania editam. Puto hunc eundem esse cum Ilacomilo prædicto.* — The only maps of WALDSEEMÜLLER (HYLACOMYLUS or ILACOMYLUS) extant are the maps of Ptolemaeus 1522, 1525, 1535 and Viennæ 1541. They are all reduced copies of the maps in the edit. 1513, with the exception of the two, of which fac-similes have been given fig. 62 and 63. In GESNER'S *Bibliotheca universalis*, fol. 501, is written: *Martini Hilacomili Instructio in chartam itinerariam, excusa Argentorati 1511 in 4:o, cum luculentiore ipsius Europæ enarratione per Ringmannum Philesium, Chartis 6.* Probably Gesner here alludes to a map of which two copies are preserved in the Bibl. Nationale at Paris. Its title is: *Das ist der Romweg von meylen zu meylen mit puncten verzeychnet von eyner stat zu der andern durch deutzsche lantt.* A monography of this geographer is anonymously published by D'AVEZAC under the title: *Martin Hylacomylus Waltzemüller, ses ouvrages et collaborateurs... par un géographe bibliophile*, Paris 1857.

Mathias Cynthius, *Hungariam, Nurenbergæ 1567.* — Hungaria: DE JUDAEIS, II: 14. In the *Brit. Mus. Cat. of printed maps* there are mentioned four maps by MATHIS ZYNDT: *Das Khynigreich Hungarn 1566; Gotta 1567; Malta 1565* and *Tabula complectens totam Belgicam, Flandriam, Brabantiam bey M. Zündten 1568.*

laid down. The Roman copper-plate does not, however, extend quite so far to the south, east, and north, as the map of 1491, or as the map described by Münster, but it is richer in geographical details. It is one of the best cartographical productions of the first half of the 16th century, and nevertheless it seems to have been hitherto altogether overlooked by geographers. An accurate full size copy of this map was engraved in copper by FERRANDO BERTELI in 1562 and inserted into Lafreri's atlas (No. 39 in my Catal.). Even the map of central Europe in Schedel's chronicle is probably a rough wood-cut copy of the Germania of Nicolaus a Cusa, though the wood-engraver was here, on account of technical difficulties, obliged to leave out a number of names occurring on the original, and to simplify the topographical details. In a letter, dated Lübeck 1574, HIOBUS MAGDEBURGUS says that he sent Cusa's Germania to Ortelius (HESSELS, p. 110). There accordingly exist at least four different editions of Nicolaus a Cusa's important map, viz. the map of 1491; the map in SCHEDEL'S *Liber cronicarum* of 1493; the Germania in Ptolemy of 1507 and 1508; the reproduction of this map in Lafreri's atlas by Ferrando Berteli in 1562. The cardinal Nicolaus a Cusa (his German name was KREBS) was one of the most prominent and influential scholars of the 15th century. He was born in 1401 at Cusa, a small place situated between Trier and Coblentz, and died in 1464.

Nicolaus Genus, *huius Tabula Regionum Septentrionalium habetur in Geographia Ptolemæi, à Girolamo Ruscelli in Italicam versa; excusa Venetiis, apud Vincentium Valgrisium.* — The map of the northern countries by Nicolo and Antonio Zeno (N. fig. 29) in the Ptolemy editions of 1561, 1562 etc. (comp. p. 57).

Nicolaus Germanus, *Huius Galliæ Chartam citat Robertus Cœnalis. Puto hunc eundem esse cum Nicolao à Cusa.* — Ortelius here evidently speaks of a map in the Ptolemy of Nicolaus Germanus, Ulmæ 1482 and 1486. On the upper border of the map in the last edition is printed: *Tabula Moderna Francie.* It would be highly interesting to know the age of the original on which this map is based. Probably it is much earlier than Cusa's map of Germany.

Nicolaus Nicolaius Delphinas, *Europam marinam; Antuerpiæ apud Ioannem Stelsium. Idem Galliæ Tabulam promittit.* — The left side of the map 11 in Ortelius 1570 is occupied by »*Caletensium et Bononiensium ditionis accurata delineatio Parisiis 1558*» by Nicolaus Nicolai. The first of the maps here mentioned by Ortelius is, probably, the reproduction of the map in the French edition of Medina's *Arte de Navegar*, translated by Nicolay and printed at Lyon in 1553. Nicolas de Nicolay was a celebrated French traveller, born in 1517, † 1583. (Didot-Hoefer; compare also Harrisse, *Cabot*, p. 239.)

[**Nicolaus Reger**, *edidit Ptolemaei tabulas gemino schemate, veteri scilicet ac recentiori, ut testatur Rob. Cœnalis in suo Galliae opere.*] — Johannes (not Nicolaus) Reger was only the factor of the editor of Ptolemy, printed at Ulm in 1486 (comp. p. 16).

Nicolaus Sophianus, *Græciæ Tabulam; Romæ [eadem postea evulgata fuit Basileæ per Oporinum.]* — Ortelius 1579, 93. According to Gesner (fol. 523) the *tabulæ Graeciæ elegantissime depictæ* of Sophianus were first printed at Rome, then, in 1543, with explanatory remarks by Nicolaus Gerbelius at Basel by Oporinus.

Orontius Fineus Delphinas, *Galliæ descriptionem; et Orbis terrarum typum, sub forma cordis humani. Idem Tabulam Regionum, quarum in sacris Bibliis fit mentio. Omnia Parisiis apud Gormontium.* — Maps by Orontius Finæus are here given on fig. 53 and T. XLI. The map of Gallia of 1525 is to be found in the Bibliothèque nationale at Paris, and editions of the same of 1561 and 1563 in the British Museum (*Catal. of Printed Maps*, p. 1336). According to Gesner Orontius' map of France of 1525 is »*circiter sex chartarum magnitudine*». Gesner also cites a double heart-shaped map by this geographer printed in 1536 by Hieronymus Gormontius, »*in tabula duabus chartis opinor*». Orontius Finæus (Finé) was born at Briançon in 1494, † at Paris in 1555.

Olaus Magnus Gothus, *Regionum Septentrionalium Tabulam, Venetiis.* — Regarding the lately rediscovered large map of the North by Olaus Magnus, see p. 60. A very exact copy, engraved at Rome in 1572, is here reproduced on a reduced scale (N. fig. 32).

Paulus Iouius, *Larij lacus tabellulam, cum, libello; Venetiis.* — Ortelius 1570, 35; 1573, 46. Regarding the map of Russia by Paulus Jovius, see page 114. *Lacus Larius* is the ancient name of the Lake of Como. Paulus Jovius was born at Como in 1483, † 1552.

Petrus Apianus, *Europam; Peregrinationem D. Pauli; et Typum universalem; omnia Ingolstadij.* — The map of Europe of Petrus Apianus is probably lost. Yet several of his maps are still extant; some of these are reproduced here: N. T. XXXVIII, fig. 57, 58, and described on p. 99.

Petrus ab Aggere, *Orbis terrarum Typum, Aquila compræhensum; Mechliniæ.*

Petrus Boekel, *Thietmarsornm Regiunculam; Antuerpiæ, apud Ioan. Liefrinck [1559].* —Thietmarsia: Ortelius 1570, 22: de Judaeis, II: 7; Quad, 74. Born at Antwerp; lived in the first part of the 16th century as a painter to the court of Mecklenburg-Schwerin. Published also a map of Denmark (Jöcher).

Petrus Coppus, *Hystriam, Venetiis.* — Lafreri's atlas. Ortelius 1573, 55. In Pietro Angelo Zeno, *Memoria de' scrittori Veneti Patritii*, Venetia 1662, is written about the family of Coppo: *1540 Pietro figliulo di Giacomo con ogni più esquisita diligenza descrisse il sito dell Istria.* A very insignificant map of the world by this geographer is reproduced fig. 65.

Petrus Laicstein, *Iudæam perlustrans eius loca descripsit, quam descriptionem Christianus Scrot in Tabulam redegit. Extat Antuerpiæ apud Hieronymum Cock 1570.* — Ortelius 1584, 97.

[**Petrus de Medina**, *Hispaniæ tabulam, Hispali per Ioannem Gutierum 1560, at valde rudem.*] — Concerning Medina's maps see above p. 110. One of them is reproduced fig. 75.

Philippus Apianus, *Bauariæ Tabulam, Ingolstadij 1568.* — Comp. p. 122.

Pyrrhus Ligorius Neapolitanus, *Regni Neapolitani; item Græciæ tabulam; Romæ, per Michaëlem Tramezinum.* — Several maps of Ligorius are inserted in Lafreri's atlas (Comp. above p. 118, No. 19 and 20; p. 122 No. 126, 134 and 138). His *Regnum Neapolitanum* was reproduced by Ortelius 1570, 37; de Judaeis, I: 19; Quad 1608, 64. Ligorius was an architect and archæologist from Naples, † 1586 (Hoefer-Didot).

Sebastianus Cabotus Venetus, *Universalem Tabulam; quam impressam æneis formis vidimus, sed sine nomine loci, et impressoris.* — At present only one copy of the large map of the world attributed to Sebastian Cabot is known. It is preserved in the Bibliothèque nationale at Paris (comp. p. 90). The great explorer Sebastian Cabot died at an advanced age shortly after the year 1557. A minute and excellent monograph: *Jean et Sebastien Cabot*, was published by Harrisse, Paris 1882.

Sebastianus Munsterus, *Germaniae Typum, Basileae, quem Tilemannus Stella emendauit et locupletauit, Wittenbergae, apud Petrum Zeitz 1567.* — Basiliensis territorii descriptio: Ortelius 1573, 38; de Judaeis, II: 24. A great number of maps of Münster are printed in his geography and cosmography (comp. p. 108). Seb. Münster was born in 1489 at Ingelheim, † 1552 at Basel.

Sebastianus à Rotenhan, *Franconiam Orientalem; Ingolstadij, Anno 1533* [1543]. — Franconia: Ortelius 1570, 24; Francia Orientalis vulgo Franconia: Quad, 22. The original is cited under the year of 1520? in *Beiträge zur Landeskunde Bayerns*, München 1884. Rotenhan was a great traveller and much employed in public service by the Emperor Charles V. He was born in Franconia 1478, † 1532.

[**Sigismundus ab Herberstein**, *Moscoviæ tabulam, in eius Commentariis Basileæ excusis, apud Joannem Oporinum.*] — Compare p. 113. A detailed biographical work about him was published by Friedrich Adelung at St. Petersburg in 1818. Herberstein was born in 1486 at Wippach in Carinthia, † at Wien in 1566.

Sta. Por. *depinxit Ducatum Oswiecimen. et Zatoriensem, Venetiis 1563.* — A map of this region, signed *Sta. Por. pinxit*, and dedicated to Sigism. Myskowski, was published »*In Venetia alla libreria del S. Marco 1563*». Size 0,321 × 0,244. It was reproduced by Quad (map 72) and Blaeu, II: p. 32 and 33.

Stephanus Geltenhofer, *Campaniae Tabulam; Suppresso tamen suo nomine, Antuerpiae.*

Thomas Geminus, *Hispaniae Tabulam; Londini.* — In the Bibliothèque nationale at Paris there is a magnificent engraved map of Spain on four leaves (total size 0,940 × 0,767) »*excusum Londini per* Thomam Geminum *1555,*» and dedicated to the King and Queen of England *Philip* and *Maria.* It is the first map I know of, that was printed in England. Not mentioned in the *British Mus. Catal. of Printed Maps.*

Tilemannus Stella Sigenensis, *Palæstinæ tabulas duas descripsit, quarum unam inscripsit Itinerarium Israëlitarum ex Ægypto; alteram, Corographia Regni Iudææ et Israëlis. Wittebergae. [Item comitatum Mansveldiensem, Coloniae apud Franciscum Hogenbergum Lutzenburgi. Item ditionem accuratissime descripsit, nondum (1595) autem edidit. Idem promittet absolutissimam totius Germaniae descriptionem.]* — Mansfeldiae Comitatus descriptio: Ortelius 1573, 28; Quad, 21; Blaeu, III: p. 113; Palestina: Ortelius 1570, 51; de Judaeis, I: 13.

Vincentius Corsulensis, *Hispaniam; Venetiis; apud Matthaeum Paganum.* — The same as Johannes Vincentius, a Spanish dominican, born in Asturia (Jöcher).

Wenceslaus Godreccius, *Poloniae Tabulam, Basileae apud Oporinum.* — Ortelius 1570, 44; Quad, 71.

Wolfgangus Lazius, *Hungariae Corographiam; Viennae. Item Austriae; Nurenbergæ. Idem Comitatum Tyrolensem; Styriam; Histriam; et Carinthiam edidit.* — Austria: Ortelius 1570, 27; de Judaeis, II: 15; Blaeu, III: p. 19; Rhetiæ alpestris descriptio in qua hodie Tirolensis Comitatus: Ortelius 1573, 40; Goritiae, Karstii, Chaczeolae, Carniolae, Histriae et Windorum Marchae descriptio: Ortelius 1573, 40; Carinthiæ ducatus & Goritiæ palatinatus: Ortelius 1573, 55; Hungaria: Ortelius 1570, 42. In Münster's cosmography, edit. 1550 and 1559, p. 682 and 683, there is a map of Vienna by W. Lazius. Wolfgang Lazius was a physician to the Emperor, councillor and historiographer at Vienna; born in 1514, † 1565.

Wolfgangus Wissenburgius Basiliensis, *Palæstinam; Argentinae apud Richelium.* — Of this map Gesner says (fol. 629): *Descriptio Terrae Sanctae per eundem* (Wolfg. Wissenburgius) *in tabula septem chartarum digesta, ita ut in parietem affigi possit, cum libello (chartae unius) eandem declarante ibidem* (Argentorati) *excusa est 1538.* The map was printed in 1532 and 1536 in Ziegler's often cited work (Gesner, fol. 367).

With due regard to the circumstance that Waldseemüller is cited twice, the names of 99 authors of maps are enumerated in the *Catalogus auctorum*. As regards 79 of these, I have been able to examine or to give references either to the original editions or to more or less accurate reproductions of their works. With respect to twelve of the remaining I have succeeded, in the literature accessible to me, in discovering at least a few biographical data, but no maps; for eight, neither biographical data nor maps could be traced. Forty six of the cited maps are reproduced in Ortelius' *Theatrum*. Hereafter several of the maps of which I have not succeeded in finding any notices, during the very incomplete researches I could devote to this question, will no doubt be discovered in the recesses of libraries. But the above remarks on the catalogue of Ortelius will suffice to show how much the history of cartography is indebted to the publisher of *Theatrum Orbis terrarum*, and this is due precisely to that want of originality for which he, from another point of view, has been justly censured. Many remarkable omissions, however, occur in his catalogue. In vain

we look in it for several of the most prominent draughtsmen and publishers of maps during the period of incunabula, such as Ruysch, Stobnicza, Battista Agnese, Petrus Apianus, Vadianus, Girava, and others, and whole groups of maps appear to have been entirely unknown to him. Thus we do not find in Ortelius even a hint of the existence of such maps as the portolanos, or of the globe-prints, or of the hand-drawn charts of the Indian Archipelago and America of which no small number is likely to have existed in the Spanish Netherlands in 1570. It is also surprising that Ortelius, as may be concluded from his remarks concerning »Nicolaus Germanus» and »Nicolaus Reger», appears to have totally overlooked the often very important new maps added to the different editions of Ptolemy's geography.

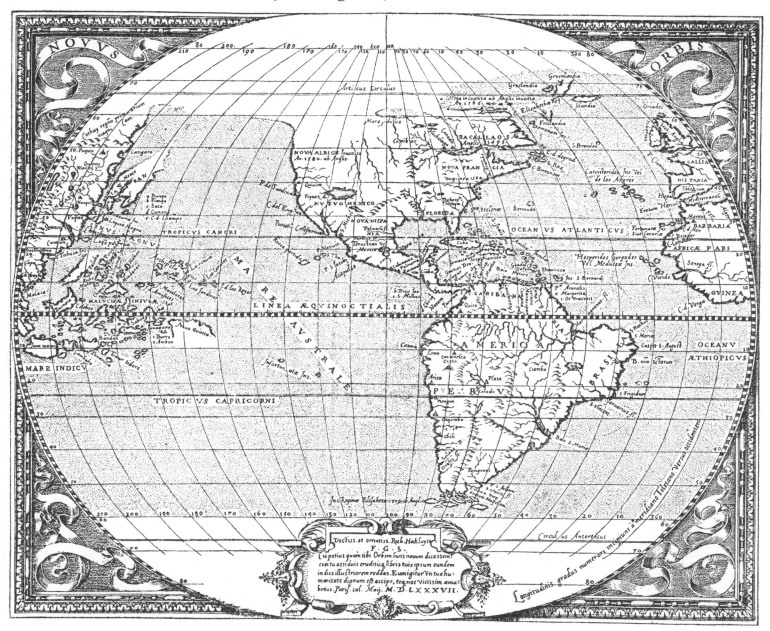

82. Map of the New World. From: Petrus Martyr, De orbe novo decades VIII, annot. Rich. Hakluyti. Paris 1587. (Orig. size 204 × 163 m. m.).

Gerard Mercator[1] was born in the small Flemish town of Rupelmonde on the 5th of March 1512.[2] When he had finished his schooling, he was sent to the University of Louvain, where he obtained an academical degree and at first

[1] The chief source for Mercator's biography is: *Vita celeberrimi clarissimique viri Gerardi Mercatoris Rupelmundani, a Domini Gualtero Ghymmio, Patricio Teutoburgensi, ac ejusdem oppidi antiquissimi Praetore dignissimo, conscripta*, introducing the first edition of *Atlas* of 1595. A monograph of Mercator based on extensive investigations in the archives, was published by Dr J. van Raemdonck in *Gerard Mercator, sa vie et ses oeuvres*, St. Nicolas 1869, and in the above mentioned works regarding Mercator's globes and his map of Flanders by the same author. Valuable contributions to his biography and to his position as a geographer have further been given by Lelewel (*Géographie du moyen âge*, Bruxelles 1852), Breusing (*Gerhard Kremer gen. Mercator*, Vortrag gehalten zu Duisburg 30. März 1869), and by others.

[2] Van Raemdonck gives the name »*Gerard Mercator ou de Cremer*», Breusing »*Gerhard Kremer genannt Mercator*». This translation of names into German or French is manifestly quite unauthorised. In his numerous writings as well as in his letters he always names himself Gerard Mercator. Ghymmius does not know of any other name. Nor is there any other to be found in the first edition of the great Atlas edited by his son Rumoldus, neither on the maps nor in any of the many introductory papers. In the French edition of 1613 only the Latin form of the name was used for Mercator himself, for his father, uncle, and children, though other names are cited in their German or Flemish form in the same work. The same is the case in the works of Ortelius, in Jöcher's *Allgemeines Gelehrten-Lexicon*, in Freherus' *Theatrum virorum eruditione singulari clarorum*, in fine in all works of the 16th—18th centuries examined by me. From this I draw the conclusion that Mercator himself never used the names Kremer or de Cremer and that he, consequently, should not be designated with any of these names. I will not deny that his ancestors may have been called Cremer or Kremer, but it is wrong, serves no purpose, and is misleading not to permit the great geographer to retain the name always used by himself, as well as a philosopher as a private man, the name by which he was exclusively designated by his contemporaries, and for which he earned such an illustrious place in the history of science.

applied himself to the study of philosophy. But after having married (1536) he began, in order to obtain the means of support for his family, to occupy himself with map-drawing, engraving in copper, and the construction of astronomical instruments. In this he had, according to the account he often gave to Ghymmius, no other guidance than some instruction by Gemma Frisius, probably in the mathematical elements of cosmography. He soon became a complete master of his new profession, a manufacturer of instruments, surveyor, drawer, and engraver of maps. His first known work is a large map of the Holy Land published in 1537, which was much admired, but which, like many other old maps printed separately, has been lost. The next year he published the map of which I have given a fac-simile on pl. XLIII (comp. p. 90 and 106). It is neither spoken of by Ghymmius, nor by any other of Mercator's earlier biographers. He was then charged by some merchants with the construction of a map of Flanders, which was already completed in 1540. This map, consisting of eight leaves, was also long considered as lost until it was lately rediscovered (comp. p. 108). In 1541 Mercator published a pamphlet on the employment of Italic letters in map-print, and his large terrestrial globe, mentioned p. 82, is dated the same year. Like some of his preceding works it is dedicated to the senator NICOLAUS PERRENOTUS GRANVELLA. Through this influential counsellor of the Emperor Mercator received the commission to make for Charles V various cosmographical instruments, which were praised as very ingenious and admirably executed. They were so much appreciated by the Emperor that he took them with him in his campaigns, during one of which they were burned by the enemy. Mercator's eminent and powerful protectors could not, however, avert from the artless and simpleminded man, whose tendency to mysticism was probably obvious to most of his acquaintants, the charge of heresy, which occasioned his imprisonment in 1544. After repeated earnest remonstrances from the authorities of the University, and when certificates of his innocence had been obtained from different quarters, Mercator was liberated. His imprisonment is supposed to have lasted about four months. Subsequently his Chronologia and the Atlas published by his son were inscribed in the *Index librorum prohibitorum*. In 1551 he finished his large celestial globe, a pendant to the above mentioned terrestrial globe, and published, as a manual for the use of them, two small tracts: *De usu globi* and *De usu annuli astronomici*. In 1552 he moved from Louvain to Duisburg, where he applied himself to his investigations with renewed energy. Here he made two small globes for Charles V, a celestial one of pure cristal, on the surface of which the stars and constellations were engraved with diamonds and indicated by a cover of gold, and a very elaborate terrestrial globe of wood. Here he also constructed for sale, but always with the utmost care, several other globes, which were sold at Nuremberg.[1] His first cartographical work accomplished at Duisburg was the large map of Europe, of which a few plates had already been engraved during his stay at Louvain. In existing collections of maps no copy of this work has been found, and this seems to be the case with a new edition of it published in 1572 (see Addenda). Of this work Ghymmius says: *Quod*

opus tantis laudibus a doctissimis quibusque viris passim effertur, ut vix simile in geographia, in lucem unquam prodiisse videatur. The first edition of »Europa« being finished, Mercator received a commission to engrave on copper an elaborate map of England, which an English friend, whose name Ghymmius does not mention, had constructed. This great work, finished in 1564, is also lost. But I suppose that the 15 maps of Britain in Mercator's atlas of 1595, signed *per Gerardum Mercatorem cum privilegio*, are slightly modified copies of these plates. About the same time he made for the government a survey of Lorraine, *»oppidatim et per singulos pagos»*, which work, according to Ghymmius, *»cum tanto vitae discrimine conjuncta fuit, adeoque vires illius debilitavit, ut parum abfuerit quin ex terrore gravissimum morbum sibi consiliasset et animi perturbationem incidisset»*. Yet no decrease in Mercator's energy can be perceived during the next few years. It therefore seems probable to me that this account, by Ghymmius, of Mercator's illness caused by fear, ought to be referred to his imprisonment for heresy, which episode is altogether omitted by the biographer. In 1568 Mercator published his comprehensive *Chronologia*, which was much praised by contemporary scholars. He had wasted much intellect and spent much labour on this work, though it has hardly had any permanent influence on science, and at present is probably very seldom consulted by chronologists. In the next year, 1569, he published his large epochal map of the world on an increasing cylindrical projection, of which I have given an account p. 96. It is one of the most original and valuable cartographical works ever published, but it was but little understood and appreciated by the author's contemporaries. It is reproduced in full size, but without some of the important legends, by Jomard, and in a much reduced scale by CHARLES P. DALY (*On the early history of cartography*, in the *Journal of American Geogr. Society.* XI, 1879). Mercator then elaborated a new edition of Ptolemy's atlas, which was published in 1578. It was received with universal approbation and was often reprinted during the next hundred and fifty years. On account of the modifications introduced on the maps of the old atlas, which were *»ad mentem Auctoris restitutae et emendatae,»* this work is at present of less value than the uncorrected and uncritical copies of the original maps, published in the 15th and the beginning of the 16th century. Yet various improvements in the map-projections were introduced in this work, concerning which I may refer to BREUSING's *Leitfaden durch das Wiegenalter der Kartographie*, Frankfurt a. M. 1883. Mercator had, according to Ghymmius, commenced the compilation of a new systematic atlas of the whole surface of the earth, long before Ortelius. The work was almost finished in manuscript, when Ortelius edited his *Theatrum*, which caused Mercator to postpone the printing of his work. Only parts of it were published, at the urgent request of his friends; namely Gallia and Germania in 1585, and Italy in 1590. A similar collection of maps of the northern countries was ready for printing at the time of Mercator's death on the 2d of Dec. 1594. In the following year his son RUMOLDUS published, at Duisburg, under the title of: *Atlas[2] sive cosmographicae meditationes de fabrica mundi et fabricati figura ... Duisburgi*

---

[1] Some interesting letters regarding the globe manufacture of Mercator are published by F. van Raemdonck in his above mentioned paper: *Les sphères de Gerard Mercator.* The globes cost 10 to 12 »florini Carolini». Such a florin contained the same quantity of silver as 4,22 francs (v. RAEMDONCK).

[2] Atlas carrying the earth on his shoulders is for the first time employed as an emblem of a collection of maps on the title-page of »Lafreri's atlas.» I have therefore employed a fac-simile of this engraving to ornament the title of the present work. Mercator gives an extremely confused explanation of the derivation of the name Atlas in the *præfatio in Atlantem*, which, besides his essay, *De mundi creatione et fabricà*, the biography of Ghymmius, a few Latin dedicatorial poems, and *Epistulae duae duorum doctissimorum virorum in laudem Atlantis conscriptae*, form the introduction to the first edition of *Atlas*. The biography of Ghymmius is a *chef d'oeuvre*, unaffected, simple, and worthy of the great and learned man whose life and deeds it describes. Yet the letters of »the two most learned men,» Mercator's *De Mundi creatione et fabrica*, and above all, his *Præfatio in Atlantem* are only a pious and benevolent, but incomprehensible nonsense. This did not prevent the name of Atlas from becoming naturalised from that time in all civilised languages, as signifying any collection of maps published in print.

Clivorum, the first incomplete edition of the whole work, evidently prepared for the press by his father. In 1602 a new edition was published at Duisburg, also in Latin, and later (1606—1636) several others much revised and augmented, in different languages, at Amsterdam by Hondius. As a practical manual of cartography they soon entirely supplanted the Theatrum by Ortelius, which was formerly so much praised. With respect to these editions the reader is referred to the often cited works of P. A. Tiele and Van Raemdonck. They do not belong to the early period of cartography.

The short review here given of Mercator's principal works may be sufficient to show what a conspicuous position he occupies in the history of cartographical science. A great part of his works belongs chronologically to the period of incunabula, — the main part, if we, like the majority of writers on the history of cartography, consider 1570 to be the

pared without disadvantage with the works of Mercator, e. g. several of the Italian maps, of which fac-similes are given above, the handsome map of the New World which was engraved on copper in Paris in 1587 for HAKLUYT's edition of Petrus Martyr (N. fig. 82), and several of Gastaldi's works. Neither, as far as I know, are any of Mercator's maps comparable with the *Bairische Landtaflen* as regards topographical accuracy. Fundamental errors also occur, as for instance his tendency, derived from medieval geographical drawings, to make rivers run right across continents from sea to sea — *a hydrographical blunder never met with on Ptolemy's maps of the second century.* He was scarcely an acute critic, and was often unlucky in the selection of his authorities, in the mapping of distant countries. His tendency to mysticism also sometimes exercised an unfortunate influence on his cartographical work. Yet with all these shortcomings

83. Hispaniola, from: CORN. WYTFLIET, Descriptionis Ptolemaicæ augmentum, Lovanii 1597. (Orig. size 282 × 228 m. m.).

limit of this period. But by the comparative accuracy and richness of details, the finish of the copper engraving, the author's insight into the mathematical principles of the art of map-drawing and into the deficiencies and merits of the different methods of projections, most of Mercator's maps have a quite modern stamp. Justice requires the admission that several maps constructed by other geographers in the 16th century may as regards finish of engraving be compared

we must fully agree with a contemporary publisher of maps, when he characterizes Mercator as *in cosmographia longe primus,* and if the genius and greatness of a philosopher is to be measured by the importance of the new and fruitful ideas he suggested, and by the quantity of useful work honestly performed in the service of science, then the master of Rupelmonde stands unsurpassed in the history of cartography since the time of Ptolemy.

The T. I—XXVII of the plates in the present work are fac-similes of the first complete atlas of the Old World, which, as is shown by the fig. on p. 1, had for this purpose been systematically divided by Ptolemy into separate regions, each represented in the atlas by a separate map. About thirteen centuries later a somewhat similar work on the New World was published under the title: *Descriptionis Ptolemaicae Augmentum, sive Occidentis notitia brevi commentario illustrata, studio et opera* CORNELII WYTFLIET, *Lovaniensis,* Lovanii MDXCVII. Of this work seven editions

were, as above (p. 29) mentioned, published between 1597 and 1611. The book is dedicated to Philip III, *Hispaniarum et Indiarum princeps,* and contains a modern account, on about 200 printed pages, of the history of the discovery, the geography, the natural history and ethnology of the New World. It contains 19 maps handsomely engraved in copper, viz. 1. *Orbis terrarum;* 2. *Australis terra et Chica;* 3. *Chili;* 4. *Plata;* 5. *Brasilia;* 6. *Peruvia;* 7. *Castillia Aurifera;* 8. *Residuum terrae firmae sive Paria et Cubagua;* 9. *Hispaniola;* 10. *Cuba et Iamaica;* 11. *Iucatana regio et Fondura;* 12. *Hi-*

spania Nova; 13. *Nova Granata et California;* 14. *Anian et Quivira;* 15. *Conibas regio;* 16. *Florida;* 17. *Virginia;* 18. *Nova Francia et Canada;* 19. *Estotilandia et Laboratoris terra.*

Space will not allow me to enter into any closer analysis of each of these maps. They do not contain any great amount of new and original material. But as the first general geography of America the text of Wytfliet's work may, at least in some degree, have contributed to dispel many of the errors regarding the New World which were adhered to in most parts of Europe at the end of the 16th century. It is still valuable as a summary of everything then known in the Spanish Netherlands about the regions to the west of the Atlantic. The following information (partly from Gomara, *Historia general de las Indias,* French edition, Book 4 chap. 14), regarding a scheme discussed before Magellan's voyage, to dig through the Isthmus of Panama, may e. g. be of some interest at the present day: *Agitatum aliquando fuisse memoriae proditum est de Isthmo Darienis perfodiendo, ut Austro mixtus Septentrio, expedita utrinque navigatione, merces transveheret et commutaret suas, qua in re cum sententiis variatum esset, negantibus plerisque committendum esse, ut refractis naturae claustris, maria inter sese committerentur; quia ingruentibus Septentrionalibus undis metuendum esset, ne in adiacentia erumpens mare, totam regionem inundaret.* Others considered it to be impossible to force a way through the high mountains, and raised warnings against any attempt to alter the arrangement of the Almighty. Such considerations delayed the commencement of the great work, until, after the discovery of the Straits of Magellan, it was altogether abandoned as useless.

In the history of early cartography the maps in Wytfliet's *Augmentum* play the same part for the New World as Ptolemy's maps do for the old hemisphere, and they give us, though chronologically belonging to a later period, a valuable summary of the early cartography of America. I shall therefore finish this fac-simile atlas with a reproduction of Wytfliet's maps. On T. LI sixteen of them are reproduced in a size reduced from about 0,290 × 0,263 to 0,135 × 0,107 m.* Of the remainder the first map *(Orbis terrarum)* is a copy of Rumold Mercator's map of the world of 1584 (N. T. XLVII), too insignificant to be worth a new reproduction. Another, the map of *Hispaniola,* which want of space has obliged me to exclude from T. LI, is reproduced on the preceding page. The third is Wytfliet's handsome map of *Australis terra et Chica,* and forms the last illustration of this work. Though long ago partly dissolved into Australian islands, and partly removed within the south polar-circle by the geographical discoveries of the last century, the large south-polar continent, laid down here, still forms a *Terra Australis incognita,* where many a problem, important to the geography and the natural history of the earth, is waiting for its solution.

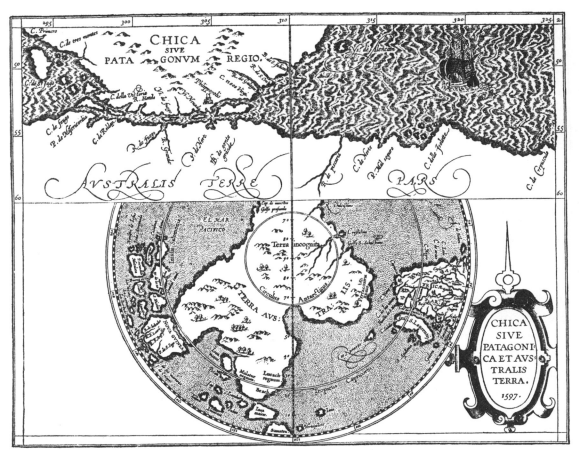

84. Australis Terra et Chica. According to Wytfliet 1597. (Orig. size 290 × 229 m. m.).

* [See Publisher's Note.]

Addenda.

P. 5. The passage »*and passing through the parallels of Rhodes and Thule,*» should be corrected to: »*Touching the earth's surface near the parallel of Rhodes, and preserving the proper ratio between the parallels of Thule and the equator*».

P. 9. During a short stay at Paris this autumn I had an opportunity of seeing the splendid codices of Ptolemy's Geography preserved at the *Bibliothèque Nationale*. Two of them are of a special interest for early cartography, viz:

1. A latin codex (No. 4802) from the middle of the 15th century.[1] It contains the prototypes or copies of the prototypes for the *Tabulae novellae* in *Berlinghieri's Septe Giornate* (comp. p. 12) but no »Tabula Novella» of the Scandinavian countries. It ends with the following sentence: *Claudii Ptolomaei cosmographie textus usq. ad tabulas filiciter finit per me Ugonem Comminelli e francia natum inter scriptores minimum.*

2. A Greek codex (No. 1401) from the end of the 14th or the beginning of the 15th century. Most of the maps in this codex are, as d'Avezac remarked, drawn on Donis' projection. If these splendid maps are not a later addition to the text, the pretension of *Dominus Nicolaus Germanus* to be the inventor of this projection is not admissible (comp. p. 14 and 86 n).[1]

P. 12. On comparing the copy of *Ptolemaeus »Bononiae 1462»* in the library of Upsala with a copy bought in Italy during the printing of the English version of my Fac-simile-atlas, I have found that two different issues exist of this remarkable edition, viz.

1. A first issue, on the maps of which the seas are generally left blank without any undulated striae.

2. A second issue, where the seas are always densely marked with such striae.

The type 1 probably contains the first maps ever published in print. In my copy of it, the tabulae 2a, 4a, 6a, 9a *Europae*, 4a *Africae* and 4a, 9a, 10a *Asiae* are wanting. It contains accordingly only 18 maps. The seas are marked by undulated striae only in the map of the world and in tab. 10a *Europae*, 2a *Africae*, and 1a, 3a and 6a *Asiae*.

P. 22, 29, and 30. *Ptolemaei Geographicae cum Eandavi annotationibus*, printed *Argentorati apud Petrum Opilionem, 1532*, is not an edition of Ptolemy's geography but identical with *Quae intus continentur* etc. of Jacobus Zieglerus Landavus (comp. p. 60), of which some copies seem to be provided with a different and misleading title, *in which Landavus is misspelt Eandavus*. The total number of authentic editions of Ptolemy is accordingly only 55, of which 32 were published before 1570.

P. 26. I have lately obtained a copy of the Italian edition of Ptolemy printed at Venice in 1564 (No. 32 of my catal.).

It corresponds to Winsor's description. The same year also a Latin edition was issued at Venice (No. 33 of my catal.).

P. 35. To the catalogue of maps printed before 1520, without any direct connection with editions of Ptolemy's Geography, should be added:

1. An edition of the map of Nicolaus a Cusa engraved in copper and printed at Eystat 1491 (comp. p. 129). A copy of this remarkable map is preserved at the British Museum (Catal. of Maps, p. 1268 and 1535). It seems to be the first map reproduced in copper-plate north of the Alps, the first and the only signed edition of the map of Central Europe by the celebrated Cardinal.

2. A map s. l. a. of central Europe from the beginning of the 16th century, which I lately had an opportunity of examining in the Bibliothèque Nationale. It gives the distances between the most important towns, and may perhaps be the *Charta itineraria of Waldseemüller, printed Argentorati 1511* (comp. p. 129).

P. 42, 71 n, and 107. In the town library at Frankfurt a. M. I have seen a copy of the *Margarita Philosophica, Basileae 1535*, containing two folding maps, one of which is identical with the map in the edition of 1503 (N. T. XXXI), the other a copy in a somewhat reduced size of the remarkable map in the edition *Argentorati 1515* (N. T. XXXVIII), but without the words *Zoana Mela*. All other copies of the edition *Basileae 1535*, which I have examined, contained only reprints of the map of 1503.

P. 45 fig. 25. This portolano of the southern parts of the Baltic and the North Sea has several times been reproduced by copper engraving. It is very exact for the time when it was published, and it is perhaps based on a Dutch original. An early edition in the Bibliothèque Nationale is signed: »*Michaelis Tramezini formis 1558. Jacobus Buschius Belga in aes incidebat.*

P. 45 n. To the portolanos in Swedish collections should be added: 4—6. Three portolanos in the collection of Skokloster viz: 4. A chart on vellum of the Mediterranean and the west coast of Europe and Africa from Iceland to Cap Bojador signed *Diego . . . puerto de S:ta Maria 1545*. 5. An atlas on vellum, containing an oval general map of the world and several special maps of the Mediterranean and Black Seas etc., by Georgio Calapoda Cretense 1552. 6. An atlas from the beginning of the 17th century.

7, 8. Two portolanos also on vellum in my private collection, viz: 7. A large and richly ornamented chart of the western parts of Europe and Africa signed: *Domingo figlio de maistre Jaume Ollives Mallorquin en Napoli Anno 1568*. 8. An atlas consisting of three maps, of which one representing the western part of Europe and Africa is signed *Augustinus Russinus* me fecit Masciliae 1590 (?). The two others representing the Mediterranean and Black Seas and the Archi-

---

[1] I am indebted to M. MARCEL, the Director of the Map Department in the Bibl. Nationale, for the following information about these codices: »Le Manuscrit latin 4802 de Ptolémée est aux armes d'Alphonse I de Naples, dont le blason se voit au bas du premier feuillet ainsi que la devise, c'est dire qu'il est antérieur à 1458, date de la mort de ce prince. Ce manuscrit qui est aux armes de Henri II est venu en France à la suite des guerres d'Italie. Le copiste Ugo Comminelli ou de Comminellis est bien connu; on lui doit un certain nombre de manuscrits dont on trouve la liste dans: BRADLEY, *A Dictionary of Miniaturists, Illuminators, Calligraphers, and Copyists*. 3 Vol. in-8:o. London 1887—1889.

Quant au manuscrit grec 1401, également aux armes de Henri II, tout ce qu'on en peut dire c'est qu'il était dans la Bibliothèque du roi à Fontainebleau, qu'il date de la fin du XV ou du commencement du XVI siècle, sans pouvoir fixer, à une trentaine d'années près, sa date certaine. Mais ce qui apparait clairement et d'une façon presque certaine, c'est qu'il a été copié par un grec — l'écriture le démontre surabondamment — et en Italie, ainsi qu'en témoigne le style de l'ornementation.»

pelago are not signed. They are probably older, and evidently drawn by other artists than the first.

P. 64 and 95 n. As to Inventio Fortunati and Nicolaus de Linna compare an elaborate paper by Rev. B. F. Costa (Journ. of The Amer. Geograph. Society XII, 1880, p. 159). The name of the author of the itinerarium to which Mercator refers for his description of the polar regions, is spelled by Gerard Mercator, on his large map of 1569, Cnoyen, by Rumoldus Mercator and Hondius, Cnoxen.

P. 76. Copies of the Mappemonde in gores described under No. 5 (N. T. XXXVII) are preserved at the Bibliothèque Nationale and in the Hauslab Collection at Vienna.

P. 85. The conical development of Ptolemy preserving the proper ratio between the parallel of Thule and the equator does not touch the earth's surface exactly at Rhodes (Lat. 36°) but at 33° 6'. If in this development also the proper ratio between the degrees of latitude and longitude at the parallel of Rhodes is maintained, the development becomes slightly intersecting.

P. 86 and 131. Professor Markgraf and Dr. A. Heyer have lately discovered, in the town-library at Breslau, some very important maps of Gerard Mercator, viz:

1. A copy of his large map on increasing cylindrical projection of 1569; 2. His large map of Europe of 1554; 3. His Angliae, Scotiae et Hiberniae nova descriptio of 1564. Of these maps, 2 and 3 have hitherto been regarded as lost, and of 1 only a single copy, preserved at the Bibliothèque Nationale in Paris, was known (*A. Heyer: Drei Mercator Karten in der Breslauer Stadt Bibliothek;* Zeitschrift für Wissenschaftlicher Geographie VII, p. 379, Weimar 1889). The description by Dr. Heyer confirms my supposition that Mercator never published any maps drawn on an unmodified intersecting conical projection.

P. 99. To the maps printed between 1520 and 1550 should be added a map of the world by »JOHANNES VESPUCCI,» of which at least two editions exist, the one undated, the other dated 1524. I am indebted to Mr. Harrisse for a photograph of the former, which unfortunately I received too late for insertion in my atlas. It is printed from a copper engraving, and forms a double planisphere on an equidistant polar projection. The southern hemisphere being divided in two parts along the great circle passing Ferro, the whole surface of the earth is here projected on a circle and two semicircles. The author was a nephew of Amerigo, Royal Pilot in Spain and one of the cosmographers consulted by the Emperor Charles V for the famous conference at Badajoz. His map, illustrating the geographical opinions in Spain before the return of the »Victoria,» is in more than one respect very remarkable. *It is the first planisphere on an equidistant polar-projection.* For special maps, that projection had already been employed in the above mentioned Bologna edition of Ptolemy's geography, and there the net of graduation is drawn with far more accuracy than on Iuan Vespucci's map, which in this respect is very defective.

P. 114 and 122. I have seen in the Bibliothèque Nationale a second edition or second issue of Gastaldi's Moscovia, signed: *per l'ecce:te M. Giacomo Gastaldo. Cosmographo in Venetia Anno MDLXIIIIII. Ferando Berteli.*

P. 118. In the Bibliothèque Nationale there exist two sets of maps corresponding to the collection here described under the name of Lafreri's atlas. Such collections, more or less complete, are also to be found in the Royal Library at Copenhagen, in the library of the University at Rostock, in the town library at Breslau?, the private Library of the King at Madrid, and probably at several other places.

INDEX.

(*a* indicates the first column, *b* the second, *n* the footnote.)

Adelung, Friedr. *113 n*, *130 b*.
Aelianus, Claudius *35 a*.
Aeneas Sylvius *29 a*, *41 a*.
Aeschler, Jacobus *20 a*, *29*, *70 b*.
Afer, Dionysius *94 a*, *112 a, b*.
Africa, connected in the south with Asia, according to Ptolemy *33 b*. — possibility of its circumnavigation *64 n*. — maps of *15*, *17*, *24 a*, *26 a*, *37 a*, *41 a*, *42 a*, *50 a*, *66 b*, *67 b*, *70 a*, *87*, *108 a*. — on Ruysch's map *64 a*.
Agathodemon *8 a*.
Ageminius, Paulus *87 n*.
Aggere, Petrus ab *130 a*.
Agnese, Battista *45 n*, *80 b*, *82 a*, *90 b*, *114 a*.
Agrippa, Marcus Vipsanius *35 a*.
Agysimba *3 b*, *4 a*.
d'Ailly, Pierre *10 a*, *37*, *38 a*, *85 a*, *93 b*, *94 n*, *100 b*.
Alantse, Lucas *99 b*.
Albano, Justus de *16 b*.
Alberti, Leandro *126 a*, *128 b*.
Albuquerque *63 n*.
Aldus Manutius *103 b*.
Alexander V, Pope *10 a*.
Alexandria *2 a*.
Algoet, Levinus *128 b*.
Aliacus, Petrus, see: Ailly.
Aloysius, Cornelius *27 a*.
Alsace (Alsatia), map of *24 a*.
Alterius, Marianus *66 a*.
Amaseo, Gregorio *128 a*.
Ambrosius, Marcus *128 b*.
America, maps of *24 a*, *26 a*, *107*, *113*, *117*, *127*, *129*, *131*, *134*. — on Ruysch's map *65 b*. — on Stobnicza's map *68 b*. — the literature on, before *1550* *62 a*. — the name *23 a*, *42 a*, *69 n*. — on maps *88 a*, *100 a*. — in book-print *100 a*.
Amoretti, Carlo *73 a*.
Anania, Lorenzo *90 a*, *94 a*.
Anaximander *35 a*.
Anian Fretum *94 b*, *120 b*.
Anselmus *29 a*.
Antilia Insula *65 a*.
Antoniszoon (Antonii, Antoniades), Cornelis *126 b*.
Apianus, Petrus *2 b*, *22 b*, *76 a*. — in the Catalogus of Ortelius *130 a*. — map of the world *1520* *6 a*, *7 a*, *88 a*, *99 a*, *101 b*, *112 a*. — map of the world *1530* *104 b*. — *Cosmographiæ introd.* *76 a*, *100 a*, *102 n*, *116 a*. — *Cosmographicus liber* *92 a*, *93 a*, *94 a*, *102 a*. — *Isagoge in typum cosmogr.* *101 a*. — projections *88 a*, *90 a*, *92 a*, *94 a*.
Apianus, Philip *122 b*, *130 a*.
Arabia Felix, map of *26 a*.
Arabian maps *43 a*.

Aristagoras *35 a*.
Aristoteles *38 b*, *71 a*. — Meteorologia *42 a*, *68 b*, *85 a*, *100 a*.
Asia Minor (Natolia), maps of *26 a*, *50 a*, *70 a*.
Assemani, Simon *89 b*.
Athos-manuscript of Ptolemy, see: Vatopedi.
Azof Sea, distance from the Baltic, according to Ptolemy *34 b*.
Aucuparius, Thomas *21 b*.
Augustinus *72 a*.
Aventinus, Johannes *24 b*, *101 b*, *128 a*.
d'Avezac *10 b*, *11 n*, *30 a*, *41 b*, *44 n*, *73 b*, *74 b*, *76 a*, *83 a*, *84 a*, *86 n*, *89 b*, *90 b*, *94 b*, *100 a*, *129 a*.
Avignon, map of *104 b*, *117 a*.

Baardson, Ivar *95 n*.
Bacalaos *26 a*, *107 b*.
Bacon, Roger *38 b*, *85 a*, *93 b*, *100 b*. — projection *93 b*, *94 n*, *102 b*.
Baduarius, Johannes *17 b*.
Bahnson, Kristian *83 a*.
Balboa, Vasco Nuñez *69 a*, *79 b*.
Baltic Sea, distance from the Sea of Azof, according to Ptolemy *34 b*. — form on the Portolanos *52 b*.
Bancroft, H. H. *63 a*.
Barentszoon (Barents), Willem *27 b*, *39*, *41*, *43*, *47 a*, *50 b*, *56 n*, *96 a*.
Barlow, Samuel L. M. *67 b*.
Bartlett, J. R. *10 n*, *11 n*, *68 a*.
Bavaria, maps of *24 a*, *108 a*, *122 b*.
Beatus Rhenanus *24 a*.
Behaim, Martin, globe *21 a*, *64 a*, *65 b*, *71 a*—*74 a*, *100 b*, *101 a*. — charts *79 a, b*.
Bellarmatus, Hieronymus *128 a*.
Belle-Foreste, François de *90 b*, *108 b*.
Benacus lacus (Lake of Garda) *111 a*.
Beneventanus, Marcus *16 b*, *18 b*, *17 a, n*, *66 a*—*67 a*.
Benincasa, Grazioso *84 b*.
Benoni-Verelst *82 b*.
Berganus, Georgius Jodocus *111 a*, *126 b*.
Bergomensis, Jacobus Philippus *40 b*.
Berlinghieri, Francesco *10 b*, *12 b* — *14 a*, *29*, *30 a*, *31 a*, *70 a*, *85 b*, *135 a*.
Bernardus Sylvanus, see: Sylvanus.
Beroaldus, Philippus *12 a*.
Bertelli, Ferrando *58 n*, *103 b*, *122 a*, *129 b*.
Bertius, Petrus *28 a*.
Bianco, Andrea *34 a*, *52 a, b*, *53 a*, *65 a*.
Blaeu (Blavius), J. *126 a*.
Blau, Jean *10 a*, *54 n*, *82 b*.
Blaurerus, Thomas *24 b*.
Boekel, Petrus *130 a*.
Boëtius *9 n*.
Bohemia, maps of *18 a*, *25, b*, *108 b*, *111*.
Bongars, J. *51 b*.

Bonus, Petrus *12 a*, *29*, *85 b*, *95 a*.
Bordone, Benedetto *36 b*, *103 b*, *117 a*, *126 a*. — projection *90 a, b*, *102 b*, *103 b*.
Bossius, Jacobus *122 a*.
Boulenger, Ludovicus *76 a*, *99 n*, *100 a*.
Brabant, map of *24 a*, *26 a*.
Brahe, Tycho *83 a*.
Brandis de Schass, Lucas *35 b*.
Brandon, St. *16 b*.
Brasil Insula *58 a*, *70 n*.
Brasiliæ Regio *26 b*, *70 b*, *77 b*, *106 b*.
Brehmer, N. H. *32 a*.
Breisgau (Brisgoia), map of *24 a*.
Brenner, O. *60 b*.
Bretagne, map of *104 a*.
Brevoort, J. Carson *90 a*.
Breusing, A. *46 b*, *48 b*, *55 n*, *88 n*, *94 b*, *118 b*, *131 n*, *132 b*.
Breydenbach, Bern. de *84 a*.
Brion, Martin. de *129 a*.
Britain, maps of *7*, *11*, *19 a*, *24 a*, *70 a*, *115 b*, *123*.
Brochard, Bonaventura *126 a*.
Brognolus, Bernardus *126 a*.
Bronchhorst, Joh. *23 b*.
Brown, John Carter *10 n*, *11 n*, *50 a*, *68 a*.
Brunet *13 n*, *22 b*, *50 a, n*.
Brusch, Caspar *126 b*.
Bruun, C. G. *126 b*.
Buchon *46 b*, *47 a*—*48 a*.
Bucius, Joh. *128 a*.
Buckinck, Arnold *14 a*, *16 a*, *29*, *31 b*.
Bulionius, Aegidius *125 a*.

Cabot, John *60 b*, *71 a*, *74 n*.
Cabot, Sebastian *90 b*, *110 a*, *130 b*.
Cabral *70 n*.
Cadamosto, Aloisius *62 b*, *103 a*.
Calameus, Johannes *128 a*.
Calapoda, Georgio *135 n*.
Calderinus, Domitius *14 a*.
Calicut, map of *26 a*.
Calvin, Jean *22 a*.
Camaroca (Madagascar) *66 b*.
Camers, Joh. *99 a*.
Camocius, Johannes Franciscus *45*, *49 b*, *50 b*, *118 n*, *122 a*.
Cantino, Alberto *63 a*.
Cão, Diogo *74 a*.
Carter Brown, see: Brown.
Cassiodorus *9 n*.
Castaldus, see: Gastaldi.
Castellani, Carlo *36 b*, *41 a*, *50 b*, *90 a*, *94 n*, *118 b*.
Castilioneus, Bonaventura *126 a*.
Ceylon, in Ruysch *64 a*. — Comp. Taprobana.
Chiauez, Hieronymus *128 a*.
Chrysoloras, Emanuel *9 a*, *10 a*.
Cimerlinus, Joh. Paulus *89 a*.

I. PTOLEMÆUS ROMÆ 1490.

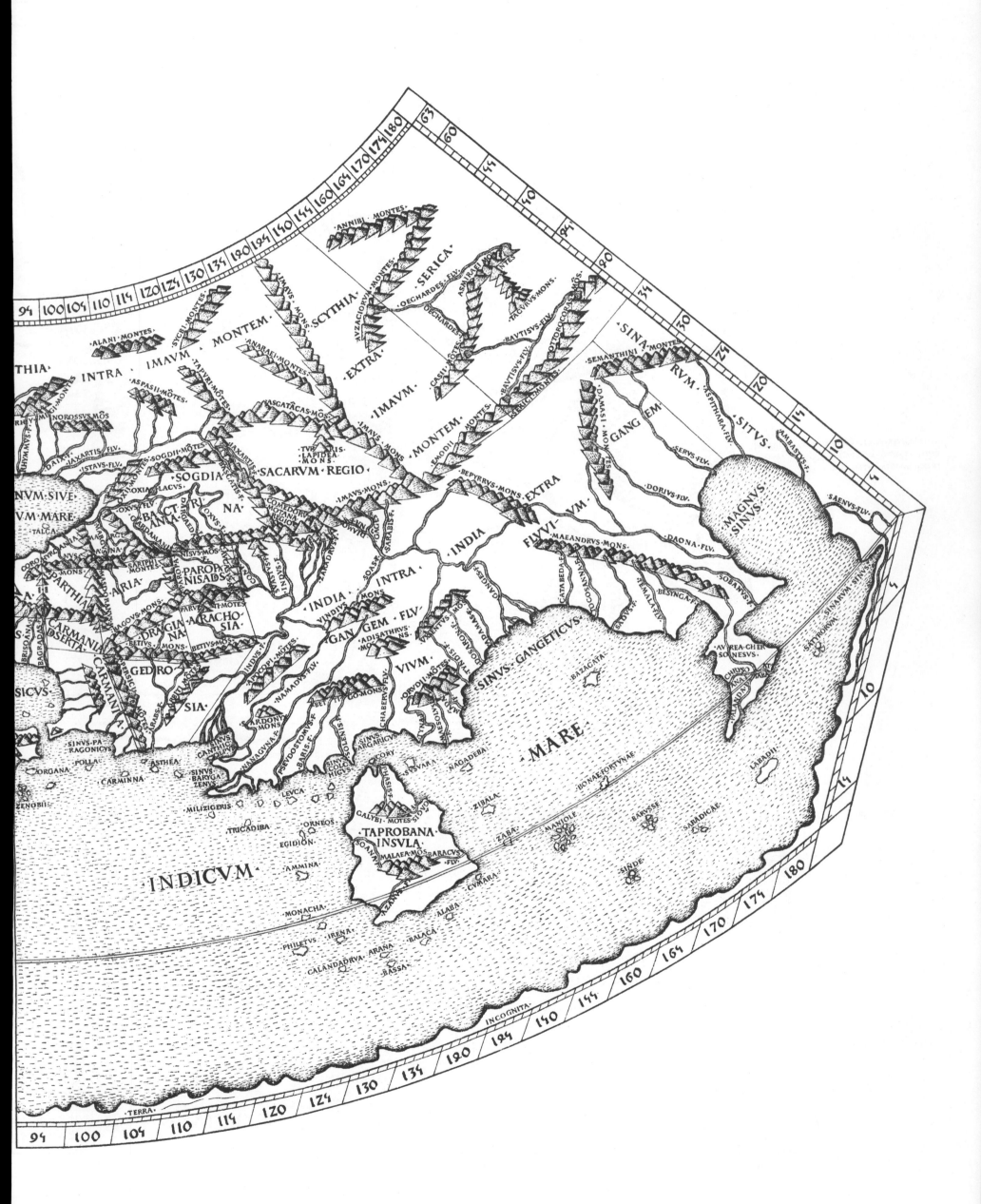

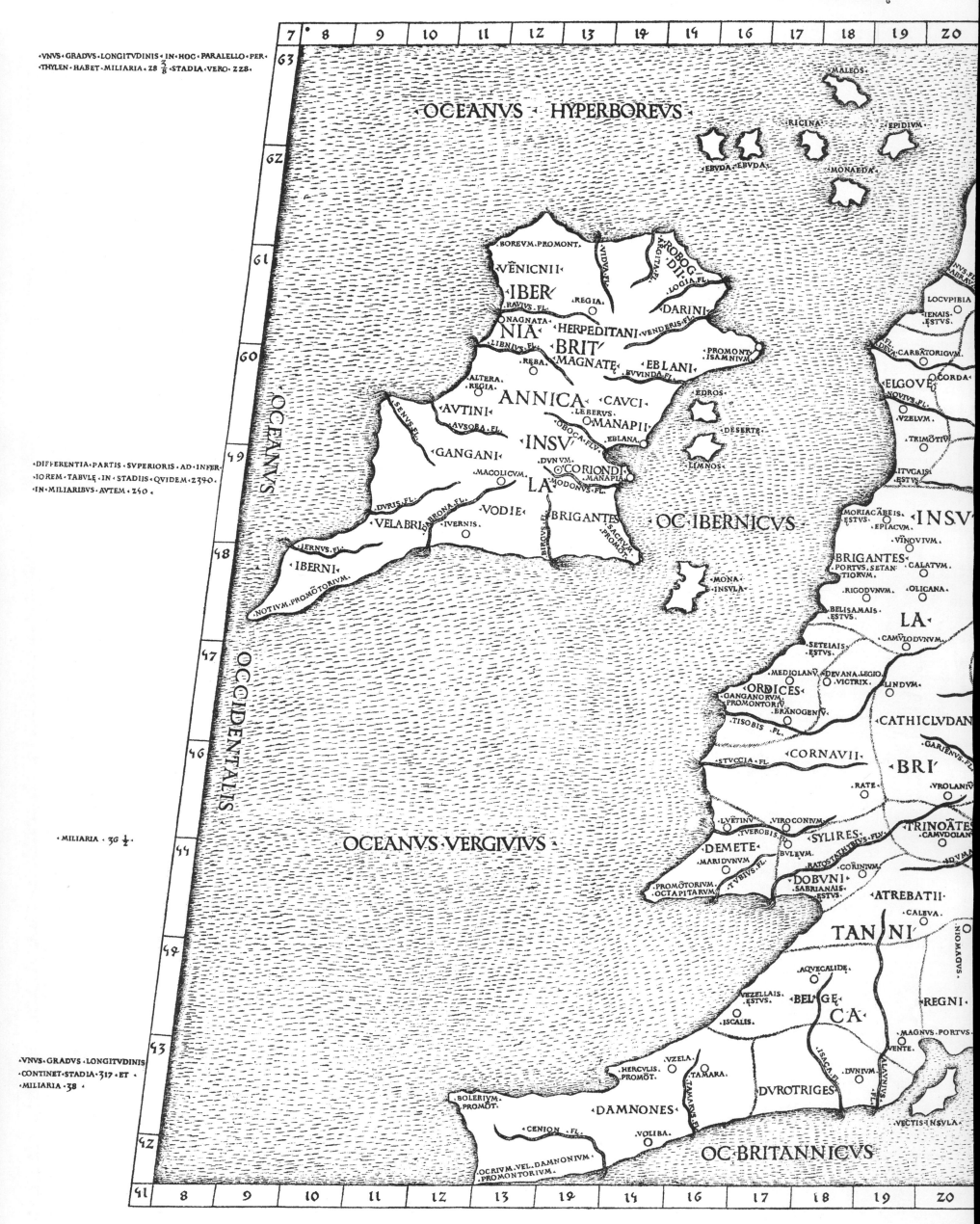

·VNVS·GRADVS·LONGITVDINIS·IN·HOC·PARALELLO·PER· ·THYLEN·HABET·MILIARIA·28 $\frac{2}{8}$·STADIA·VERO·228.

·DIFFERENTIA·PARTIS·SVPERIORIS·AD·INFER· ·IOREM·TABVLE·IN·STADIIS·QVIDEM·2390· ·IN·MILIARIBVS·AVTEM·240·

· MILIARIA · 36 $\frac{1}{2}$ ·

·VNVS·GRADVS·LONGITVDINIS· ·CONTINET·STADIA·317·ET· ·MILIARIA·38·

OCEANVS · HYPERBOREVS ·

MALEOS

RICINA   EPIDIVM

EBVDA EBVDA   MONAEDA

BOREVM·PROMONT.
VIDVA·FL
VENICNII   ROBOG·
                  DII·   LOCVPIBIA
IBER   ARGITA·FL
·RAVIVS·FL   ·REGIA   LOGIA·FL   IENAIS·
                           DARINI   ESTVS·
NAGNATA HERPEDITANI VENDERIS·FL   FL·DEVA·CARBATORIGVM·
NIA   CORDA
LIBNIVS·FL BRIT
         REBA   MAGNATE   EBLANI   ELGOVE·
                        EVVINDA·FL   NOVIVS·FL·
ALTERA·   VZELVM·
REGIA·   EDROS
ANNICA· CAVCI   TRIMOTIV·
AVTINI· LEBERVS·   DESERTE·
AVSOBA·FL OBOCA·FLV·MANAPII   ITVGAIS·
GANGANI· INSV EBLANA·   ESTVS·
                DVNVM·   INSV
MACOLICVM· CORIONDI LIMNOS·   MORIACEBEIS·
              MANAPIA   ESTVS·
LA MODONVS·FL   EPIACVM·
                                         VINOVIVM·
DVRIS·FL· VODIE·   BRIGANTES OC·IBERNICVS BRIGANTES
VELABRI BARONA·FL   SACRVE·   PORTVS·SETAN·
        IVERNIS·   BIRGVS·FL PROMOT·   TIORVM· CALATVM·
                                           RIGODVNVM· OLICANA·
IERNVS·FL·   MONA·
IBERNI·   INSVLA   BELISAMAIS·
                                            ESTVS·   LA·
NOTIVM·PROMOTORIVM·   CAMVLODVNVM·
                                            SETEIAIS·
                                            ESTVS·
                                         MEDIOLANV· DEVANA·LEGIO·
                                         ORDICES· VICTRIX·
                                         GANGANORVM·   LINDVM·
                                         PROMONTORIV·
                                         BRANOGENV·
                                         TISOBIS·FL·   CATHICLVDAN·
                          OCEANVS · VERGIVIVS ·   CORNAVII·   BRI·
                                         STVCCIA·FL·
                                         RATE·   VROLANIV·
                                         LVETINV· VIROCONIVM·
                                         TVEROBIS·FL·   TRINOATE·
                                         SYLIRES·FLV·   CAMVDOLAN·
                                         DEMETE· BVLEVM·   ADVA·
                                         MARIDVNVM·
                                         RATOSTATHYBIVS·FL·
                                         PROMOTORIVM· CORINIVM·
                                         OCTAPITARVM· DOBVNI·
                                         T·TVBIVS·FL   SABRIANAIS· ATREBATII·
                                                      ESTVS·   CALEVA·
                                                                   NOMAGVS·
                                         TAN   NI·
                                         AQVECALIDE·
                                                      BEL GE·   REGNI·
                                         VEZELLAIS·   CA·
                                         ESTVS·   MAGNVS·PORTVS·
                                         ISCALIS·   VENTE·
                                         HERCVLIS· VZELA·
                                         PROMOT· TAMARA· DVNVM·
                          BOLERIVM·   ISACA·FL
                          PROMOT·   DVROTRIGES·
                                                         VECTIS·INSVLA·
                             CENION·FL· VOLIBA·
                                 DAMNONES·
                          OCRIVM·VEL·DAMNONIVM·   OC·BRITANNICVS·
                          PROMONTORIVM·

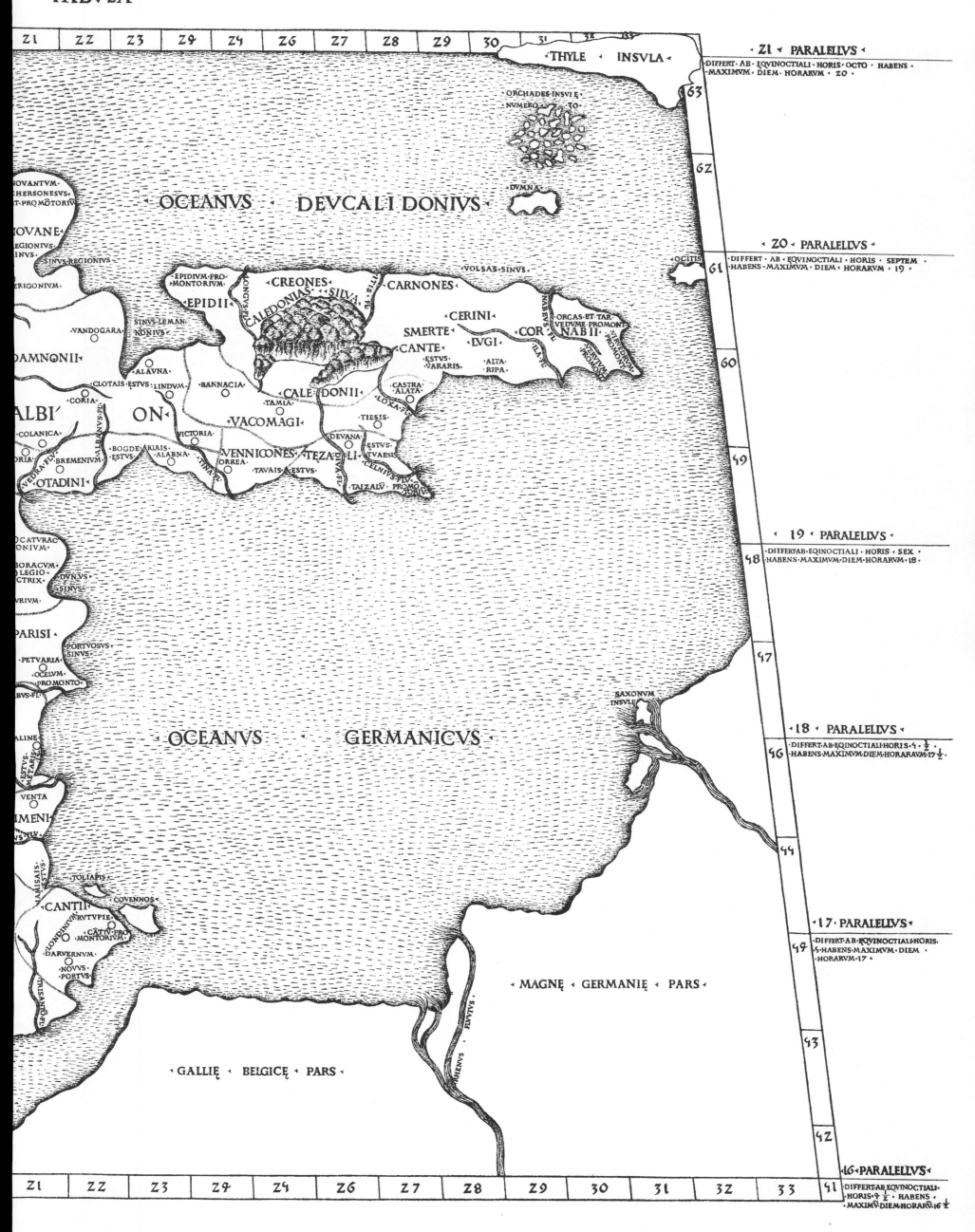

ZI	ZZ	Z3	Z9	Z4	Z6	Z7	Z8	Z9	30	31	3Z	133

·THYLE · INSVLA ·

·DIFFERT · AB · EQVINOCTIALI · HORIS · OCTO · HABENS ·
·MAXIMVM · DIEM · HORARVM · ZO ·

63

ORCHADES·INSVI E
·NVMERO· TO·

6Z

·OVANTVM·
·HERSONESVS·
·T·PROMOTORIV·

OCEANVS · DEVCALI DONIVS ·

·DVMNA·

· ZO · PARALELLVS ·

·DIFFERT · AB · EQVINOCTIALI · HORIS · SEPTEM ·
HABENS · MAXIMVM · DIEM · HORARVM · 19 ·

·OVANE·
·LEGIONVM·
·SINVS·

·SINVS·REGIONIVS·

·VOLSAS·SINVS·

·OCITIS·

61

·ERIGONIVM·
·VANDOGARA·

·EPIDIVM·PRO·
·MONTORIVM·

·CREONES·

·CARNONES·

·EPIDII·

CALEDONIAS · SILVA·

·CERINI·

·SINVS·LEMAN
NONIVS·

·SMERTE·

·ORCAS·ET·TAR
VEDVME·PROMONT
NAB II

DAMNONII·

·ALAVNA·

·CANTE·

·LVGI·

·COR·

60

·CLOTAIS·ESTVS· ·LINDVM·

·BANNACIA·

·ESTVS·
·VARARIS·

·ALTA·
·RIPA·

ALBI·

·CORIA·

ON·

·CALE DONII·

·CASTRA·
·ALATA·

·COLANICA·

·TAMIA·

·LOXA·F·

49

·ORIA· ·BREMENIVM·
·VETRA·FLV·

·BOGDE·ARIAIS·
·ALABNA·

·VICTORIA·

·VACOMAGI·

·TIESIS·

·DEVANA·

·OTADINI·

·ESTVS·

·TINAE·

·ORREA·

·VENNICONES· TEZA·LI·

·ESTVS·
·TVAESIS·

·TAVAIS·ESTVS·

·TAIZALV· ·PROMO·
·TORIV·

·CELNIVS·FLV·

· 19 · PARALELLVS ·

·DIEFERTAB·EQINOCTIALI·HORIS · SEX ·
HABENS·MAXIMVM·DIEM·HORARVM·18·

48

·OCATVRAC·
·TONIVM·

·BORACVM·
·LEGIO·
·CTRIX·

·VRIVM·

·DVNVS·
·SINVS·

47

·PARISI·

·PORTVOSVS·
·SINVS·

·PETVARIA·

·OCELVM·
·PROMONTO·

·BVS·FL·

·SAXONVM·
·INSVLE·

· 18 · PARALELLVS ·

·DIFFERT·AB·EQINOCTIALI·HORIS·4·½·
HABENS·MAXIMVM·DIEM·HORARVM·17·½·

46

OCEANVS · GERMANICVS ·

·ALINE·

·ESTVS·
·METARIS·

·VENTA·

44

·IMENI·

·VS·FLV·

·MISAIS·
·ESTVS·

·TOLIAPIS·

· 17 · PARALELLVS ·

·CANTII·
·RVTVPIE·

·COVENNOS·

·LONDINIV·
·CATIV·PRO·
·MONTORIVM·

·DIFFRT·AB· EQVINOCTIAL·HORIS·
·5·HABENS·MAXIMVM·DIEM·
·HORARVM·17·

49

·DARVERNVM·
·NOVVS·
·PORTVS·

·TRISANTO·

· MAGNE · GERMANIE · PARS ·

43

·RHENVS· FLV·

· GALLIE · BELGICE · PARS ·

4Z

· 16 · PARALELLVS ·

ZI	ZZ	Z3	Z9	Z4	Z6	Z7	Z8	Z9	30	31	3Z	33	41

·DIFFERT·AB·EQVINOCTIALI·
·HORIS·4·½· HABENS·
·MAXIMV·DIEM·HORARV·16·½·

·VNVS·GRADVS·LONGITVDINIS·VALET·MILIARIA·42·

·MILIARIA·44·¼·

·MILIARIA·47·

·MILIARIA·40·

III.   PTOLEMÆUS ROMÆ 1490.

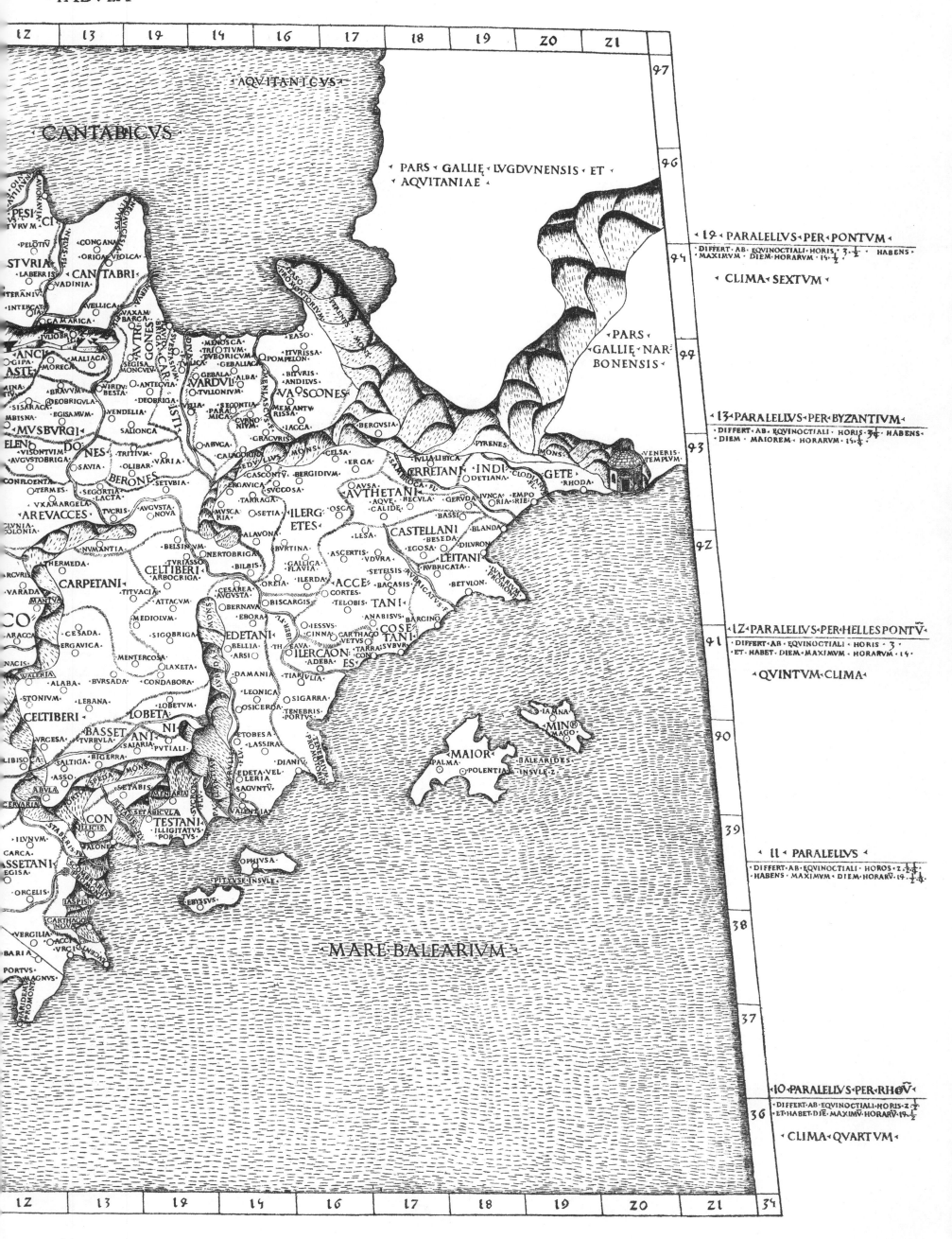

The map shows the following labels:

AQVITANICVS

CANTADICVS

· PARS · GALLIE · LVGDVNENSIS · ET · AQVITANIAE ·

PARS GALLIE NARBONENSIS ·

· 19 · PARALELLVS · PER · PONTVM ·
· DIFFERT · AB · EQVINOCTIALI · HORIS · 3 · ½ · HABENS ·
· MAXIMVM · DIEM · HORARVM · 19 · ½ ·

· CLIMA · SEXTVM ·

· 13 · PARALELLVS · PER · BYZANTIVM ·
· DIFFERT · AB · EQVINOCTIALI · HORIS · 3 ¼ · HABENS ·
· DIEM · MAIOREM · HORARVM · 15 · ¼ ·

· 12 · PARALELLVS · PER · HELLESPONTV ·
· DIFFERT · AB · EQVINOCTIALI · HORIS · 3 ·
· ET · HABET · DIEM · MAXIMVM · HORARVM · 14 ·

· QVINTVM · CLIMA ·

· 11 · PARALELLVS ·
· DIFFERT · AB · EQVINOCTIALI · HOROS · 2 ¾ ·
· HABENS · MAXIMVM · DIEM · HORARV · 14 ¾ ·

· MARE · BALEARIVM ·

· 10 · PARALELLVS · PER · RHOV ·
· DIFFERT · AB · EQVINOCTIALI · HORIS · 2 ½ ·
· ET · HABET · DIE · MAXIMV · HORARV · 14 ½ ·

· CLIMA · QVARTVM ·

Regions and places: STVRIA, CANTABRI, ANCI, ASTE, MVSBVRGI, ELENO, NES, BERONES, AREVACCES, CARPETANI, CO, CELTIBERI, BASSET, ANI, CON, TESTANI, ASSETANI, VARDVLI, VASCONES, ERRETANI, AVTHETANI, ILERGETES, CASTELLANI, LEITANI, ACCETANI, COSETANI, ILERCAONES, EDETANI, GETE, INDI, MAIOR, PALMA, POLENTIA, MINO MAGO, BALEARIDES INSVLE 2, OPHIVSA, PITYVSE INSVLE, EBYSSVS

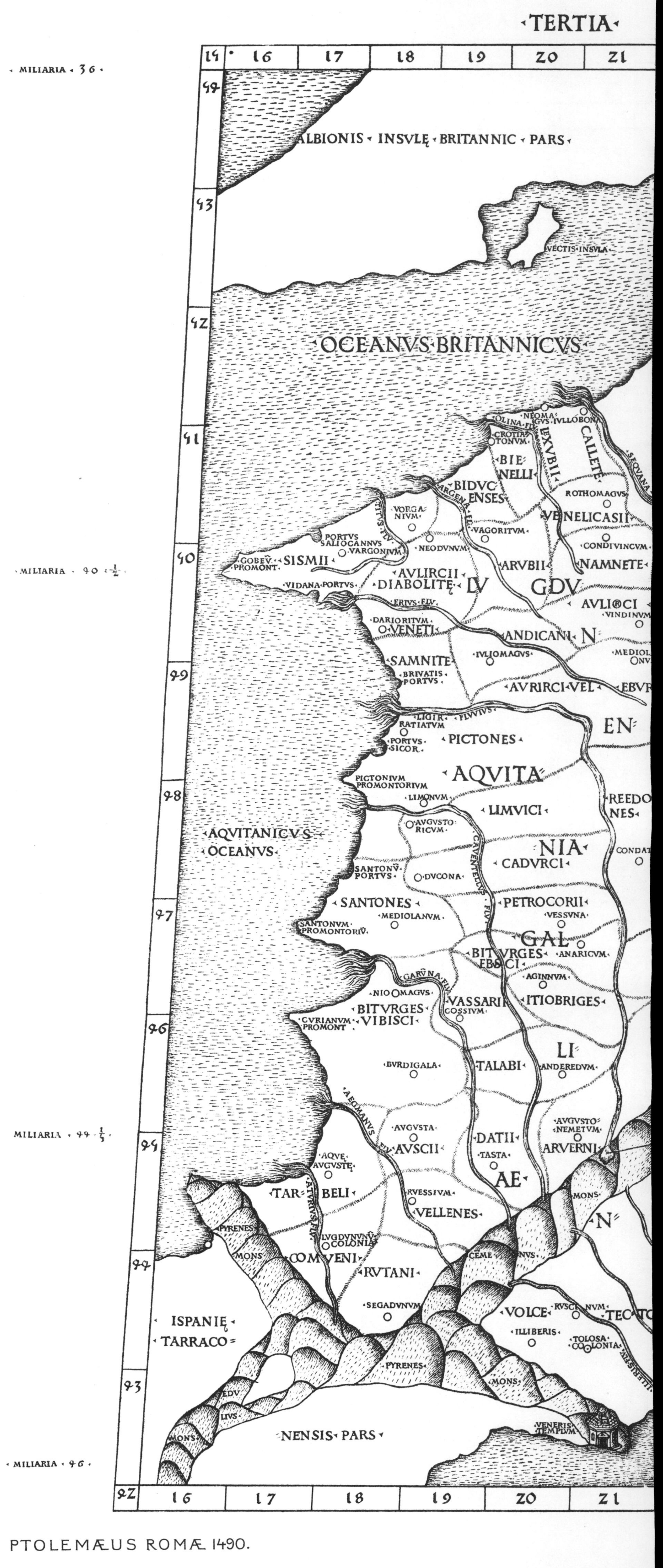

·MILIARIA · 36·

19 16 17 18 19 20 21

44

·ALBIONIS · INSVLÆ · BRITANNIC · PARS·

43

·VECTIS·INSVLA

42

OCEANVS·BRITANNICVS·

41

NEOMA
GVS·IVLLOBON
OLINA·FL
CROTIA
TONVM
LEXVBII·
BIE
NELLI·
BIDVC
ENSES·
ROTHOMAGVS·
VENELICASII·

·MILIARIA · 90 · ½·

40

VORGA
NIVM
NEODVNVM·
VAGORITVM
CONDIVINCVM·

PORTVS
SALIOGANNVS·
VARGONVM
SISMII·
GOBEV
PROMONT.
AVLIRCII·
DIABOLITÆ·
LV
GDV
NAMNETE·

·VIDANA·PORTVS·
AVLIOCI·
VINDINVM·

ERIVS·FLV·
DARIORITVM
O·VENETI·
ANDICANI·
N·
MEDIOL
ONV·

SAMNITE·
BRIVATIS·
PORTVS·
IVLIOMAGVS·

49

AVRIRCI·VEL·
EBVR

·LIGIR·
RATIATVM
·FLVVIVS·
EN·

PORTVS
·SICOR·
PICTONES·

PICTONVM
PROMONTORIVM
AQVITA·

48

·LIMONVM·
LIMVICI·
REEDO
NES·

AQVITANICVS·
OCEANVS·
AVGVSTO
RICVM·
CONDAT

NIA·
CADVRCI·

SANTONV
PORTVS·
O·DVCONA·

47

·SANTONES·
MEDIOLANVM·
PETROCORII·
VESSVNA·

SANTONVM
PROMONTORIV·
GAL
BITVRGES
EBOICI·
ANARICVM·

GARVNA·FL
AGINNVM·

NIO·OMAGVS
VASSARII·
ITIOBRIGES·
BITVRGES·
COSSIVM·

46

CVRIANVM
PROMONT·
VIBISCI·
LI·
ANDEREDVM·

·BVRDIGALA·
·TALABI·

AEGNANVS·
AVGVSTO
NEMETVM·

·MILIARIA · 44 · ⅓·

45

AVGVSTA
DATII·
ARVERNI·

AQVE
AVGVSTÆ·
AVSCII·
TASTA·

AE·

TAR·BELI·
RVESSIVM·
MONS·

VELLENES·
N·

PYRENES·
LVGDVNVM
COLONIA·
CEME
NVS·

MONS·

44

COM VENI·
RVTANI·

ISPANIÆ·

SEGADVNVM·
VOLCE·
RVSCINVM·
TECTO

·TARRACO·
ILLIBERIS·
TOLOSA
COLONIA·

PYRENES·

43

EDV
LIVS·
MONS·

NENSIS·PARS·
VENERIS·
TEMPLVM·

·MILIARIA · 46·

42 16 17 18 19 20 21

IV. PTOLEMÆUS ROMÆ 1490.

ZZ	Z3	Z9	Z4	Z6	Z7	Z8	Z9	

GERMANICVS

·GESORIACVM·
·NAVALE·
·ITIVM·PRO-
MONTORIV·

·MORINI·
·TARVANA·

·ATVACVTV·

·MOSA·FLV·

·LVGODVNVM·

·RHENVS·FLV·

43

·AMBIANI·

·SAMAROBRIGA·

·CASTELLVM·

MENAPII·

·BVTAVO·
DVRVM·

42

ATRIBATII·

·METACVM·

BELGI· CA·

·CESAROMAGVS·
·BELVACI·

NERVSII·

BAGACVM·

GAL·

·VEGERRA·

·LVPPIA·

·AGRIPPI·
NENSIS·

·FLVDIS·FLV·

·BATAVI·

·BONNA·

41

L·

ROMANDI·

I·

·VBANECTI·

·ROTOMAGVS·

·AVGVSTA·
·ROMANDORVM·

·TRAIANA·
LEGIO·

·MOGONTIA·
CVM·

40

AE·

INGENA·

ABRINCAV·

·OBRICVS·FLVVIVS·

·NEOMAGVS·

·RVFINIANA·

TEL·
CENOMAN·

TRIBERI·

·AVGVSTA·
·TRIBERORVM·

NEMETI·

·VICI·

·VESSONES·

·AVGVSTA·
·VESSONVM·

·REMI·

·DVROCOT·
TORVM·

·BORRETO·
MAGVS·

SIS·

VANGIONES·

·AVTRICVM·

·LVCESIA·VEL·
·PARISIVS·

PARISII·

·ARGENTO·
RAGVM·

·BREVCO·
MAGVS·

·CAENVTE·

·CENABVM·

·AVGVSTO·
BONA·

GERMANIA · SVPERIOR ·

TRIBONI·

·ELCEBVS·

98

GALLIE·

·TRICCASII·

MEDIOMATRICES·

·ARGENTOVARIA·

·AVGVSTA·
·RAVRICVM·

·SENNONES·
O·AGEDECVM·

·DIVODVRVM·

RAVRICI·

97

·CEMENORVM·

·MONTES·

·TVLLIVM·

LEVCI·

·NASIVM·

·ADVLAS·
MONS·

·CESARODVNVM·

TVRVPII·

·AVGVSTODV·
NVM·

·IVRAS·
SVS·

·ANDVMANTVN·

·LONGONES·

·GAN·NODVRV·

EGVSIACI·

·NIOMAGVS·

·MO·
NS·

·VISONTIVM·

ELVETII·

·FORVM·
·TYBERII·

·SALVTH·SEND·

96

FORVM·SE·
STATVM·

MELDE·
IATINVM·

·RVDV·
MNA·

·CAEVLLINVM·

VADICASSI·

·DIATAVVM·

·IVRAS·
SVS·

·MONS·

EDORVM·
GENS·

SEQVANI·

·EQVESTRIS·

·AVATICVM·

·LVGDVNVM·
METROPOLIS·

·ARAR·FLV·

·DVBIS·F·

·RHODANVS·FLVVIVS·

·LEMANVS·LACVS·

·NEOMAGVS·

·VIENA·

·ALLOBRIGES·

·FORVM·NE·
RONIS·

·TRICASTINI·

·DINIA·

·VINDOMAGVS·

·ISABA·FLV·

VOCONTII·

·SENTII·

·NEMAV·
SVM·

·VALENNTIA·
COLONIA·

·ACVSIA·
COLONIA·

MEMINI·

A· R·

·VASION·

SEGALAVNI· CAVARII·

BONENSIS · GALLIAE·

·ISSERO·

VOLCE·ARICOMI·

·AVENION·
COLONIA·

·DRVENTIA·FLVVIVS·

·ARAVSION·

·CARCASO·

·MASSILIA·
GRECA·

ELVCOTI·

·ADVLAS·
MONS·

99

SACES·

·TARVSCON·

·ERNAGINVM·

·GLANVM·

·ALBAVGVSTA·

·DECIATII·

·CHAETIRAE·

·SALIES·

·ARELATVM·
COLONIA·

·COLONIA·
·MARITIMA·

·AQVE·
·SEXTIE·

ANATIALI·
MASSILIA

·ANTIPOLIS·

·ALPES·
·MONS·

·TVROE·
TIVM·

·CITHARISTA·
PROMONTORIVM·

·OLBIA·

·FORVM·
·IVLIVM·

· ITALIAE ·
· PARS· 93

·AGATA·POLIS·

·ARAVRIS·FLV·

·AGATHA·

·BLASCON·

·SCICADES·

·VERO·
·INSVLA·

MARE

GALLICVM

ZZ	Z3	Z9	Z4	Z6	Z7	Z8	Z9	92

GERMANIA · INFERIOR ·

· MAGNE · GERMANIE · PARS ·

SEQVANA · FLVVIVS ·

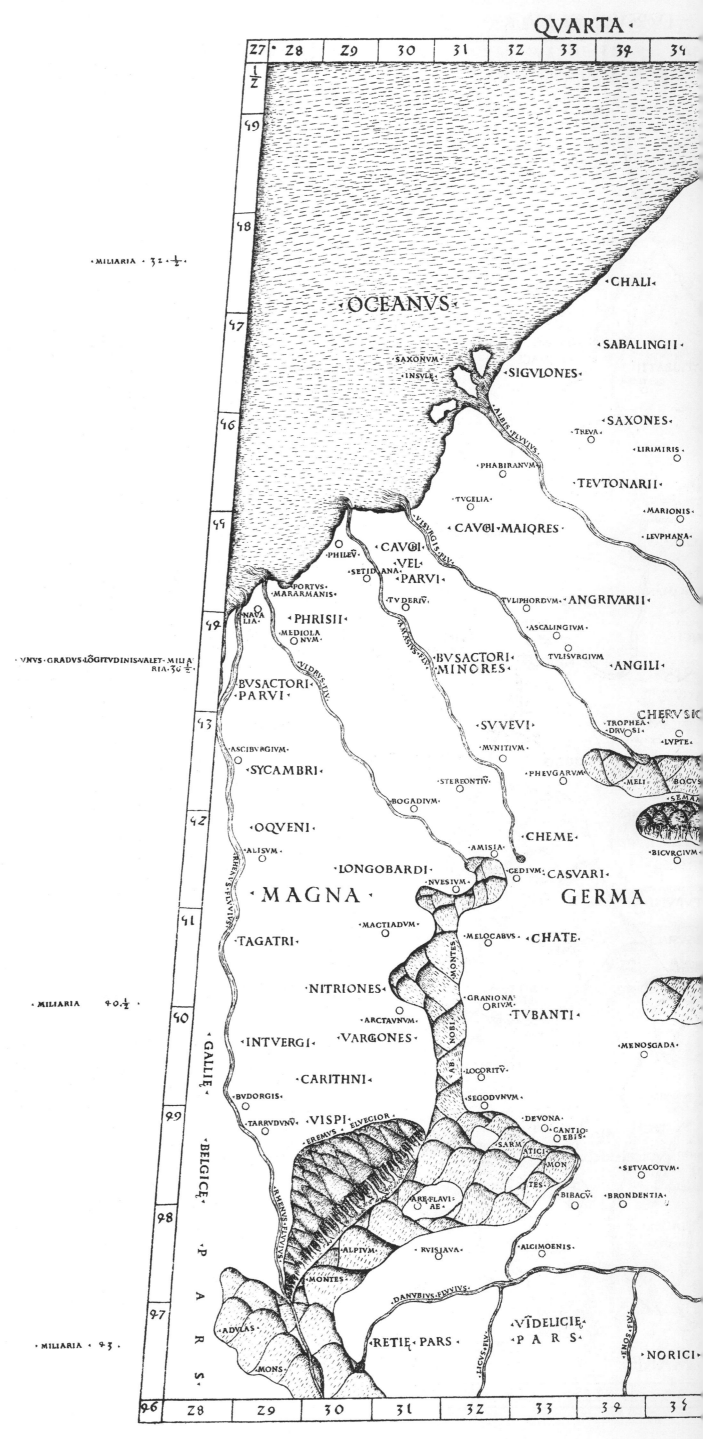

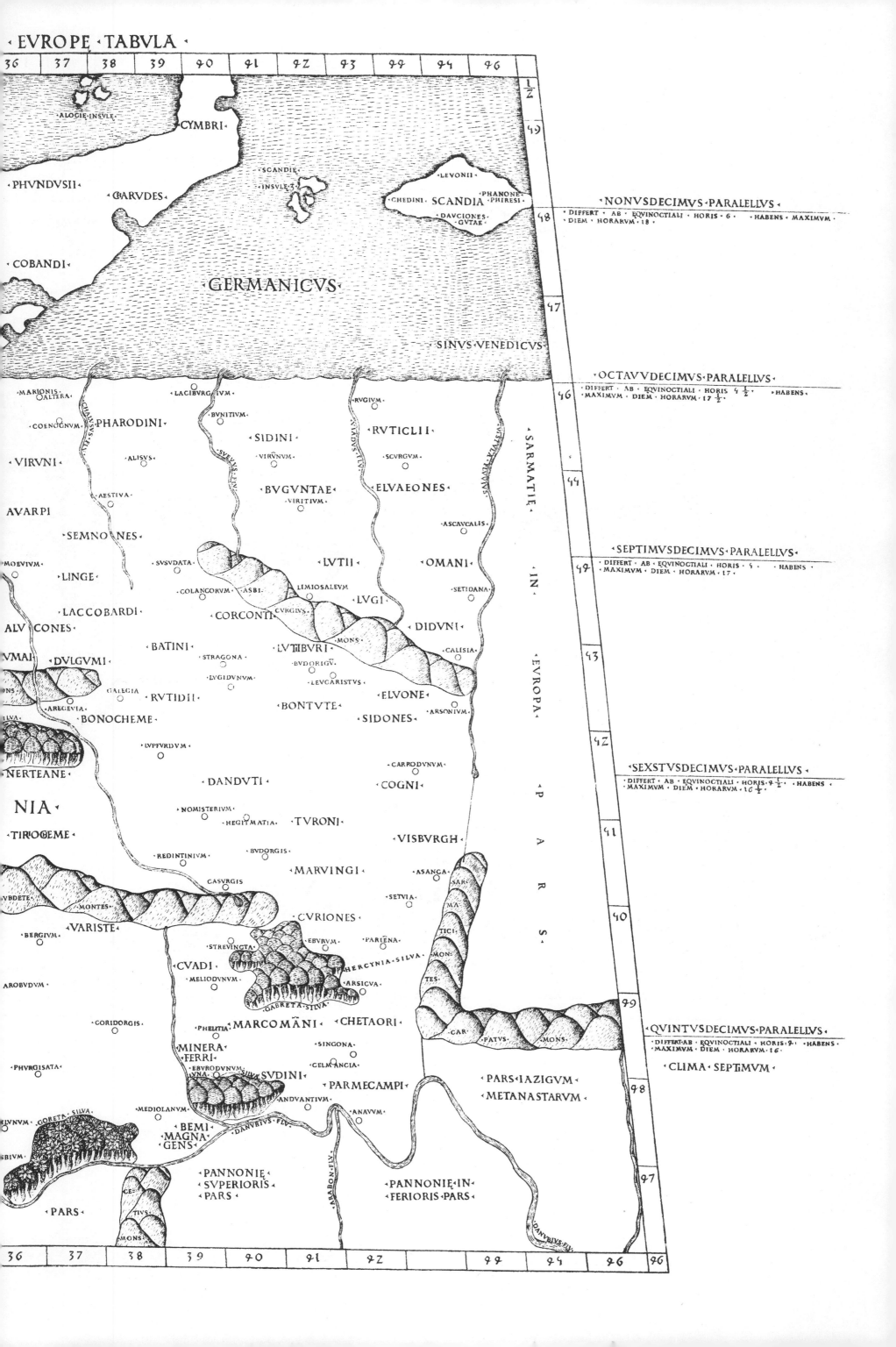

· ALOGIE · INSVLE ·

· CYMBRI ·

· PHVNDVSII ·

· GARVDES ·

· SCANDIE ·
· INSVLE · 3 ·

· LEVONII ·
· PHANONE ·
· CHEDINI · SCANDIA · PHIRESI ·
· DAVCIONES ·
· GVTAE ·

· COBANDI ·

GERMANICVS ·

· SINVS · VENEDICVS ·

· NONVSDECIMVS · PARALELLVS ·
· DIFFERT · AB · EQVINOCTIALI · HORIS · 6 · HABENS · MAXIMVM ·
· DIEM · HORARVM · 18 ·

· MARIONIS ·
· ALTERA ·

· LACIBVRGIVM ·

· RVGIVM ·

· CHALVSVS · FLV ·
· COENOGNVM ·

PHARODINI ·

· BVNITIVM ·

· SIDINI ·

RVTICLII ·

· SVEVVS · FLV ·

· VISTLA · FLVVS ·

· SARMATIE ·

· OCTAVVDECIMVS · PARALELLVS ·
· DIFFERT · AB · EQVINOCTIALI · HORIS · 4 ½ · HABENS ·
· MAXIMVM · DIEM · HORARVM · 17 ½ ·

· VIRVNI ·

· ALISVS ·

· VIRVNVM ·

· SCVRGVM ·

· AESTIVA ·

BVGVNTAE ·

ELVAEONES ·

· AVARPI ·

· VIRITIVM ·

· SEMNONES ·

· ASCAVCALIS ·

· IN ·

· SEPTIMVSDECIMVS · PARALELLVS ·
· DIFFERT · AB · EQVINOCTIALI · HORIS · 4 · HABENS ·
· MAXIMVM · DIEM · HORARVM · 17 ·

· MOEVIVM ·

· LINGE ·

· SVSVDATA ·

· LVTII ·

· OMANI ·

· SETIDANA ·

· LACCOBARDI ·

· COLANCORVM · ASBI ·

· LIMIOSALEVM ·

· LVGI ·

· EVRGIVS ·

· EVROPA ·

· ALVCONES ·

CORCONTI ·

· DIDVNI ·

· VMAI ·

· BATINI ·

· LVTIIBVRI ·

· MONS ·

· CALISIA ·

· DVLGVMI ·

· STRAGONA ·

· BVDORIGV ·

· GALEGIA ·

· LVGIDVNVM ·

· LEVCARISTVS ·

· RVTIDII ·

· ELVONE ·

· AREGEVIA ·

BONOCHEME ·

· BONTVTE ·

· ARSONIVM ·

· SIDONES ·

· P ·

· LVPFVRDVM ·

· NERTEANE ·

· CARRODVNVM ·

· DANDVTI ·

· COGNI ·

· A ·

NIA ·

· NOMISTERIVM ·

· HEGITMATIA ·

· TVRONI ·

· R ·

· SEXSTVSDECIMVS · PARALELLVS ·
· DIFFERT · AB · EQVINOCTIALI · HORIS · 3 ½ · HABENS ·
· MAXIMVM · DIEM · HORARVM · 16 ½ ·

· TIRIOGEME ·

· REDINTINIVM ·

· BVDORGIS ·

· VISBVRGH ·

· S ·

· CASVRGIS ·

MARVINGI ·

· ASANGA ·

· SAR ·

· VBDETE ·

· SETVIA ·

· MA ·

· MONTES ·

· CVRIONES ·

· TICI ·

· VARISTE ·

· BERGIVM ·

· STREVINGTA ·

· EBVRVM ·

· PARIENA ·

· MON ·

· CVADI ·

HERCYNIA · SILVA ·

· TES ·

· MELIODVNVM ·

· ARSICVA ·

· AROBVDVM ·

· GABRETA · SILVA ·

· QVINTVSDECIMVS · PARALELLVS ·
· DIFFERT · AB · EQVINOCTIALI · HORIS · 3 · HABENS ·
· MAXIMVM · DIEM · HORARVM · 16 ·

· CORIDORGIS ·

· PHELITIA · MARCOMANI ·

· CHETAORI ·

· CAR ·

· PATVS ·

· MONS ·

MINERA
FERRI ·

· SINGONA ·

· CLIMA · SEPTIMVM ·

· PHVRGISATA ·

· EBVRODVNVM · SILVA ·

· CELMANCIA ·

· VNA · SVDINI ·

PARS · IAZIGVM ·

· MEDIOLANVM ·

· ANDVANTIVM ·

· PARMECAMPI ·

METANASTARVM ·

· LVNVM · GORETA · SILVA ·

· BEMI ·
MAGNA ·
GENS ·

· DANVBIVS · FLV ·

· ANAVVM ·

· SBIVM ·

· ARABON · FLV ·

· PANNONIE ·
· SVPERIORIS ·
· PARS ·

· PANNONIE · IN ·
· FERIORIS · PARS ·

· CE ·

· PARS ·

· TIVS ·

· MONS ·

· DANVBIVS · FLV ·

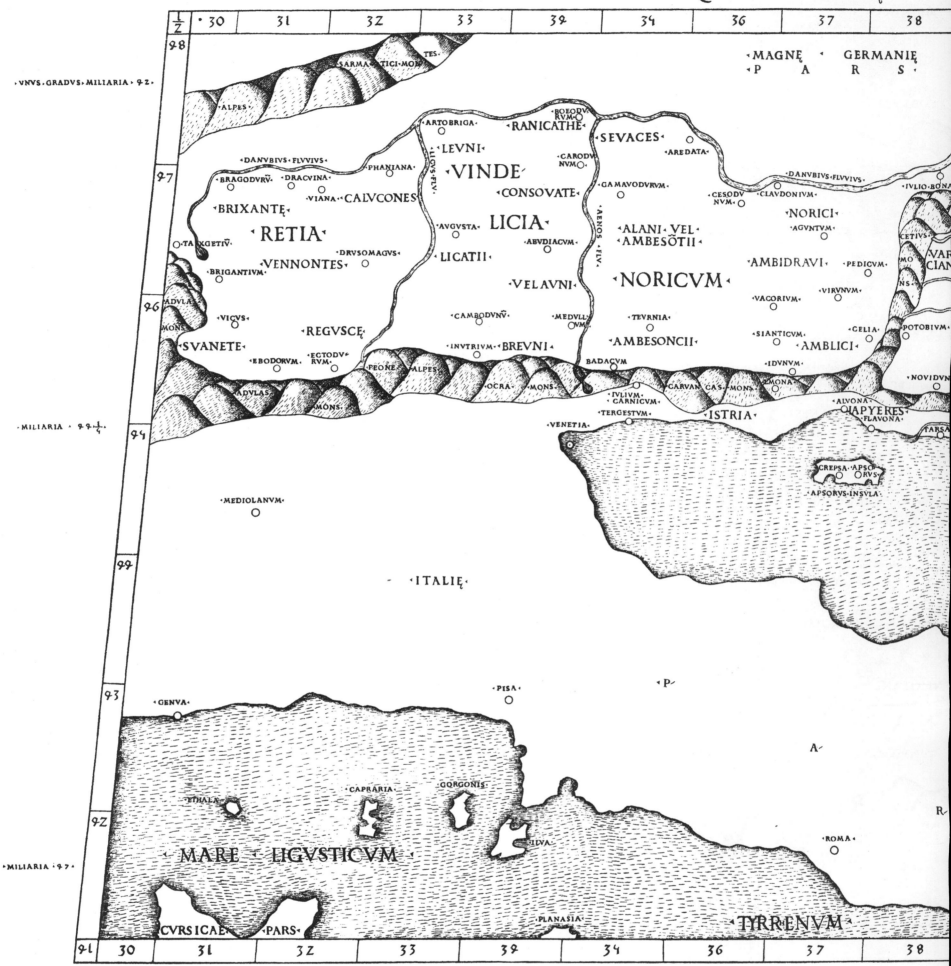

· VNVS · GRADVS · MILIARIA · 42 ·

· MILIARIA · 44 1/4 ·

· MILIARIA · 47 ·

MAGNE · GERMANIE
· P   A   R   S ·

·SARMA·TICI·MON·TES·

·ALPES·

·ARTOBRIGA· ·BOEORV·RVM·
·RANICATHE·
·LEVNI· ·SEVACES·
·DANVBIVS·FLVVIVS· ·PHANIANA· ·ILLVS·FLV· ·CARODV·NVM· ·AREDATA· ·DANVBIVS·FLVVIVS· ·IVLIO·BONA·
·BRAGODVRV· ·DRACVINA· VINDE ·GAMAVODVRVM· ·CESODV·NVM· ·CLAVDONIVM· ·NORICI· CETIVS·
·VIANA· ·CALVCONES· ·CONSOVATE· ·AGVNTVM· MO VAR
BRIXANTE ·AVGVSTA· LICIA ·ALANI·VEL· CIA
·RETIA· ·ABVDIACVM· ·AMBESOTII· ·AMBIDRAVI· ·PEDICVM· NS
·TA XGETIV· ·DRVSOMAGVS· ·AENOS·FLV· ·VIRVNVM·
·BRIGANTIVM· VENNONTES· ·LICATII· NORICVM ·VACORIVM·
·ADVLA· ·VELAVNI· ·TEVRNIA· ·SIANTICVM· ·CELIA· POTOBIVM
MONS ·VICVS· REGVSCE ·CAMBODVNV· ·MEDVLL·VM· ·AMBESONCII· ·AMBLICI·
·SVANETE· ·EBODORVM· ·ECTODV·RVM· ·INVTRIVM· BREVNI ·BADACVM· ·IDVNVM· NOVIDVM
·PEONE· ·ALPES· ·EMONA·
ADVLAS ·OCRA· ·MONS· ·CARVAN·CAS·MONS· ·ALVONA· IAPYERES
·MONS· ·IVLIVM·CARNICVM· ISTRIA ·FLAVONA·
·TERGESTVM· TARSA
·VENETIA·

CREPSA·APSO·RVS·
·APSORVS·INSVLA·

·MEDIOLANVM·

·ITALIE·

·P· A·

·PISA·

·GENVA· R

·CAPRARIA· ·GORGONIS·
·ETHALA· ·ROMA·
·ILVA·
· MARE = LIGVSTICVM · ·PLANASIA· · TYRRENVM ·
·CVRSICAE· ·PARS·

VI.   PTOLEMÆUS ROMÆ 1490.

39   90   91   92   93   94   94   96   97

CLIMA·SEPTIMVM·

28

CARPIS·
SALVA·
AQVI·CV·

·ARAVISCI·

IAZIGVM·ME
TANASTAV·PARS·

CARPA
TVS· MONS

·CHERT
OBALVS·
·PHLEXVM·

AMÃTINI

SALI
NVM·

27

·SACARBAN
TIA·
·CARNVS·

CYTNI

VLLINA·

·CVRTA·

LVSSONI
V.M.

·SAVARIA·
·AZALI·
·RISPIA·

PANNO
NIA·SVPE
RIOR· IASSII·

HERCVNIATE

VAGONTIVM·
LVGIO
NVM

PANNONIA·

DACIE·

26

·SALA·
·MOROELA·
·BOEI· ·SAVARIA· FLV·
ROS·FLV· ·CORRODVNV·

PTORIVM·

BOLENTIVM·
MAGNIANA·

BERBIS· IVOLLVM·

ANDIANTES·

MVRSELLA·
MVSIA
COLNIA·

IN

TEVTOEVR
GIVM·

TIBISCVS·FLV·

PARS·

·BONONIA·
COLETIANI
VDRIA·

·LENTVDVM·

ODERIATES·
OLIMACVM·

SOGORA·
SERBINVM·

CERTISSA·

ACVMIN
CVM·

BIBALI·

SCISCIA·LATOBICI·SISOPA·
MESEI·
NIVM·

VISONTIVM·
BE·BII

BREVCI·

MORSONA·

FER·

SAVS·FLV·

SALLIS·

CORNACVM·

I·

PER·PONTVM·
19·PARALELLVS·

SCORDISCI·

ALBA
NVS· MONS·

ARVCCIA·
ARDOTIV·

MON TES·

SIRMIVM·

OR·

RITTIVM·

DIFFERT·AB·EQVINOCTI
ALI·HORIS·3½·HABENS·
DIEM·MAIOREM·HORARV·15½

24

CLIMA·SEXTVM·

MAEZAEI·
VS·FLV·VOLCE·
RA· SENIA· LOPSICA·

AVSANCALA·

DITIONES·

ADRA·

SIDRONA·

BASSIANA·
TAVRV·
RVM·

MYSIAE·

DANVBIVS·FLV·

PEDANIVS·P·

CVRVM·

DINDARII·

BLANONA·

FVLFI
NVM· CVRICTA·
CVRICV· INSVLA·

ORTOPV·
LA·
VETIA·

ILLYRIS·
SALVIA·

ARGIRV·
TVM· VARVARIA·

CERAVNII·
ARAVZONA·

SEV·
ASSESIA·

BVRNVM·

LYBVR
N·
I·

NEDINVM·

DRINI·FLV·

HERONA·

DELMI
NIVM·

SVPERIORIS·

MORICLVS·FLV·

PARS·

23

PER·BYZÃTIVM
13·PARALELLVS·

CORINI
VM· ENONVM·

DERII·

OVPORVM·

A·

COMENII·

VARDEI·

DIFFERT·AB·EQVI
NOCTIALI·HORIS·
3¾·HABES·DIE·MAIOR
EM·HORARV·15¾·

ARBA·

IADER
COLONIA·

SCARDONA·INSVLA·
COLENTVM·

SCARDONA·

ANDECRIVM·

DAVRSII·

DA·

EQVM·
COLONIA· SALONIANA·

L·
SARDOATE·

SINVS·HADRIATICVS·

TICIS·FLV·

SICVM·

ISSA·

SALONA
COLONIA·

ALETA·

NARENSII·

SARDO
NIVS·

DIOMEDEE
INSVLAE·4·

EPETVM·

NARBONA·
COLONIA·

MA·

MONS·

TRAGVRIVM·

PIGVN
TIVM·

ENDERVM·

DVCLEATE·

22

PHARIA·

SICVLOTE·
ONEVM·

NARBON·
FLV·

CHINNA·

T·
EPIDAVRVS·

DOCLEA·

SIPARVN
TVM·

S·

CORCIRA·
NIGRA·

RIS·
INV·

ASTRVIVM·

RIZANA·

SINVS·RISO·
NISCVS·

BVLVA·

TERME
DA·NA·

PIRVSE·

SCAR·
DVS·

I·

SCIRTONES·

MO·

MELIGINA·

VLCINIVM·

A·

NEAPOLIS·

DRILON·
LISSVS·

EPICARIA· EMINACIVM·

·MACEDONIE·PARS·

CLIMA·4·

39   90   91   92   93   94   94   96   97   91

VII. PTOLEMÆUS ROMÆ 1490.

Map labels (top margin coordinates): 36 37 38 39 90 91 92 93

· ILLYRIDIS · SIVE · LYBVRNIE · PARS ·

SINVS·hADRIATICS

TR IA
FORMION·FLV.
PIQVEN
TVM
ALVVM
PARENTIVM
POLA
APSORVS
CVRICTA
SCHADONA

AVRVM·FANVM
SENAGALLICA
SEM NONES
SVASA
OSTRA
TRAIANA
ANCO
FORVMSE
PRONII
NV
MANA
VRSA
BALVIA
SEPTE
PEDA
VMBRI
AISSIS
POTENTIA
ISVLVM
TVFICVM
CAMARINVM
PICENI
CVPRA
MO.TANA
CVPRA
MARITIMA
DIOMEDEE
INSVLE
ISSA

PERVSIA
CETINA
FIRMVM
HADRIA
ASCVLVM
CASTRVM
SISIVM
NVCERIA
SPOLETVM
SABINI
NVRSIA
CLITEN
VAEX
SEV·CA
PRA
MARVCINI
PREGVTII
BERETRA
PINNA
AVIA
TEATEA
ANGOLVS
ORTON
FORVM
FLAMINII
AMERIA
MEVANIA
VILVBRI
CALSIO
LE
MARSI
ALFABVCE
LIS
INTERA
MNIA
VESTINI
CVRFELINIV
SVLMO
APENN
FERENTIA
ESOLVM
TVRDER
EQVICVLI
OBRICOLVM
AMITERNV
PELIGNI
SVDERNVM
VICVS·ELBII
TVSCI
TIBVR
L
APENNINVS
NVMENTVM
ATINA
BOIANVM
CARACENI
SARVS·FLV.
ANXANVM
TARQVINE
BLERA
FALERIVM
LATINI
PNESTE
TVSCVLVM
FIDENE
AQVINV
MONS
SET·NVM
ESERNIA
SAMNITES
TVTICV
AVFIDENA
ISTONIVM
BVBA
GRAVISCE
ONEPETA
FORVM
LANVRIVM·TREBA
FRVSIN
MINTVRN
SORA
TEANVM
CALES
ALIFA
TELESIA
LARINVM
FRENTANI
CASTRV·CLAVDII
NOVVM
PIRGI·ALSIVM
ROMA
ARICIA
ANANIA
FERENTINV
TREBVLA
SVESSA
CASILINV
FORVM
CAPVA
BENEVEN
TVM
CLVSIV
ATERNVS·FLV.
AP
HYRIVM
OSTIA
ARDEA
VELIPSVM
VELITRE
PRIVERNV
FVNDI
VENA
FRVM
SECTA
SOESSA
POPILII
ABELLA
ARPE
NVCERIA
APPVLORV
VIBRA
NAC.
GARGANVS
MONS
ANTI
VM
CLOSTRA
CIRCEVM
PROMONTORIV.
TARRA
CINE
FORME
CAMPANI
LITERNV
MISENV
CVME
ATELLA
NEAPOLIS
PVTE
OLI
SARNVS·FLV.
ABELLINV
AECVLANVM
APENESTE
MONS
PER·HELLESPÕTV
IZ·PARALELLVS
PONTIA
PANDATORIA
HITHECVSA
PROCYTA
PERTENOPE
NOLA
SVRENTVM
MINERVEPRO
MONTORIV
PICEN
TINI
SALERNV
IRPINI
FRATVOLVM
NVCERIA
TEANV
P
ARDONIA
V
SIPA
SALPIE
CAPREA
SILARVS·FLV.
COMPSA·POTE
TIA
AQVILONIA
CANVSIVM
AVFIDV
FLV.
BARIVM
BLANDA
VLCI
VENV
SIA
GELIA
AETOLI
EGNACIA
TYRRENVM
LVCANI
PESTVM
GRVMENTVM
VELIE
SYRENV·INSVLE
BVXENTVM
MAGNA
GRECIA
CROTON
TAREN
RVDIA·BRVND
ISIVM
CALA
LVPIE
STRONGYLE
LAVS·FLV.
THVRIVM
SINVS
TARANTINVS
VAS·VERETVM
OTVM
BRIA
OTVRNA
DIDYME
HICESIA
NVMISTRVM
LACINIVM
PROMONT
BAVROTA
SALETINI
NERITV
ALETIV
HYDRA
AEOLII
TEMPSA
COSETIA
PETILIA
BRVTII
SALENTINVM
PROMONTORIV
ERICODES
LIPARA
EVONIMOS
SCOPVLVS
SCVLATIV
VSTICA
PHOERICODES
VVLCANI
SINVS
ISPONIASTA
SINVS
SCILACEVS
BRVTII
SCYLEVM
PROMONT
REGIVM·IV
MESENA
ZEPHIRIV
PROMONT
LOCRIS
HERCEPETRA
PROMONTORIV
HADRIATICVM·MARE
· SICILIE · INSVLE · PARS ·

Map labels (bottom margin coordinates): 36 37 38 39 90 91 92 93

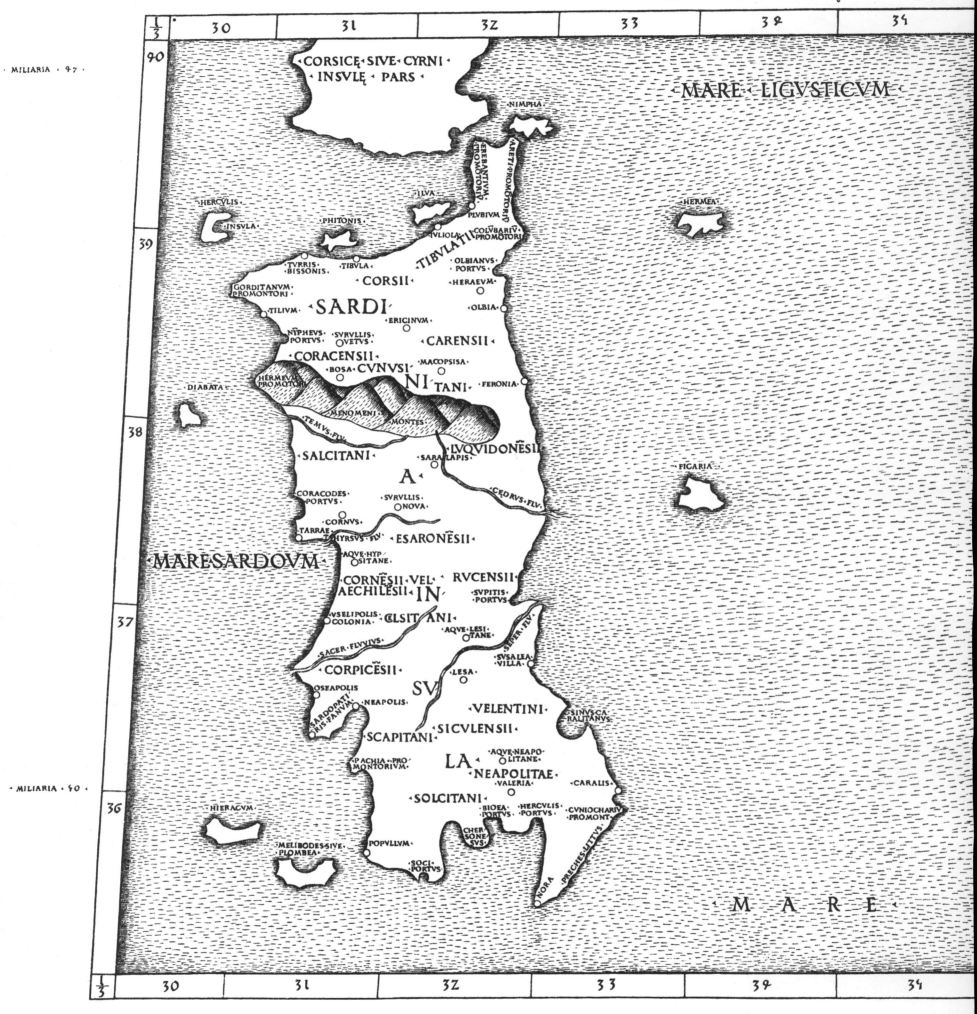

· MILIARIA · 47 ·

· MILIARIA · 40 ·

VIII. PTOLEMÆUS ROMÆ 1490.

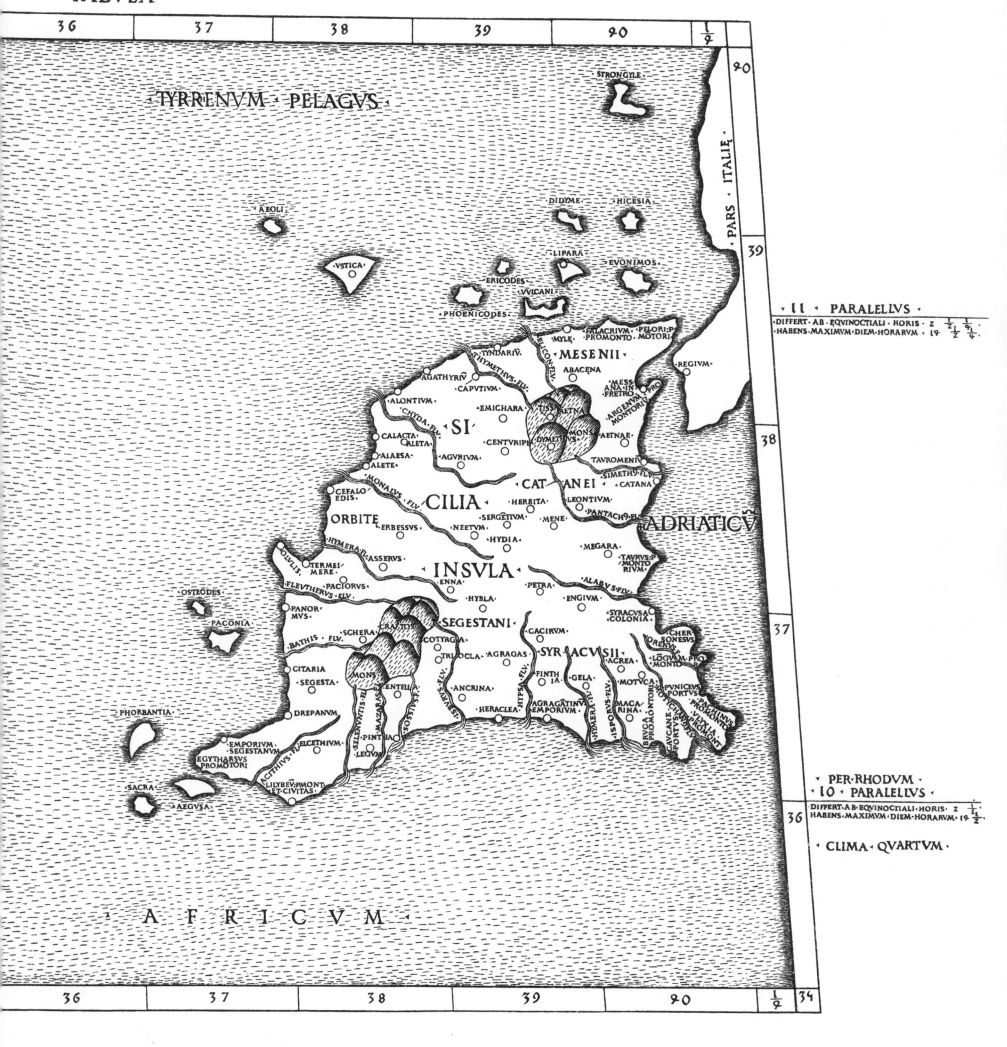

| 36 | 37 | 38 | 39 | 90 | ¼ | 90 |

TYRRENVM · PELAGVS ·

· STRONGYLE ·

· PARS · ITALIE ·

· 39

· AEOLI ·

· DIDYME · · HICESIA ·

· VSTICA · · LIPARA ·
· EVONIMOS ·

· ERICODES ·
· VVLCANI ·
· PHOENICODES ·

· FALACRIVM · · PELORI · P ·
· MYLE · PROMONTO · MOTORI ·
· TYNDARIV · MESENII ·
· THYMETHVS · FLV · ABACENA ·
· AGATHYRIV · · REGIVM ·
· CAPVTIVM · ETNA · · MESS ·
· ALONTIVM · · EMICHARA · TISA · ANA · IN ·
SI · MONS · ARGENTA · FRETRO ·
· CHYDA · FLV · DYMETHVS · ARGENTA · PRO ·
· CALACTA · · CENTVRIPE · AETNAE · MONTORI ·
· ALETA · · AGVRIVM · · TAVROMENIV ·
· ALAESA · CICILIA · · SIMETHV · FLV ·
· ALETE · · CATANEI · · CATANA ·
· CEFALO · · MONALVS · FLV · · HERBITA · · LEONTIVM ·
· EDIS · ORBITE · · SERGETIVM · · MENE · · PANTACH9 · FLV · · ADRIATICV ·
· ERBESSVS · · NEETVM · · MEGARA ·
· HYMERA · FL · · ASSERVS · · HYDIA · · TAVRVS ·
· OVLIS · INSVLA · · MONTO ·
· TERMEI · ENNA · · PETRA · RIVM ·
· MERE · · ALABVS · FLV ·
· OSTEODES · · FLEVTHERVS · ELV · PACIORVS · · HYBLA · · ENGIVM ·
· PANOR · · SYRACVSA ·
· PACONIA · MVS · SEGESTANI · · COLONIA ·
· SCHERA · · COTYRGA · · CACIRVM · · CHER ·
· BATHIS · FLV · · CITARIA · · TRIOCLA · · AGRAGAS · SYRACVSII · SONESVS ·
· SEGESTA · · MONS · · ACREA · · LOGIVM · PRO ·
· PHORBANTIA · · ENTELLA · · ANCRINA · · FINTH · · GELA · · MONTO ·
· IA · · MOTVCA · · PVNICEVS ·
· EMPORIVM · · HERACLEA · · AGRAGATINV · · MACA · PORTVS ·
· SEGESTANVM · · EMPORIVM · RINA · · PACHINIS · PROMONTO ·
· EGYTHARSVS · · ELCETHIVM · · BRVCA · · VVCHMI ·
· PROMOTORI · · PINTHIAC · · PROMONTORI · PORTVS ·
· SACRA · · LEGVM · · PAVCANE ·
· LILYBEV · PMONT · · PORTVS ·
· AEGVSA · ET · CIVITAS ·

· A F R I C V M ·

| 36 | 37 | 38 | 39 | 90 | ½ | 34 |

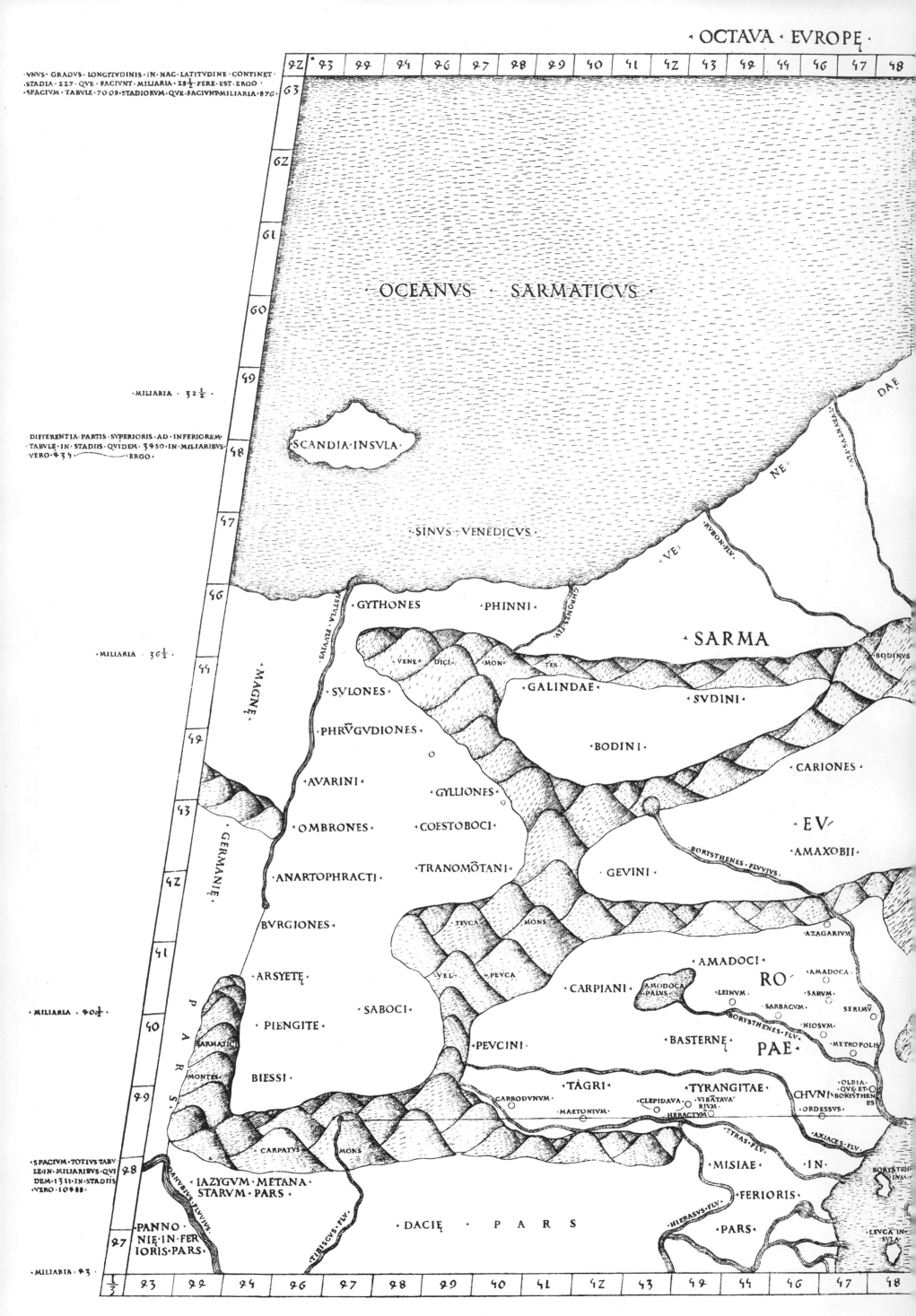

92 · 93 94 94 96 97 98 99 40 41 42 43 44 44 46 47 48

· VNVS · GRADVS · LONGITVDINIS · IN · HAC · LATITVDINE · CONTINET · ·STADIA · 227 · QVE · FACIVNT · MILIARIA · 28½ · FERE · EST · ERGO · ·SPACIVM · TABVLE · 7008 · STADIORVM · QVE · FACIVNT · MILIARIA · 876 ·

63

62

· MILIARIA · 32½ ·

61

60

OCEANVS · SARMATICVS ·

59

DIFFERENTIA · PARTIS · SVPERIORIS · AD · INFERIOREM · TABVLE · IN · STADIIS · QVIDEM · 3450 · IN · MILIARIBVS · VERO · 434 ·   ERGO ·

58

DAE

TVRVNTVS · FLV

NE

57

SCANDIA INSVLA

RVBON · FLV

VE

56

SINVS · VENEDICVS ·

CHRONVS · FLV

· GYTHONES ·   · PHINNI ·

SARMA

BODINVS

· MILIARIA · 30½ ·

55

MAGNE

VISTVLA · FLVVIVS

VENE   DICI   MON   TES

· GALINDAE ·

· SVDINI ·

54

· SVLONES ·

· CARIONES ·

53

· PHRVGVDIONES ·

· BODINI ·

GERMANIE

· AVARINI ·

· GYLLIONES ·

EV

52

· OMBRONES ·

· COESTOBOCI ·

· AMAXOBII ·

BORYSTHENES · FLVVIVS

51

· ANARTOPHRACTI ·

· TRANOMŌTANI ·

· GEVINI ·

· BVRGIONES ·

50

TEVCA   MONS

AZAGARIVM

· AMADOCI ·

RO

AMADOCA

49

· ARSYETE ·

VEL   PEVCA

· CARPIANI ·

AMODOCA PALVS

LEINVM

SARVM

· MILIARIA · 40½ ·

48

· PIENGITE ·

· SABOCI ·

SARBACVM

SERIMV

P

A

R

S

SARMATICI

· PEVCINI ·

· BASTERNE ·

NIOSVM

METROPOLIS

MONTES

· BIESSI ·

PAE

47

· TAGRI ·

· TYRANGITAE ·

OLBIA QVE · ET · BORYSTHEN ES

CARRODVNVM

CLEPIDAVA   VIBATAVA RIVM

CHVNI

MAETONIVM

HERACTVM

ORDESSVS

TYRAS · FLV

AXIACES · FLV

CARPATVS   MONS

· MISIAE ·

· IN·

BORYSTEN INS

· SPACIVM · TOTIVS · TABV· LE · IN · MILIARIBVS · QVI· DEM · 1311 · IN · STADIIS · VERO · 10488 ·

46

IAZYGVM METANA STARVM · PARS

DANVBIVS FLVVIVS

· FERIORIS ·

LEVCA IN· SVLA ·

45

PANNO NIE · IN · FER IORIS · PARS

TIBISCVS · FLV

· DACIE ·   PARS

· PARS ·

HIERASVS · FLV

· MILIARIA · 43 ·

⅓ 93 94 96 97 98 99 40 41 42 43 44 44 46 47 48

49 | 60 | 61 | 62 | 63 | 64 | 64 | 66 | 67 | 68 | 69 | 70 | 71 | 72 | ½

· VIGESIMVSPRIMVS · PARALELIVS ·

· DIFFERT · AB · EQVINOCTIALI · HORIS · 8 · HABENS · MAXIMVM · DIEM · HORARVM · 20 ·

63

62

· VIGESIMVS · PARALELIVS ·

· DIFFERT · AB · EQVINOCTIALI · HORIS · 7 · HABENS · DIEM · MAIORĒ · HORARVM · 19 ·

61

· CARBONES ·

· CAREOTAE ·

· SALI ·

· AGATHYRSI ·

· SARMATIĘ ·   · ASIATICĘ · PARS ·

60

· AORSI ·

· PAGYRITĘ ·

· VELTĘ ·

· SAVARI ·

49

· BORVSCI ·

· HOSSII ·

· NONVSDECIMVS · PARALELIVS ·

· DIFFERT · AB · EQVINOCTIALI · HORIS · 6 · HABENS · MAXIMVM · DIEM · HORARVM · 18 ·

48

· CHERSINVS · FLV ·

· ACIBI ·

·NASCI·

· OPHLONES ·

47

MONTES

RIPHEI

ALEXANDRI · ARĘ

CAESARIS · ARĘ

·VIBIONES·

· TANAITĘ ·

· OCTAVVSDECIMVS · PARALELIVS ·

· DIFFERT · AB · EQVINOCTIALI · HORIS · 5 ½ · HABENS · MAXIMVM · DIEM · HORARVM · 17 ½ ·

46

· IDRĘ ·

·STVRNI·

· TIA ·

· OSYLI ·

· TANAIS · FLVVIVS ·

ALAVNVS   MONS

44

· ALAVNI ·

· SARMATIĘ ·

TANAIS

· SCYTHĘ ·

· SEPTIMVSDECIMVS · PARALELIVS ·

· DIFFERT · AB · EQVINOCTIALI · HORIS · 5 · HABENS · MAXIMVM · DIEM · HORARVM · 17 ·

48

·STAVANI·

· ROXOLANI ·

PORTVS · FLV

CAROEA · VILLA

ALOPECIA
QVAE · ET ·
· TANAIS ·

43

· REVCANALI ·

· HYGREIS ·

· ASIATICĘ ·

· SARGATII ·

42

· SEXTVSDECIMVS · PARALELIVS ·

· DIFFERT · AB · EQVINOCTIALI · HORIS · 4 ½ · HABENS · MAXIMVM · DIEM · HORARVM · 16 ½ ·

· EXOBYGITĘ ·

· LYCVS · FLV ·

· AMADOCI · MONTES ·

41

· LVCVS · DEI · SALTVS ·

· AGARVS · FLV ·

· IAZYGES ·

· PALVS · MEOTIS ·

· PARS ·

40

·NA· VARI·

· TORRECCADĘ ·

· AGARI · PRO · MONTORI ·

· NAVBARVM ·

· CNEMORVM ·

· TRACANA ·

BYCIS · FLV ·

· GERVS · FLV ·

39

· PASTRIS ·

· HERCABV ·

· BYCIS · PALVS ·

·TORROCCA·

· CVRSVS ·

ACRA

· 18 · PARALELIVS ·

· DIFFERT · AB · EQVINOCTIALI · HORIS · 4 · HABENS · MAXIMVM · DIEM · HORARVM · 16 ·

·HIPENES·FEL·

· ACHILLEIS ·

· CARCINA ·

· PATACYS · FLV ·

· NOVA · MENIA ·

ZENONIS · CHERSO · NESVS

· PARTHENIV ·

· CIMMERIVM ·

· BOS ·

· TAVRO · SCYTHĘ ·

· SINVS · CARCI · NITVS ·

TAPHR

· HERACLIVM ·

· CA ·

· CHER ·

· MYRMECIV · PROMONT ·

· CIMMERVM · PROMONTORIV ·

· CLIMA · 7 ·

38

· BONVS · PORTVS ·

TAV

PAROSTA

CIMERIV · FLV ·

RI

TARONA

· POSTIGIA ·

SO

· IVRATV ·

· PANTI · CAPAE ·

· ACHILLEVM · PROMONTORIV ·

CORAX

MONS

· NEMVS · CEPHALO · DIANE · NESVS ·

EVPATO

ORIA

SAT

PORTAGRA

· ARCHE ·

· CYTEVM ·

· BOEON ·

· TYRIC · TADA ·

· NE ·

· TAZVS ·

· NYMPHE ·

· SACRV · MYSARIS · MONTO · PROMO ·

SYMBOLORV · PORTVS ·

BADATIV

· SVS ·

· THEODO · SIA ·

· TABANA ·

· SINVS · CER · CECIDIS ·

PARTHENIVS · PORTVS ·

CTENIS

· CHARAX · LAGIRA ·

· COLCHIDIS · PARS ·

37

ARIESTON · PROMONTORIV ·

49 | 60 | 61 | 62 | 63 | 64 | 64 | 66 | 67 | 68 | 69 | 70 | 71 | 72 | ½ | ⅓

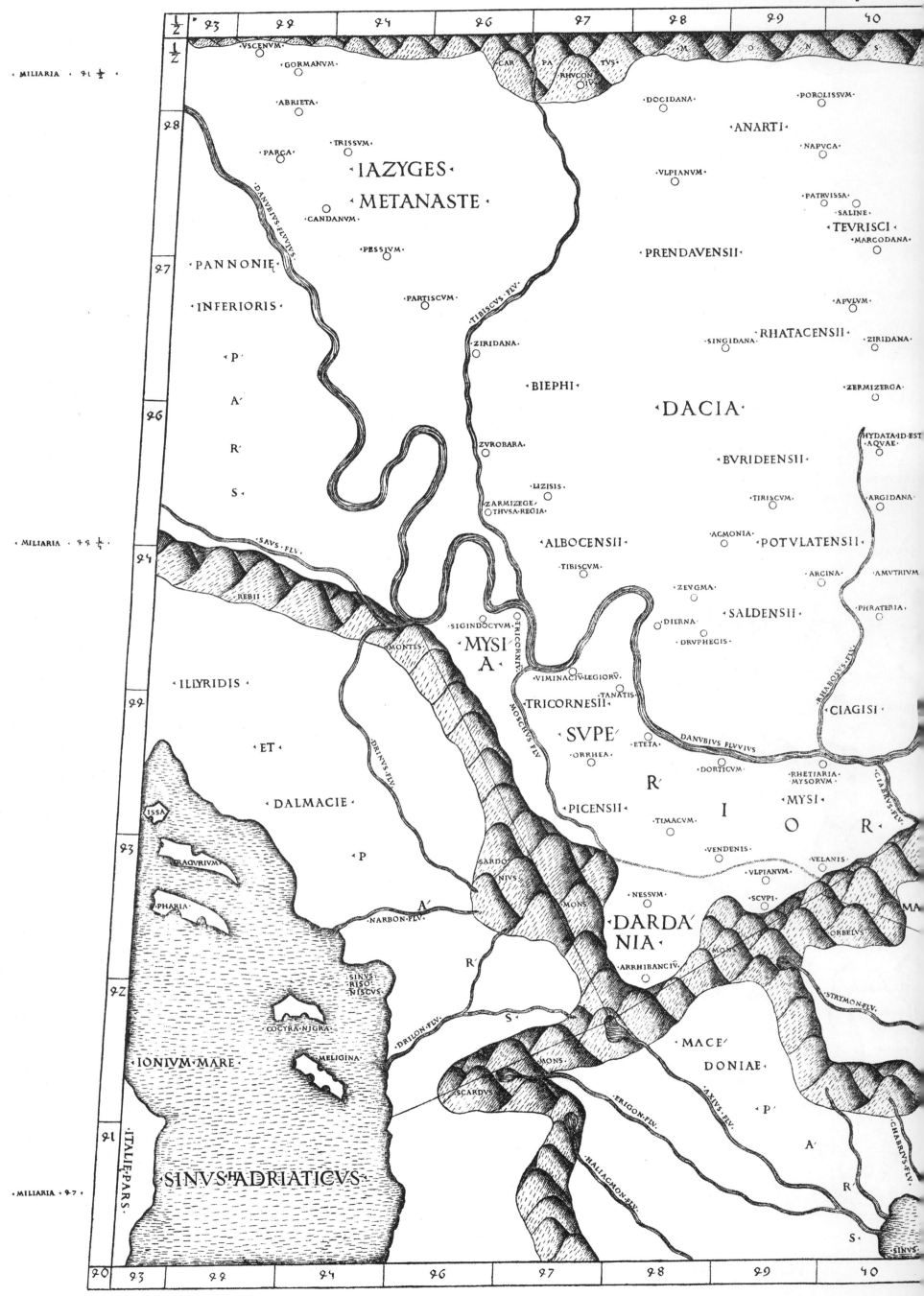

X. PTOLEMÆUS ROMÆ 1490.

· TABVLA ·

41 42 43 44 45 46 47 48

14· PARALELLVS· PER· BORYSTHENE
·DIFFERT· AB· EQVINOCTIALI· HORIS· 4·
·HABENS· MAXIMVM· DIEM· HORARVM· 16·

· CLIMA· SEPTIMVM·

·TRIPHVLVM·
·ARCOBADARA·
·CARSIDANA·
·PATRIDANA·
·ARPII·
·AXIACES· FLV·
·NICONI-
VM·
·OPHIVSSA·
·BORYSTHENES· 48

·CISTOBOCI·
·VTIDANA· ·ZARGIDANA·
· MYSIA ·
·TYRAS·
FLV·
·TYRAS· ·PHYSCA·

·SANDANA·
·TAMISIDANA·
·HERMONAC
TVS·

·ANGVSTIA· ·PETRODANA·
· BRITTOLAGE·
·HARPIS·

·PRAETORIA
·AVGVSTA·
·PALODA·
·HIERASVS· FLV·
·PIROBO
RIDANA·
· IN ·
·THIAGVLA· PALVS·
THIAGOLA·
OSTIVM·
·LEVCA· INSVLA·
ACHILLIS· 47

·CAVCOENSII·
· PEV·
·AQVILONIVM·

·COMIDANA·
·DINOGETIA·
CI·
·PSEVDOSTOMVM·

·RHAMIDANA·
·ZVSIDANA·
·OBVLENSII·
·NVIODV
NVM·
·SITIOEVIA
·PEVCA
INSVLA·
NI·
·CALOSTOMVM·
·INARIATIVM·

·TRISMIS·
· F·
·SACRVM· VEL· PEVCA· 46

·PIRVM·
·SINSII·
·TIBISCA·
· TROGLODYTE·
·ISTROPOLIS· ·PTERVM
PROMONTO·

·COTENSII·
·TIASVM·
·NENTIDANA·
·CARSCV
·SVCIDANA·
·TOMI·
·AXIVM·
CIVITAS·

·TROMA
RISCA·
·CALLATIS·

·DOROSTOLVM
LEGIO·
E·
TENSII·
DIONYS·
OPOLIS·
· MARE· PONTICVM· VEL·

·SORNVM·
·TIRISTA·
·TRIMANIVM·
· CRYBYZI·
·TRISTRIS· PR
OMOTORIVM·
·ODYSSVS·
· PONTVS· EVXINVM·

·PIEPHISI·
R·
·DAVSDANA·
·PANYSVS· FLV·
·MESEMBRIA·

·PINVM·
·DANVBIVS· FLVVIVS·
·NOVE·
I·
· PIARENSII·
·ANCHIALVS·

·DIACVM·
O·
·DEVELTVS·
COLONIA·
·APOLLOIA·
·CYANEIS·

·TRIBALLI·
N·
·OPISINA·
·PRAETVRA·
SELLETICA·
·TONZVS·
·PEROTICV·

·OESCVS· TRIBALLORVM·
·VALLA·
· PRAETVRA·
· BIZYA·
·THINIAE· PRO
MONTORI·

·DIMENSII·
R·
M·
·NICOPOLIS·
·VSDICESICA·
·ORCELLIS·
·HALMIDISSV
LITVS·

·PRASIDIVM·
·TONZOS·
·PHILIA· PRO
MONTORIVM·

·PRAETVRA· SARDICA·
·ARZOS·
·CARPVDAEM·
·OS· PONTI
EXTERIVS·

·PRAETVRA·
DANTHE
LETICA·
·TEITA·
·OSTAPHOS·
·RHODOPE·
·MONS·
CIA·
·BERGVLA·
·BATHYNIVS· FLV·
·PHINOP
OLIS·

·SARDICA·
· THRA·
·HADRIANOPOLIS·
·VEL· TRIMVNTIVM·
·SEV· PHILIPPOPOLIS·
·PLOTINOPOLIS·
·ATHYRA· FLV·
·BYZANTIVM·
· PON·

·BESICA·
· BENNICA·
·PERGAMV·
·DIVSIPARA·
·OS· PONTI
INTERIVS·
TI·

·ANTALIA·
·PRAETVRA·
· SAMAICA·
·CYPSELLA·
· SELYMBRIA·
ET·

·DROSICA·
·NICOPOLIS·
·TRAIANOPOLIS·
·PRAETVRA·
CAENICA·
·PERINTH
OS·
·PROPONTIDIS·
BITHY·

·COELETICA·
·TOPIRIS·
· SAPAICA·
·APROE·
COLONIA·
·ARZVS· FLV·
·BOS·
PHO·
·PROECONESVS·
NIE·

·CORPIA· LICA·
·BISANTHA·
RVS·
P·

·ABDERA·
·DYMA·
·APHRODI·
SIA·
·HERACLIA·
·THRA·
CI·
A·

·MARONIA·
·LYSIMACHIA·
·LONGVS·
MVRVS·
VS·
R·

·AENOS·
·PACTIA·
·HELLES
PON·
· ASIE·
S·

·THALASSIA·
·SAMOTHRACE·
·SINVS· MELANE·
·CALLI·
POLIS·
TV·

·SINVS· EDONICVS·
·IMBROS·
·CHERSO·
·MADI·
·SESTOS·
S·
· PROPRIE·

·HEFESTIA·
·CARDIA·
·CRITHEA·
NE·
·QVILA·
P·

· LEMNOS·
·MASTVSA
ORCONTORIVM·
SVS·
·SCHOE·
·IV· PROM·
A·

·MYRINA·
·PELEVS·
·TENEDON·
R·

·SINVS· SINOTICVS·
S·

·SINVS· TORO
NAICVS·
· MARE· AEGEVM ·
·LESBOS·
PARS·

·HERMAICVS·

41 42 43 44 45 46 47 48

· PER· PONTVM·
19· PARALELLVS·
·DIFFERT· AB· EQVINOCTIALI· HORIS·
3 1/2· HABENS· MAXIMVM· DIEM·
HORARVM· 15 1/2·

· CLIMA· SEXTVM·

44

·THYNIAS· 44

· PER· BYZANTIVM·
13· PARALELLVS·
·DIFFERT· AB· EQVINOCTIALI·
HORIS· 3 3/4· HABENS·
MAXIMV· DIEM· HORARV· 15 1/4·

43

42

· PER· HELLESPOTV·
12· PARALELLVS·
·DIFFERT· AB· EQVINOC·
TIALI· HORIS· 3·
HABENS· MAXIMVM·
DIEM· HORARV· 15·

· CLIMA· 5·

41

50

· THRACIĘ · PARS ·

THRACIE PARS

NESVS · FLV

GASORVS · PHILIPVS
EDONI · DON
OBSYMA · NEAPOIS
SINVS · EDONICVS
TANTIRA
PANORMVS
ACATHVS STRATONICE
BALCICICA ANTRAEVM
SINGVS
AVGAEA
CHAETAE AMPELVS · PROM
PAR · MORTILVS TORONE ONTO
DERRIS · PROM
AXIA ANTIG OONA
CASANDRIA
PATALENES

THALASSIA

SARMOTHRACE

IMBROS

PROECONNESVS

PROPONTIDIS

HELLESPON T V S

ASIĘ · ET · PHRIGIĘ PARS

CHER SO NESVS

HEPHAESTIA
LEMN O OS
MYRINA

TENEDOS

LESBOS INSVLA

MARE · AEGAEVM

PEPARETHOS

ON COLONIA
PHER
ANSSA
OSSA MONS
TI
MELITARA
CORONIA
PAGASAE
PHTHIOTIDA
DEMETRIAS
LAMIA
ECHINVS
SPERCHIVS FLV
ERETRIA SPERCHIA
THEBAE
AEVM MONS
BRAGR SCARPHIAS
CEPHISVS FLV
LOCRI
DAVLIS
ELATIA
DELPHI PYTHIA AEGOSTHENIA
BVLIA A
PHO CIDA BOETA
CRISSA ANTICYRRHA
INVS CORINTHIAGVS

MAGNESA PROMOT
PREPA
AEANTIVM
IOLCOS
SINVS PELASGICVS
POSIDIVM
SCIATHOS

SCOPELOS

SCYROS

CENAEVM PRO MONTORI
PHIASIA
NESVM PROM SORENS
ATALAM TES
AEDIPSVS
CYNVS
PHOCAE ANTHEDON
CRAPHIA
 LEBADIA
THEBE ISMENVS FLV
CORONIA DELIVM
THESPIAE COPAE
SIPHAE HYAPOLIS
CREVSA
MEGARA
NISAEA RI
DIS
SINVS
ATHENAE
ELENSIS
OENOA
RHAMNVS
MARATHON
PIREVS ANAPHLYSTVS
ATTICA DIANE SACRV
MVNICHIAE PORTVS
PANORMVS PORTVS
HIPHORS PORTVS

DIANA SACRVM
EV BOEA
CERITHVS
CHALCIS
ERETRIA
CAVARE PROMON
GERESTVS PORTVS
AMARYN THVS
CARYS TVS
CALA ACTIA

INSVLE CICLADES

ANDRI

TENI

SCYRI

RHENA
MIGON
DELI
CARS SVS
OEVS
CTIM NVS
NAXI
SERI THVM SVS
PARI
CARTHAEA INSV
SIPHNVS
PHILO SICIN CAN DRVA
OLEARVS

CORIN
SICYON
CENCHREA SINV
LECHAE CORINTHVS
SARO NICVS
THIA BVCEPHALE DV PORTVS
ATHENIENSV VM PORTVS
EGINA
SALAGIS
SICYOIA
EL SPIREV PRO MONTORI
NEMEA
ARGOS ARGIA
EPIDAVR
CHERSO NESVS
THE RA
OEA
ELEVSIN

THERASIA
MIRTHOVM MARE

MELOS

CRETICVM · MARE ·

DIA

PEL
LILAEA
STIMPHA MEGALE
DIA CLITOR
PO
LA CYPHANT
THALAMA
LERNA
PRASIA
CYTHANTA PORTVS
BLEMINA ARGO
ZARENX LICVS
CEDAEMON
GERENIA
THVRIVM
EPIDA VRVS
ICA

AEGIRA
PELLENA
PHLIVS
THIA
ASINE MYCENAE
INACHVS
NAVPLIA
TROE ZENA
N
HERMIONE
ARGOS
NEMEA

EPLA
CYTHERA

AMPHIMALIS SINVS
DION CITAEVM
ELEVTHERAE
ZEPHRIVM PROM
HERACLI CHERSO
OLVLIS CAXA
PANNONA MVS
PANOR MVS
SAMONI PORTVS

C RETA INSVLA
DICTAMVS PROMONTORI
DICTANV
CISAMVS
CISAMVS
RHANVS FLV
LYRRHENIA
ARTACINA
SVBRITA
GORTYNA
LAPPA
MASALIA
PSYCHIVM
MATALIA
CHERSO NESVS
INACHO RIVM
POECILA
TERMOS PROM
PHOENIX
LETOA

CIMOLIS

GLAVDVS

CARPATHICVM · MARE ·

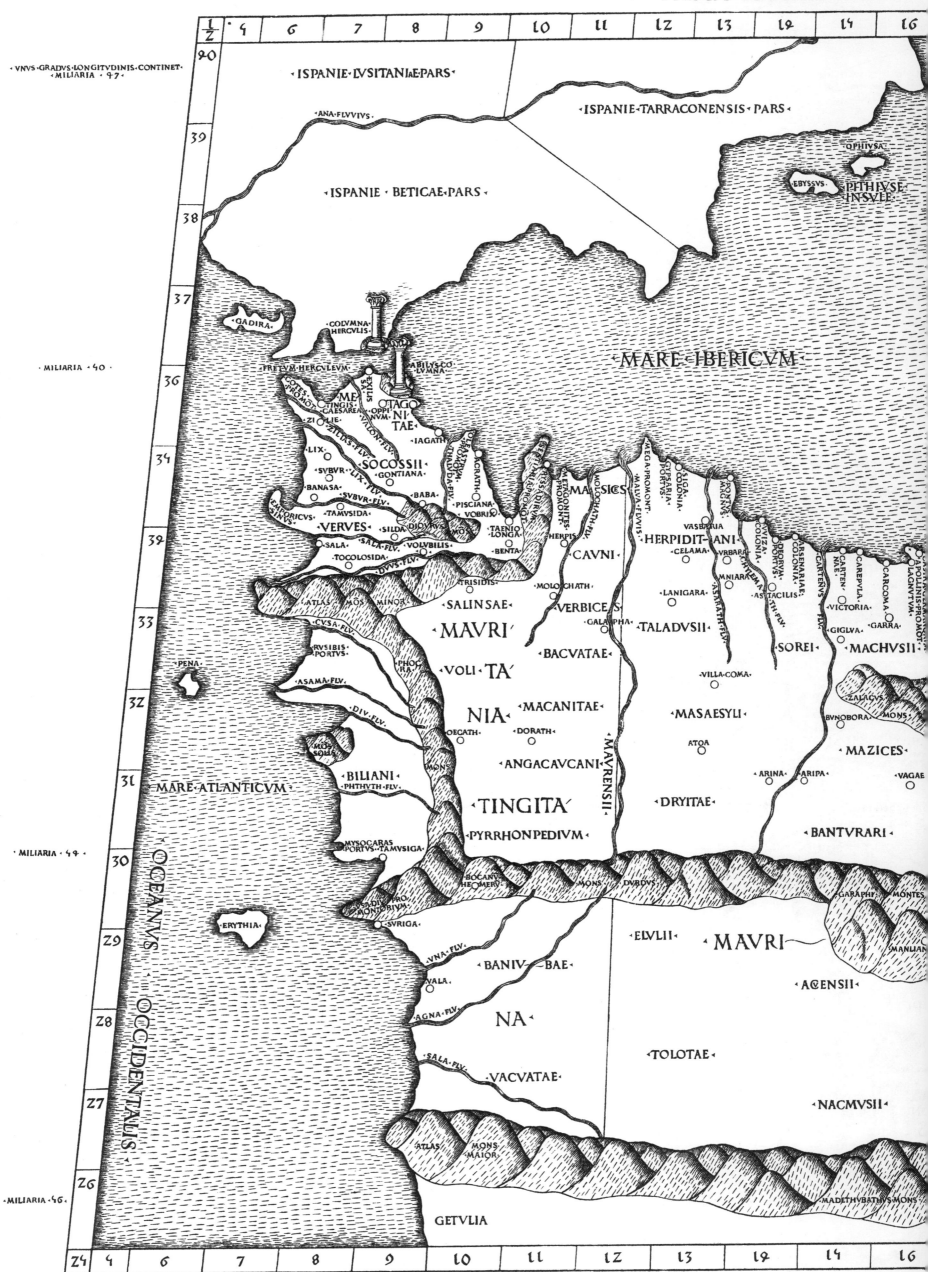

XII. PTOLEMÆUS ROMÆ 1490.

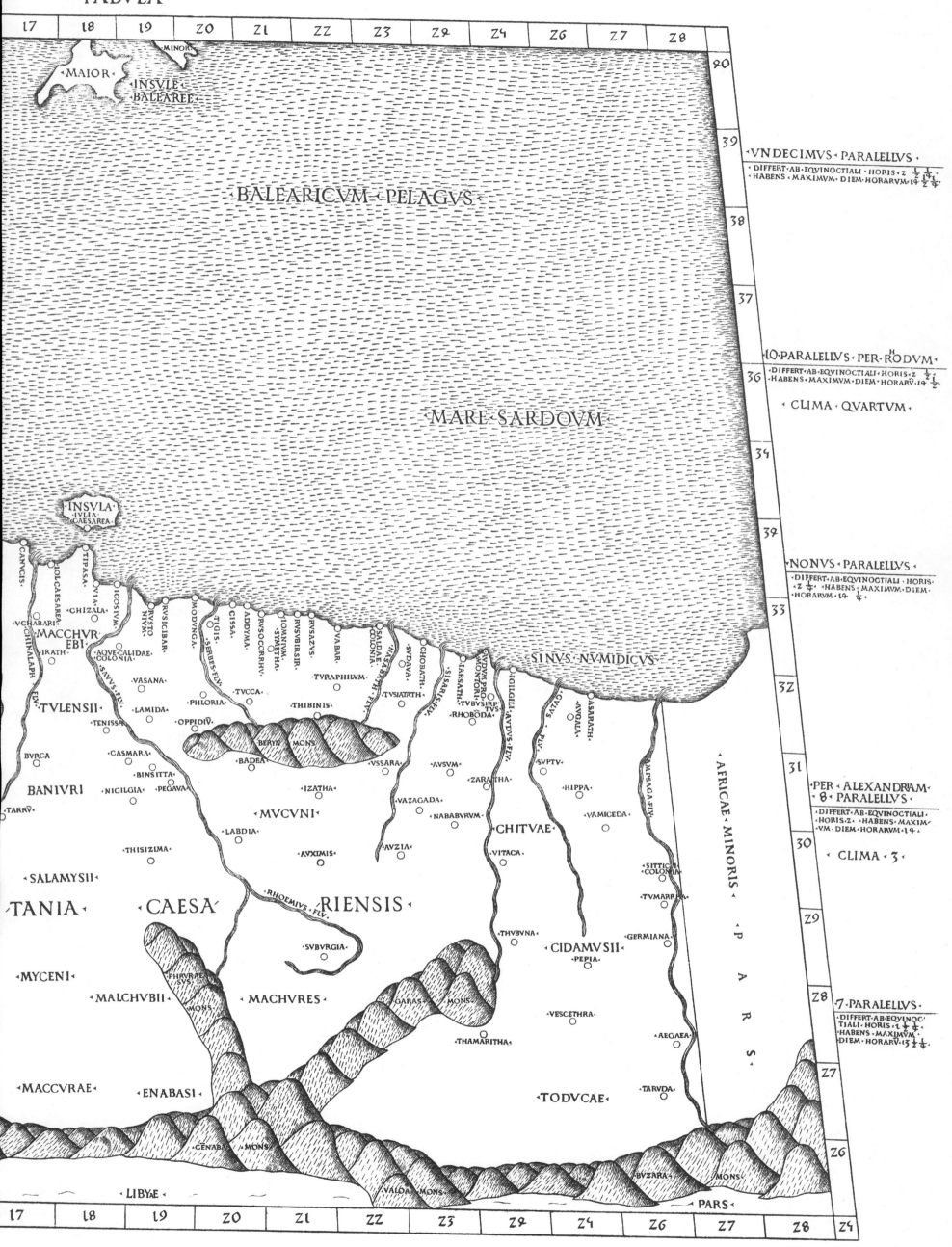

·TABVLA·

17 18 19 20 21 22 23 24 25 26 27 28

·MINOR·

·MAIOR· ·INSVLE·
·BALEAREE·

40

39

BALEARICVM · PELAGVS

38

37

·11·PARALELLVS·PER·RODVM·
·DIFFERT·AB·EQVINOCTIALI·HORIS·2 ½ ⅛·
·HABENS·MAXIMVM·DIEM·HORARV·14 ½ ⅛·

36

MARE · SARDOVM ·

·CLIMA·QVARTVM·

35

·INSVLA·
·IVLIA·
·CAESAREA·

34

·NONVS·PARALELLVS·
·DIFFERT·AB·EQVINOCTIALI·HORIS·
·2 ⅛·HABENS·MAXIMVM·DIEM·
HORARVM·14 ⅛·

·CARVCIS· ·TIPASA· ·VTA·
·IOLCAESAREA· 33

·VCHABARI· ·CHIZALA· ·ICOSIVM· ·RVSTO·
·CHINALAPH· ·NIVM· ·RVSICIBAR· ·CISSA· ·ADDYMA· ·VABAR· ·NASTARRHA· ·SVDVA· SINVS · NVMIDICVS

·MACCHVR·
·EBI· ·IRATH· ·AQVE·CALIDAE· ·SERES·FLV· ·RVSOCORRHV· ·IOMNYM· ·RVSVBIRSIR· ·SADAE· ·CHOBATH· ·AVDVM·PRO· 32
·COLONIA· ·TLIGIS· ·SYNETHA· ·COLONIA· ·MONTORI·
·SAVVS·FLV· ·VASANA· ·TVCCA· ·TVRAPHILVM· ·TVSIATATH· ·TVBVSIRP· ·RHOBODA·
·TVLENSII· ·LAMIDA· ·PHLORIA· ·THIBINIS· ·TVS·
·TENISSA· ·OPPIDIV· ·BERYL·MONS· ·VSSARA· ·SISARATH·

·BVRCA· ·CASMARA· ·BADEA· ·AVSVM· ·SVPTV· 31
·BINSITTA· ·ZARATHA· ·HIPPA·
·BANIVRI· ·NIGILGIA· ·PEGAVA· ·IZATHA· ·VAZAGADA· ·VAMICEDA· ·PER·ALEXANDRIAM·
·TARRV· ·MVCVNI· ·NABABVRVM· ·CHITVAE· ·8·PARALELLVS· 30
·LABDIA· ·AVZIA· ·VITACA· ·DIFFERT·AB·EQVINOCTIALI·
·THISIZIMA· ·AVXIMIS· ·HORIS·2·HABENS·MAXIM·
·VM·DIEM·HORARVM·14·
·SALAMYSII· ·RHOEMIVS· ·SITTIC·
·FLV· ·COLONIA· ·CLIMA·3· 29
·TANIA· ·CAESA· ·RIENSIS· ·TVMARRH·
·SVBVRGIA· ·THVBVNA· ·GERMIANA·
·MYCENI· ·CIDAMVSII· 28
·MALCHVBII· ·MACHVRES· ·PEPIA· ·7·PARALELLVS·
·MONS· ·GARAS· ·MONS· ·VESCETHRA· ·DIFFERT·AB·EQVINOC·
·PHVRAE· ·TIALI·HORIS·1 ½ ⅛·
·SVS· ·AEGAEA· ·HABENS·MAXIMVM·
·THAMARITHA· ·DIEM·HORARV·13 ½ ⅛· 27
·MACCVRAE· ·ENABASI· ·TODVCAE· ·TARVDA·
26
·CENABA·MONS· ·BVZARA· ·MONS·
·LIBYAE· ·VALDA·MONS· ·PARS· 25

17 18 19 20 21 22 23 24 25 26 27 28 29

MARE·TYRRHENVM·

MARE·SARDOVM·

·SARDINIA·
·INSVLA·

·MILIARIA·50·

M   A   R   E   ·   A   F

·CALATH
A·

·LARVNE·
SIAE·2·

·HYDRAS·

·DRACONTINVS·

·AEGIMIVS·

SINVS·NVMIDICS·

·CRVSICADA·
·TRETON·PROMON·
·COLLOPS·MAGN·
·AVS·
·VZICATH·
·SINVS·OIO·
·ACHITES·
·COLLOPS·
·PARVS·
·TACATVA·
·ASPVGA·
·STVR·PORTVS·
·HIPPI·PROMÔTORIVM·
·STOBOR·PROMONTOR·
·APHRODISIS·COLONIA·
·HIPPON·REGI
·HVBRICATV·FLV·
·THABRACA·
·COLONIA·
·SVLLA·
·REA·
·SACRVM
·APOLLINIS·
·ITYCA·
·NEPTVNI·
·MAX·
·HIPPON·
·TISPAN·
·THINISSA·
·VZAN·
·THISICA·
·CORNELIICAS·
·TRAIEETACIO·
·CIPIPA·
·CARTHAGO·
·MAGNA·
·MAXVLA·
·VETVS·
·AVOL·
·THIMISA·
·MAXVLA·
·ORBITA·
·VTICNA·
·HERMAEAPROS·
·CLIPEA·
·NEAPOLIS·
·SIAGV·
·APHRODISIVM·
·NISA·
·CARPIS·
·ALMENA·
·ADRVMET·
·COLONIE·
·MACHYNI·

·CIRTESII·
·VAGA·
·NVMI·
·DIA·
·THVBVRNICA·
·COLONIA·
·MIDENI·
·MADVRVS·
·THEVDALI·
·CATADA·
·TVRZA·
·GISIRA·
·ZVRM
·CARTA·
·GONII·

·CIRTA·
·IVLIA·
·CAENA·
·BATRAE·
·CVLCVA·
·COLONIA·
·SIMISTHI·
·TVCCA·
·AF·
·MIAEDII·
·CVINA·
·COLONIA·
·VLIZIBIRRA·
·MEDICCARA·
·DIOS·

·IONTII·
·GAVSAPHNA·
·THIEBA·
·SICCA·VENE·
·RIA·
·ASSVRVS·
·AMAEDARA·
·SISARA·
·PALVS·
·AVITTA·
·TOBROS·
·ABDIRA·
·AVGVSTV·
·LIBYPHOENICES·
·LEAE·

·LARES·
·THVNVDROMV·
·THVBVRSICA·
·NARANGARA·
·NOVA·
·LAMBAESA·
·VCIBI·
·THEBESTA·
·GEDNA·
·MVSVNI·
·THANVTADA·
·CIRNA·
·MONS·
·DABIA·
·ILICA·
·THVBVRBO·
·TVCMA·
·BVLLAMEN·
·SA·
·CERBICA·
·AVIDVS·
·SASRVA·
·BAZACITIS·
·REGIO·
·BVNTHV·
·VBA·
·RI·

·MIREVM·
·APARI·
·MISVLAMI·
·AVDVS·
·MOS·
·GAZACVPADA·
·VAZVA·
·NVROLI·
·BENDENA·
·TVCCA·
·CILMA·
·VEPILLIV·
·ZVTAE·
·TICELIA·
·NENSA·
·TEMISVA·
·THABBA·
·TICHASA·
·TISVR·

·NATTABVTAE·
·THVBVTIS·
·CEROPHAEI·
·AQVAE·
·CALIDAE·
·ZAMAMIZON·
·TVSCVBIS·

·AZAMA·
·THAM·
·MES·
·MONS·
·ZIGIRA·
·THASIA·
·TIMICA·
·NEGETA·

·NISIBES·
·SABVRBVRES·
·THVNVBA·
·MVSTA·
·MAMPSARI·

·BVZARA·
·HALIARDI·
·MAMP·
·SARVS·

·MONS·
·SITTAPHIVS·CAMPVS·
·MOTVTVRII·
·MONS·

·MILIARIA·56·
·BAGRADAS·FLV·

·LIBYAE·
·INTERIORIS·

·MAVRITANIAE·CAESARIENSIS·PARS·
·AMPSAGA·FLV·
·BAGRADASFLVVVS·

XIII.   PTOLEMÆUS ROMÆ 1490.

38 39 40 41 42 43 44 45 46 47

· ITALIAE · PARS ·

· EPIRI · PARS ·

39

· VNDECIMVS · PARALELLVS ·
· DIFFERT · AB · EQVINOCTIALI · HORIS · 2 $\frac{1}{4}\frac{1}{8}$ · HABENS · MAXIMVM · DIEM · HORARVM · 14 $\frac{1}{2}\frac{1}{4}$ ·

38

· MARE · ADRIATICVM ·

· CORCYRA · INSVLA ·

37

· SICILIA · INSVLA ·

· 10 · PARALELLVS · PER · RODVM ·
· DIFFERT · AB · EQVINOCTIALI · HORIS · 2 $\frac{1}{2}$ · HABENS · MAXIMVM · DIEM · HORARVM · 14 $\frac{1}{2}$ ·

36

· CLIMA · QVARTVM ·

35

· GLAVCONIS ·
· MELITA ·
· IVNONIS · TEMPLVM ·
· COSYRA ·
· HERCVLIS · TEMPLV ·

R   I   C   V   M ·

32

· NONVS · PARALELLVS ·
· DIFFERT · AB EQVINOCTIALI · HORIS · 2 $\frac{1}{4}$ · HABENS · MAXIMVM · DIEM · HORARV · 14 $\frac{1}{4}$ ·

· LOBADVSA ·
· AETHVSA ·

33

· CERCINNA ·

· LEPTIS · PARVA ·
· THAPSVS ·
· ACHOLA ·
· OVZECIA ·
· RVSPE ·
· NSVLA ·
· TAPHRVRA ·
· USVDRVS ·

· SYRTIS · PARVA ·

32

· THEAENAL ·
· ASICA ·
· SETIENSIS ·
· CARAGA ·
· MACOD · AMA ·
· IOVIS · MOS ·
· BYZACINA ·

· MINIX ·
· LOTOFAGES · GIRAPOLIS ·
· HEDAPH · THA ·
· ZETHA PRO / MONTORI ·
· SABRATHA ·
· PISINDON · PORTVS ·
· HEDA ·
· GARAPHA · PORTVS ·
· NEAPOLIS · EL · LEPTIS ·
· SYDDENIS ·

· BARATH LAC ·
· CINSTERNE ·
· CINYPHVS · FL ·

31

· PER · ALEXANDRIAM · OCTAVVS · PARALELLVS ·
· DIFFERT · AB · EQVINOCTIALI · HORIS · 2 · HABENS · MAXIMVM · DIEM · HORARV · 9 ·

· MARGARV ·
· ZVGAR ·
· SICHTHIS ·
· NIGI ·
· SVMVCIS ·
· SABRATA ·
· TIMI ·
· AMMONIS ·
· TRIEORORVM · PROMONT ·
· MACOMA · CA · VILLA ·
· GERISA · ISCINA ·
· MISY · NVS ·

· ARARVS ·
· TRITON · F ·
· TACAPA ·
· CINII ·
· CHVZIS ·
· ASTACVRES ·
· HAMVCVLV ·
· ASPIS ·
· SICAPHA ·
· LOTOPHAGI ·
· PONTIA ·
· SYRTIS · MAGNA ·

30

· CLIMA · TERCIVM ·

· MACHRYES ·
· CAPSA ·
· TRITO · PALVS ·
· NITIS ·
· SIGIPLOSII ·
· GILIVS ·
· MONS ·
· EROPAEI ·
· SACAMA · ZA · VILLA ·
· SAMAMVCII ·
· PYRGOS · EVPHRATA ·
· GEA ·

· PV · TEA ·
· PALLAS · PALVS ·
· ACHAEMENES ·
· DAMENSII ·
· GALYBA ·

· GEPHES ·
· LIBYA · PALVS ·
· C ·
· MVSTA · VILLA ·
· VDDITA ·
· PHARAXA · VILLA ·
· OESPORIS · VILLA ·
· HIPPI · PROMONT ·
· PHILENI · VILLA ·

29

· MIMACES ·
· DOLPES ·
· BVTTA ·
· NIGBENI ·
· NYCPII ·

· VSALAETVS ·
· MONS ·
· MVTVRGVRES ·
· THIZIBIVS · MOS ·
· A ·
· ELAEONES ·

28

· 7 · PARALELLVS ·
· DIFFERT · AB · EQVINOCTIALI · HORA · 1 $\frac{1}{2}\frac{1}{4}$ · HABENS · MAXIMVM · DIEM · HORARVM · 13 $\frac{1}{2}\frac{1}{4}$ ·

· VZARA ·
· MVCHTHVSII ·
· TEGA ·
· MACEI · SYRITAE ·

27

· EREBIDAE ·
· DVRGA ·
· MONTES ·
· VELPAE ·

· DESERTAE · LIBYAE · PRINCIPIVM ·
· ZVCH ·
· ABARVS ·
· MONS ·
· ACHABE · FONS ·

26

· DESERTA · LIBYA ·

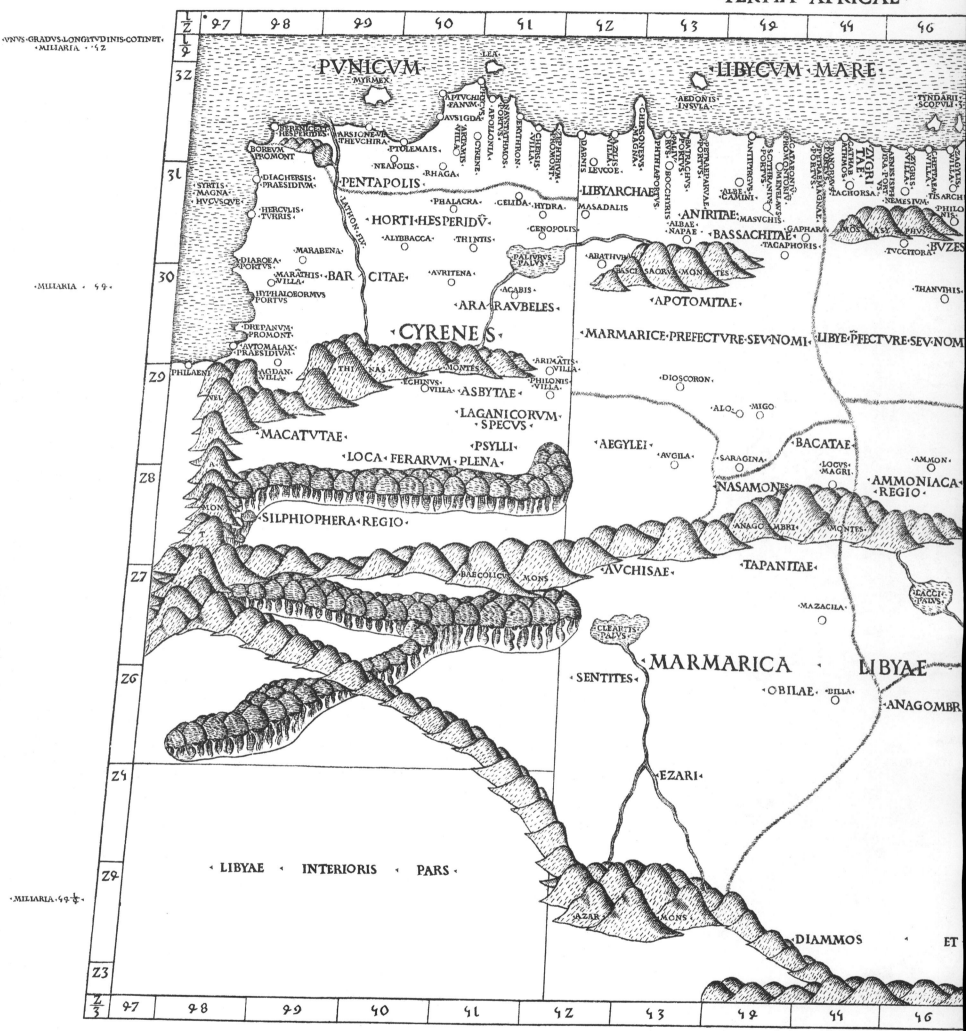

XIV. PTOLEMÆUS ROMÆ 1490.

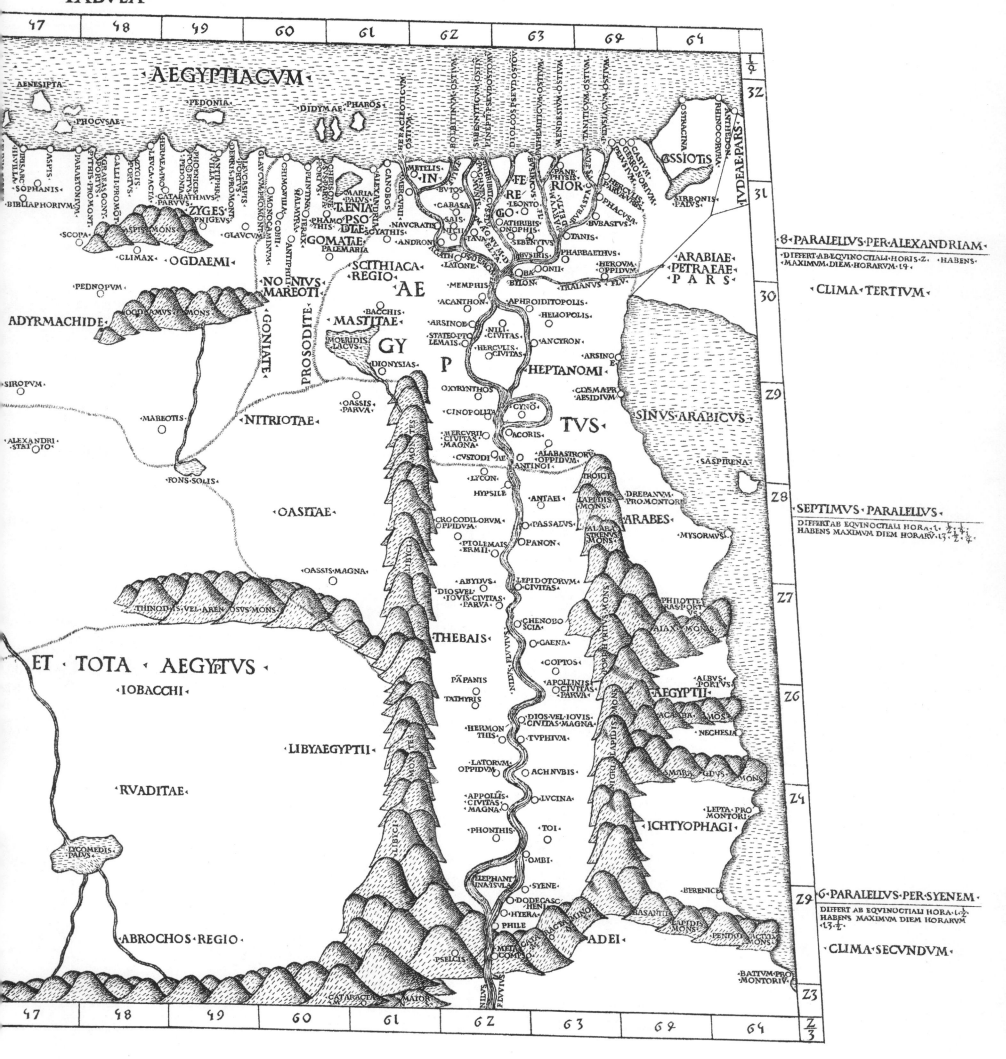

| 47 | 48 | 49 | 60 | 61 | 62 | 63 | 64 | 64 | |

·AEGYPTIACVM·

AENESIPTA

PHOCVSAE
ASPIS
PEDONIA
DIDYMAE · PHAROS·

CHIVILLA·
SOPHANIS·
BIBLIAPHORIVM·
SCOPA·

ADYRMACHIDE

SIROPVM·

ALEXANDRI·
STATIO

LEVCA·ACTA·
APIS·PORTVS·
PHILES·
PHOENICIS·
ZYGIS·
MONOCRISE·
CATABATHMVS·
CHIMOMA·
ALEXANDRIA
NO·NIVS
MAREOTI
ADYRMACHIDE

OGDAEMI
NITRIOTAE·
PEDNOPVM·
OGDAMVS·MONS
MARBEOTIS
FONS·SOLIS
OASITAE·

SCITHIACA·
REGIO·
MASTITAE
MOERIDIS·
LACVS·
DIONYSIAS·

ZYGES
PNIGEVS·
GLAVCVM·
CLIMAX·
GOMATAE·
PALEMARIA·
PROSODITE·
GONIATE·
BACCHIS·
OASSIS·
PARVA

AE
GY
P
OXYRINTHOS·
CINOPOLITA·
MERCVRII·
CIVITAS·
MAGNA·
CVSTODI·AE·
LYCON·
HYPSILE·

METELIS·IN
CANOBOS
BVTOS
CABASA·
NAVCRATIS·
SAIS·
ANDRON
LATONE·
MEMPHIS·
ACANTHON·
ARSINOE·
STATEO·PTO·
LEMAIS·
HERCVLIS·
CIVITAS·
NILI·
CIVITAS·

FE
RE
GO
LEONTO·
OATHRIBIS·
ONOPHIS·
SEBENTVS·
BVSIRIS·
OBA
BYLON·
APHROIDITOPOLIS·
HELIOPOLIS·
ANCYRON·

RIOR·
TANIS·
PHARBAETHVS·
HEROVM·
OPPIDVM·
TRAIANVS·FLV·
ARSINO
E·
CLISMA·PR·
AESIDVM·

CSSIOTIS
SIRBONIS·
PALVS·
PHACVSA·
BVBASTVS·

ARABIAE·
PETRAEAE·
PARS

SINVS·ARABICVS·

SASPIRENA·

HEPTANOMI
TVS

ACORIS·
CYNO
ALABASTRORV·
OPPIDVM·
ANTINOI·
THOIG·
LAPIDIS·
MONS·
ANTAEI·
PASSALVS·
ALABA
STENNA·
MONS·
OPANON·
ARABES·

DREPANVM·
PROMONTORIE·
MYSORMVS·

OASSIS·PARVA

CROCODILORVM·
OPPIDVM·
PTOLEMAIS·
ERMII·

THINODIS·VEL·AREN·OSVS·MONS·
OASSIS·MAGNA·

ABYDVS·
DIOSVEL·
IOVIS·CIVITAS·
PARVA·
LEPIDOTORVM·
CIVITAS·
CHENOBO
SCIA·
CAENA·
COPTOS·

PHILOTTE
RAS·PORT·
AIAX·MONS·

ET·TOTA·AEGYPTVS·
·IOBACCHI·
LIBYAEGYPTII·
·RVADITAE·
THEBAIS
PAPANIS·
TATHYRIS·
HERMON
THIS·
LATORVM·
OPPIDVM·
APPOLLIS·
CIVITAS·
MAGNA·
PHONTHIS·
APOLLINIS·
OCIVITAS·
PARVA·
DIOS·VEL·IOVIS·
CIVITAS·MAGNA·
TVPHIVM·
ACHNVBIS·
LVCINA·
TOI·

ALBVS·
PORTVS·
AEGYPTII·
ACABA·
MOS·
NECHESIA·
SMARAGDVS·MON·
LEPTA·PRO
MONTORI·
ICHTYOPHAGI·

DICOMEDIS·
PALVS·
OMBI·
ELEPHANT
INA·INSVLA·
SYENE·
DODECASC
HENI·
HIERA·
PHILE·
BERENICE·

ABROCHOS·REGIO·
PSELCIS·
METACO
COMPSO·
ADEI·
ASANTE·
LEPIDIS·
MONS·
PENTA
MONS·
BATTVM·PRO·
MONTORIV·

CATARACTA
MAIOR·

| 47 | 48 | 49 | 60 | 61 | 62 | 63 | 64 | 64 | |

Z/3

32	
31	
30	
Z9	
Z8	
Z7	
Z6	
Z4	
Z9	
Z3	

IVDEAE·PARS·

·8·PARALELLVS·PER·ALEXANDRIAM·
·DIFFERT·AB·EQVINOCTIALI·HORIS·Z· HABENS·
·MAXIMVM·DIEM·HORARVM·19·

·CLIMA·TERTIVM·

·SEPTIMVS·PARALELLVS·
DIFFERTAB EQVINOCTIALI HORA·I· HABENS MAXIMVM DIEM HORARV·13·

·6·PARALELLVS·PER·SYENEM·
DIFFERT AB EQVINOCTIALI HORA·1·
HABENS MAXIMVM DIEM HORARVM·13·

·CLIMA·SECVNDVM·

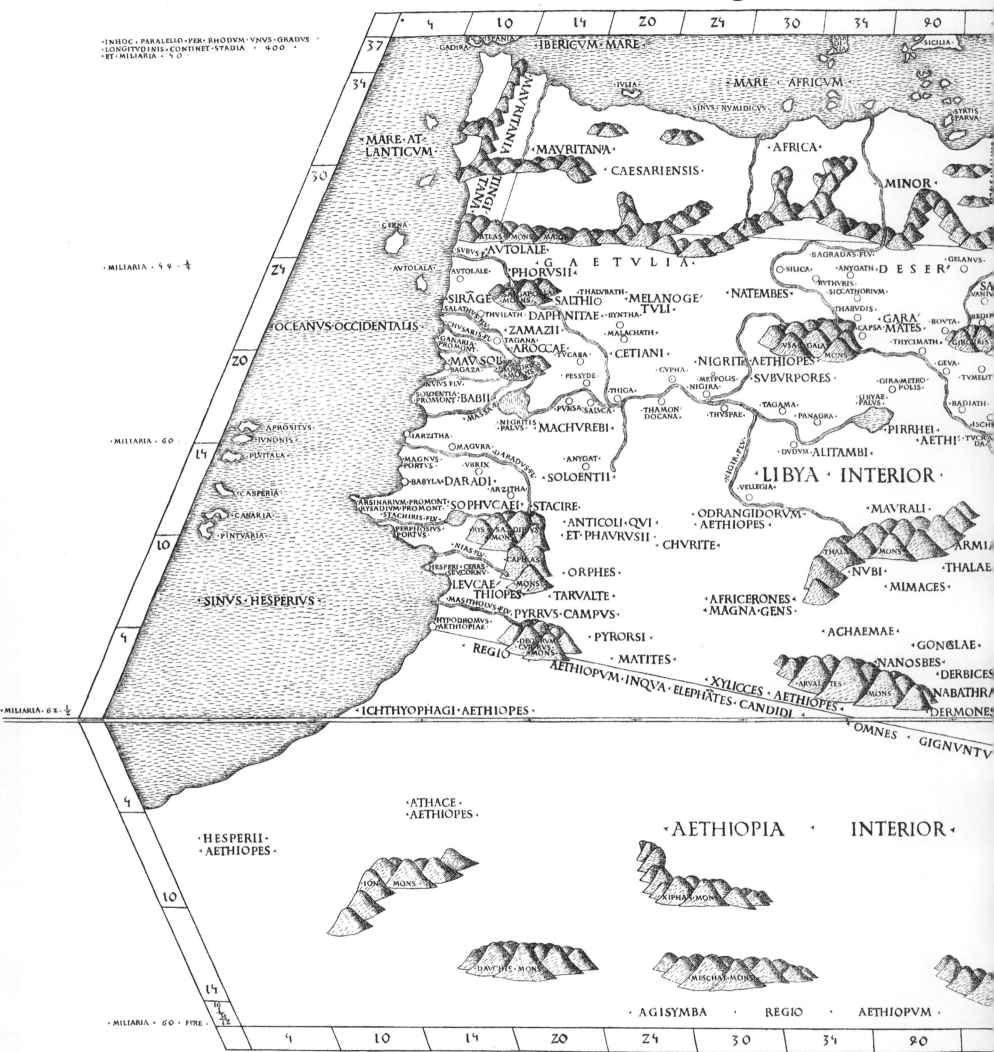

·INHOC · PARALELLO · PER · RHODVM · VNVS · GRADVS ·
·LONGITVDINIS · CONTINET · STADIA · 400 ·
·ET · MILIARIA · 50 ·

IBERICVM · MARE ·

GADIRA
HISPANIA
SICILIA

· MARE · AFRICVM ·

IVLIA ·
· SINVS · NVMIDICVS ·
SYRTIS PARVA

· MARE · AT
LANTICVM ·

MAVRITANIA TINGITANA

· MAVRITANA ·

· AFRICA ·

· CAESARIENSIS ·

MINOR

CERNA ·

ATLAS · MONS

GAETVLIA

DESER

SVBVS F
· AVTOLALE ·

BAGRADAS · FLV
GELANVS

AVTOLALA ·
AVTOLALE ·
· PHORVSII ·
SALAP MONS
THALVBATH
SILICA
ANYGATH
BVTHVRIS ·
SIOCATHORIVM

· NATEMBES ·

SAT
VANIV

MILIARIA · 49 · ½

OCEANVS · OCCIDENTALIS ·

SIRAGE
SALATHV FLV
THVILATH
DAPHNITAE ·
SALTHIO
· MELANOGE
TVLI ·
BYNTHA
THABVDIS ·
· GARA
CAPSA
MATES ·
BOVTA

CHVSARIS FLV
TAGANA
· ZAMAZII ·
MALACHATH
THYCIMATH
GIRRI

GANARIA
PROMONT
AROCCAE ·
TVCABA
· CETIANI ·

USAR GALA
MONS
GEVA ·
TVMELIT

MAVSOL
BAGAZA
MADVRE
AMON
PESSYDE ·
CVPHA ·
· NIGRITE · AETHIOPES
METPOLIS ·
NIGIRA ·
· SVBVRPORES ·
LIBYAE
PALVS ·
GIRA · METRO
POLIS ·
BADIATH

INIVS FLV
THIGA ·
· PIRRHEI
· AETHI
TVCA

MILIARIA · 60 ·

SOLDENTIA
PROMONT
BABII
MASSA
NIGRITIS
PALVS
· PVNSA · SALVCA ·
· THAMON
DOCANA ·
· THVSPAE ·
TAGAMA ·
· PANAGRA ·

APROSITVS
IARZITHA
· MACHVREBI ·
· DVDVM · ALITAMBI ·

IVNONIS
PLVITALA
OMAGVRA
DARADVS FLV
ANYGAT ·
NIGIR · FLV
VELLEGIA ·

LIBYA · INTERIOR

MAGNVS
PORTVS
VBRIX
· SOLOENTII ·
· ODRANGIDORVM ·
AETHIOPES ·
· MAVRALI ·

CASPERIA
BABYLA
DARADI
ARZITHA
· ANTICOLI · QVI
· ET · PHAVRVSII ·
THALA
MONS
ARMIA

CANARIA
ARSINARIVM · PROMONT
RYSADIVM · PROMONT
STACHIRIS · FLV
SOPHVCAEI
STACIRE
NVBI ·
THALAE ·

PINTVARIA
PERPHOSIVS
PORTVS
NIAS FLV
SALADIRVS
MONS
· CHVRITE ·
· MIMACES ·

SINVS · HESPERIVS ·

HESPERI · CERAS
SEVCORNV
CAPRA
· ORPHES ·
· AFRICERONES ·
· MAGNA · GENS ·
· ACHAEMAE ·

LEVCAE
THIOPES
MONS
· TARVALTE ·
· GONGLAE ·

MASITHOLVS FLV
· PYRRVS · CAMPVS ·
NANOSBES ·
· DERBICES

HYPODROMVS
AETHIOPIAE
· PYRORSI ·
ARVALTES ·
NABATHRA

DEORVM
CVRRVS
MONS
REGIO
· MATITES ·
MONS
DERMONES

· AETHIOPVM · INQVA · ELEPHATES · CANDIDI ·
· XYLICCES · AETHIOPES ·

MILIARIA · 62 · ½

· ICHTHYOPHAGI · AETHIOPES ·

· OMNES · GIGNVNTV

· ATHACE ·
· AETHIOPES ·

AETHIOPIA · INTERIOR ·

· HESPERII ·
· AETHIOPES ·

ION · MONS
XIPHAS · MONS

DAVCHIS · MONS
MESCHAE · MONS

MILIARIA · 60 · FERE ·

· AGISYMBA · REGIO · AETHIOPVM ·

XV. PTOLEMÆUS ROMÆ 1490.

4 · 40 · 44 · 60 · 64 · 70 · 74 · 80 · 84

PELOPONESVS
CRETA · RHODVS · ISSICVS
MESO POTAMI
P A R S · AE · 37
DECIMVS · PARALELLVS · HABET · DIEM · MAXIMVM · HORARVM · 19 ⅓

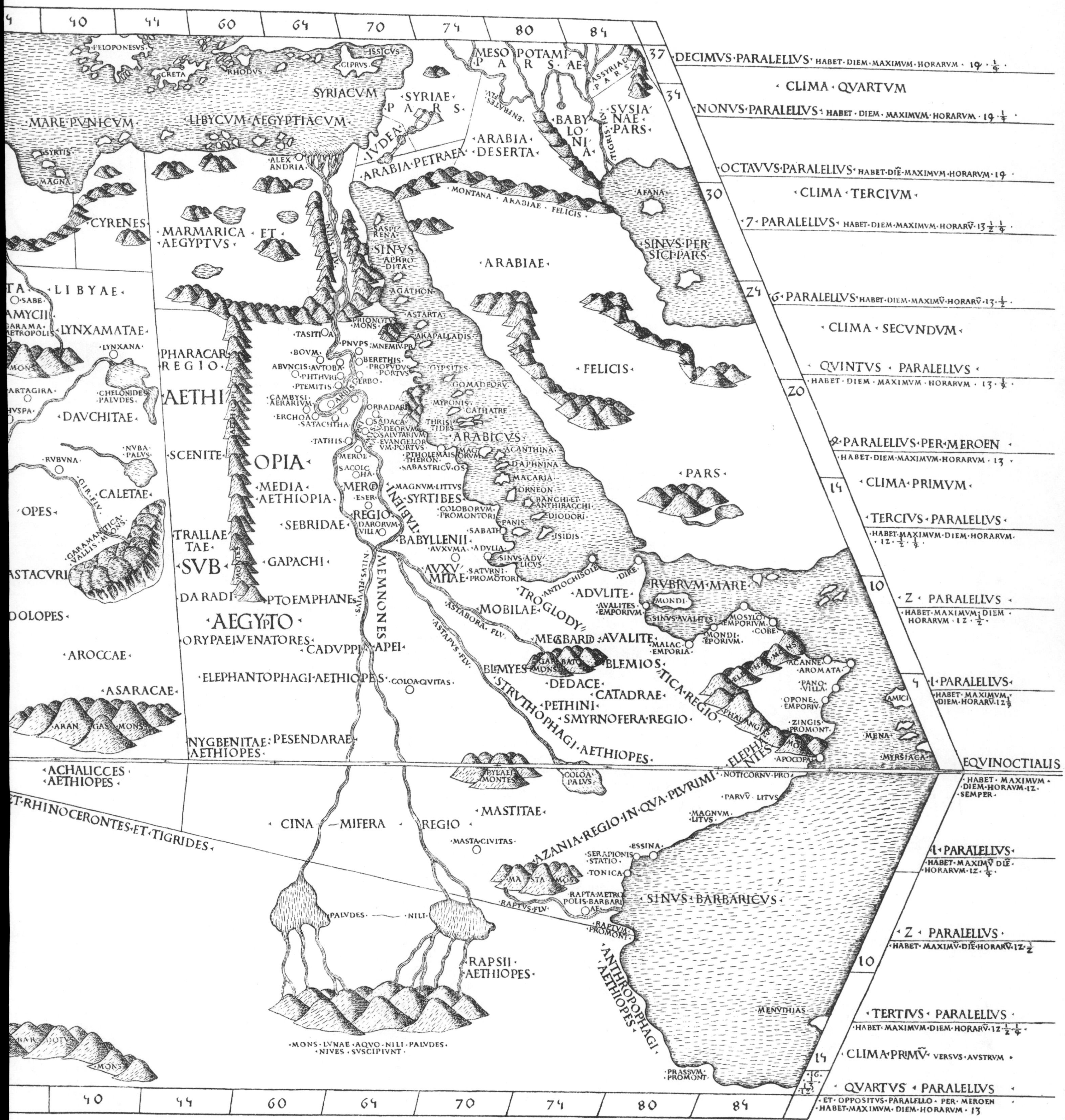

SYRIACVM
SYRIAE
PARS
ASSYRIAE
PARS
34
· CLIMA · QVARTVM ·

MARE · PVNICVM
LIBYCVM · AEGYPTIACVM
IVDEA
ARABIA
DESERTA
EVPHRATES
BABY
LO
NIA
SVSIA
NAE
PARS
TIGRIS
NONVS · PARALELLVS · HABET · DIEM · MAXIMVM · HORARVM · 19 ⅓

SIRTIS
MAGNA
ALEX
ANDRIA
ARABIA · PETRAEA
OCTAVVS · PARALELLVS · HABET · DIE · MAXIMVM · HORARVM · 19

CYRENES
MARMARICA · ET ·
AEGYPTVS ·
MONTANA · ARABIAE · FELICIS ·
AFANA
30
· CLIMA · TERCIVM ·

SINVS
ACHRO
DITA
ARABIAE
SINVS · PER
SICI · PARS
7 · PARALELLVS · HABET · DIEM · MAXIMVM · HORARV · 13 ½ ⅓

TA
LIBYAE
SABE
AGATHON
24
6 · PARALELLVS · HABET · DIEM · MAXIMV · HORARV · 13 ⅓

AMYCII
GARAMA
METROPOLIS
LYNXAMATAE
LYNXANA
ASTARTA
· CLIMA · SECVNDVM ·

MONS
ARTAGIRA
CHELONIDES
PALVDES
PHARACAR
REGIO ·
TASITIOA
PNVPS · MNEMIV · PR
AKAPALLADIS
FELICIS ·
QVINTVS · PARALELLVS
HABET · DIEM · MAXIMVM · HORARVM · 13 ½

HVSPA
DAVCHITAE
AETHI
BOVM
ABVNCIS · AVTOBX
PHTHVRI
BERETHIS
PROPVDVS
PORTVS
GYPSITES
20

RVBVNA
NVBA
PALVS
PTEMITIS
GERBO
GOMADEORV
MYPONIS
CATIATRE
2 · PARALELLVS · PER · MEROEN
HABET · DIEM · MAXIMVM · HORARVM · 13

OPES
GIR · FLV
CALETAE
OPIA
CAMBYSI
AERARIVM
ERCHOAC
SATACHITHA
ARVIS
ORBADARI
SADACA
DEORVM
THRISI
TIDES
ARABICVS
ACANTHINA
18
· CLIMA · PRIMVM ·

GARSMANTICA
VALLIS · MONS
MEDIA
AETHIOPIA ·
TATHIS
SALVTARIVM
VM · PORTVS
EVANGELOR
DAPHNINA
TERCIVS · PARALELLVS
HABET · MAXIMVM · DIEM · HORARVM · 12 ½ ⅓

ASTACVRI
SEBRIDAE ·
MERO
SACOIC
OHA
MEROE
EXER
PTOLEMAIS
THERON ·
SABASTICV · OS
MAGI
MACARIA
TORNEON
2 · PARALELLVS
HABET · MAXIMVM · DIEM · HORARVM · 12 ½

OOLOPES
TRALLAE
TAE
SVB ·
GAPACHI
REGIO
DARORVM
VILLA
MAGNVM · LITTVS
COLOBORVM
PROMONTORI
RANCHI · ET ·
ANTHIBACCHI
DIODORI
SINVS · ADV
10

DA RADI
PTOEMPHANE
SABATH
EPANIS
ISIDIS ·

AROCCAE
AEGYPTO ·
ORYPAEI · VENATORES ·
AVXV
MITAE
AVXVMA
ADVLIA
SATVRNI
LICVS
ANTIOCHISOLI
TROGLODY
DIRE
ADVLITE
RVBRVM · MARE
I · PARALELLVS
HABET · MAXIMVM ·
DIEM · HORARVM · 12 ⅔

ASARACAE
CADVPPI
APEI
ASTABORA · FLV
ASTAPVS · FLV
MOBILAE ·
MEGBARI
AVALITES · EMPORIVM
MONDI
MOSYLI
EMPORIVM
COBE
ACANNE
AROMATA

ARAN
GIAS · MONS
ELEPHANTOPHAGI · AETHIOPES ·
COLOA · CIVITAS
BLEMYES
SARBATA
MONS
AVALITE
MALAC
EMPORIA
MONDI
EPORIVM
PANO
VILLA

NYGBENITAE · PESENDARAE ·
AETHIOPES ·
STRVTHOPHAGI · AETHIOPES ·
DEDACE ·
CATADRAE
PETHINI
SMYRNOFERA · REGIO
TICA · REGIO
ELEPHAS · MONS
OPONE
EMPORIVM
ZINGIS
PROMONT
TAMICI
MENA

ACHALICCES ·
AETHIOPES ·
PYLAEI
MONTES
COLOA
PALVS
NOTICORNV · PRO
ELEPHA
NTES
APOCOPA
MYRSIACA
EQVINOCTIALIS
· HABET · MAXIMVM ·
DIEM · HORARVM · 12 ·
SEMPER ·

ET · RHINOCERONTES · ET · TIGRIDES ·
CINA — MIFERA
REGIO ·
MASTITAE
AZANIA · REGIO · IN · QVA · PLVRIMI ·
PARVV · LITVS
MAGNVM
LITVS
I · PARALELLVS ·
HABET · MAXIMV · DIE ·
HORARVM · 12 ⅓

MASTA · CIVITAS
ESSINA
SERAPIONIS
STATIO
TONICA
2 · PARALELLVS ·
HABET · MAXIMV · DIE · HORARV · 12 ½

PALVDES ·
NILI
RAPTVS · FLV
RAPTA · METRO
POLIS · BARBARI
AE
RAPTVM
PROMONT
SINVS · BARBARICVS
ANTHROPOPHAGI
AETHIOPES
10

RAPSII ·
AETHIOPES ·
Z · PARALELLVS ·
HABET · MAXIMV · DIE · HORARV · 12 ½

MENVTHIAS
TERTIVS · PARALELLVS ·
HABET · MAXIMVM · DIEM · HORARV · 12 ½ ⅓

MARPADOT
MONS
MONS · LVNAE · AQVO · NILI · PALVDES ·
NIVES · SVSCIPIVNT ·
14
· CLIMA · PRIMV · VERSVS · AVSTRVM ·

PRASSVM
PROMONT
G
12
QVARTVS · PARALELLVS ·
ET · OPPOSITVS · PARALELLO · PER · MEROEN
HABET · MAXIMVM · DIEM · HORARVM · 13

40 · 44 · 60 · 64 · 70 · 74 · 80 · 84

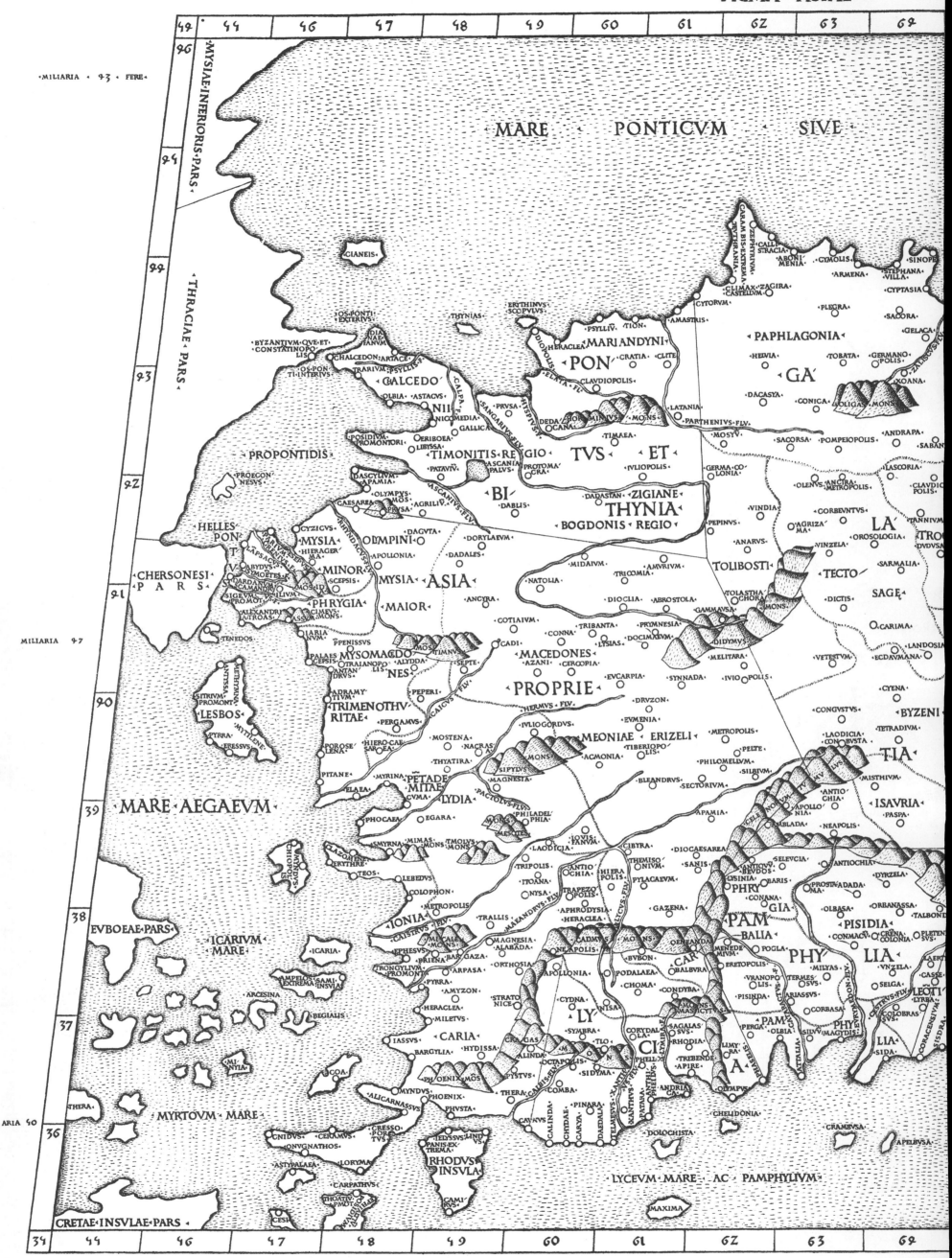

XVI. PTOLEMÆUS ROMÆ 1490.

| 64 | 66 | 67 | 68 | 69 | 70 | 71 | 72 | 73 |

· COLCHIDIS · P A R S ·

26

PONTVS · EVXINVS ·

·DIFFERT·AB·EQVINOCTIALI·HORIS· 3· 2 1
·HABENS·MAXIMVM·DIEM·HORARVM· 14·½

25

SEBASTO
POLIS

· CLIMA · SEXTVM ·

24

APSO
RVS

XYLINA

ARCHABIS·FLV

CISSA·FLV

·ASIBA· CISSI

CERASVS·

·PHARNACIA·

HYSSI·
PORTVS

·TRAPEZVS·

·CAMVRESARB

·13·PER·BYZANTIVM·

ISCHOPOLIS

PONTVS·
CAPPADOCIAE

·CORDVLA·

·DIFFERT·AB·EQVINOCTIALI·HORIS· 3· 3 1
·HABENS·MAXIMVM·DIEM·HORARV· 14·¾

23

·COCALIA·

·ORBALISSENA·

ARME

·AZA·

·SINIBRA·

CALORVM·
·AMISSVS·

·EVSENA·
·TITVA·

LEVCO
SYRI

CAP

LEVCOSTRORVM·
CVBITVS·

HERCVLISPO
PRONTORIVM·

THEMISCYRA

THERMODON·FLV

IRIS·FLVVIS

·AMONTORIVM·
ARGYRIA

CYTEORVM·
HERMONASSA

POLEMONIVM
GENONIA

GASALENA

NIA

·SATALA· ·TAPVRA·

·DOMANA·

·AZIRIS·

·CHARAX·

22

·ANDROSIA·

·AMASIA·

PONTVS·POLEMONIACVS

TEBENDA·

·BARBA
NISSA·

·NICOPOLIS·

·CHORSABIA·

·LADANA·

·CARISSA·

PA

·EVDIPHVS·

CARVANIS·

·ABLATA·

AETVLANA·

·BOENASSA·

·CHOLOGI·

·NEOCAESAREA·

·PIDA·

·PIALA·

·SERMVGA·

COMANA·
PONTICA·

MEGALVLA·

·SISMARA·

MI

·ETHONIA·

·PLEVRAMIS·

SEBASTO
POLIS

ZELA·

·DANATI·

DO

·DAGONA·

·CALTIORISSA·

MI

·ANALIBLA·

·SELEOBERIA·

ARMENIAE

·RASTIA·

·ZAMA·

·GADASENA·

PRAETVRA

PHIARA·

MEGA
LOSSVS

AERETHICA

·PISINGARA·

·GODASA·

MAIO

·SADAGENA·

ARIATHIRA·

·SAVRANIA·

ANTI·TAVRVS·

·MASORA·

·ZIMARA·

ROSER
MENI
AE

TENESSVS·

PRAEFECTVRA·CHA
MANESIS·

SARGAVRASENA

·SEBASTIA·

·GAVRAENA·

·SARVENA·

·ANDRACA·

·ODOGA·

SABALASSVS·

·MAROGA·

HORSENA·

·DASCVTA·

ANTI·TAVRVS·

RIS

21

SARALVS·

ZEPHYRIVM·

·ISPA·

CLIMA · QVINTVM ·

VADATA·

ARGEVS

SIVA·

EBAGENA·

C

ARCHALLA

ORBISENA

N

·PHVPHENA·

·ARANA·

·DAGVSA·

EVPHRATES·FLVVIVS·

·CARAPA·

PHREATA·

MOS·

PRAEFECTVRA·CILICIAE·

PHVPHATENA·

·OROMADVS·

·ZOPARISTVS·

·SINISCO
LONO·

PARS

20

APERTA·

PRAETVRA·GAR

NANESSVS·

MVSTILIA·

CAMPE·

·MARDARA·

·PHVSIPARA·

ARCHELAIS·

MAZA·QVE·ET·
CAESAREA

SOBAR·

·VARSAPA·

·IASSVS·

·TITARISSVS·

MELITENA·

ASADA·

·DIOCAESAREA·

NELOS·FLV

·CIANICA·

·ADOPISSVS·

DAV

SALAMBRIAE

TETRAPYRGIA·

CYZISTRA·

·HORSA·

MELI

·EVSIMARA·

TENA

·CARMALA·

CRETA

I

·SINDITA·

CIACIS·

O

·CORNA·

YCAONIA

PRAEFECTVRA·TYA

BAZIS·

CHOTAENA·

ZOROPASSVS·

·LEVCAESA·

·METITA·

DRATE·

NIDIS·

TYANA·

SIAL

LADOENERIS·

·SEMISVS·

39

CANNA·

CHASBIA·

PRAEFETVRA·MVRIANNA

·NYSA·

·ZIZOATRA·

·CLAVDIA·

·11·PARALELLVS·

ICONIVM·

TAVRVS·

M·O·N·S·

CABASSVS·

·CARNALIS·

·PASARNA·

·DIFFERT·AB· EQVINOCTIALI
·HORIS· 2· 3 1 ·HABENS·
·MAXIMVM· DIE·HORARV·
19· 2 1

PARALAIS·

BARAT
THA·

TYNNA

ARASAZA·

GARNA·CA·

·CIZARA·

·NOSALENA·

IVLIOPO
LIS·

A

TIRALLIS·

·SABAGENA·

ARA
SERASTERA

·MONS·

DERBA·

ANTI

CORNA·

CYBISTRA·

PRAEFECTVRA·CATAONIA

PADIANDVS·

·COMANA·

PRAEFECTVRA·
LANIANA

VENA

LACRIASSVS·

·BARZALO

38

LARANDA·

OLBASA·

MVSBANDA·
OANA·

CLAVDIO
POLIS·

DALISADVS·

TAVRVS·MOS·

TANADARIS·

·LEANDIS·

R

·LAVTASA·

ENTELIA·

TAV

RVS

·OLBASA·

NINICA·

FLAVIOPO
LIS·

LACANI
TIDIS·

AVGVSTA·

·NICOPOLIS·

MESOPOTAMIAE

37

·CETIDA·

LALASSIDA·

CHARACIA

DOMICIO
POLIS·

PHILADE
PHIA·

DIOCAESAREA

IRINOPOLIS·

LAMO
TIDIS·

BRYELICES

CILICIAE·

CI

STRA
THIA·

CAYSTRVS·

OLMEA

PINARVS·FLV

ORIMAGLA

ARSINOE

PHRODI
SIA·

SELEV
CIA

CALDVS·FLV

SARPEDON·

CORICVS·

SEBASTE

CI

CAESAR
EA

MOPSIESTIA·

CASTABALA·

A

PROPRIAE

PARS

36

CILI

ANTIOCHIA·

SELENVS·

ANEMVRIVM

CELENDERIS·

SARDENIS

TARSOS·

CYDNVS·FLV·

SARVS·FLV·

ADANA

PYRAMVS·FLV·

MALLVS·

EPIPHANIA·

·10·PER·RHODV

NEPHE
LIS·

PIRAIVS·FLV·

PIERIVS

EGEE·

AMANVS·MONS·

·DIFFERT·AB·EQVI
NOCTIALI·HORIS·
2· 4 1 ·HABENS·
·MAXIMVM· DIE·
HORARVM· 14·½

ISSVS·

·PORTAE·VEL·
PYLAE·AMANICAE·

·ANGVSTIAE · CILICIAE ·

·SINVS·ISSICVS·

· SYRIAE · PARS ·

· CLIMA · 4 ·

· CYPRI · INSVLAE · PARS ·

CLIDES·

| 64 | 66 | 67 | 68 | 69 | 70 | 71 | 72 | 73 | 34 |

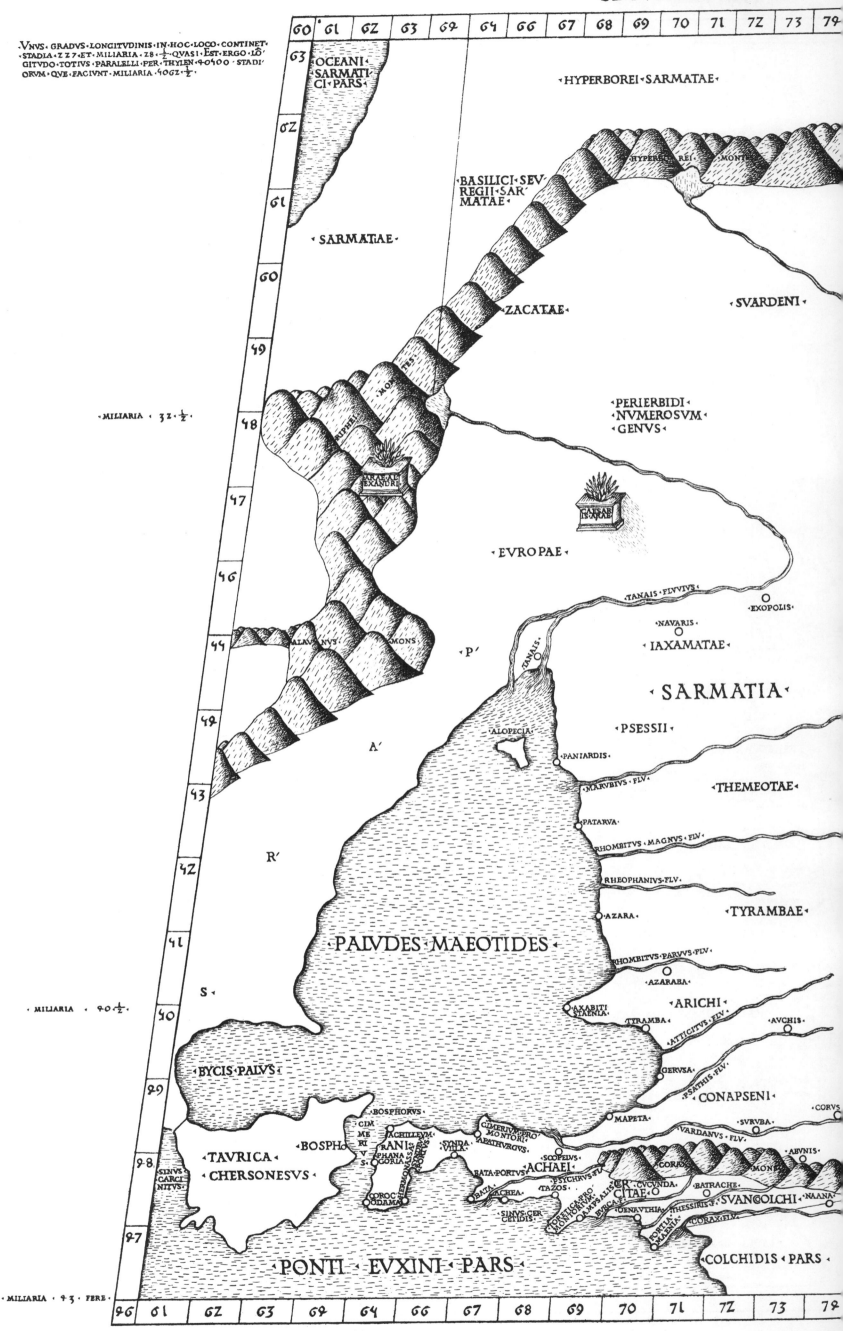

·VNVS· GRADVS· LONGITVDINIS · IN· HOC· LOCO· CONTINET·
·STADIA· Z Z7· ET· MILIARIA· Z8· ½· QVASI· EST· ERGO· LŌ
GITVDO· TOTIVS· PARALELLI· PER· THVLEN· 4O4OO· STADI
ORVM· QVE· FACIVNT· MILIARIA· 4O6Z· ½·

HYPERBOREI · SARMATAE ·

OCEANI
SARMATI
CI· PARS

·BASILICI· SEV·
·REGII· SAR·
·MATAE·

HYPER   REI   MONT

·SARMATAE·

·SVARDENI·

·ZACATAE·

·MILIARIA· 3 Z ½·

·PERIERBIDI·
·NVMEROSVM·
·GENVS·

·EVROPAE·

·TANAIS· FLVVIVS·

O
·EXOPOLIS·

ALA   NVS·   MONS

·NAVARIS·
O
·IAXAMATAE·

·P'

·SARMATIA·

·ALOPECIA·

·PSESSII·

·PANIARDIS·

·THEMEOTAE·

·MARVBIVS· FLV·

·A'

·PATARVA·

·RHOMBITVS· MAGNVS· FLV·

·R'

·RHEOPHANIVS·FLV·

·AZARA·

·TYRAMBAE·

·RHOMBITVS· PARVVS· FLV·

·PAIVDES· MAEOTIDES·

O
·AZARABA·

·S

·MILIARIA· 4O ½·

·AXABITI·
·STAENIA·

·ARICHI·

·TRAMBA·

·AVCHIS·

·ATTICITVS· FLV·

·EYCIS· PALVS·

·GERVSA·

·PSATHIS· FLV·

·CONAPSENI·

·BOSPHORVS·

·MAPETA·

·CORVS

·VARDANVS·
·SVRVBA·
FLV·

·TAVRICA·
·CHERSONESVS·

·BOSPHO

CIM
ME
RI
V
S

ACHILLEVM

RANIS
PHANA
GORIA

SYNDA
VILLA

CIMERIVM PRO
MONTORI
APATHVRGVS·

·SCOPELVS·
·CORA

·ABVNIS·

MONS

·SINVS·
·CARCI·
NITVS·

COROC
ODAMA

HERMO

ACHAEI
RATA PORTVS·
RATA   ACHEA

PSYCHRVS· F·
·TAZOS·   CITAE
OENIA

·SINVS· CER·
CETIDIS·

CVCVNDA·
SER   O
CA
OAMPSALIS·
·FORETICA· PRO   CORAX
MONTORIVM·
OENAVTHIA·
PORTVS

BATRACHE

SVANCOLCHI

NAANA

THESSIRIS· F·
CORAX· FLV·

AZARIA·

·PONTI· EVXINI· PARS·

·COLCHIDIS· PARS·

·MILIARIA· 4 3· FERE·

·TABVLA·

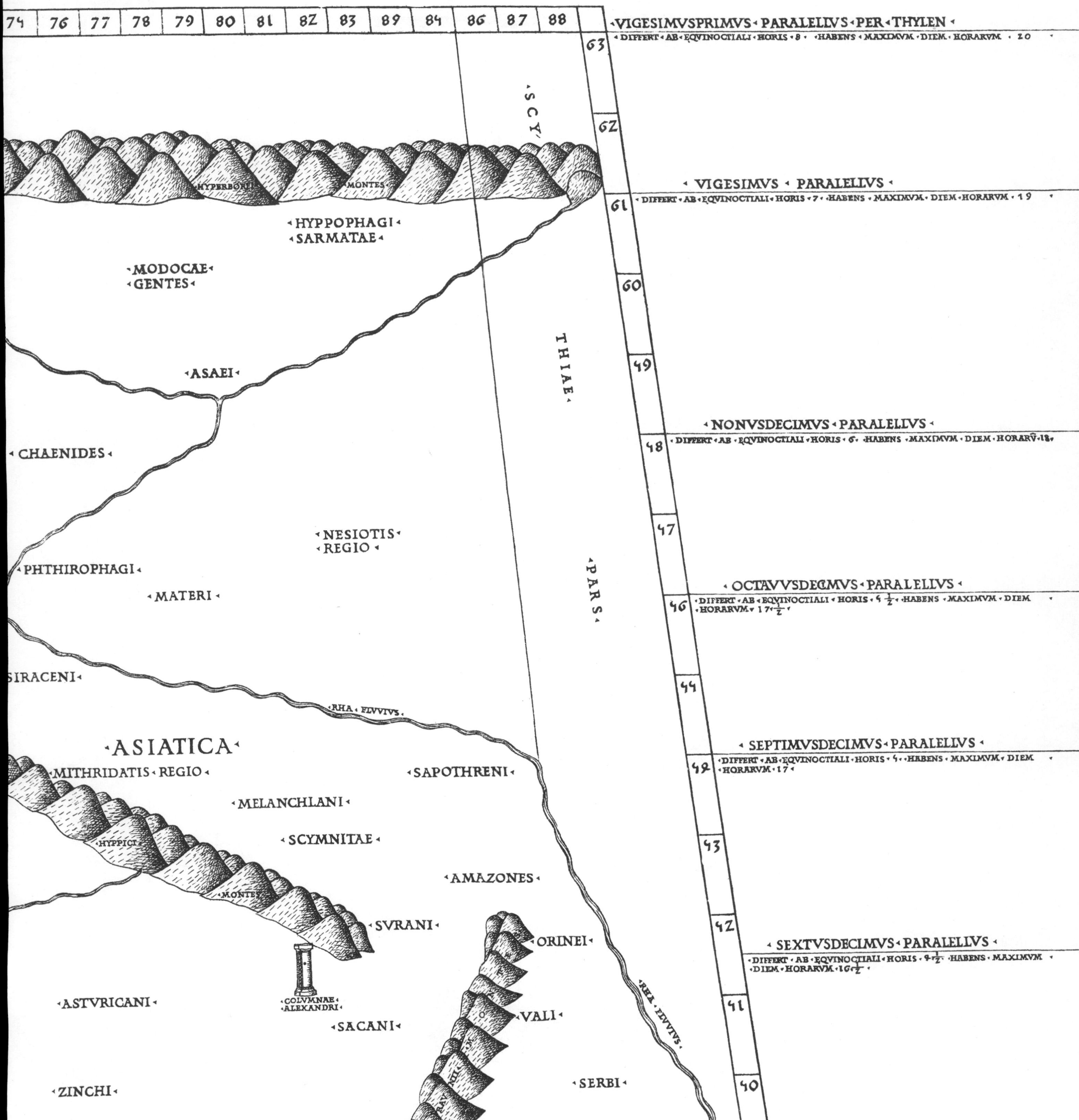

74	76	77	78	79	80	81	8Z	83	89	84	86	87	88

·VIGESIMVSPRIMVS·PARALELIVS·PER·THYLEN·
·DIFFERT·AB·EQVINOCTIALI·HORIS·8·HABENS·MAXIMVM·DIEM·HORARVM·20

63

6Z

·SCY·

·VIGESIMVS·PARALELIVS·
·DIFFERT·AB·EQVINOCTIALI·HORIS·7·HABENS·MAXIMVM·DIEM·HORARVM·19·

61

·HYPPOPHAGI·
·SARMATAE·

·MODOCAE·
·GENTES·

60

·ASAEI·

THIAE·

49

·NONVSDECIMVS·PARALELIVS·
·DIFFERT·AB·EQVINOCTIALI·HORIS·6·HABENS·MAXIMVM·DIEM·HORARV·18·

48

·CHAENIDES·

47

·NESIOTIS·
·REGIO·

PARS·

·PHTHIROPHAGI·

·OCTAVVSDEGMVS·PARALELIVS·
·DIFFERT·AB·EQVINOCTIALI·HORIS·4½·HABENS·MAXIMVM·DIEM·
HORARVM·17½·

46

·MATERI·

44

SIRACENI·

·RHA·FLVVIVS·

·SEPTIMVSDECIMVS·PARALELIVS·
·DIFFERT·AB·EQVINOCTIALI·HORIS·4·HABENS·MAXIMVM·DIEM·
HORARVM·17·

4Z

·ASIATICA·

·MITHRIDATIS·REGIO·

·SAPOTHRENI·

·MELANCHLANI·

43

·HYPPICI·

·SCYMNITAE·

·AMAZONES·

·MONTE·

4Z

·SVRANI·

·ORINEI·

·SEXTVSDECIMVS·PARALELIVS·
·DIFFERT·AB·EQVINOCTIALI·HORIS·4½·HABENS·MAXIMVM·
DIEM·HORARVM·16½·

·COLVMNAE·
·ALEXANDRI·

·ASTVRICANI·

·VALI·

41

·SACANI·

·ZINCHI·

·SERBI·

40

·TVSCI·

·VDAE·

·METIBI·

·DIDVRI·

·VDON·FLV·

·SALAGA·FLVVIVS·

Z9

·SERACA·

·QVINTVSDECIMVS·PARALELIVS·
·DIFFERT·AB·EQVINOCTIALI·HORIS·9·HABENS·MAXIM·
VM·DIEM·HORARVM·16·

·EBRIAPA·

·PORTAE·
·SARMATICAE·

·AGORITAE·

·OLONDAE·

Z8

ASVNIA· MIA·

·CLIMA·SEPTIMVM·

·ALEVS·FLV·

·CAV CASVS·

·SANARAEI·

·ISONDAE·

·SOANA·FLV·

·GERRI·

Z7

·PORTAE·MONS·
·ALBANIAE·

·CAV CASVS·

·IBERIAE·
·PARS·

·ALBANIAE·PARS·

·HYRCANI·MARIS·PARS·

74	76	77	78	79	80	81	8Z	83	84	84	86	87	88	Z6

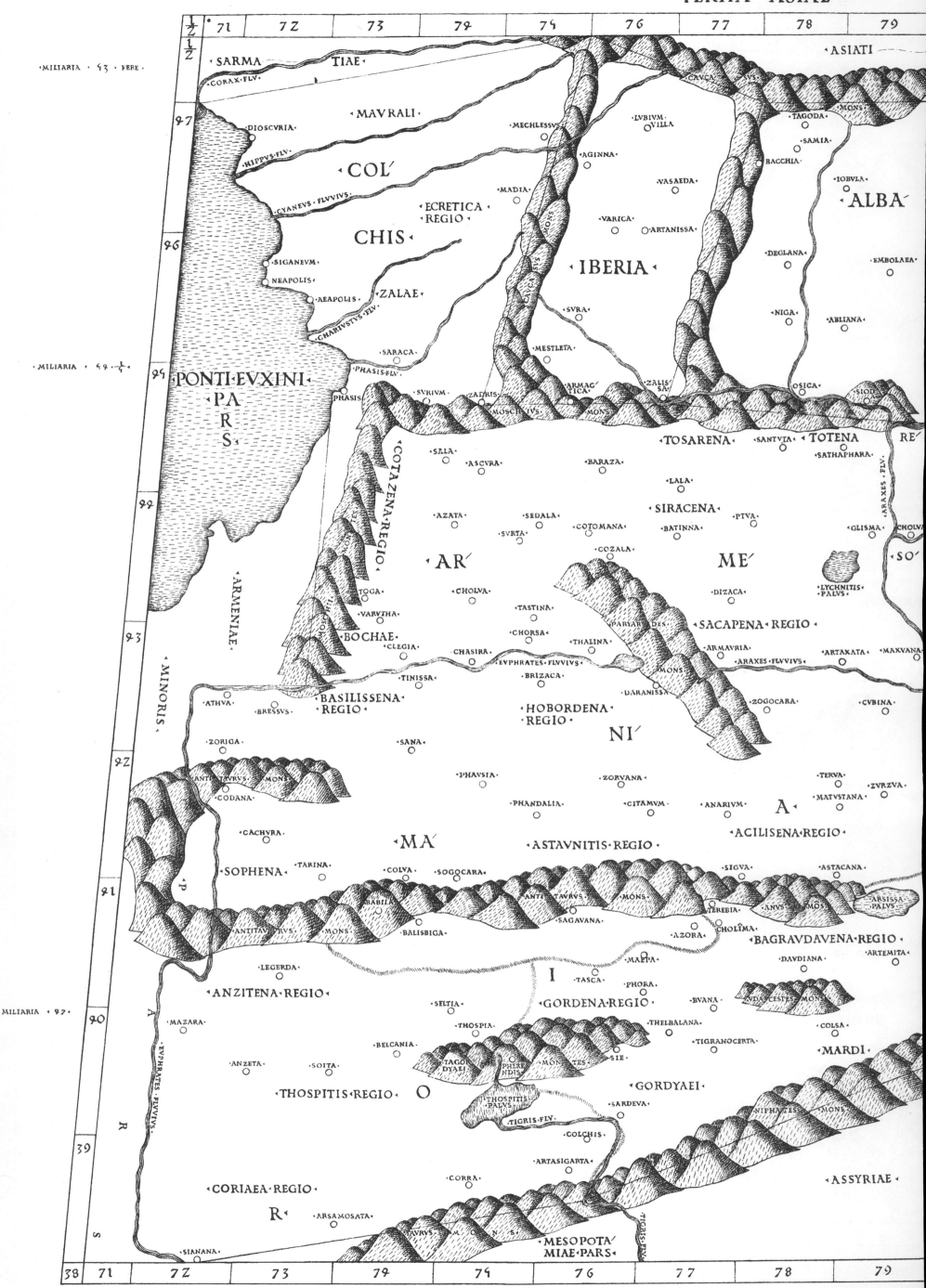

XVIII. PTOLEMÆUS ROMÆ 1490.

80   81   8Z   83   8٩   8٤   86   87   ½

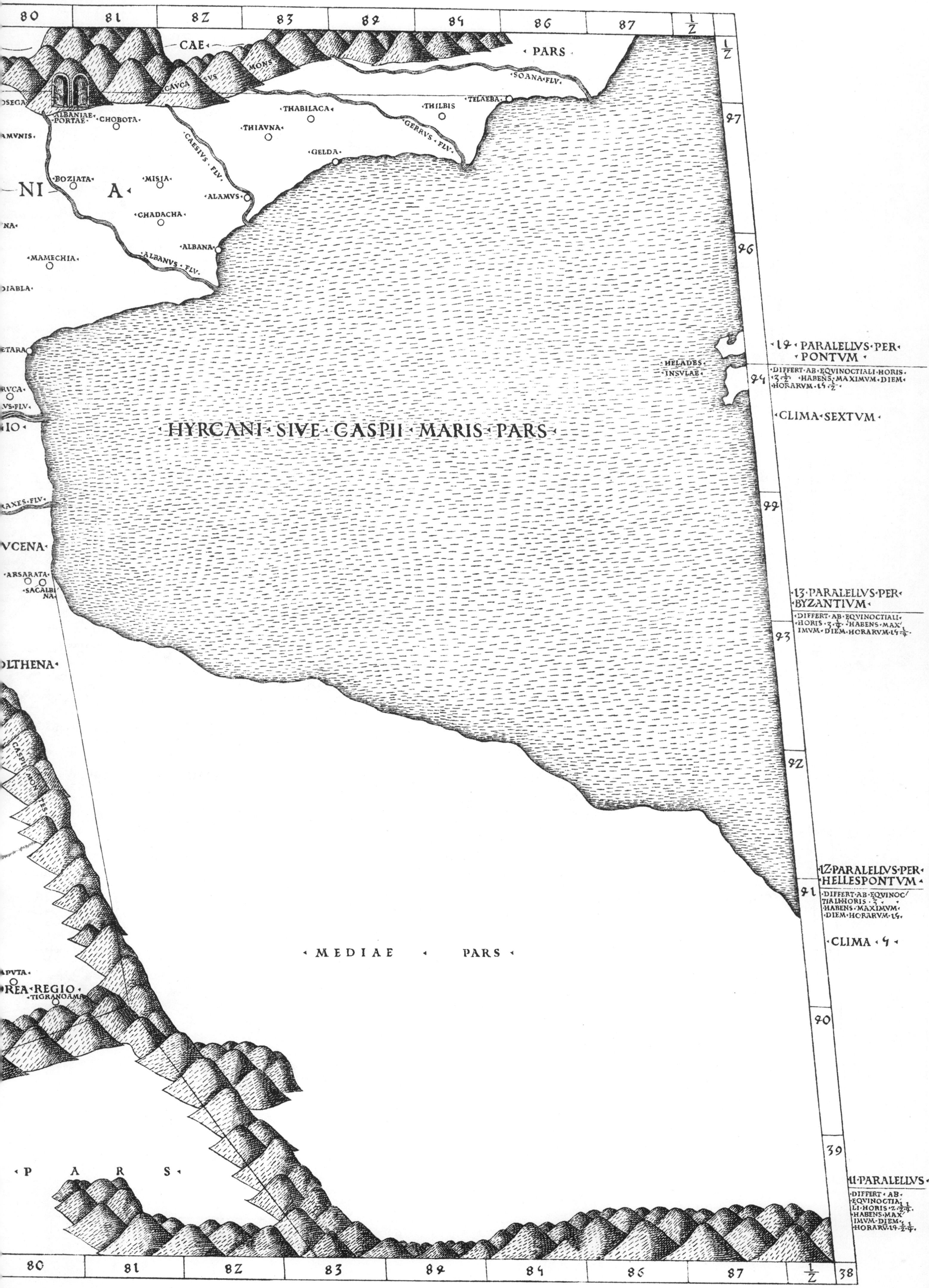

· CAE ·

· PARS ·

MONS

· CAVCA · SVS ·

· SOANA · FLV ·

OSEGA

· ALBANIAE · PORTAE ·   CHOBOTA ·

· THABILACA ·

· THILBIS ·

· TELAEBA ·

AMVNIS ·

· THIAVNA ·

· GERRVS · FLV ·

· CAESIVS · FLV ·

NI   A

· BOZIATA ·

· MISIA ·

· GELDA ·

· ALAMVS ·

NA ·

· CHADACHA ·

· MAMECHIA ·

· ALBANA ·

· ALBANVS · FLV ·

DIABLA ·

ETARA ·

RVCA ·

VS · FLV ·

IO ·

HYRCANI · SIVE · CASPII · MARIS · PARS

· HELADES · INSVLAE

· 1٩ · PARALELLVS · PER · PONTVM ·

· DIFFERT · AB · EQVINOCTIALI · HORIS · 3 ¾ · HABENS · MAXIMVM · DIEM · HORARVM · 1٩ ½ ·

84

· CLIMA · SEXTVM ·

RAXES · FLV ·

VCENA ·

· ARSARATA ·

· SACALBI NA ·

· 13 · PARALELLVS · PER · BYZANTIVM ·

· DIFFERT · AB · EQVINOCTIALI · HORIS · 3 ½ · HABENS · MAX · IMVM · DIEM · HORARVM · 1٩ ¼ ·

93

OLTHENA ·

CASPII PROGRESS

92

· 1Z · PARALELLVS · PER · HELLESPONTVM ·

· DIFFERT · AB · EQVINOC · TIALI · HORIS · 3 · HABENS · MAXIMVM · DIEM · HORARVM · 1٩ ·

91

· CLIMA · ٩ ·

APVTA ·

REA · REGIO ·

· TIGRANOAMA ·

· MEDIAE · PARS ·

90

39

· PARS ·

· 11 · PARALELLVS ·

· DIFFERT · AB · EQVINOCTIA · LI · HORIS · Z · ¾ · HABENS · MAX · IMVM · DIEM · HORARV · 1٩ ¼ ·

80   81   8Z   83   8٩   8٤   86   87   ½   38

XIX. PTOLEMÆUS ROMÆ 1490.

| 73 | 79 | 74 | 76 | 77 | 78 | 79 | 80 | 1/2 |

1/2

· MAIORIS · PARS ·

· OSARARA

· ANTHEMVSIA ·

· DORBE
TA

38

· BITHIAS · O
· SACANA O
· AMMAEA O
· BITHIGA ·

· SAPPHA ·

TIGRIS·FLVVIVS ·

· ASSY — RIAE · · P ·

· CAPRVS·FLV·

· GORGOS·FLV·

· PORSICA · O
· EDESSA · O
· RHISINA · O
· SYNNA ·
· MAMBVTA ·
· NISIBES ·
· DEBA · O

· CHALCITIS ·
REGIO ·

· SVMA · O
CASIVS · O
· GIZAMA
· ARSAMA ·
· BAZALA O
· SYNGARA O
· BETOVM · O

· EVPHRATES·FLV·

· OMBRAEA · O

37

· ME
· OLIBERA · O

· ELIIA · O
· ZAMA O
· SYNNA · O

· P ·

· EVROPVS ·
· CAECILIA ·
· ANIANA O
· AVLADIS · O
· BALLATHA ·
· GAVZANI — TIS·REGIO ·
· THENGVBIS ·

· ACABENA ·
REGIO ·
· LAMBANA O
· BIRTHA O
· A

· MARIA
· BARSAMPSA O
· GARRAE ·
· SO ·

· SYN CARA ·
M.
· GORBATHA

· CARTHARA O

· SARNVCA · O
· BERSIMA O
· CHABORAS·FLV·
· TIRITTHA · O
· ACRABA · O
· ORTHAGA · O
· INGENA·REGIO ·

· PO ·
· DABAVSA · O
· BARIANA · O
· MANCHANA · O

36

10·PARALELLVS ·
HABES·MAXIMVM·
DIE·HORARV·14.1/2

· SSVS·
· ATHIS ·
· SVRA · O
· RHESAENA · O
· PELIALA · O
· SIPPHARA · O
· SELEVCIA ·

CLIMA·
QVARTV ·

· NITIS ·
· ALALIS · O
· MAVBE · O
· APHADANA ·
· ALVANIS · O
· BIMATRA ·
· NAARDA · O
· TERIDATA · O
· THELBENCANA ·

· T

· R

· DERRIMA O
· NAGVDA
· TAVSACVS ·
NICEPHORIV ·
· CABORA ·
· BA ·
· AVCHANITIS ·
REGIO ·
· MAARSARES·FLV·
· REGIVS·FLV·
· BABYLON ·

· A

35

· RHESAPHA O
· ALAMATHA O
· BIRTHA · O
· ANCOBARITIS·REGIO ·
· DAREMMA ·
· ASCORAS·FLV·
· PACORIA · O
· THACCONA · O
· VOLGAESIA O
· BY ·
· EVPHRATES·FLV·
· TEVSCHA
PHA ·
· TIGRIS·FLVVIVS ·

· S ·

· HOLLA · O
· GADIRTHA · O
· THELDA · O
· APHPHADANA ·
· BARSITA ·
· APAMIA ·

· ORIZA O
· ADADA O
· AVZARA O
· BANABA O
· ZITHA O
· ADDAEA · O
· DVRABA O
· BIBLA · O

34

· AVDATTHA O
· BETHAV
NA O
· A

· ADACHA O
· DADARA O
· RHESCI
PHO
· MIA ·
· EVDRAPA O
· DIDIGVA O

· S ·

· BATHANAEI ·
· BALAGAEA O
· PHARGA ·
· AGAMNA O
· CHVDVGA · O

· CAVCHABENI ·
· EVPHRATES·FLVVIVS ·
· LO ·
· CHVMANA O

· ARABIA ·
· COLARINA O
· BELGYNAEA O
· IDICARA O
· PVNDA O

9·PARALELLVS ·
HABES·MAXIMVM·
DIE·HORARV·14.1/4

· RHEGANA O
· CHALDAEA·REGIO ·
· BETHTHANA O

· SAVA · O
· BARATHENA O
· CAESA O
· ORCHOA · BEANA O

33

· GHOTA · O
· AVRANA O
· GABARA O
· DE ·
· ESITAE ·
· ORTHIDA O
· BIRANDA O
· MARDO — CEA·REGIO ·
· BATRACHARTA O

· THALATHA ·

· AGRAEI ·
· THELMA O
· NI ·

32

· ALACA · O
· MARTENI ·
· STROPHADES ·

· TEMMA ·
O
· IAMBA O
· RHAGIA O
· ALTHA O

· ERVPA · O
· SER ·
· MASANI ·
· TEREDON O

· AGVBENI ·
· LVMA ·
O
· A ·

31

· ODAGANA O
· CHIRIPHA O

· THAVBA · O
· SEVIA · O
· DAPHA · O
· TEDIVM O

· SINVS·MESANITES ·

8·PARALELL ·
HABES·MAXI
MVM·DIEM·
HORARV·19

· SORA O
· RAHABENI ·
· TA ·
· AMMAEA O

· ARTEMITA O
· ARRADA O
· ZAGMAIS O

30

CLIMA·
TERTIV ·

· BANACHA O
· HORCHENI ·

· SINVS·PERSI
CI·PARS
· ARA BIAE ·
· DVMAETHA O

· BERA O
· CALATHVA O
· SALMA O
· IDICARA O

· FELICIS ·
· IVCARA O

29

· P
· A
· R
· S ·

1/2

1/2

| 73 | 79 | 74 | 76 | 77 | 78 | 79 | 80 | 1/2 |

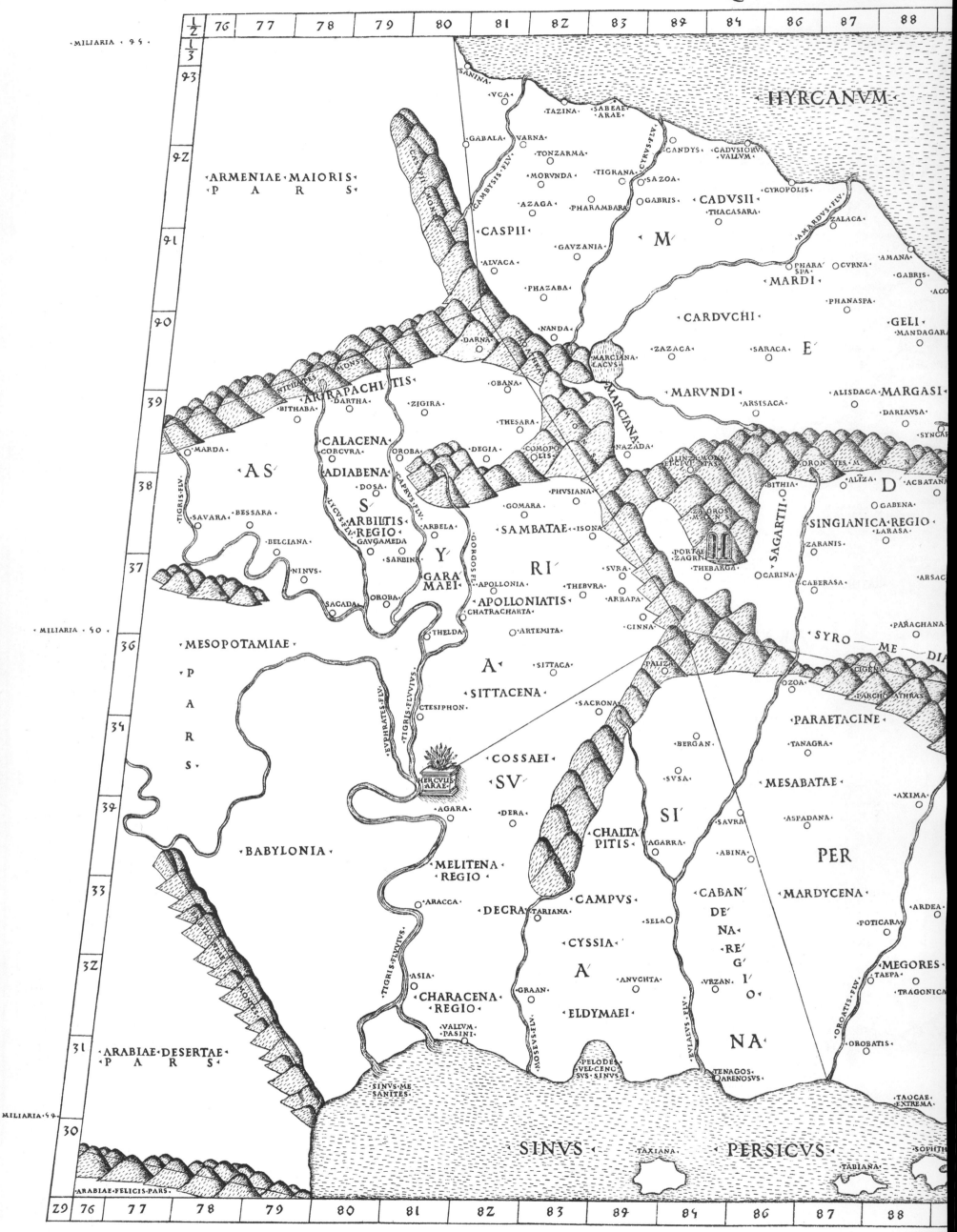

XX. PTOLEMÆUS ROMÆ 1490.

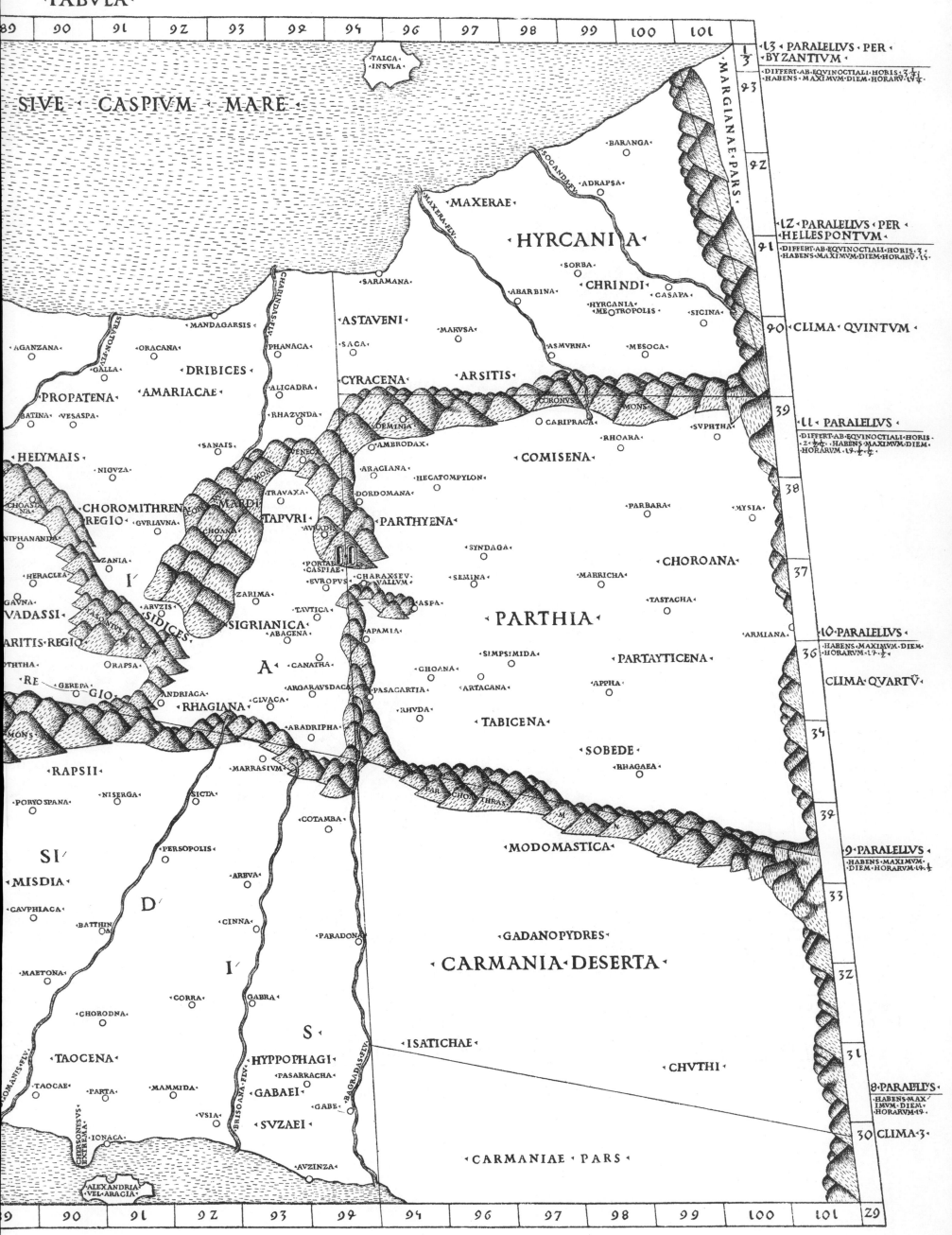

·L3 · PARALELLVS · PER ·
BYZANTIVM ·
· DIFFERT · AB · EQVINOCTIALI · HORIS · 3 ·
HABENS · MAXIMVM · DIEM · HORARV ·

· SIVE · CASPIVM · MARE ·

·TALCA·
·INSVLA·

·BARANGA·

·ADRAPSA·

·MAXERAE·

·HYRCANIA·

·LZ · PARALELLVS · PER ·
HELLESPONTVM ·
· DIFFERT · AB · EQVINOCTIALI · HORIS · 3 ·
HABENS · MAXIMVM · DIEM · HORARV · LS ·

·SORBA·
·CHRINDI·
·SARAMANA·
·ABARBINA·
·HYRCANIA·
·MEOTROPOLIS·
·CASAPA·
·SICINA·

·ASTAVENI·

·MARVSA·
·SACA·
·ASMVRNA·
·MESOCA·

·MANDAGARSIS·
·PHANACA·

·90 · CLIMA · QVINTVM ·

·ORACANA·
·DRIBICES·
·ALICADRA·
·CYRACENA·
·ARSITIS·
·CORONVS·
·MONS·

·39·

·PROPATENA·
·AMARIACAE·
·RHAZVNDA·
·CARIPRACA·
·RHOARA·
·SVPHTHA·
·LL · PARALELLVS ·
· DIFFERT · AB · EQVINOCTIALI · HORIS ·
Z · HABENS · MAXIMVM · DIEM ·
HORARVM · LS ·

·BATINA· ·VESASPA·
·OEMINIA·
·AMBRODAX·
·COMISENA·

·SANAIS·
·VENEC·
·ARAGIANA·
·HECATOMPYLON·
·38·

·HELYMAIS·
·MONS·
·DORDOMANA·

·NIGVZA·
·TRAVAXA·
·PARBARA·
·MYSIA·

·CHOROMITHREN·
·MARDI·
·TAPVRI·
·PARTHYENA·

·CHOASTA·
REGIO·
·GVRIAVNA·
·CHOANA·
·AVIADII·
·SYNDAGA·
·CHOROANA·
·37·

·NIPHANANDA·
·PORTAE·
·CASPIAE·
·SEMINA·
·MARRICHA·

·HERACLEA·
I·
·ZARIMA·
·CHARAX SEV·
VALLVM·
·TASTACHA·

·VADASSI·
·JASONIS·SE·
·SIDICES·
·EVROPVS·
·ASPA·
·PARTHIA·
·ARMIANA·
·LO · PARALELLVS ·
· HABENS · MAXIMVM · DIEM ·
HORARVM · L7 ·

·ARITIS·REGIO·
·ARVZIS·
·SIGRIANICA·
·LAVTICA·
·APAMIA·

·OTHTHA·
·ORAPSA·
·ABACENA·
·SIMPSIMIDA·
·PARTAYTICENA·
·36·

·RE·
·GEREPA·
GIO·
·CANATHA·
·CHOANA·
·APPHA·
CLIMA · QVARTV ·

·ANDRIACA·
·ARGARAVSDACA·
·PASACARTIA·
·ARTACANA·

·RHAGIANA·
·GLVACA·
·RHVDA·

·MONS·
·ARADRIPHA·
·TABICENA·
·34·

·RAPSII·
·MARRASIVM·
·SOBEDE·

·PORVOSPANA·
·NISERGA·
·SICTA·
·PARCHOVTHIRAS·
·RHAGAEA·
·32·

·COTAMBA·
·M·
·9 · PARALELLVS ·
· HABENS · MAXIMVM ·
DIEM · HORARVM · L9 ·

·SI·
·PERSOPOLIS·
·MODOMASTICA·
·33·

·MISDIA·
·ARBVA·

·GAVPHIACA·
D·
·CINNA·
·GADANOPYDRES·

·BATTHIN·
OA·
·PARADONA·
·CARMANIA·DESERTA·
·3Z·

·MAETONA·
I·

·CORRA·
·GABRA·

·CHORODNA·
S·
·ISATICHAE·
·31·

·TAOCENA·
·HYPPOPHAGI·
·CHVTHI·

·PASARRACHA·
·8 · PARALELLVS ·
· HABENS · MAX·
IMVM · DIEM ·
HORARVM · L9 ·

·TAOCAE·
·PARTA·
·MAMMIDA·
·GABAEI·
·GABE·

·VSIA·
·SVZAEI·
·30 CLIMA · 3 ·

·IONACA·

·CHERSONESVS·
EXTRMA·
·AVZINZA·

·ALEXANDRIA·
VEL·ARACIA·
· CARMANIAE · PARS ·

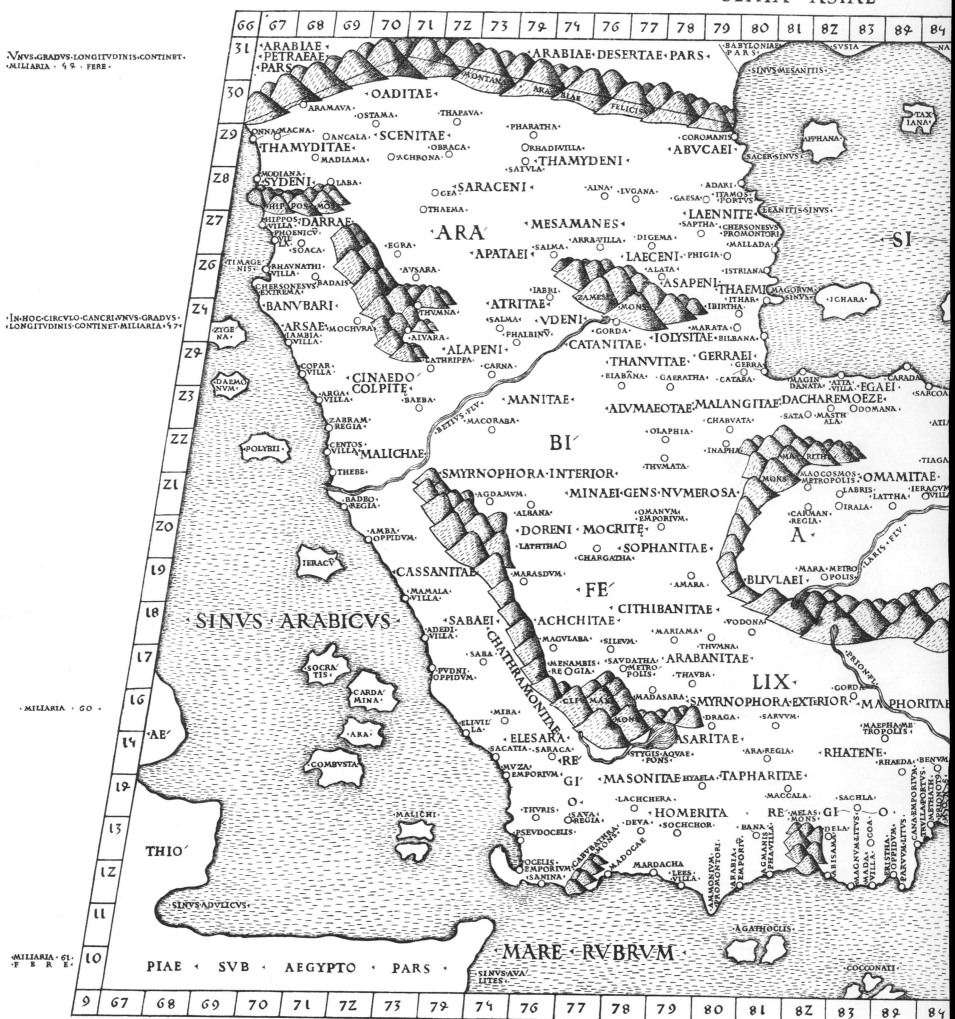

XXI.   PTOLEMÆUS ROMÆ 1490.

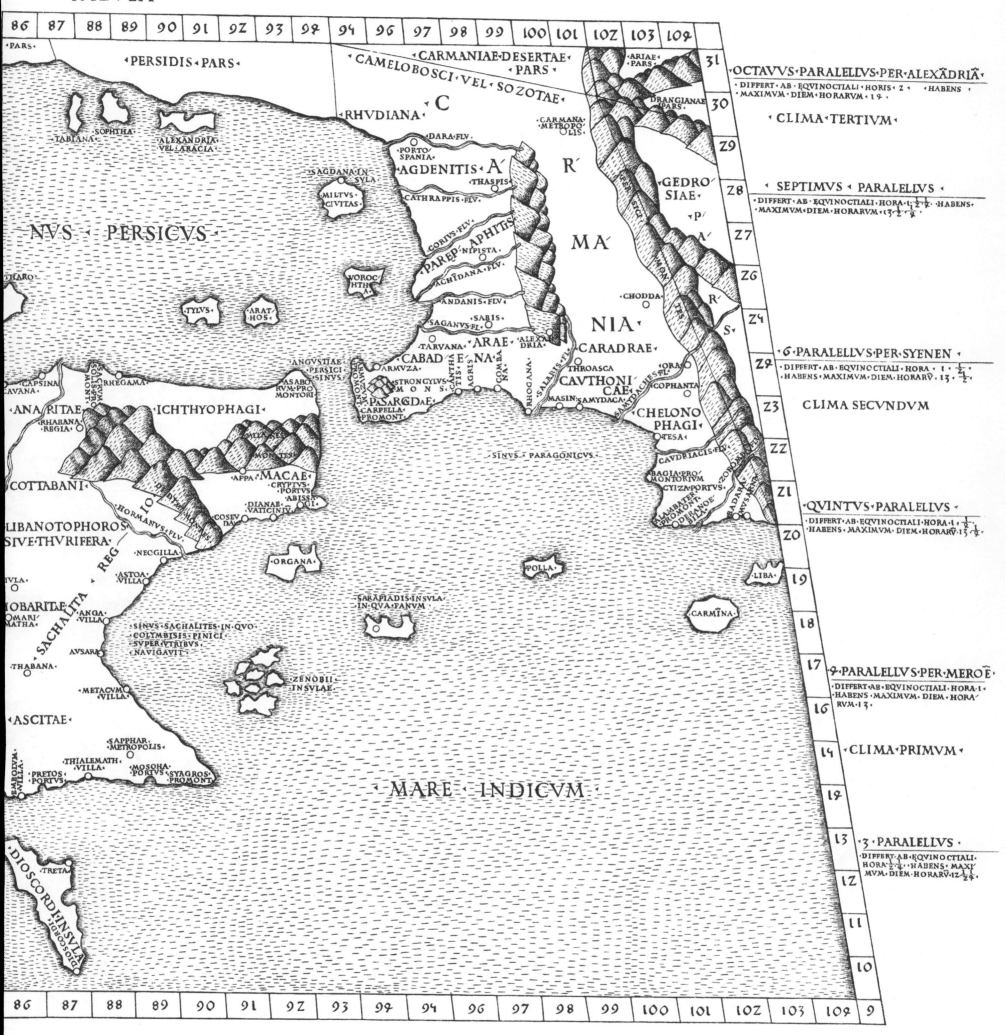

| 86 | 87 | 88 | 89 | 90 | 91 | 9Z | 93 | 9⁴ | 9⁴ | 96 | 97 | 98 | 99 | 100 | 101 | 10Z | 103 | 10⁴ |

· PARS ·

· PERSIDIS · PARS ·

· CARMANIAE · DESERTAE ·
· PARS ·

· CAMELOBOSCI · VEL · SOZOTAE ·

· ARIAE ·
· PARS ·   31

· RHVDIANA ·   C

· SOPHTHA ·

· TABIANA ·   · ALEXANDRIA ·
· VEL · ARACIA ·

· DARA·FLV ·

· DRANGIANAE ·
· PARS ·   30

· CARMANA ·
· METROPO ·
LIS ·

· OCTAVVS · PARALELLVS · PER · ALEXĀDRIĀ ·
· DIFFERT · AB · EQVINOCTIALI · HORIS · Z ·   · HABENS ·
· MAXIMVM · DIEM · HORARVM · 19 ·

· CLIMA · TERTIVM ·

· PORTO ·
SPANIA ·

· SAGDANA·IN ·
SVLA ·

AGDENITIS · A´

R´   Z9

· THASPIS ·

· MILTVS ·
CIVITAS ·

· CATHRAPPIS · FLV ·

MA´

· GEDRO ·
SIAE ·   Z8

· SEPTIMVS · PARALELLVS ·
· DIFFERT · AB · EQVINOCTIALI · HORA · ¹·Z·⁷ · HABENS ·
· MAXIMVM · DIEM · HORARV̄ · 13·¹·⁷ ·

NVS · PERSICVS ·

· CORIVS · FLV ·

PAREP · APHITIS

· NIPISTA ·

· P ·   Z7
A´

· THARO ·

· VOROC ·
HTHA ·

· ACHIDANA · FLV ·

R ·   Z6

· TYLVS ·   · ARAT ·
HOS ·

· ANDANIS · FLV ·

· SABIS ·

· SAGANVS·FL ·

ARAE
NA

· ALEXA ·
DRIA ·

MA
NIA ·

· CHODDA ·

S ·   Z⁴

· 6 · PARALELLVS · PER · SYENEN ·
· DIFFERT · AB · EQVINCTIALI · HORA · 1·¹·⁴ ·
· HABENS · MAXIMVM · DIEM · HORARV̄ · 13·¹·⁴ ·

· ANGVSTIAE ·
· PERSICI ·
SINVS ·

· TARVANA ·

CABAD

· ASABO ·
RVM·PRO ·
MONTORI ·

· STRONGYLVS ·
MONS ·

PASARGDAE ·

· CARPELLA ·
PROMONT ·

· RHOGANA ·

· CARADRAE ·

· THROASCA ·

· CAVTHONI ·
CAE ·

· MASIN · SAMYDACA ·

· ORA ·

· COPHANTA ·

Z9   CLIMA · SECVNDVM ·

Z3

· CAPSINA ·
· CAVANA ·

· RHEGAMA ·

· ICHTHYOPHAGI ·

· SINVS · PARAGONICVS ·

· CHELONO ·
PHAGI ·

· TESA ·

· CAVDRIACIS · FLV ·

ZZ

· ANA ·
RITAE ·

· RHABANA ·
REGIA ·

· APPA · MACAE ·

· CRYPTVS ·
· PORTVS ·
· ABISSI ·

· BAGIA·PRO ·
MONTORIVM ·

· CYIZA·PORTVS ·

Z1

· QVINTVS · PARALELLVS ·
· DIFFERT · AB · EQVINOCTIALI · HORA · 1·¹·⁴ ·
· HABENS · MAXIMVM · DIEM · HORARV̄ · 13·¹·⁴ ·

· COTTABANI ·

· LIBANOTOPHOROS ·
· SIVE · THVRIFERA ·

· DIANAE ·
· VATICINIV ·

· HORMANVS·FLV ·

· COSEV ·
DA ·

· ALAMBATES ·
PROMONT ·

· DEIANOE ·
VILĀE ·

· BADARA ·
AVS·ANA ·

Z0

· NECGILLA ·

· ORGANA ·

· POLLA ·

· LIBA ·   19

· OBARITE ·

· OMARI ·
MATHA ·

· ANGA ·
VILLA ·

· ASTOA ·
VILLA ·

· SARAPIADIS · INSVLA ·
· IN·QVA · PANVM ·

· CARMINA ·   18

· THABANA ·

· AVSARA ·

· SINVS · SACHALITES · IN · QVO ·
· COLYMBISIS · PINICI ·
· SVPER · VTRIBVS ·
· NAVIGAVIT ·

17   · 9 · PARALELLVS · PER · MEROĒ ·
· DIFFERT · AB · EQVINOCTIALI · HORA · 1 ·
· HABENS · MAXIMVM · DIEM · HORA ·
RVM · 13 ·

· METACVM ·
VILLA ·

· ZENOBII ·
INSVLAE ·

16

· ASCITAE ·

· SAPPHAR ·
METROPOLIS ·

1⁴   · CLIMA · PRIMVM ·

· THIALEMATH ·
VILLA ·

· MOSOHA ·
PORTVS · SYAGROS ·
PROMONT ·

· PRETOS ·
PORTVS ·

· MARE · INDICVM ·

1⁴

13   · 3 · PARALELLVS ·
· DIFFERT · AB · EQVINOCTIALI ·
· HORA · ²·³ · HABENS · MAXI ·
MVM · DIEM · HORARV̄ · 12·¹·⁴ ·

· DIOSCORIDIS·INSVLA ·
· IGNOSOGI ·

· TRETA ·

1Z

11

10

| 86 | 87 | 88 | 89 | 90 | 91 | 9Z | 93 | 9⁴ | 9⁴ | 96 | 97 | 98 | 99 | 100 | 101 | 10Z | 103 | 10⁴ | 9 |

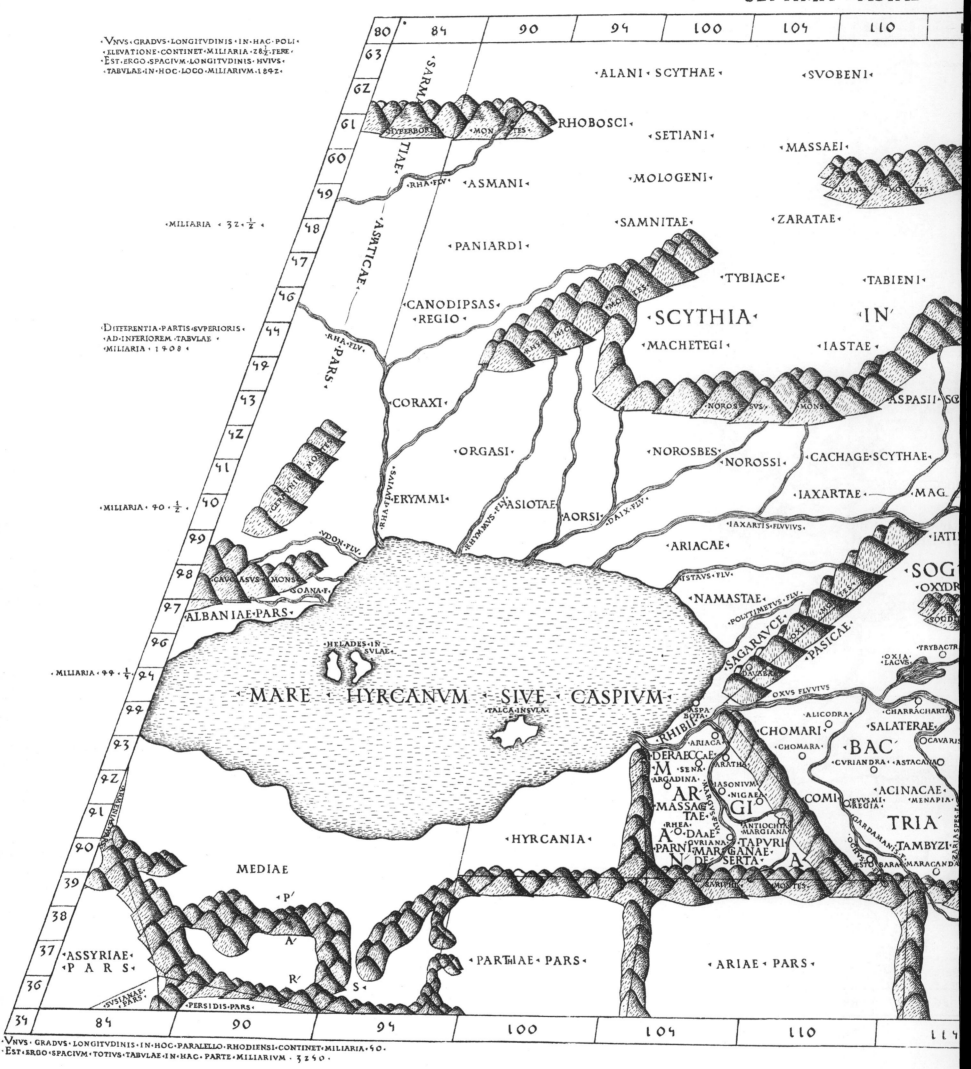

·VNVS·GRADVS·LONGITVDINIS·IN·HAC·POLI·
·ELEVATIONE·CONTINET·MILIARIA·28½·FERE·
·EST·ERGO·SPACIVM·LONGITVDINIS·HVIVS·
·TABVLAE·IN·HOC·LOCO·MILIARIVM·1842·

80   84   90   94   100   104   110

63

62

61
·RHOBOSCI·
60

59

·MILIARIA·32·½·   48

47

46

·DIFFERENTIA·PARTIS·SVPERIORIS·   44
·AD·INFERIOREM·TABVLAE·
·MILIARIA·1908·   42

43

42

41

·MILIARIA·40·½·   40

39

38

37

36

·MILIARIA·44·⅕·   34

33

32

31

30

39

38

37   ·ASSYRIAE·
·PARS·
36   ·SVSIANAE·
·PARS·   ·PERSIDIS·PARS·

34   84   90   94   100   104   110

·ALANI · SCYTHAE·   ·SVOBENI·

·SETIANI·

·MASSAEI·

·ASMANI·   ·MOLOGENI·

·SAMNITAE·   ·ZARATAE·

·PANIARDI·

·TYBIACE·   ·TABIENI·

·CANODIPSAS·
·REGIO·   SCYTHIA·   IN·

·MACHETEGI·   ·IASTAE·

·CORAXI·   ASPASII·SC

·NOROSBES·   ·CACHAGE·SCYTHAE·

·ORGASI·   ·NOROSSI·

·IAXARTAE·   ·MAG·

·ERYMMI·   ·ASIOTAE·
·AORSI·

·IAXARTIS·FLVVIVS·   IATI·

·ARIACAE·   SOG·
OXYDR·

·NAMASTAE·

·SAGARAVCE·   ·PASICAE·

MARE · HYRCANVM · SIVE · CASPIVM·   ·CHOMARI·   BAC·
·SALATERAE·

·RHIBII·   ·ALICODRA·   ·CHARRACHARTA·

·DERAECCAE·   ·ACINACAE·
·COMI·   ·MENAPIA·

·HYRCANIA·   AR·GI·   TRIA·
TAPVRI·   TAMBYZI·

MEDIAE   ·MARGANAE·
SERTA·   A·

·PARTHIAE·PARS·   ·ARIAE·PARS·

·VNVS·GRADVS·LONGITVDINIS·IN·HOC·PARALELLO·RHODIENSI·CONTINET·MILIARIA·50·
·EST·ERGO·SPACIVM·TOTIVS·TABVLAE·IN·HAC·PARTE·MILIARIVM·3250·

XXII.   PTOLEMÆUS ROMÆ 1490.

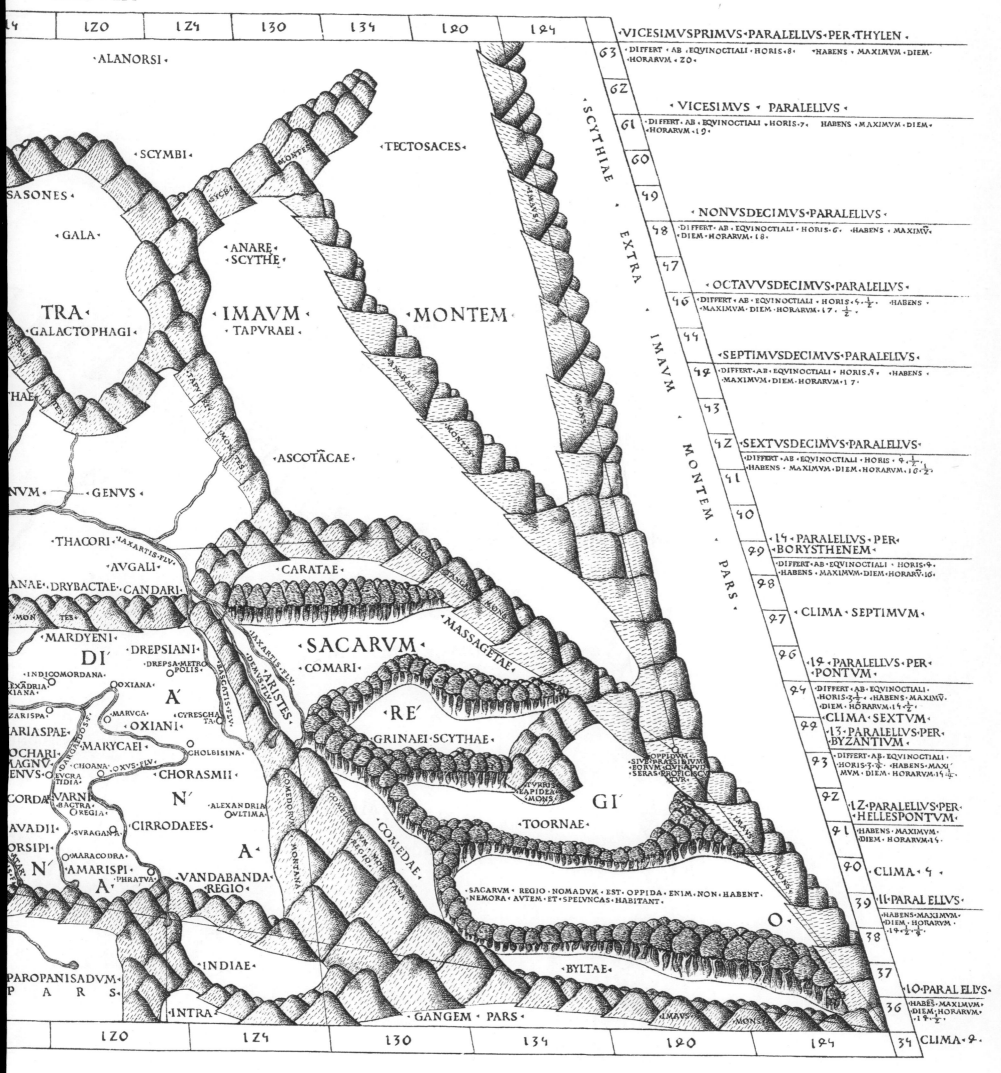

	14	120	124	130	134	120	124

· VICESIMVSPRIMVS · PARALELLVS · PER · THYLEN ·

63 · DIFFERT · AB · EQVINOCTIALI · HORIS · 8 ·   · HABENS · MAXIMVM · DIEM · HORARVM · 20 ·

· ALANORSI ·

62

· VICESIMVS · PARALELLVS ·

61 · DIFFERT · AB · EQVINOCTIALI · HORIS · 7 ·   HABENS · MAXIMVM · DIEM · HORARVM · 19 ·

· SCYMBI ·   · TECTOSACES ·

60

49

· NONVSDECIMVS · PARALELLVS ·

48 · DIFFERT · AB · EQVINOCTIALI · HORIS · 6 ·   · HABENS · MAXIMV̄ · DIEM · HORARVM · 18 ·

· SASONES ·

47

· GALA ·

· ANARE · SCYTHE ·

· OCTAVVSDECIMVS · PARALELLVS ·

46 · DIFFERT · AB · EQVINOCTIALI · HORIS · 5 ½ ·   · HABENS · MAXIMVM · DIEM · HORARVM · 17 ½ ·

· TRA ·   · IMAVM ·   · MONTEM ·
· GALACTOPHAGI ·   · TAPVRAEI ·

44

· SEPTIMVSDECIMVS · PARALELLVS ·

42 · DIFFERT · AB · EQVINOCTIALI · HORIS · 5 ·   · HABENS · MAXIMVM · DIEM · HORARVM · 17 ·

· THAE ·

43

· ASCOTACAE ·

42 · SEXTVSDECIMVS · PARALELLVS ·

· DIFFERT · AB · EQVINOCTIALI · HORIS · 4 ½ · HABENS · MAXIMVM · DIEM · HORARVM · 16 ½ ·

41

· NVM ·   · GENVS ·

40

· THACORI · · IAXARTIS · FLV ·

39 · 14 · PARALELLVS · PER · BORYSTHENEM ·

· AVGALI ·

· DIFFERT · AB · EQVINOCTIALI · HORIS · 4 · HABENS · MAXIMVM · DIEM · HORARV · 16 ·

· CARATAE ·

· ANAE · DRYBACTAE · CANDARI ·

38

· MON   TES ·

37 · CLIMA · SEPTIMVM ·

· MARDYENI ·

· DI' ·   · DREPSIANI ·   · SACARVM ·   · MASSAGETAE ·

36

· DREPSA · METRO · POLIS ·   · COMARI ·

· 14 · PARALELLVS · PER · PONTVM ·

· INDICOMORDANA ·

24 · DIFFERT · AB · EQVINOCTIALI · HORIS · 3 ½ · HABENS · MAXIMV̄ · DIEM · HORARVM · 15 ½ ·

· EXADRIA · XIANA ·   · OXIANA ·

· A' ·

· CLIMA · SEXTVM ·

· ZARISPA ·

· RE' ·

· CYRESCHA · TA ·

· 13 · PARALELLVS · PER · BYZANTIVM ·

· ARIASPAE ·   · MARVCA ·
· OXIANI ·

· GRINAEI · SCYTHAE ·

23 · DIFFERT · AB · EQVINOCTIALI · HORIS · 3 ⅓ · HABENS · MAXI · MVM · DIEM · HORARVM · 15 ⅓ ·

· OCHARI · MAGNV · GENVS ·   · MARYCAEI ·   · CHOLBISINA ·

· CHOANA · OXVS · FLV ·

· CHORASMII ·

· SIVE · OPPIDVM · EORVM · QVIBVS · EORVM · CVM AD · SERAS · PROFICISCV ·

· 12 · PARALELLVS · PER · HELLESPONTVM ·

· CORDA · VARNI · BACTRA · REGIA ·

· N' ·

· ALEXANDRIA · OVLTIMA ·

· GI' ·

22

· AVADII ·   · SVRAGANA ·

· CIRRODAEES ·

· TOORNAE ·

21 · HABENS · MAXIMVM · DIEM · HORARVM · 15 ·

· ORSIPI ·   · MARACODRA ·

· A' ·

20 · CLIMA · 4 ·

· N' ·   · AMARISPI ·
· PHRATVA ·   · VANDABANDA · REGIO ·

· COMEDAE ·

39 · 11 · PARALELLVS ·

· A' ·

· SACARVM · REGIO · NOMADVM · EST · OPPIDA · ENIM · NON · HABENT · NEMORA · AVTEM · ET · SPELVNCAS · HABITANT ·

· O' ·

· HABENS · MAXIMVM · DIEM · HORARVM · 14 ⅔ ·

38

· INDIAE ·

· BYLTAE ·

37

· PAROPANISADVM · PARS ·

· 10 · PARALELLVS ·

36 · HABES · MAXIMVM · DIEM · HORARVM · 14 ½ ·

· INTRA ·   · GANGEM · PARS ·   · IMAVS ·   · MONS ·

	120	124	130	134	120	124	34

CLIMA · 4 ·

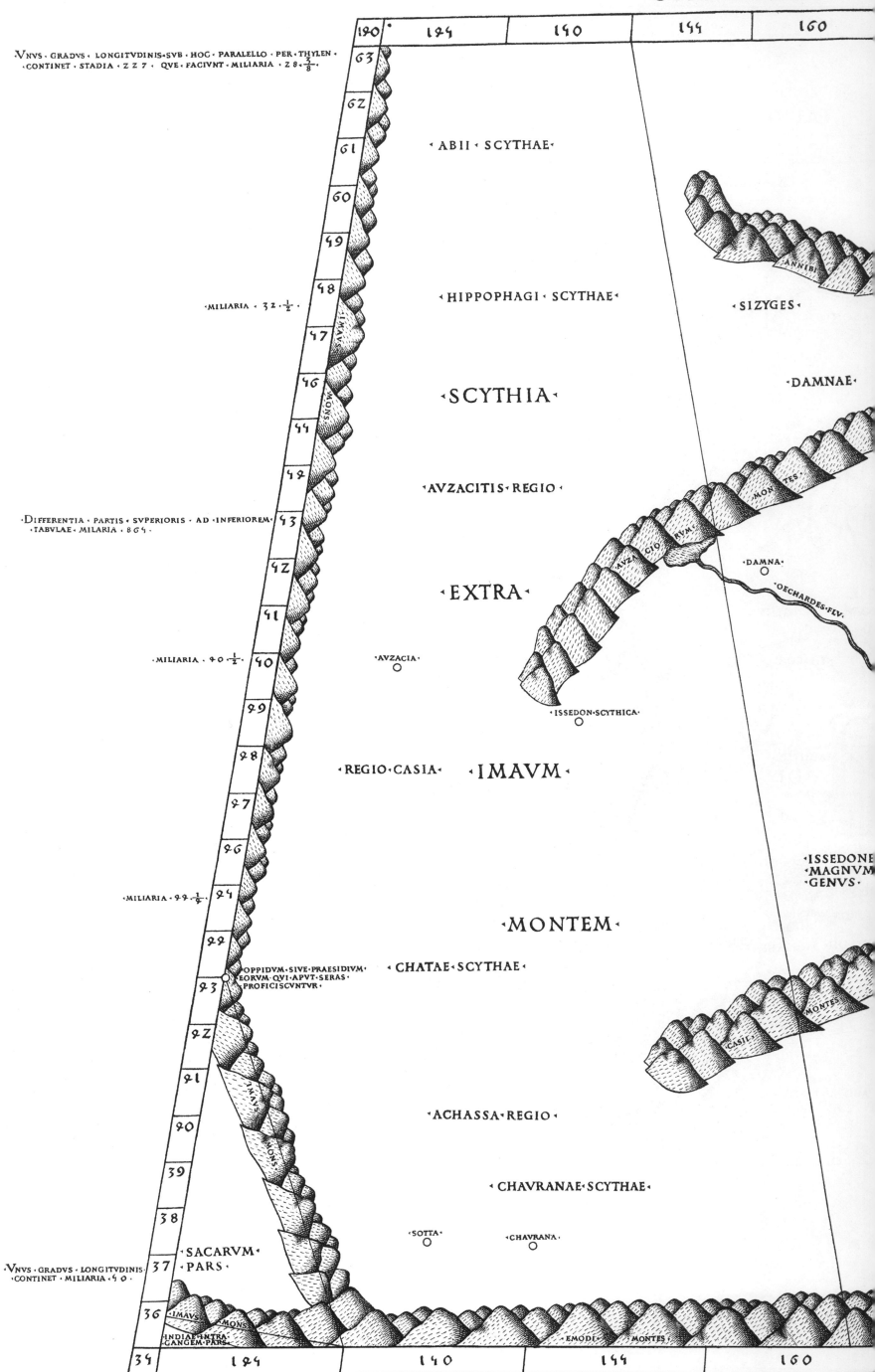

190 185 180 175 160

·VNVS·GRADVS·LONGITVDINIS·SVB·HOC·PARALELLO·PER·THYLEN·
·CONTINET·STADIA·ZZ7·QVE·FACIVNT·MILIARIA·Z8·3/8·

63

6Z

61 ·ABII·SCYTHAE·

60

59

58

·MILIARIA·3Z·1/2· 57 ·HIPPOPHAGI·SCYTHAE· ·SIZYGES·

56

55 ·SCYTHIA· ·DAMNAE·

54

53

·AVZACITIS·REGIO·

·DIFFERENTIA·PARTIS·SVPERIORIS·AD·INFERIOREM· 53
·TABVLAE·MILIARIA·864·

5Z

51 ·EXTRA· ·DAMNA·

·OECHARDES·FLV·

·MILIARIA·40·1/2· 50 ·AVZACIA·
○

49 ·ISSEDON·SCYTHICA·
○

48

·REGIO·CASIA· ·IMAVM·

57

56 ·ISSEDONE
·MAGNVM·
·GENVS·

·MILIARIA·44·1/2· 55

53

·MONTEM·

52

·OPPIDVM·SIVE·PRAESIDIVM· ·CHATAE·SCYTHAE·
53 EORVM·QVI·APVT·SERAS·
PROFICISCVNTVR· ○

5Z

51

50 ·ACHASSA·REGIO·

39

38 ·CHAVRANAE·SCYTHAE·

·SOTTA· ·CHAVRANA·
○ ○

·SACARVM·
37 PARS·

·VNVS·GRADVS·LONGITVDINIS·
·CONTINET·MILIARIA·50·

36 ·IMAVS·MONS·

·INDIAE·INTRA·
GANGEM·PARS· ·EMODI·MONTES·

35 185 140 155 160

| 164 | 170 | 174 | 180 | |

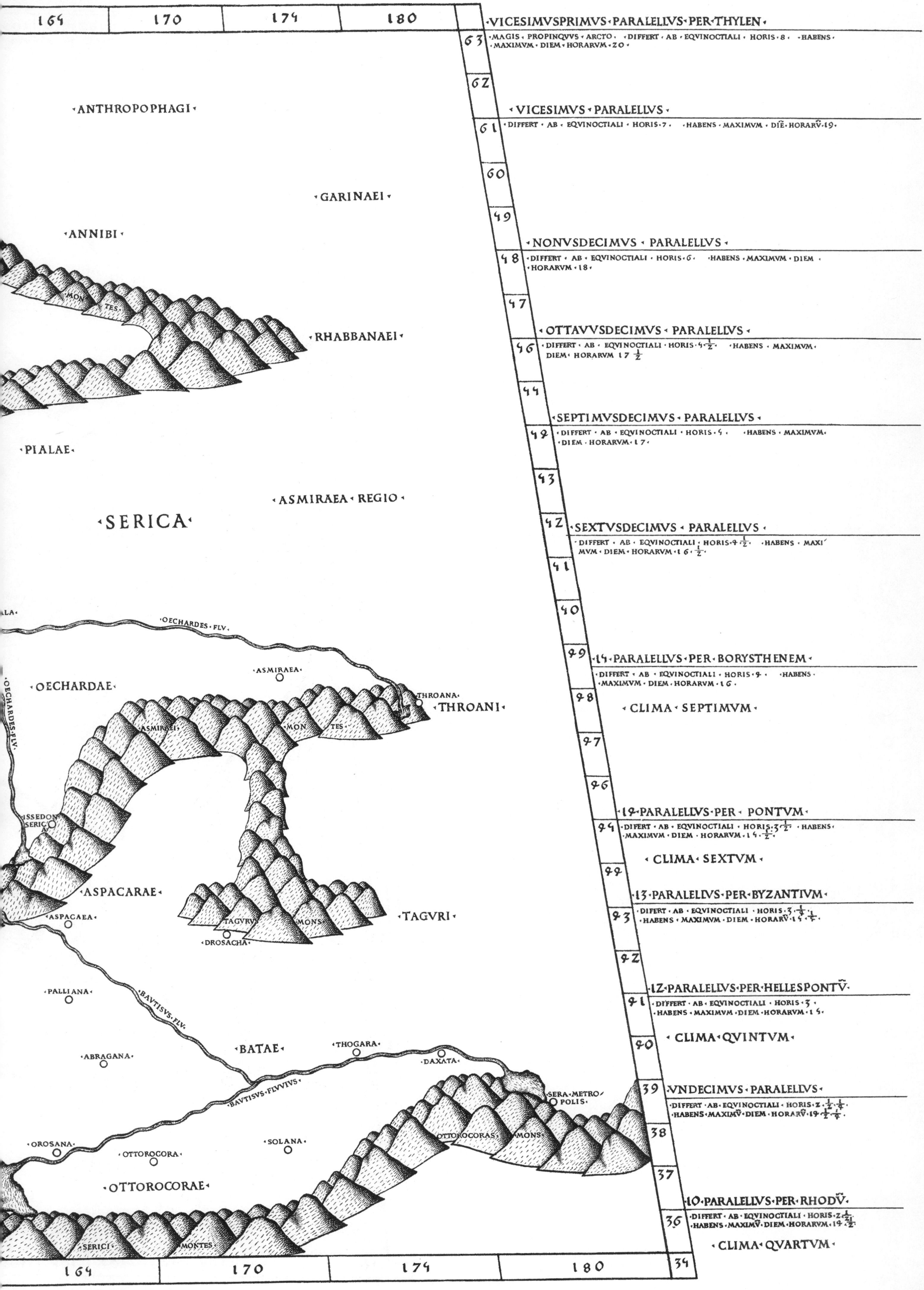

·VICESIMVSPRIMVS·PARALELLVS·PER·THYLEN·

63 · ·MAGIS· PROPINQVVS· ARCTO· ·DIFFERT· AB· EQVINOCTIALI· HORIS·8· ·HABENS· ·MAXIMVM· DIEM· HORARVM· 20 ·

6Z

·ANTHROPOPHAGI·

·VICESIMVS· PARALELLVS·

61 ·DIFFERT· AB· EQVINOCTIALI· HORIS·7· ·HABENS· MAXIMVM· DIE· HORARV·19·

60

·GARINAEI·

49

·ANNIBI·

·NONVSDECIMVS· PARALELLVS·

48 ·DIFFERT· AB· EQVINOCTIALI· HORIS·6· ·HABENS· MAXIMVM· DIEM· ·HORARVM· 18·

47

·MON TES·

·RHABBANAEI·

·OTTAVVSDECIMVS· PARALELLVS·

46 ·DIFFERT· AB· EQVINOCTIALI· HORIS·4 1/2· ·HABENS· MAXIMVM· DIEM· HORARVM 17 1/2

44

·SEPTIMVSDECIMVS· PARALELLVS·

·PIALAE·

42 ·DIFFERT· AB· EQVINOCTIALI· HORIS·5· ·HABENS· MAXIMVM· DIEM· HORARVM· 17·

43

·ASMIRAEA· REGIO·

4Z ·SEXTVSDECIMVS· PARALELLVS·

·SERICA·

·DIFFERT· AB· EQVINOCTIALI· HORIS·4 1/2· ·HABENS· MAXI' MVM· DIEM· HORARVM· 16 1/2·

41

40

·OECHARDES· FLV.

49 ·14·PARALELLVS·PER·BORYSTHENEM·

·DIFFERT· AB· EQVINOCTIALI· HORIS·4· ·HABENS· MAXIMVM· DIEM· HORARVM· 16·

·OECHARDAE·

·ASMIRAEA·

·THROANA·

·THROANI·

48 ·CLIMA· SEPTIMVM·

·ASMIREII· ·MON TES·

47

46

·ISSEDON SERICA·

·19·PARALELLVS·PER· PONTVM·

44 ·DIFERT· AB· EQVINOCTIALI· HORIS·3 1/2· ·HABENS· ·MAXIMVM· DIEM· HORARVM· 14 1/2·

·CLIMA· SEXTVM·

4Z

·ASPACARAE·

·13·PARALELLVS·PER·BYZANTIVM·

·ASPACAEA·

·TAGVRV MONS·

·TAGVRI·

43 ·DIFERT· AB· EQVINOCTIALI· HORIS·3 1/4· ·HABENS· MAXIMVM· DIEM· HORARV·15 1/4·

·DROSACHA·

4Z

·1Z·PARALELLVS·PER·HELLESPONTV·

·PALLIANA·

·BAVTISVS·FLV.

41 ·DIFFERT· AB· EQVINOCTIALI· HORIS·3· ·HABENS· MAXIMVM· DIEM· HORARVM· 15·

·BATAE·

·THOGARA·

40 ·CLIMA· QVINTVM·

·ABRAGANA·

·DAXATA·

·BAVTISVS·FLVVIVS·

39 ·VNDECIMVS· PARALELLVS·

·SERA·METRO POLIS·

·DIFFERT· AB· EQVINOCTIALI· HORIS·Z 1/2 1/3· ·HABENS·MAXIMV·DIEM·HORARV·14 1/2 1/4·

·OROSANA·

·SOLANA·

·OTTOROCORAS MONS·

38

·OTTOROCORA·

37

·OTTOROCORAE·

·10·PARALELLVS·PER·RHODV·

36 ·DIFFERT· AB· EQVINOCTIALI· HORIS·Z 1/4· ·HABENS·MAXIMV·DIEM·HORARVM·14 1/2·

·SERICI· MONTES·

·CLIMA· QVARTVM·

| 164 | 170 | 174 | 180 | 34 |

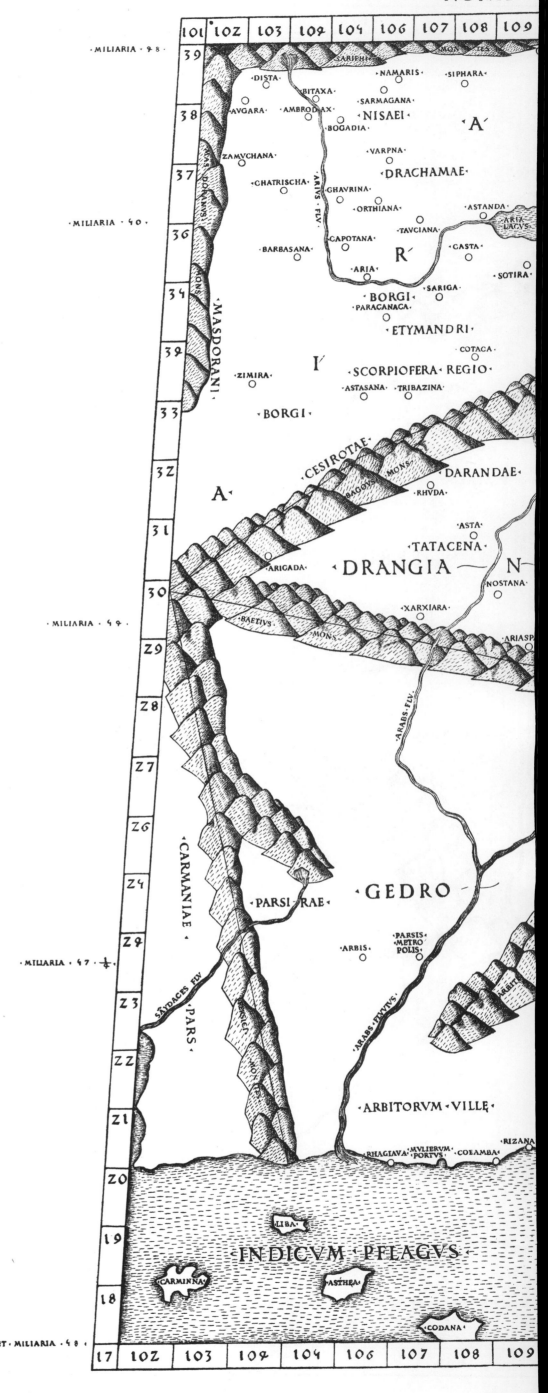

| 101 | 102 | 103 | 104 | 104 | 106 | 107 | 108 | 109 |

·MILIARIA·98· 39

·DISTA· ·SARIPH· ·NAMARIS· ·SIPHARA·
·BITAXA·
38 ·AVGARA· ·AMBROD AX· ·SARMAGANA· NISAEI ·A·
·BOGADIA·

·ZAMVCHANA· ·VARPNA·
37 DRACHAMAE
·CHATRISCHA· ·GHAVRINA·
·ORTHIANA· ·ASTANDA· ARIA LACVS
·MILIARIA·40· 36 ·TAVCIANA·
·BARBASANA· ·CAPOTANA· ·CASTA·
R· ·SOTIRA·
·ARIA·
34 ·SARIGA·
BORGI
·PARACANACA·
ETYMANDRI

32 ·COTACA·
I· SCORPIOFERA·REGIO·
·ZIMIRA· ·ASTASANA· ·TRIBAZINA·
33 BORGI

·CESIROTAE·
32 A· ·BAGOX·MONS· DARANDAE
·RHVDA·

31 ·ASTA·
·TATACENA·
·ARIGADA· DRANGIA N
30 ·NOSTANA·

·XARXIARA·
·MILIARIA·44· 29 ·BAETIVS· ·MONS· ·ARIASP·

28

·ARABS·FLV·
27

26

CARMANIAE· 24 PARSI RAE· ·GEDRO·

·PARSIS· METRO POLIS·
29 ·ARBIS·
·MILIARIA·47·⅓·

23 ·SVDACES FLV· ·PARS·

22 ·ARABS·FLVVIVS·

21 ·ARBITORVM·VILLE·

·RHAGIAVA· ·MVLIERVM· ·COEAMBA· ·RIZANA·
PORTVS·
20

19 ·LIBA·
INDICVM·PELAGVS·
18 ·CARMINNA· ·ASTHEA·

·CODANA·

XXIV. PTOLEMÆUS ROMÆ 1490.

110	111	112	113	114	114	116	117	118	119	1/2

·TAVA·

·RHAVGARA·

·ASTAVENI·

·GODANA·
○

·PHORAVA·
○

·PARVTI·

·PAROPANISVS·MONS·

· BOLITAE ·

·BACTRIANAE·PARS·

·PARSIANA·

·GARDAMANIS·FLV·

·BARZAVRA· ·ARTOARTA· ·BABORANA·
○

·GATISA·
○

39

38

·ARTIGAVDNA·
○

·ALEXANDRIA·
IN·O·ARIA·

·ORBITANA·
○

·NISIBIS·
○

·OBARES·

·DARCAMA·

·ARISTOPHYLI·

PA

ROPANISA DES·

·PARSII·

·PARSIA·
○

·NIPHANDA·
○

·DRASTOCA·AMBAVTE·
○

·GAVZACA·
○

·GANGA·

·NAVLIBIS·
○

·CABVRA·
○

·DOROACANA·
○

37

36

34

·PARSI E TAE·

·TARBACANA·

·CHOLARNA·
○

·BAGARDA·
○

·ARGVDA·
○

·MOS·

34

·PROPHTHASIA·
○

·AZOLA· ·PARGYETAE·
○

PAR VETI

·MON LES·

·PHOCLIS·
○

33

INDIAE·INTRA·GANGEM· P· A·

32

·ARACHOSIA·

·ARICACA·
○

·RHIZANA·
○

·ARBACA·
○

31

·BATRII·

·SYDRI·

·ALEXANDRIA·
○

·ROPLVTAE·

·CHOASPA·
○

·ARACHOTVS·
○

30

·PHARAZANA·
○

·BIGIS·
○

·SIGARA·
○

·FONS·
○

·ARACHOTOS·

·MALLIANA·
○

29

A·

·ASIAGA·
○

·GAMMACA·
○

·DAMMANA·
○

·EORITAE·

28

ARAN

BAETIVS

·MON·

SPATIA

S·

·CVNI·
○

·MVSARNA·
○

·COTTOBARA·
○

27

·BADARA·
○

·MVSARNAEI·

·OSCANA·
○

26

·PARADENA·

24

SIA ·

·INDVS·FLVVIVS·

24

·MON·

·PARISINA·

23

·OMIZA·

·RHAMNAE·

·INDIAE·INTRA·
·GANGEM·PARS·

22

21

20

19

18

110	111	112	113	114	114	116	117	118	119	1/2	17

·SINVS·CAN·
THICOLPVS·

BARACA

· VNDECIMVS · PARALELLVS ·
·DIFFERT · AB · EQVINOCTIALI · HORIS·2·1/2·1/4· · HABENS · MAXIMVM · DIEM · HORARVM · 19 ·1/2·1/4·

· DECIMVS · PARALELLVS·PER·RHODVM ·
·DIFFERT · AB · EQVINOCTIALI · HORIS·2·1/2· · HABENS · MAXIMVM · DIEM · HORARVM · 19 ·1/2·

· CLIMA · QVARTVM ·

· NONVS · PARALELLVS ·
·DIFFERT · AB · EQVINOCTIALI · HORIS·2·1/4· · HABENS · MAXIMVM · DIEM · HORARV̄ · 19 ·1/4·

· OCTAVVS · PARALELLVS · PER · ALEXANDRIAM ·
·DIFFERT · AB · EQVINOCTIALI · HORIS·2· · HABENS · MAXIMVM · DIEM · HORARV̄ · 19 ·

· CLIMA · TERTIVM ·

· SEPTIMVS · PARALELLVS ·
·DIFFERT · AB · EQVINOCTIALI · HORA·1·1/2·1/4· · HABENS · MAXIMVM · DIEM ·
·HORARVM · 13 ·1/2·1/4·

· SEXTVS · PARALELLVS · PER · SYENEN ·
·DIFFERT · AB · EQVINOCTIALI · HORA·1·1/2· · HABENS · MAXIMVM · DIEM ·
·HORARVM · 13·1/2·

· CLIMA · SECVNDVM ·

· QVINTVS · PARALELLVS ·
·DIFFERT · AB · EQVINOCTIALI · HORA·1·1/4· · HABENS · MAXIMVM · DIẼ ·
·HORARVM · 13·1/2·

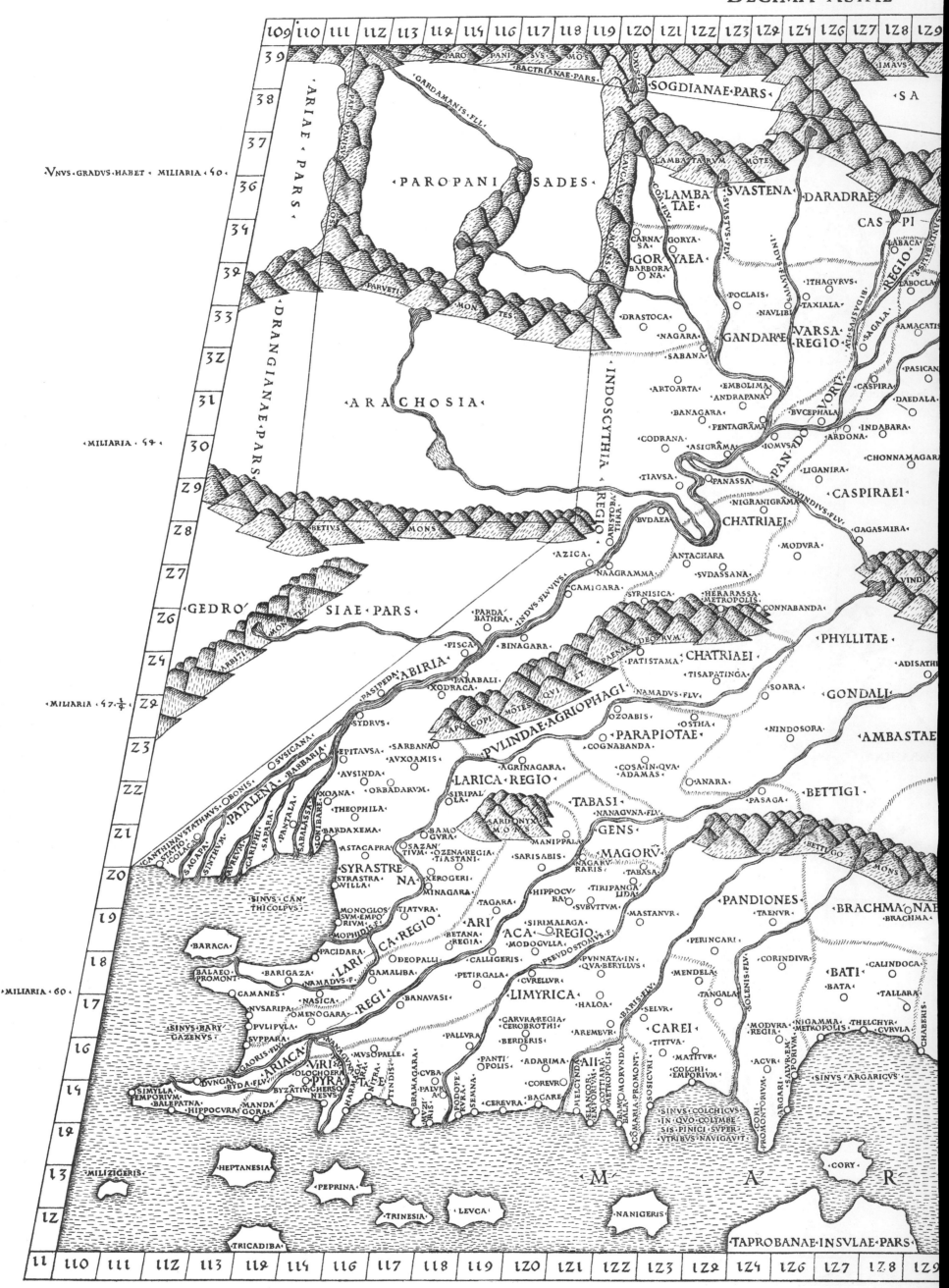

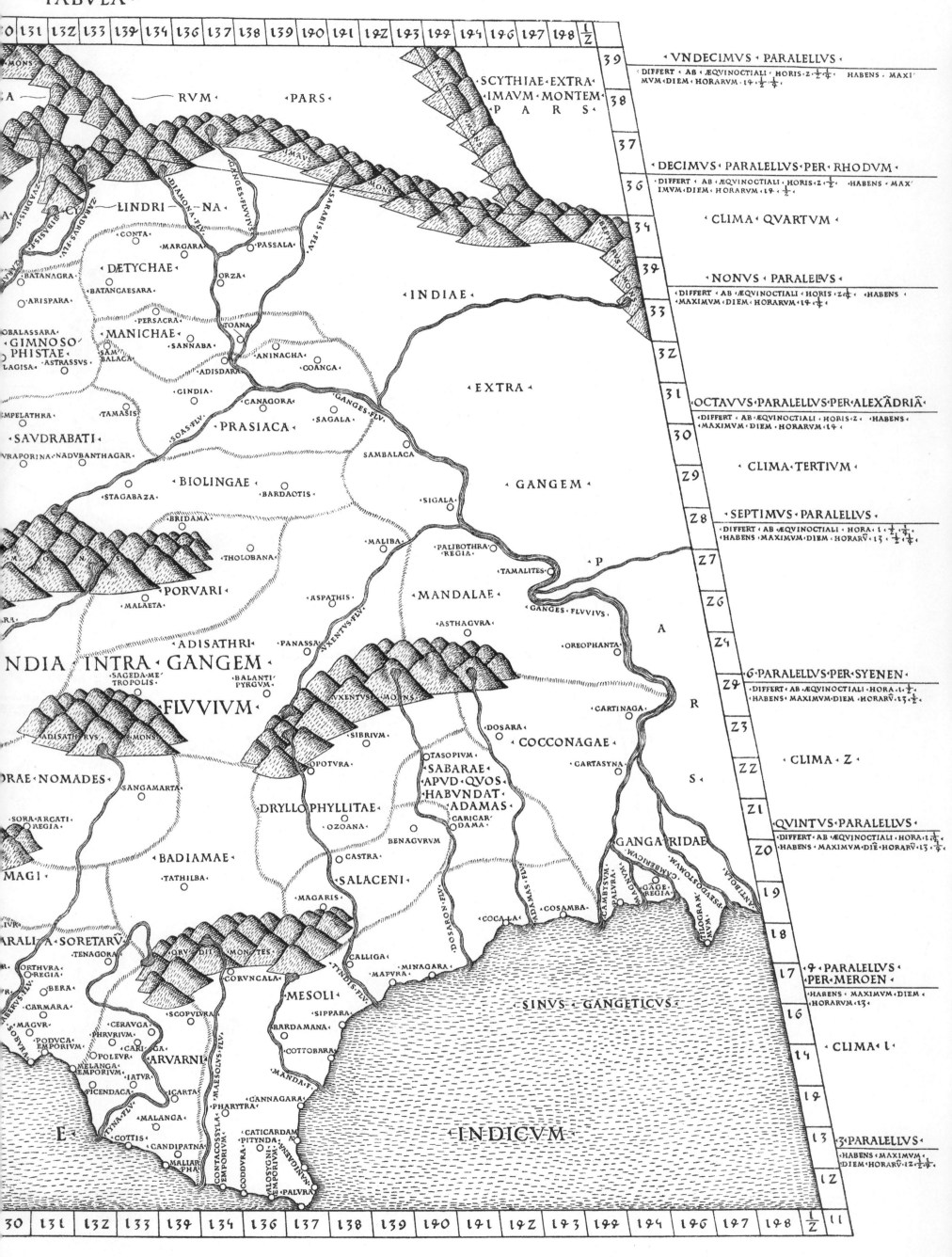

| 30 | 131 | 132 | 133 | 134 | 134 | 136 | 137 | 138 | 139 | 190 | 191 | 192 | 193 | 199 | 199 | 196 | 197 | 198 | 1/2 |

MONS

RVM ·   PARS

SCYTHIAE · EXTRA IMAVM · MONTEM P A R S

39

38

37

ZARBADISTE
CU
ZARADVS FLV.
LINDRI — NA

36

DIAMONA FLV.
GANGES FLVVIVS
MAV
MONS

· CONTA
· MARGARA
· PASSALA

34

· CLIMA · QVARTVM ·

BATANAGRA
· DÆTYCHAE ·
ORZA
· BATANCAESARA
· INDIAE ·
SARAIS FLV.
MONS

34

· ARISPARA
PERSACRA
TOANA
33

OBALASSARA
· MANICHAE ·
· SANNABA
· EXTRA ·
32

GIMNOSO PHISTAE
SAM BALACA
ANINACHA
· COANGA
· LAGISA
· ASTRASSVS
ADISDARA

31

· GINDIA
· CANAGORA
GANGES FLV.
30

· SAVDRABATI
· TAMASIS
SOAS FLV.
· PRASIACA ·
SAGALA

29

· CLIMA · TERTIVM ·

VRAPORINA · NADVBANTHAGAR

SAMBALACA
· GANGEM ·

· BIOLINGAE ·
· BARDAOTIS
· SIGALA
28

· STAGABAZA
· BRIDAMA

MALIBA
PALIBOTHRA REGIA
27

· MONS
· THOLOBANA
· TAMALITES
· P
26

· PORVARI ·
· ASPATHIS
· MANDALAE ·
GANGES · FLVVIVS
25

· MALAETA
VXENTVS FLV.
· OREOPHANTA
· A
24

NDIA · INTRA · GANGEM ·
· ASTHAGVRA
· PANASSA
24

SAGEDA · ME TROPOLIS
· BALANTI PYRGVM ·
· R
23

· FLVVIVM ·
VXENTVS · MONS
· CARTINAGA
· S ·
22

· CLIMA · 2 ·

ADISATHRVS MONS
· SIBRIVM
· DOSARA
· COCCONAGAE ·
21

DRAE · NOMADES ·
· OPOTVRA
· TASOPIVM
· CARTASYNA
20

· SANGAMARTA
SABARAE APVD · QVOS HABVNDAT ADAMAS
· GANGA · RIDAE
19

· SORA · ARCATI REGIA
· DRYLLO PHYLLITAE ·
CARICAR DAMA
CAMBYSVM PALVRA
CAMBERICVM
MAMOLA
18

MAGI
· OZANA
BENAGVRVM
· PSEVDOSTOMVM
17

· BADIAMAE ·
· CASTRA
MANDAGORA
ANTIBOLA
16

· TATHILBA
· SALACENI ·
GAGE REGIA
· SINVS · GANGETICVS ·
14

ARALIA · SORETARV ·
· MAGARIS
· COCALA
COSAMBA
· CLIMA · 1 ·

TENAGORA
ORV · DII · MONTES
CALLIGA
13

· ORTHVRA REGIA
CORVNCALA
· MINAGARA
12

· BERA
· TYNDIS FLV.
MAPVRA

CARMARA
· MESOLI ·
SIPPARA
· SINVS · GANGETICVS ·

MAGVR
· SCOPVLVRA
BARDAMANA

· CERAVGA
PHRVRIVM
· ARVARNI ·
COTTOBARA

SOBVRA
CARI
OPOLEVR
14

PODVCA EMPORIVM
MELANGA EMPORIVM
IATVR
MANDA · F.
13

PICENDACA
· ICARTA
CANNAGARA

E
TYNA FLV.
· MALANGA
PHARYTRA

COTTIS
· CANDIPATNA
CONTACOSSYLA EMPORIVM
CATICARDAMA PITYNDA
· INDICVM ·
13

MALIAR PHA
CODDVRA
ALOSYGNI EMPORIVM
PALVRA
12

| 30 | 131 | 132 | 133 | 134 | 134 | 136 | 137 | 138 | 139 | 190 | 191 | 192 | 193 | 199 | 199 | 196 | 197 | 198 | 1/2 | 11 |

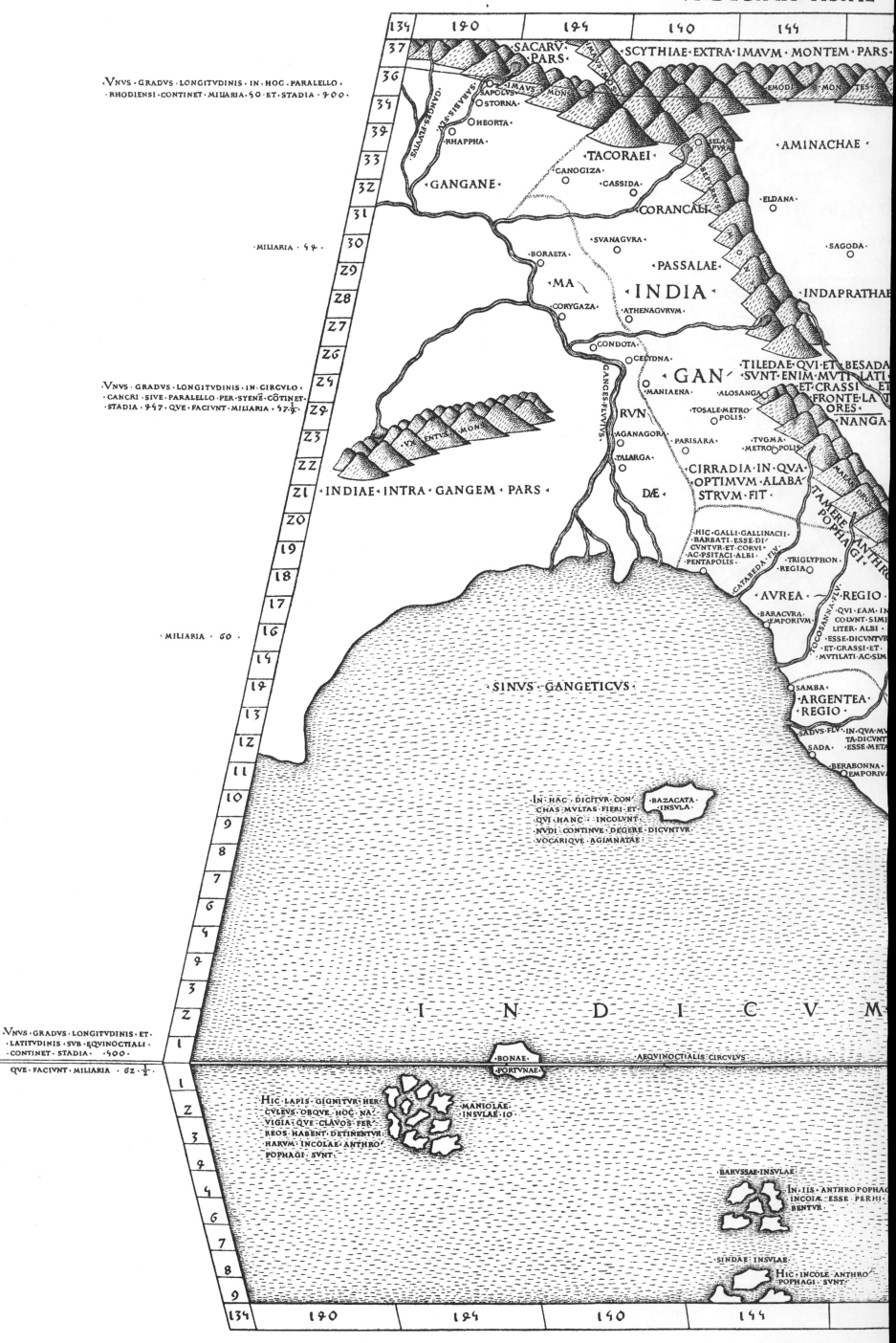

XXVI. PTOLEMÆUS ROMÆ 1490.

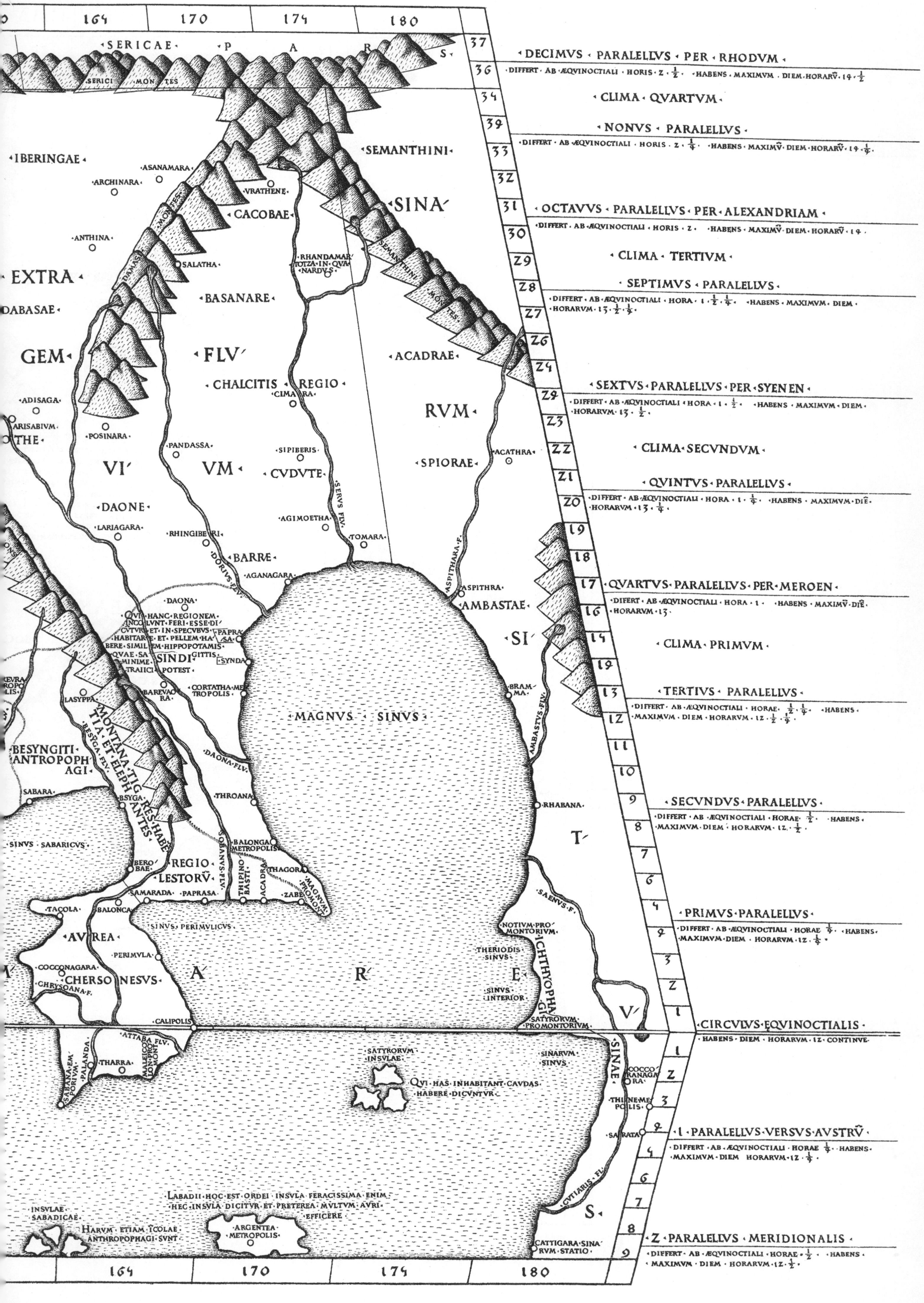

164 170 174 180

SERICAE · P · A · R · S · | 37

· DECIMVS · PARALELLVS · PER · RHODVM ·

SERICI · MONTES | 36

· DIFFERT · AB · ÆQVINOCTIALI · HORIS · Z · ½ · HABENS · MAXIMVM · DIEM · HORARV̄ · 19 · ½ ·

· CLIMA · QVARTVM · | 34

· IBERINGAE · | 39

· NONVS · PARALELLVS ·

ASANAMARA · | SEMANTHINI · | 33

· DIFFERT · AB · ÆQVINOCTIALI · HORIS · Z · ¼ · HABENS · MAXIMV̄ · DIEM · HORARV̄ · 19 · ¼ ·

ARCHINARA · | VRATHENE · | 3Z

· SINA · | 31

· OCTAVVS · PARALELLVS · PER · ALEXANDRIAM ·

· CACOBAE · | 30

· DIFFERT · AB · ÆQVINOCTIALI · HORIS · Z · HABENS · MAXIMV̄ · DIEM · HORARV̄ · 19 ·

· ANTHINA · | Z9

· CLIMA · TERTIVM ·

· SALATHA · | RHANDAMAR TOTZA · IN · QVA · NARDVS · | Z8

· SEPTIMVS · PARALELLVS ·

· EXTRA · | Z7

· DIFFERT · AB · ÆQVINOCTIALI · HORA · 1 · ½ · ¼ · HABENS · MAXIMVM · DIEM ·

DABASAE · | BASANARE · | Z6

· HORARVM · 13 · ½ · ¼ ·

· GEM · | · FLV · | · ACADRAE · | Z4

· SEXTVS · PARALELLVS · PER · SYENEN ·

· ADISAGA · | CHALCITIS · REGIO | Z4

· DIFFERT · AB · ÆQVINOCTIALI · HORA · 1 · ½ · HABENS · MAXIMVM · DIEM ·

· CIMARA · | RVM · | Z3

· HORARVM · 13 · ½ ·

· ARISABIVM · | POSINARA · | ACATHRA · | ZZ

· CLIMA · SECVNDVM ·

THE · | PANDASSA · | SPIORAE · | Z1

· QVINTVS · PARALELLVS ·

· VI · | VM · | CVDVTE | SIPIBERIS · | ZO

· DIFFERT · AB · ÆQVINOCTIALI · HORA · 1 · ¼ · HABENS · MAXIMVM · DIE ·

· HORARVM · 13 · ¼ ·

· DAONE · | 19

· AGIMOETHA · | 18

· LARIAGARA · | 17

· QVARTVS · PARALELLVS · PER · MEROEN ·

· RHINGIBE · RI · | TOMARA · | 16

· DIFFERT · AB · ÆQVINOCTIALI · HORA · 1 · HABENS · MAXIMV̄ · DIE ·

· DORIVS · RV · | BARRÆ · | ASPITHARA · F. | 14

· HORARVM · 13 ·

· AGANAGARA · | ASPITHRA · | 19

· CLIMA · PRIMVM ·

· AMBASTAE · | 13

· DAONA · | SI · | TERTIVS · PARALELLVS ·

QVI · HANC · REGIONEM · INCOLVNT · FERI · ESSE · DI CVTVR · ET · IN · SPECVBVS · HABITARE · ET · PELLEM · HA BERE · SIMILEM · HIPPOPOTAMIS · QVAE · SA MINIME TRAIICI · POTEST | PAPRA SA · | BRAM MA · | 1Z

· DIFFERT · AB · ÆQVINOCTIALI · HORAE · ½ · ¼ · HABENS ·

· SINDI · GITTIS · | SYNDA | · MAXIMVM · DIEM · HORARVM · 1Z · ½ · ¼ ·

· CORTATHA · ME TROPOLIS · | AMBASTVS · FLV. | 11

· BAREVACORA · | 10

· LASYPPA · | MAGNVS · SINVS · | 9

· SECVNDVS · PARALELLVS ·

BESYNGITI ANTROPOPH AGI · | MONTANA · TIG TIA · ET · ELEPH ANTES | RES · HABE BESGCA · FLV. | RHABANA · | 8

· DIFFERT · AB · ÆQVINOCTIALI · HORAE · ½ · HABENS ·

· SABARA · | T · | · MAXIMVM · DIEM · HORARVM · 1Z · ½ ·

· BSYGA · | DAONA · FLV. | 7

· SINVS · SABARICVS · | THROANA · | 6

· BERO BAE · | REGIO LESTORV̄ | SOBANVS · FLV. | 4

· PRIMVS · PARALELLVS ·

· SAMARADA · | PAPRASA · | THIPINO BASTI · | THAGORA | SAENVS · F. | 9

· DIFFERT · AB · ÆQVINOCTIALI · HORAE · ¼ · HABENS ·

· TACOLA · | BALONCA · | ACADRA | ZABE | NOTIVM · PRO MONTORIVM · | 3

· MAXIMVM · DIEM · HORARVM · 1Z · ¼ ·

· AVREA · | SINVS · PERIMVLICVS · | MAGNV̄ PROMON. | THERIODIS · SINVS · | Z

· COCCONAGARA · | PERIMVLA · | A · | R · | E · | ICHTHYOPHA GI · | 1

· CHRYSOANA · F. | CHERSO NESVS · | SINVS · INTERIOR · | V ·

· CIRCVLVS · ÆQVINOCTIALIS ·

CALIPOLIS · | SATYRORVM · PROMONTORIVM · | · HABENS · DIEM · HORARVM · 1Z · CONTINVE ·

· ATTABA FLV. | SINARVM · SINVS · | SINAE · | 1

· THARRA · | COCCO RANAGA RA · | Z

· SATYRORVM · INSVLAE · | THINE ME POLIS · | 3

QVI · HAS · INHABITANT · CAVDAS · HABERE · DICVNTVR · | SARATA · | 4

· 1 · PARALELLVS · VERSVS · AVSTRV̄ ·

· 4 | · DIFFERT · AB · ÆQVINOCTIALI · HORAE · ¼ · HABENS ·

· 6 | · MAXIMVM · DIEM · HORARVM · 1Z · ¼ ·

LABADII · HOC · EST · ORDEI · INSVLA · FERACISSIMA · ENIM HEC · INSVLA · DICITVR · ET · PRETEREA · MVLTVM · AVRI · EFFICERE | 7

· INSVLAE · SABADICAE · | S · | 8

HARVM · ETIAM · ICOLAE ANTHROPOPHAGI · SVNT | ARGENTEA · METROPOLIS · | · Z · PARALELLVS · MERIDIONALIS ·

CATTIGARA · SINA RVM · STATIO · | 9 | · DIFFERT · AB · ÆQVINOCTIALI · HORAE · ½ · HABENS ·

· MAXIMVM · DIEM · HORARVM · 1Z · ½ ·

164 170 174 180

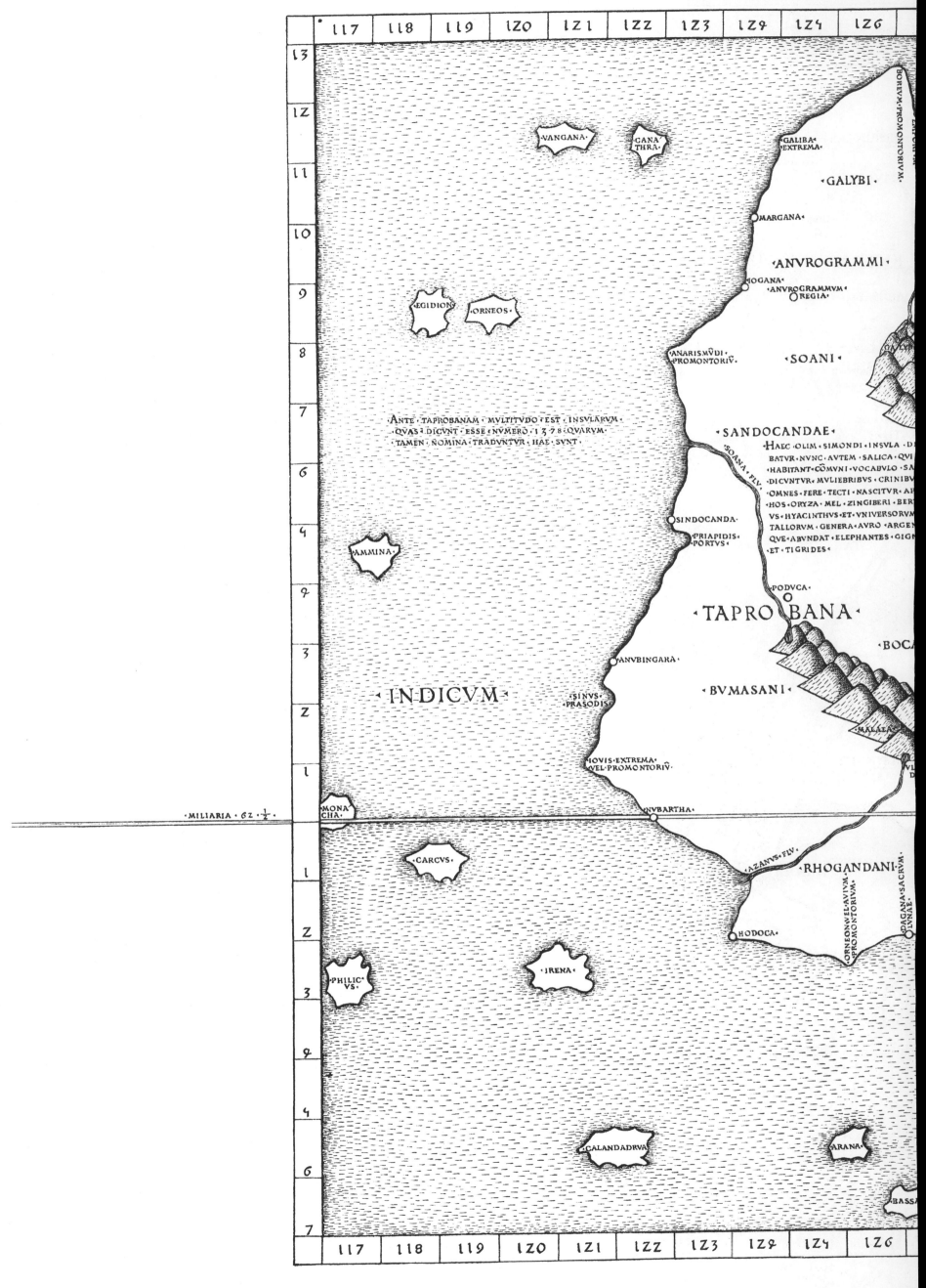

Ante·Taprobanam·Multitudo·est·insularum·
Quas·dicunt·esse·numero·1378·Quarum·
Tamen·nomina·traduntur·hae·sunt·

TAPROBANA·

·INDICVM·

·MILIARIA·62·½·

XXVII. PTOLEMÆUS ROMÆ 1490.

₂7	1Z8	1Z9	130	131	13Z	133	139	134

13

·TERTIV⁹ PARALELLVS·

·DIFFERT · AB · AEQVINOCTIALI · HORAE $\frac{1}{2}$ $\frac{1}{9}$·
·HABENS · DIEM · MAXIMVM · HORARVM · 1Z · $\frac{1}{2}$ · $\frac{1}{4}$·

1Z

·MODVRGI·
EMPORIVM·

·SVSVARA·

11

·MVDVNTI·

10

·ANVBINGARA·

·NANAGADIBI·

9

·NAGADE
BA·

·SECVNDVS ·PARALELLVS·

·DIFFET · AB · AEQVINOCTIALI · HORA · $\frac{1}{2}$·
·HABENS · DIEM · MAXIMVM · HORARVM · 1Z · $\frac{1}{2}$·

·SINVS·
PATI·

·NAGADINA·

8

·SPATANA·
PORTVS·

·MAAGRAMMVM·
METROPOLIS·

·OXIA·PRO·
MONTORIV·

·GANGIS·FLV·

7

·RHIZALA·
PORTVS·

·SEMNI·

6

·PROCVRI·

·ADISAMVM·

4

·MAGNVM·
LITTVS·

·TARACHI·

·PRIMVS · PARALELLVS·

·DIFFERT · AB · AEQVINOCTIALI · HORAE · $\frac{1}{9}$·
·HABENS · DIEM · MAXIMVM · HORARVM · 1Z · $\frac{1}{4}$·

·ZIBALA·

·SOLIS·
PORTVS·

8

·INSVLA·

·ABARATHTHA·

3

·MORDVLI·

·MORDVLAE·
PORTVS·

·MARE·

Z

·BARACVS·FLV·

·BOCANA·

1

·NACADVMA·

·AEQVINOCTIALIS·

·HABENS · DIEM · HORARVM · 1Z · SEMPER ·

·ZABA·

·CETAEVM·PRO·
MONTORIVM·

1

·NANIGIRI·

·DIONYSI·SEV·BAC·
CHI·OPPIDVM·

·CVMARA·

Z

·CORCOBARA·

3

·ALABA·

·PRIMVS·PARALELLVS·AVSTRALIS·

·DIFFERT · AB · AEQVINOCTIALI · HORAE · $\frac{1}{4}$·
·HABENS · DIEM · MAXIMVM · HORARVM · 1Z · $\frac{1}{4}$·

9

4

·BALACA·

6

7

1Z8	1Z9	130	131	13Z	133	139	134

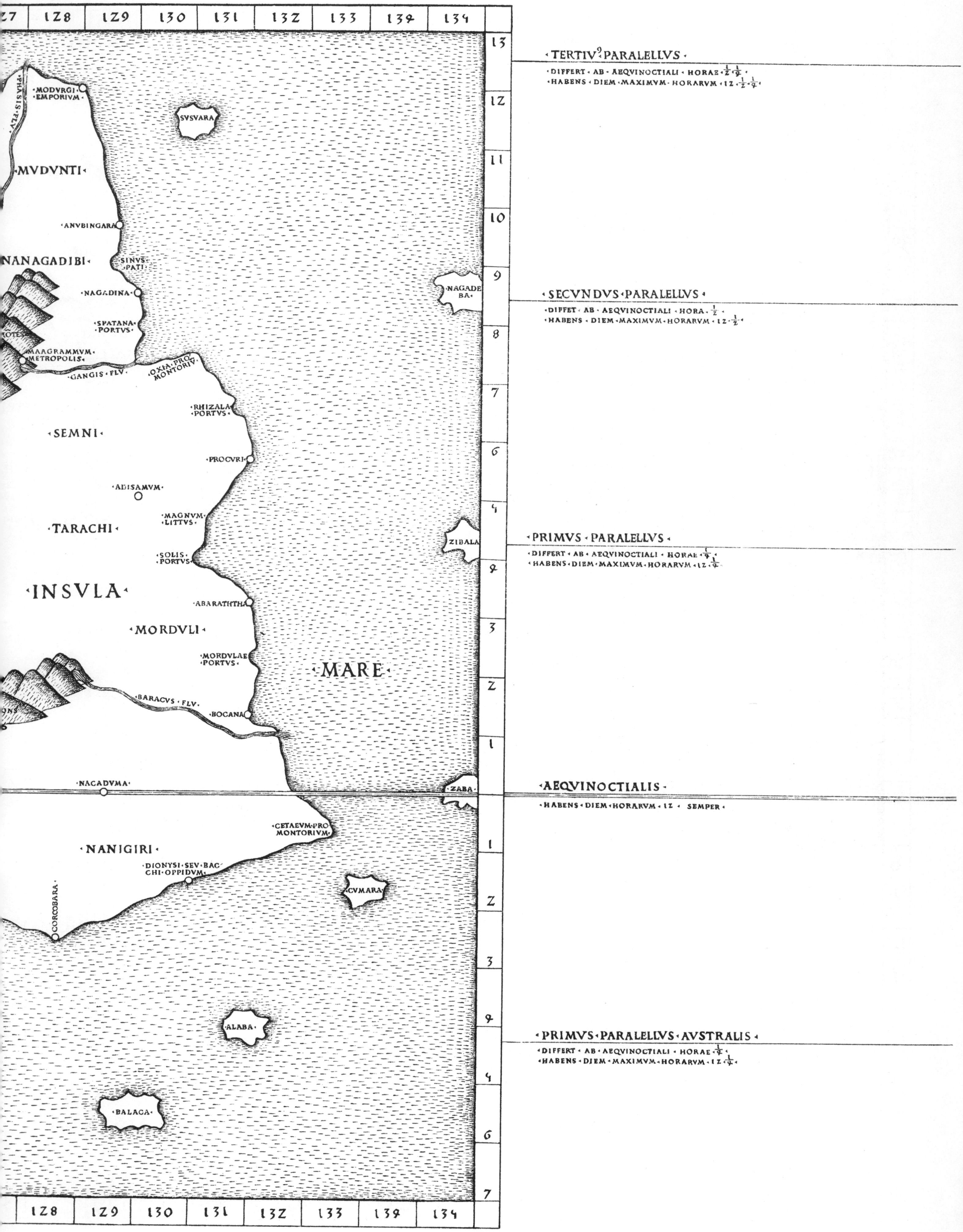

TABVLA QVA
HISPANIA
SARDOO PELAGO
SARDINIA
TYRENO
SICILIA
ADRIATICO
IBERICO MARE
MAVRITANIA TING
APHRICANO PELAGO
OCEANO OCCIDENTALE
ATLANTICO MARE
MAVRITANIA CESARENSE
APHRICA
CINYPHO FIV
BAGRADA F
GETVLIA
LIBYA DE
MERNA
AVTOLALE
ISOLA DI GIVNONE QVALE ET AVTOLALA
GANARIA PO
MAVSOLI
PHARVSII
MELANO GETVLI
NATEBRE
LYNXAMATE
GARAMA METPI
GIR GYRE
BABII
MALCOE
NIGRITE ETHIOPI
TAMODOCANA
APROSITO
ISOLA DI GIVNONE
PLVITANA
CASPERIA
CANARIA
PINTVARIA
RYSSADIO
MACHVREBI
SOLENTII
ARATICOLI
LIBYA PALVDE
ALITABI
MANRALI
ARMIE
THALE
DOLOPI
ASTACVRI
NVBI
PYRREI E
CHVRITE
ODRANGIDI ETHIOPI
LIBYA INTER
MIMACE
AECHEME
PYRROCAMPO
ORPHEI
TARVALTE
MATITE
APHRICERONI MAGNA GETE
PERORSI
GONGALE
NANOSBEI
HESPERIO SENO
AGANGINE ETHIOPI
NABATHRE
DERMONI
DERBI
XYLICI ETHIOPI
ICHTHYOPHAGI ETHIOPI
MAGNA REGIONE DELLI ETHIOPI INQVALE
PHATI CADIDIA TVTTI ET RINOCEROTI ET TIGRI
ETHIOPIA INT
HESPERII ETHIOPI
ANTHACE ETHIOPI
ZIPHA
DAVCHI
MESCHA
AGISYMBA MAGNA
TERRA INC

XXVIII. GEOGRAPHIA DI FRANCESCO BERLINGHIERI

(FIRENZE CIRCA 1478.)

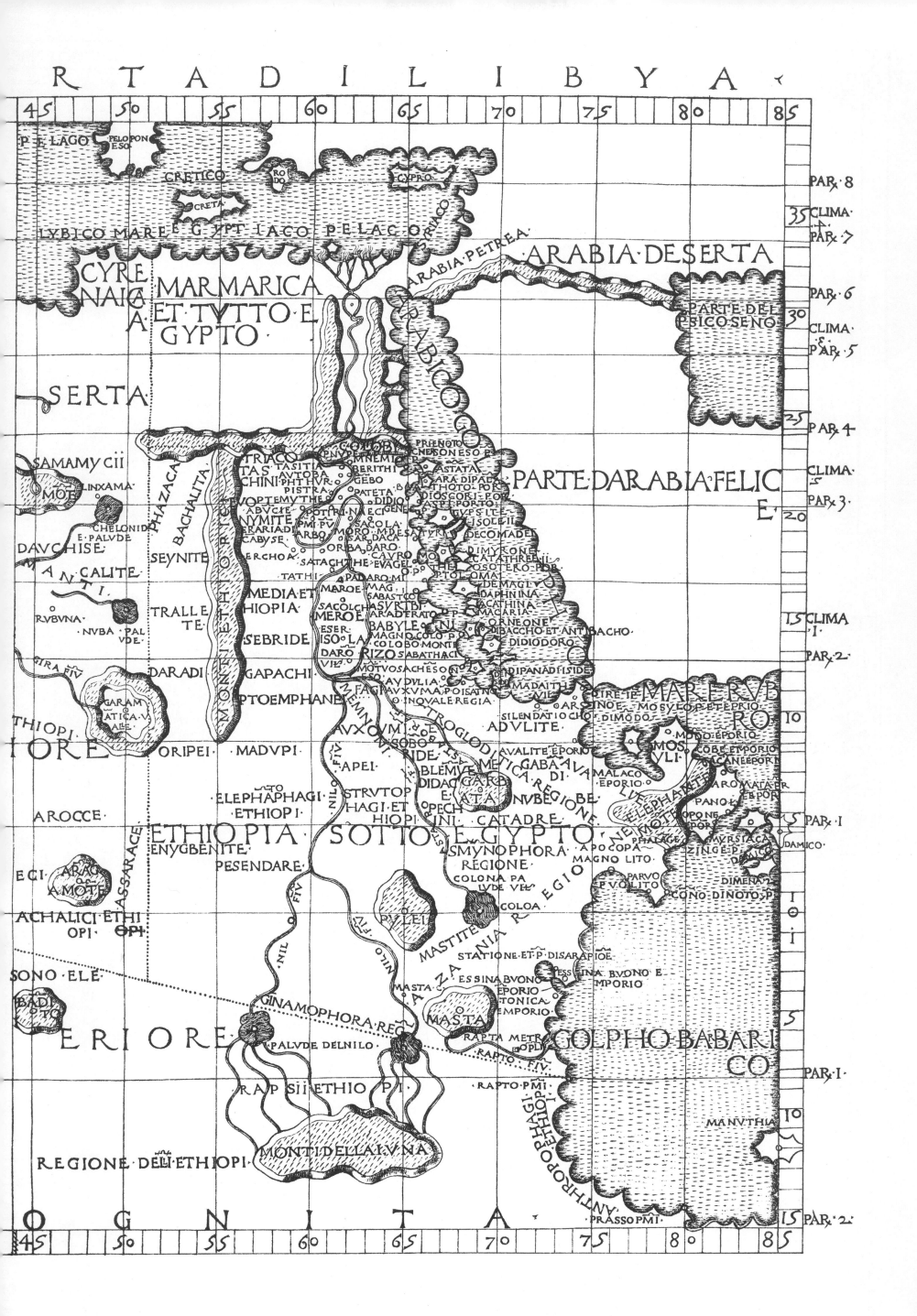

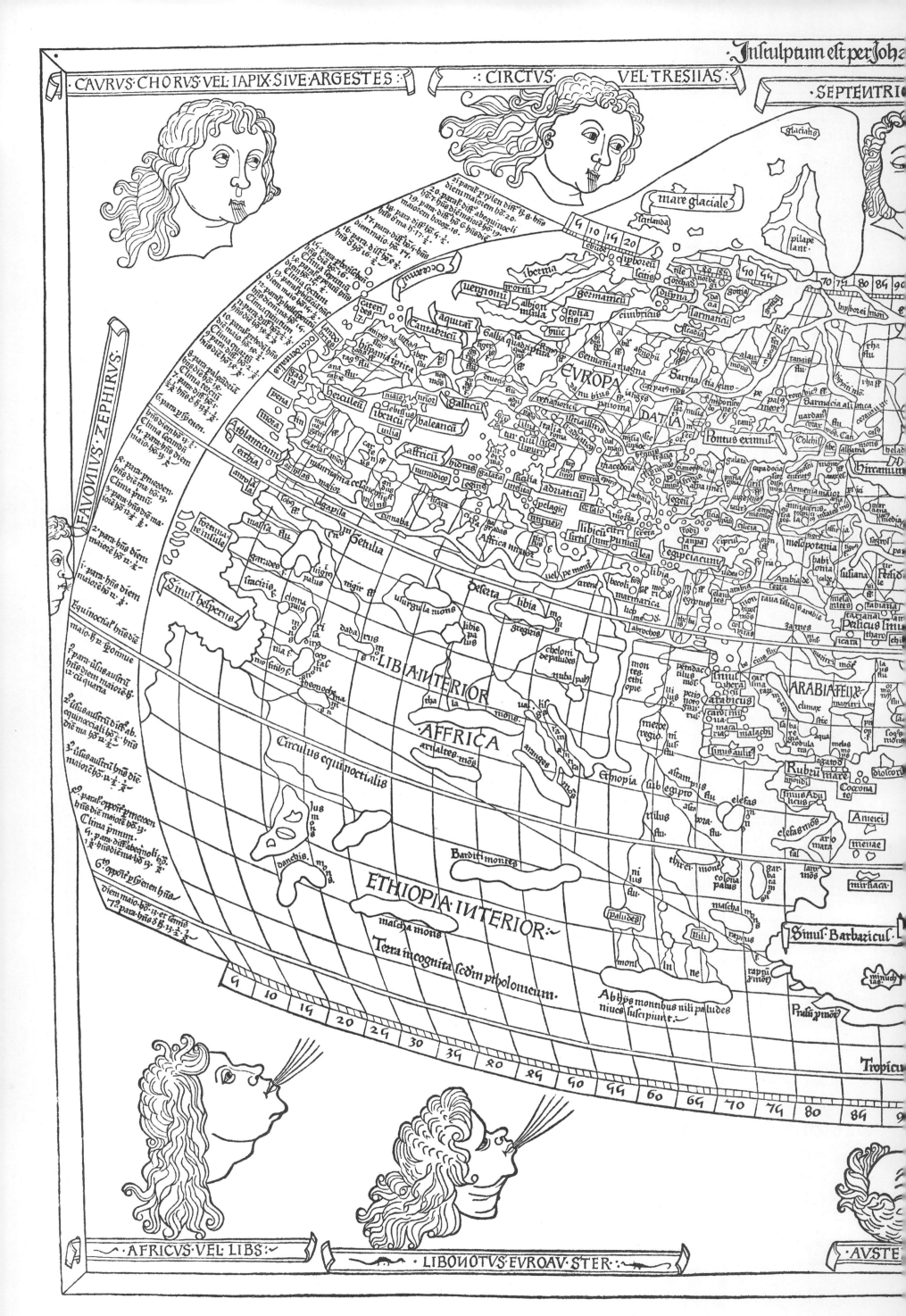

XXIX. PTOLEMÆUS ULMÆ 1482.

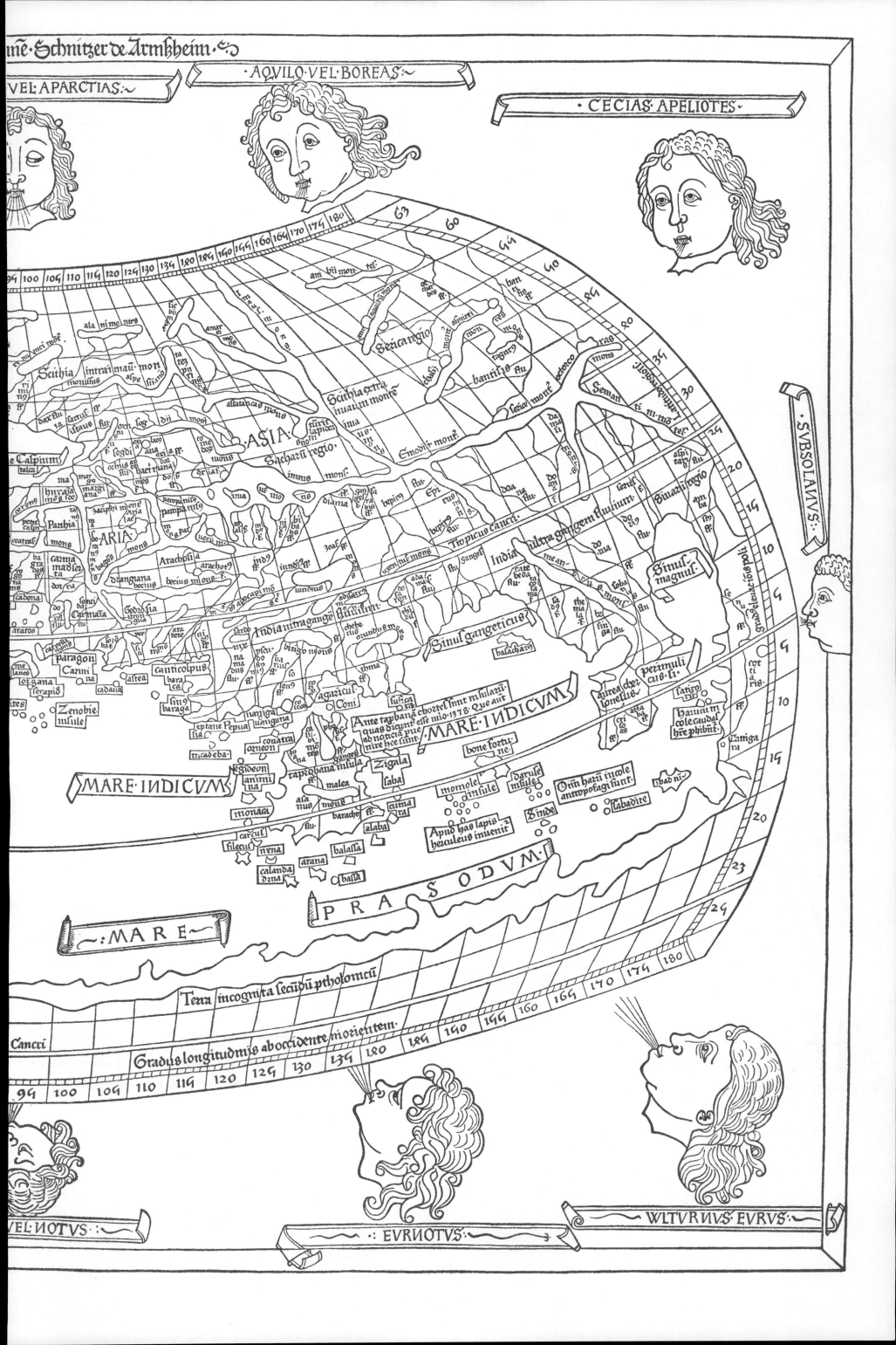

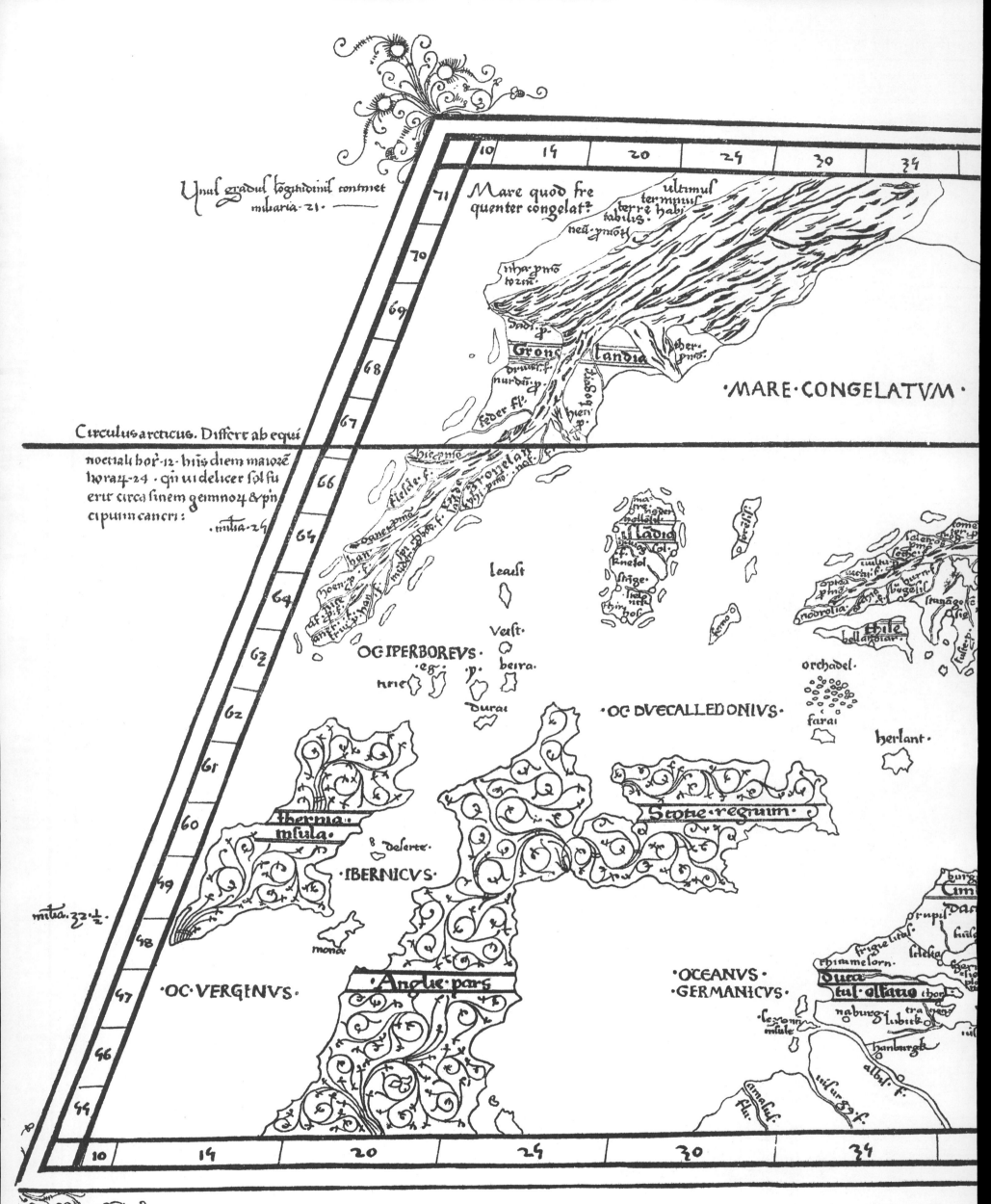

Ynus gradus logitudinis contmet miliaria · 21 ·

Mare quod fre quenter congelat?

ultimus terminus terre habi neū pnoss

·MARE·CONGELATVM·

Circulus arcticus. Differt ab equi noctiali hor· 12· huis diem maiorē hora 4·24 · qn ui delicet sol fu erit circa finem geminoq & pri cipium cancri :   · mulia· 24

Gronelandia

iha · pnō tozin

OC IPERBOREVS·

·OC DVE CALLEDONIVS·

leauk

veauk

beira.

Duras

Lladia

orchadel.

farai

Hile bellatorsi

herlant.

·Scone · regnium·

thernia infula

8 Deferte

·IBERNICVS·

monai

·OC·VERGINVS·

·Anglie· pars

·OCEANVS· ·GERMANICVS·

Cim

hanburg

albis f.

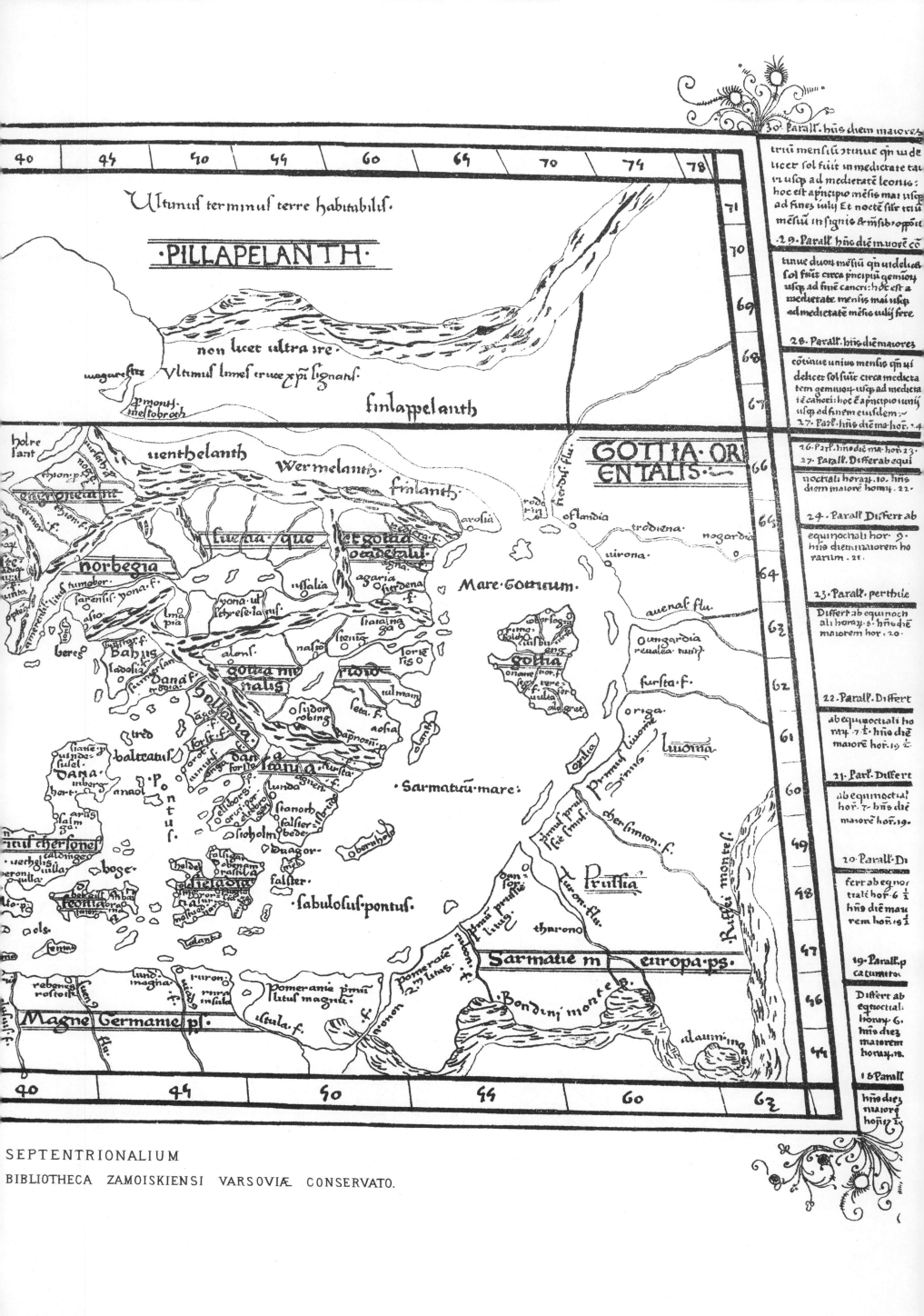

(1) JOHANNES ESCHUIDUS, SUMMA ANGLICANA
VENETIIS 1489.

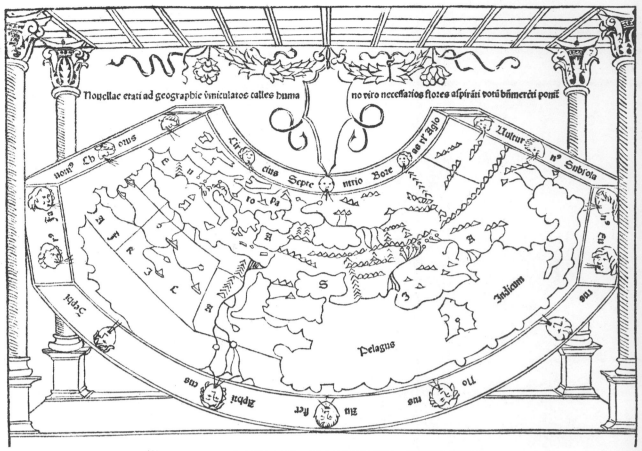

(2) POMPONII MELLAE COSMOGRAPHI GEOGRAPHA
VENETIIS 1482

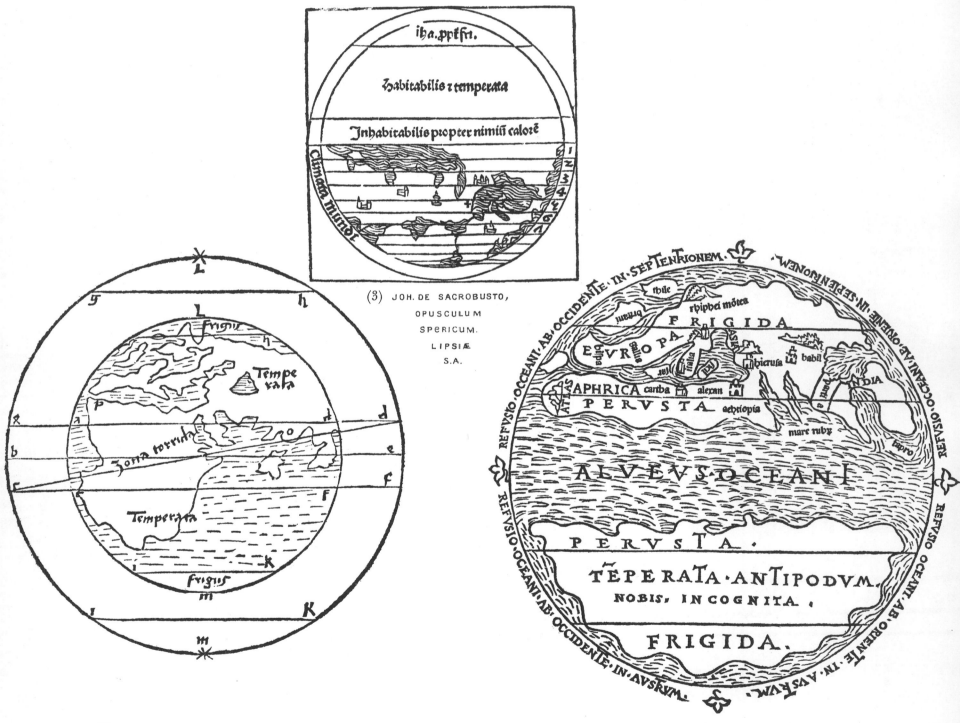

(3) JOH. DE SACROBUSTO,
OPUSCULUM
SPERICUM.
LIPSIÆ
S.A.

(4) METEOROLOGIA ARISTOTELIS
ELEGANTI JACOBI FABRI STAPULENSIS PARAPHRASI EXPLANATA.
NORINBERGÆ 1512.

(5) MACROBII IN SOMNIUM SCIPIONIS EXPOSITIO,
BRIXIÆ 1483.

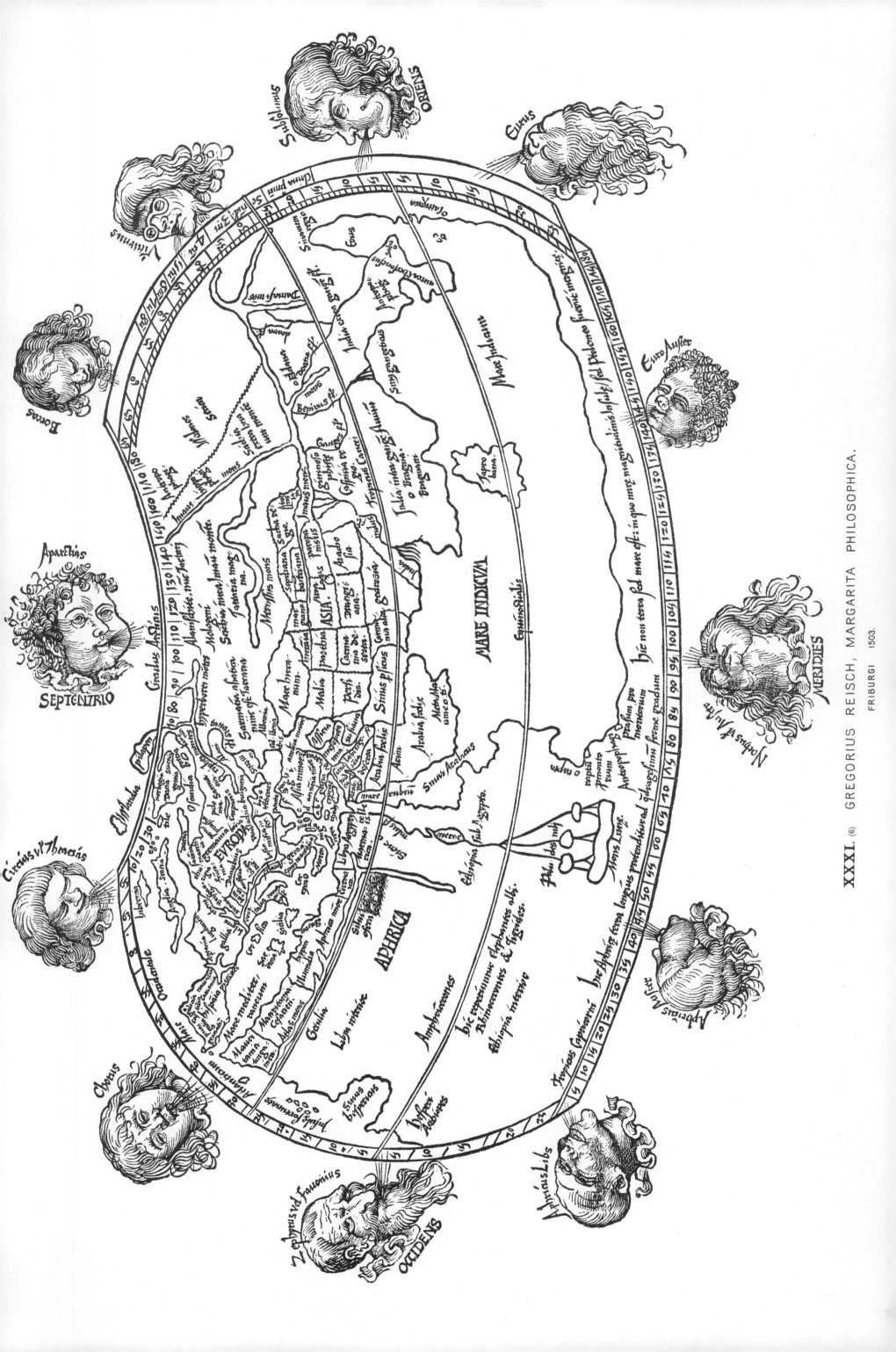

XXXI. (6) GREGORIUS REISCH, MARGARITA PHILOSOPHICA.
FRIBURGI 1503.

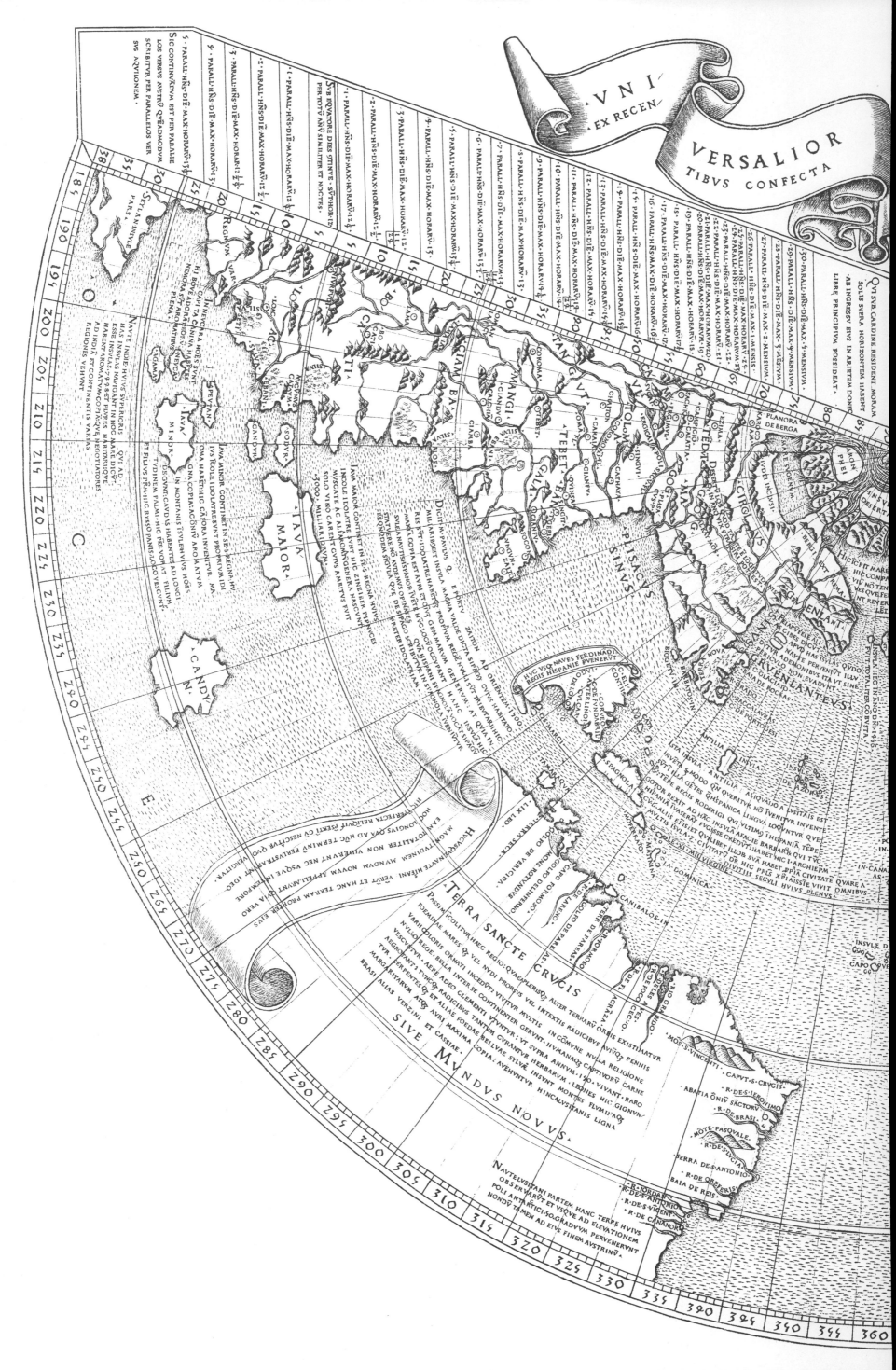

XXXII. RUYSCH, UNIVERSALIOR COGNITI ORBIS TABULA

PTOLEMÆUS

EX RECENTIBUS CONFECTA OBSERVATIONIBUS
ROMÆ 1508

FAVO NIVS

CYRCI NVS

OCCIDENS

SEPTEN

ZEPHYRVS

CLIMA.I.

CLIMA.II.

CLIMA.TER.

CLIMA.QVAR.

LYBS

AFRICVS

LIBO

NOTVS

terra cube

ifpa
niaɛ
infu

regalis domus

terra labora

toruɜ

iuernia.inf.

iflandia
congelatum.m.

albion.i.

aquila oceā

magna
germa

fortunatæ
infule

SCY
THI

dalma

P.

lia
lia

m pont.

hyrca

agepo
mai
madrus
atlas

iberi
mauri
catari
croli
mazia

gallicū
africū
fardo

ARABIA

PERSI
CVS.SI

LIBYA

GETVLIA

AFRI CA

puniuɜ

cyrenaica
regio

mamari
capre

libycapeɜ lacī

mö tana arabiɛ

FELIX

INTERI

caphas.m.

cheloni
des.pal.

ptus

arabias

merce

Si.

tli
gis.aq.

LIBYADE SERTA

libyae.pa.

garamanti
ca uallis

AETHI
SVB
AEGYP.

TRO

mare rubruɜ

alcabora

GLO

hyppodromus

Nilus.flu.

alcapusɜ

OR

DITICA.RE.

diofc
ridis.

AETHI
ins.m.

OPIA IN
ziptha.mös.

barditi.m.

melinde

barbari
cus.Si.

BARBAR
CVM.PELA
GVS

AGYSIN BA

paludes

nili

AZANIA

menutias.inf
BREVE

MAGNVS
PERIVS

ETHES
SINVS

dauchis.mö.

niger mons.

mons
pala
REGIO

lune
des niues

aquo nili
fufcipiunt

MARE

lapidū

OCEA NVS

TERI OR

lapidis.m.

caputbonɛ
ipei

currentiū

primum
promon

comorbina.in

OCEA NVS MERI

AVSTER

NVS

SAN CTAE CRVCIS OCEA NVS

TERRA

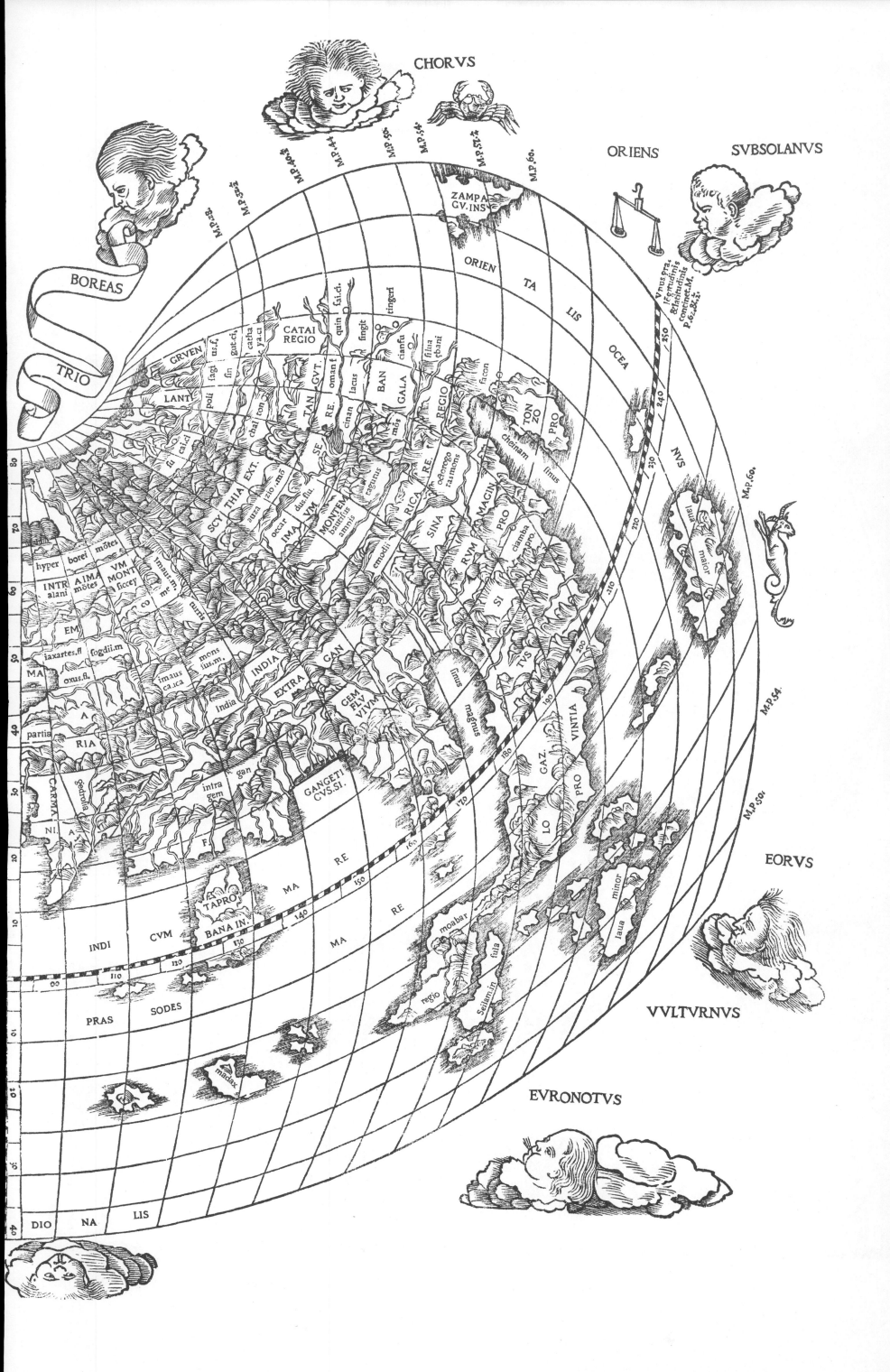

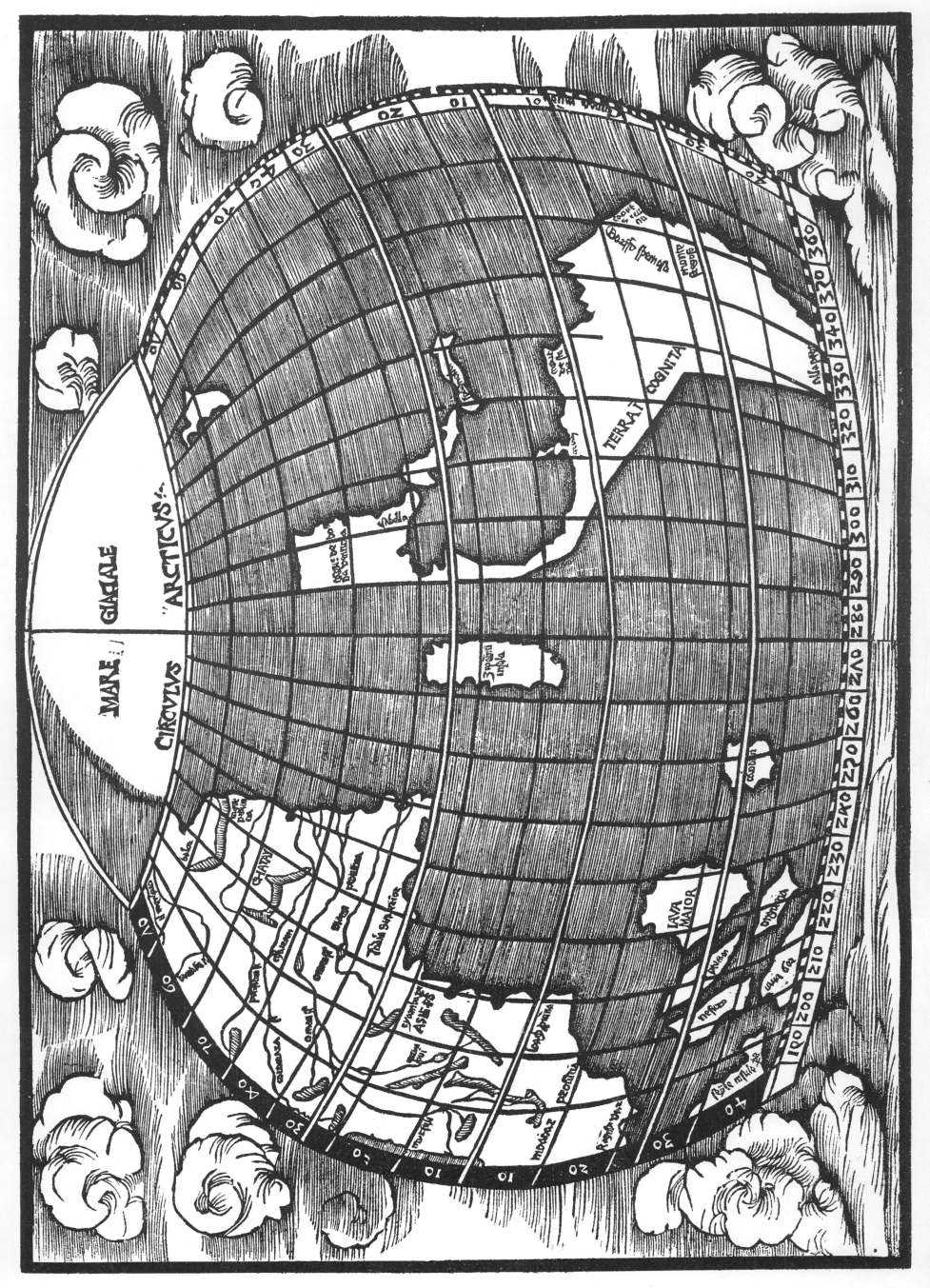

XXXIV. JOANNES DE STOBNICZA, INTRODUCTIO
CRACOUIE

IN PTHOLOMEI COSMOGRAPHIAM

1512.

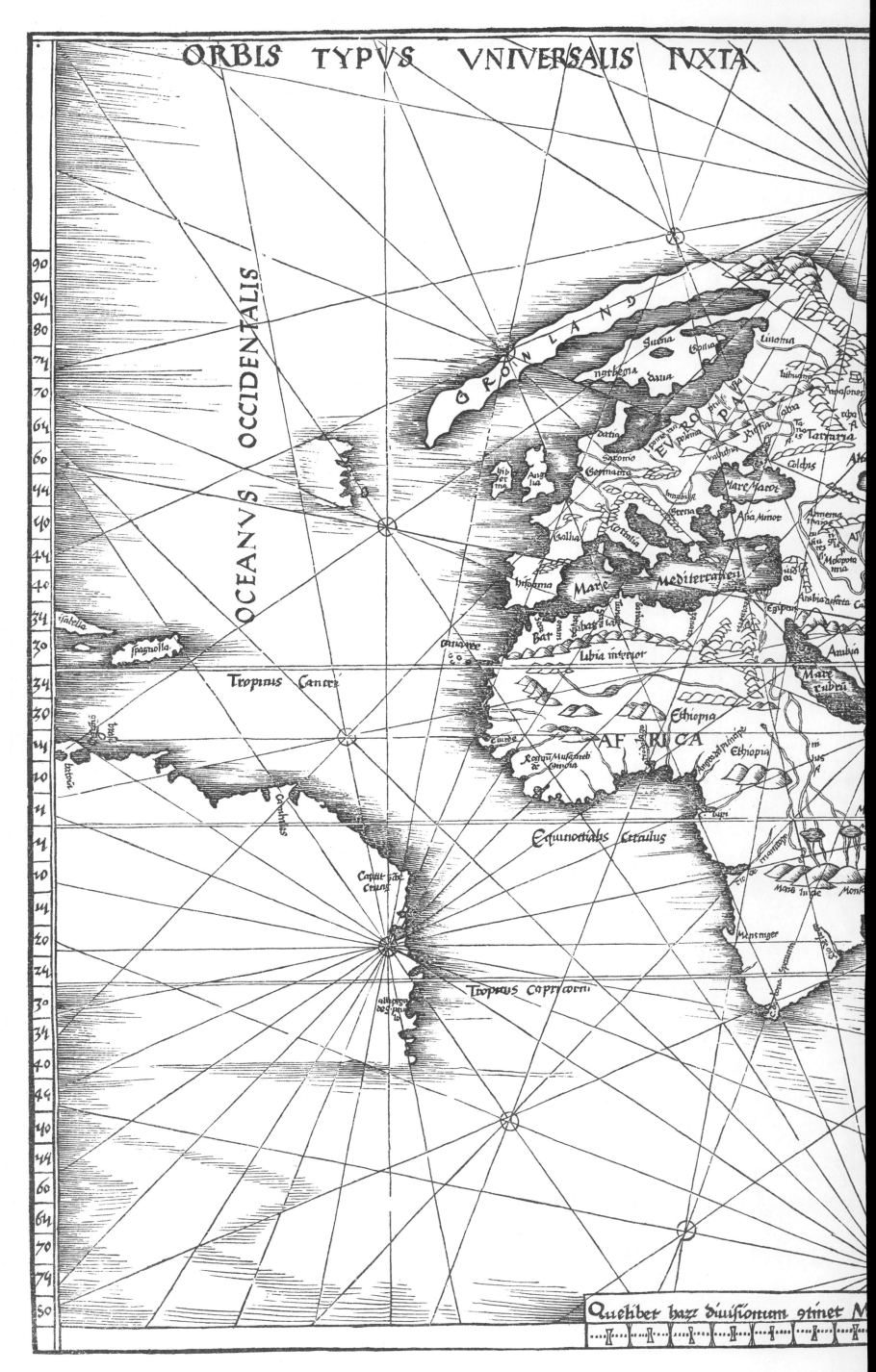

XXXV. HYDROGRAPHIA SIVE CHARTA MARINA

PTOLEMÆUS, ARGENTINÆ 1513.

ASIA

MARE INDICVM

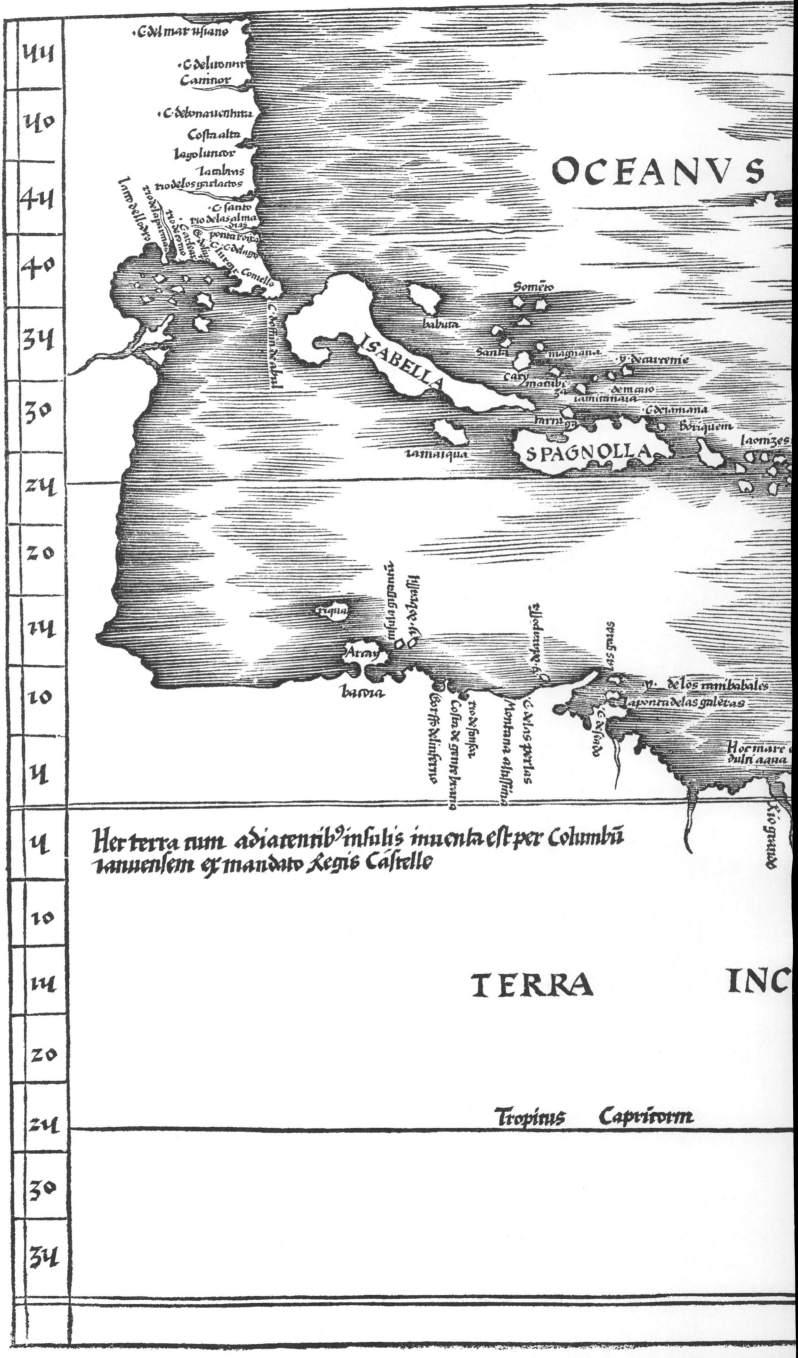

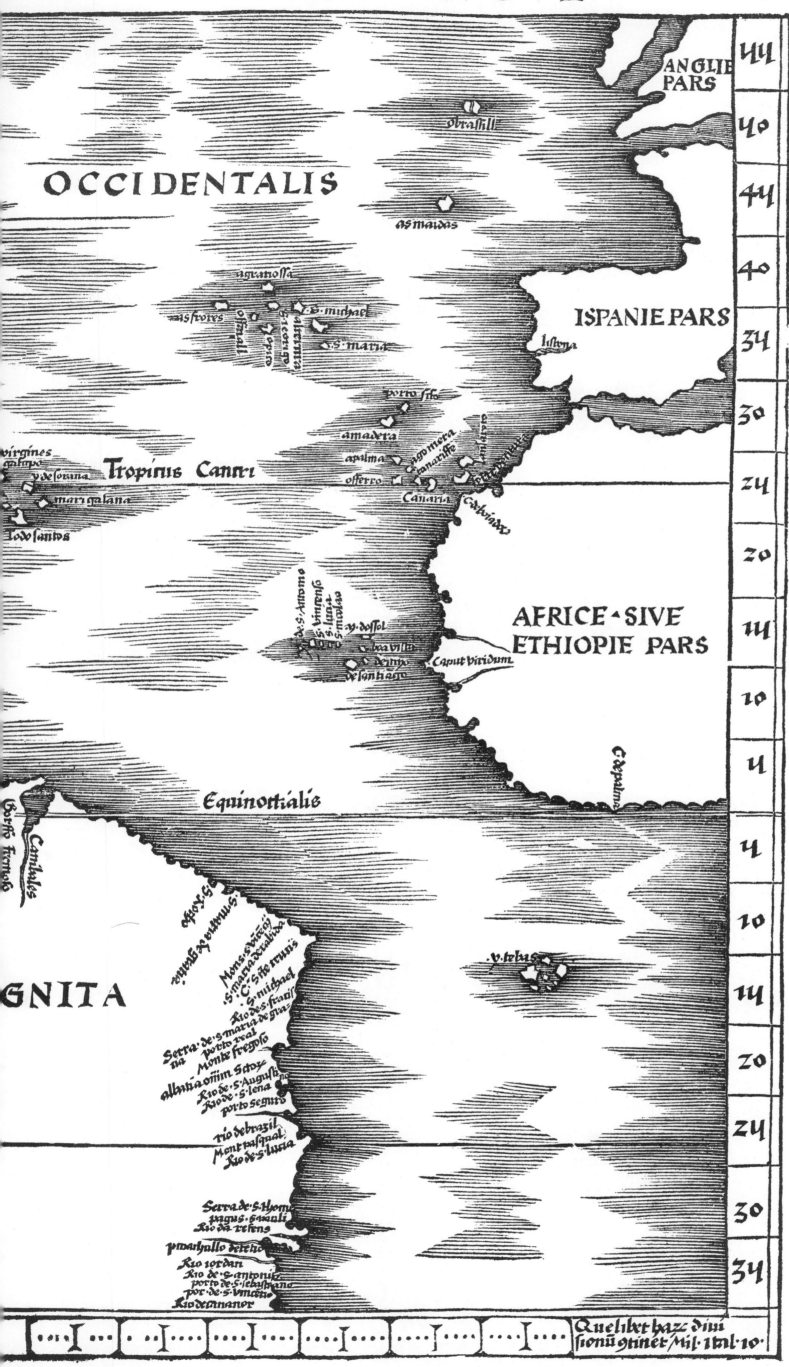

OCCIDENTALIS

44
40
44
40

ANGLIE PARS

ISPANIE PARS

34
30
24
20
14
10
4

AFRICE · SIVE
ETHIOPIE PARS

Tropirus Cancri

Equinottialis

4
10
14
20
24
30
34

GNITA

OCCIDENTALIS SEU TERRÆ NOVÆ.
ARGENTINÆ 1513

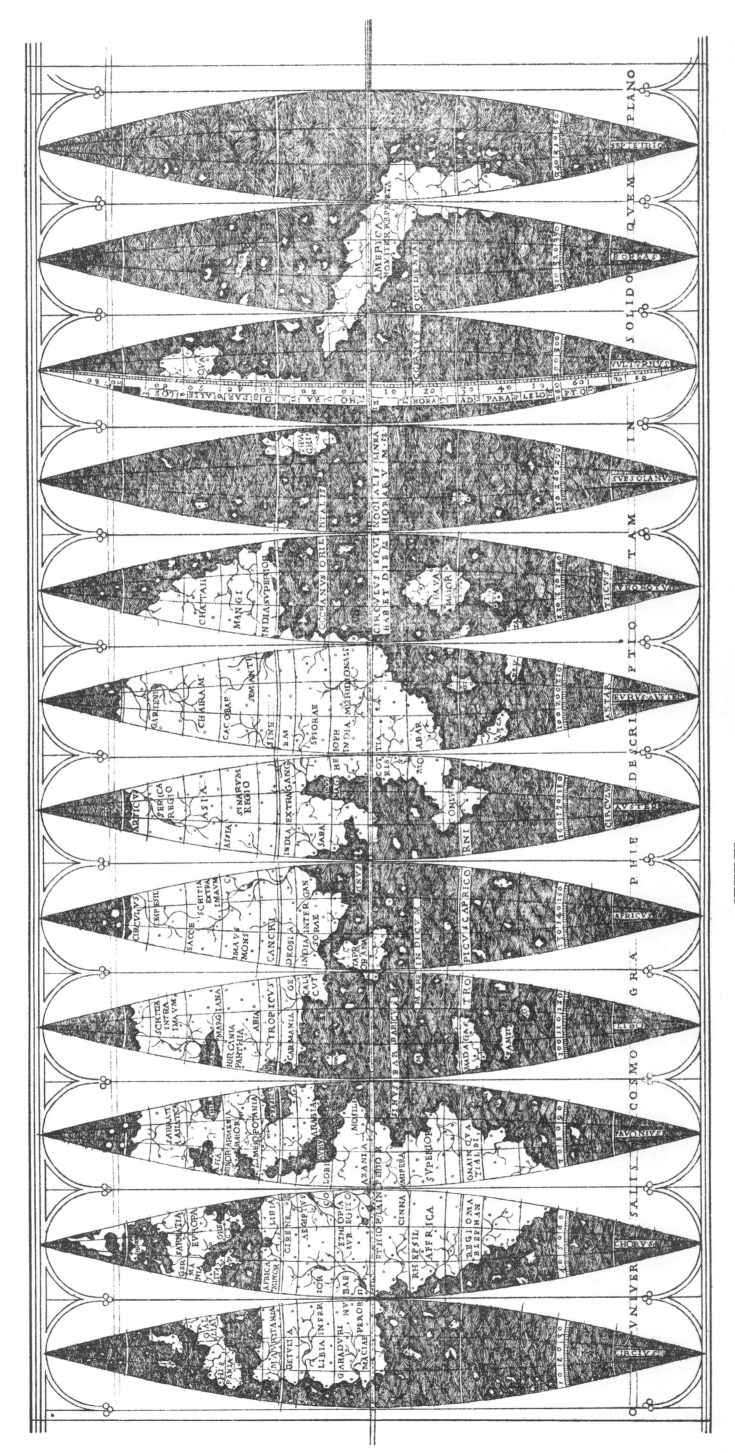

XXXVII. (1) LUDOVICUS BOULENGER 1514.

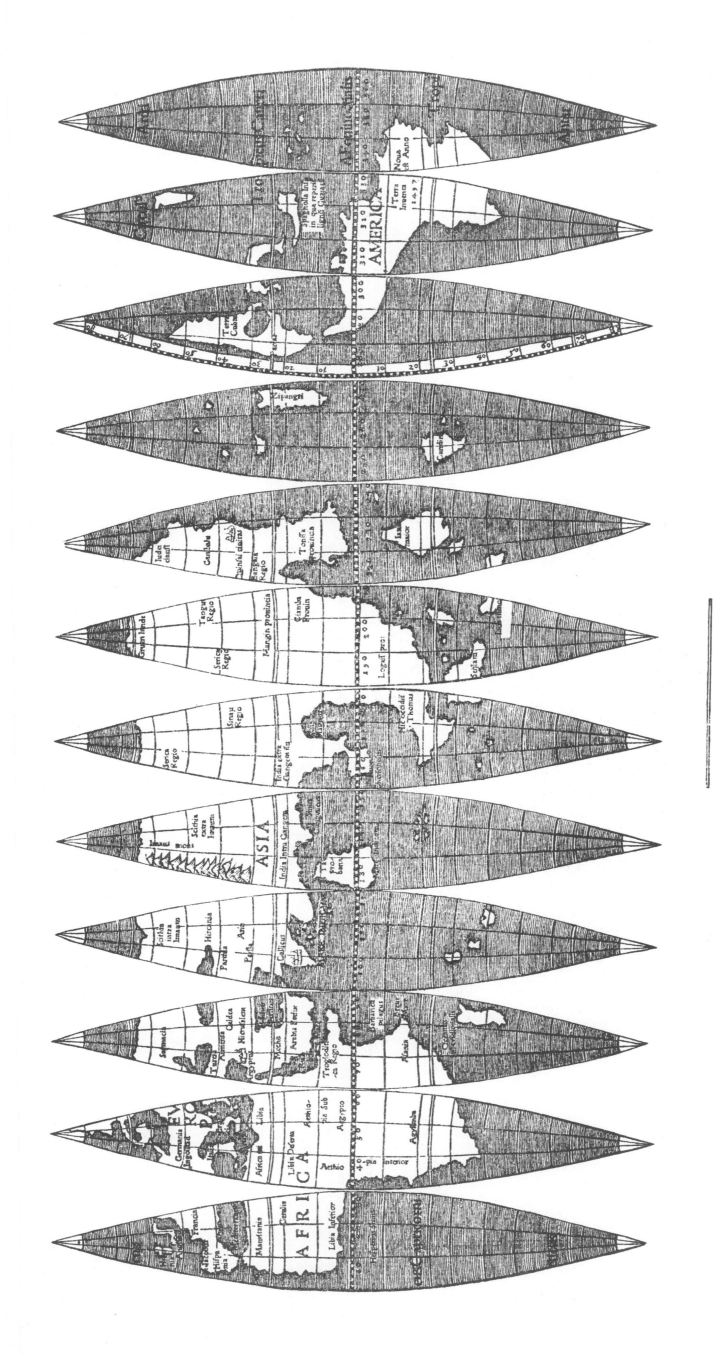

XXXVII. (2) MAPPA MUNDI AD GLOBUM INDUCENDUM

LUSTRO TERTIO SECULI XVI. IN LIGNUM INCISA.

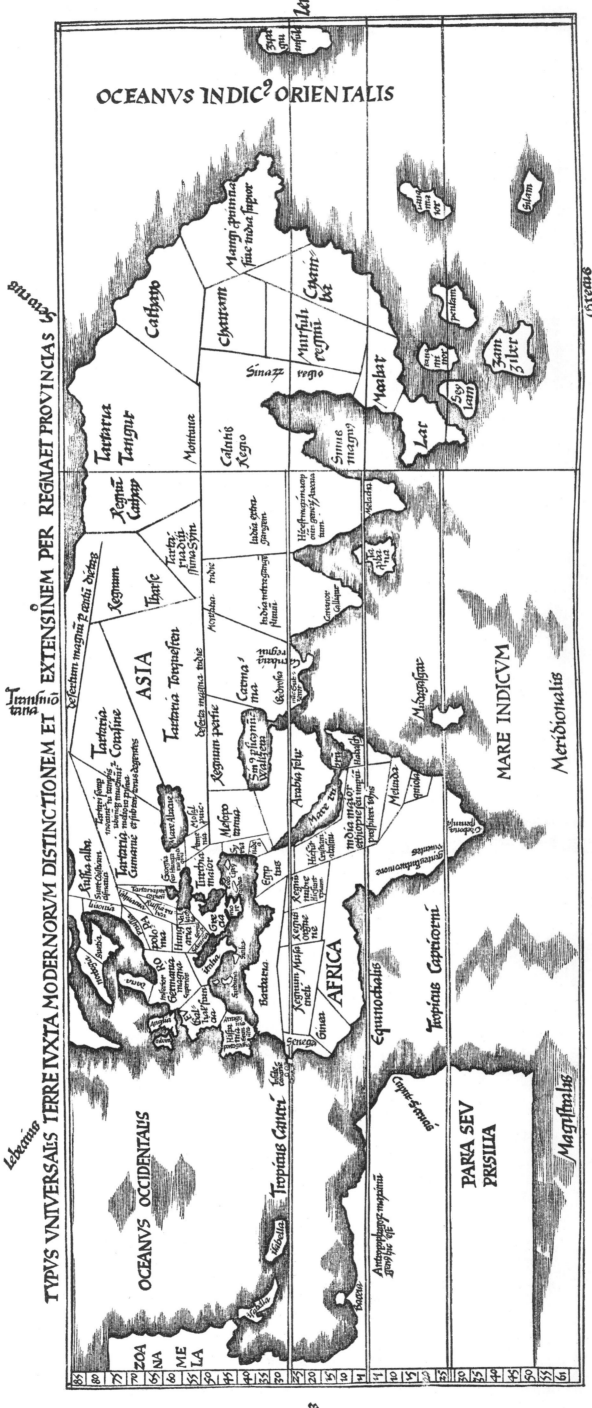

XXXVIII. (1) GREGORIUS REISCH, MARGARITA PHILOSOPHICA NOVA. STRASSBURG 1515.

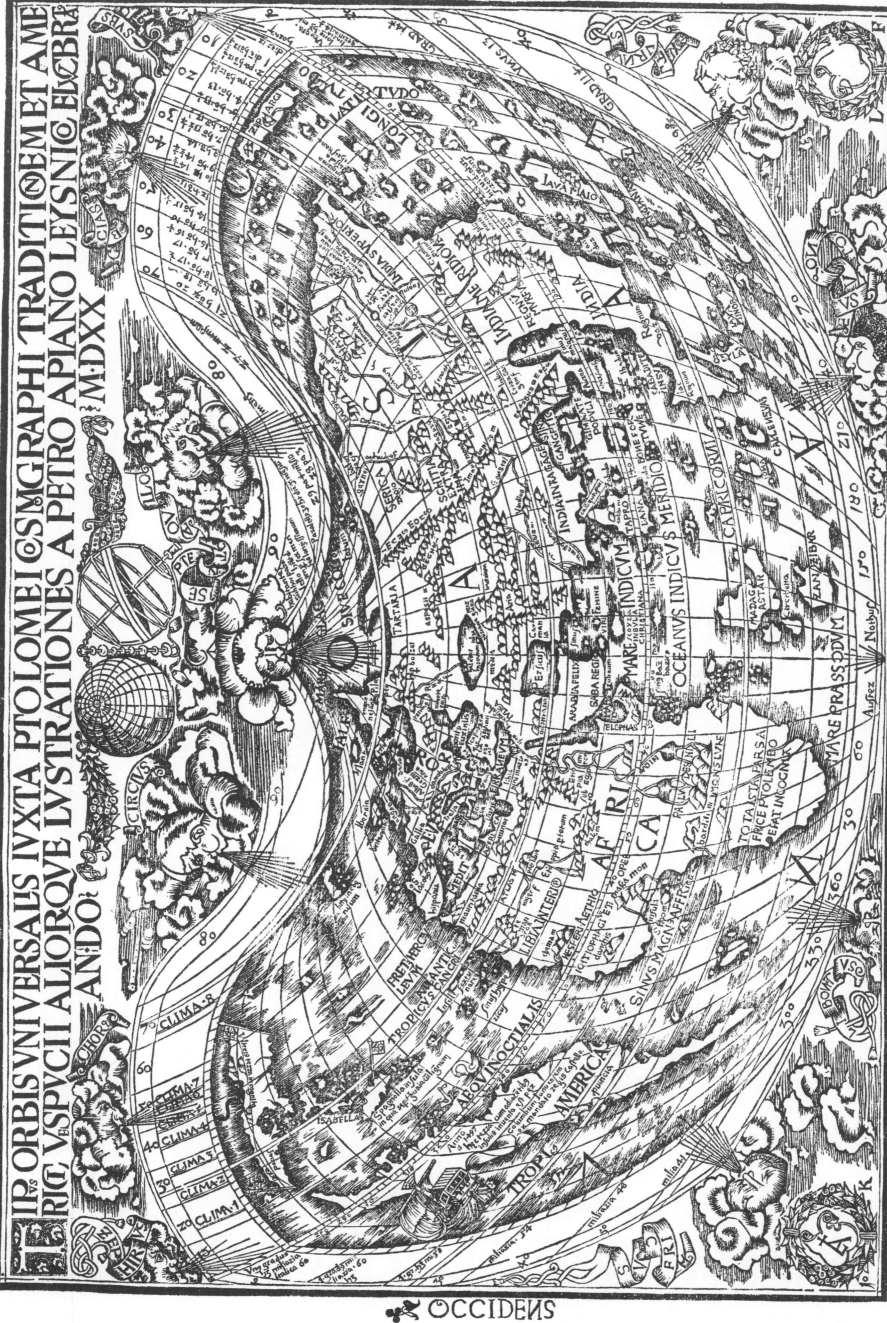

ORIENS

SEPTENTRIO

MERIDIES

OCCIDENS

TYPVS ORBIS VNIVERSALIS IVXTA PTOLOMEI COSMOGRAPHI TRADITIONEM ET AMERICI VESPVCII ALIORVQVE LVSTRATIONES A PETRO APIANO LEYSNICO ELVCVBRAT ANNO MDXX

LONGITVDO

LATITVDO

OCEANVS INDICVS MERIDIONALIS

AMERICA

AFRICA

INDIA

XXXVIII. (2) PETRUS APIANUS 1520.

JOANNIS CAMERTIS MINORITANI IN C. JULII SOLINI ΠΟΛΙΣΤΩΡΑ ENARRATIONES VIENNÆ. 1520.

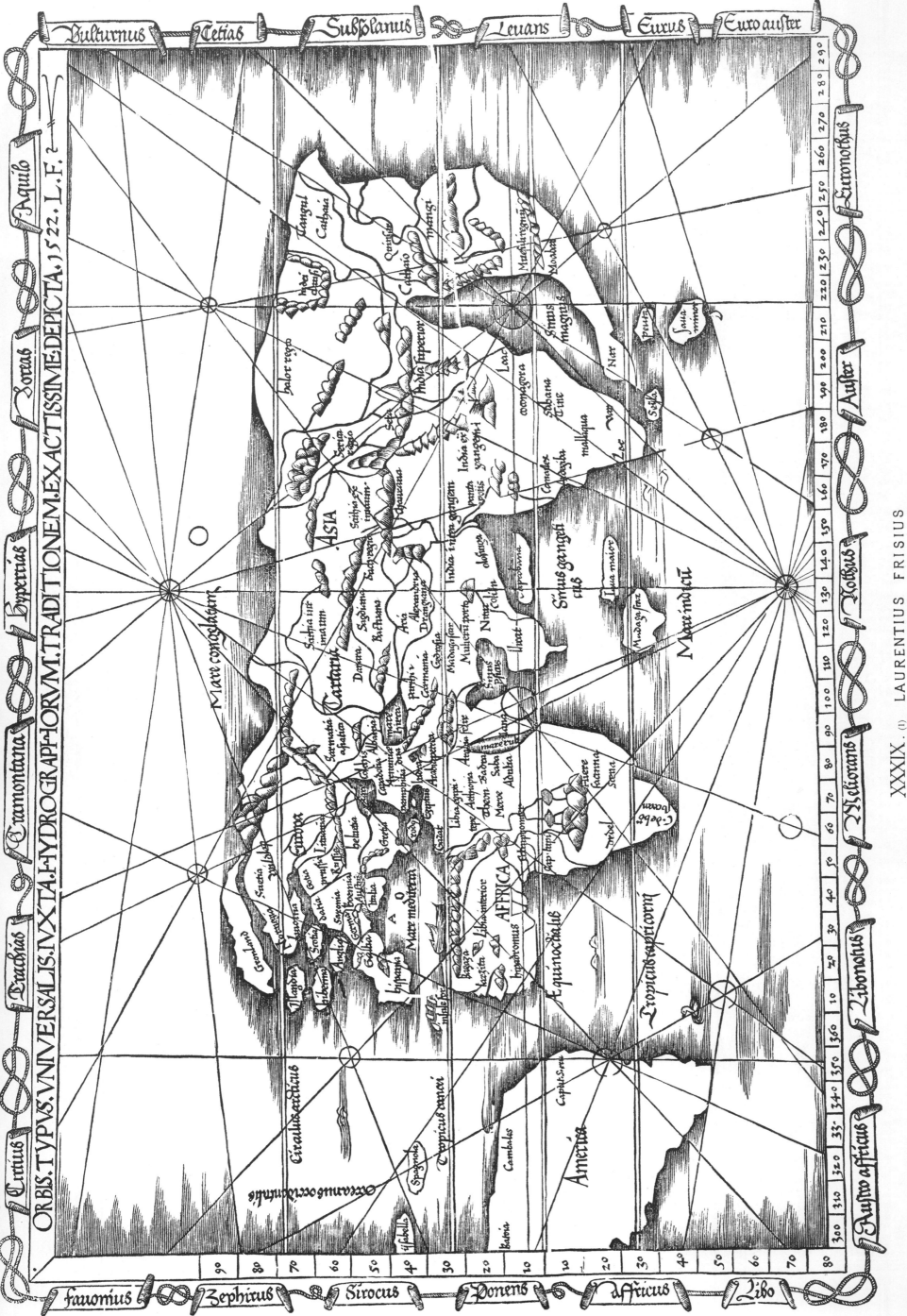

XXXIX. (1) LAURENTIUS FRISIUS
PTOLEMÆUS, ARGENTORATI 1522.

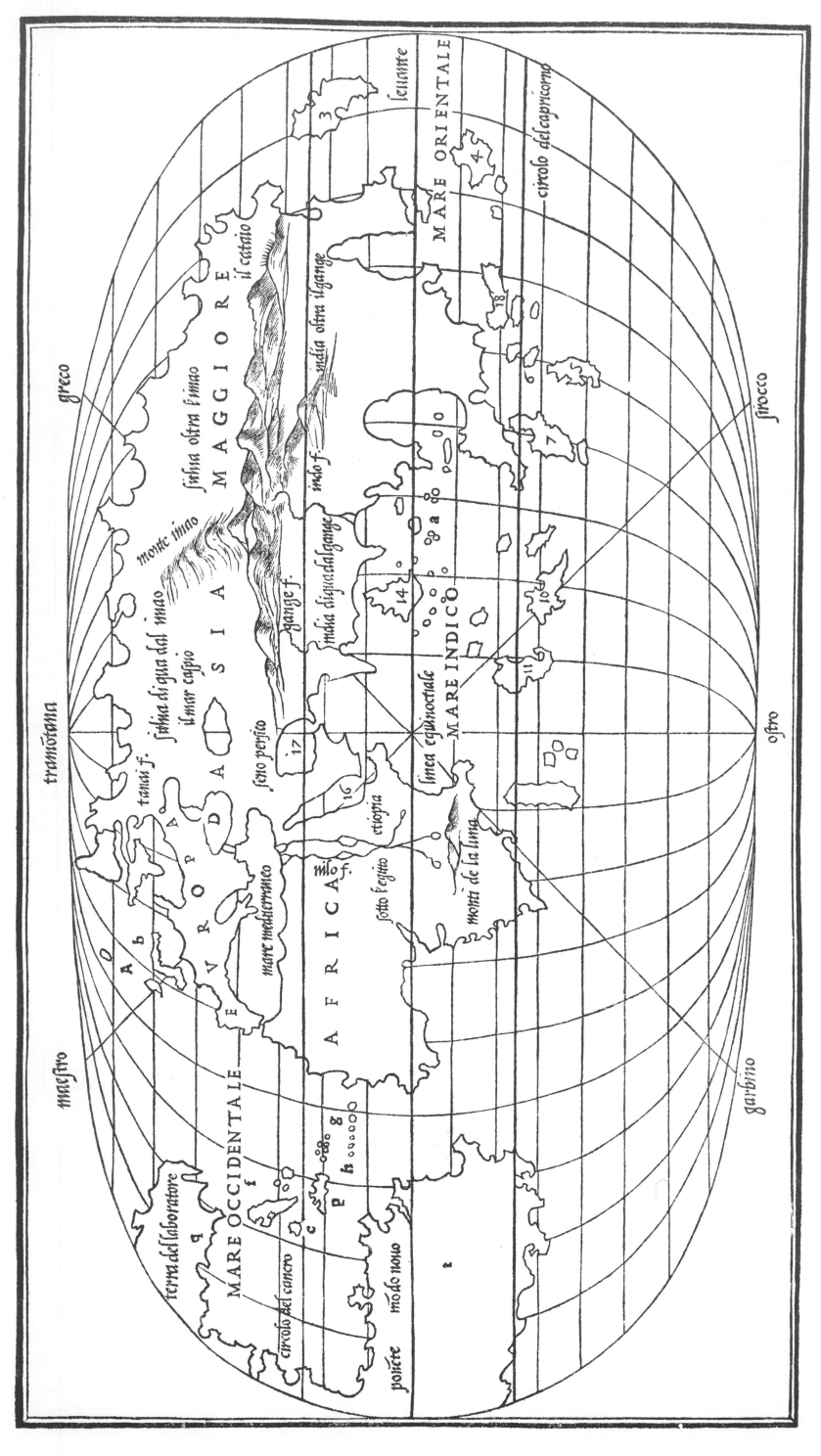

XXXIX . (2) LIBRO DI BENEDETTO BORDONE
VINEGIA 1528.

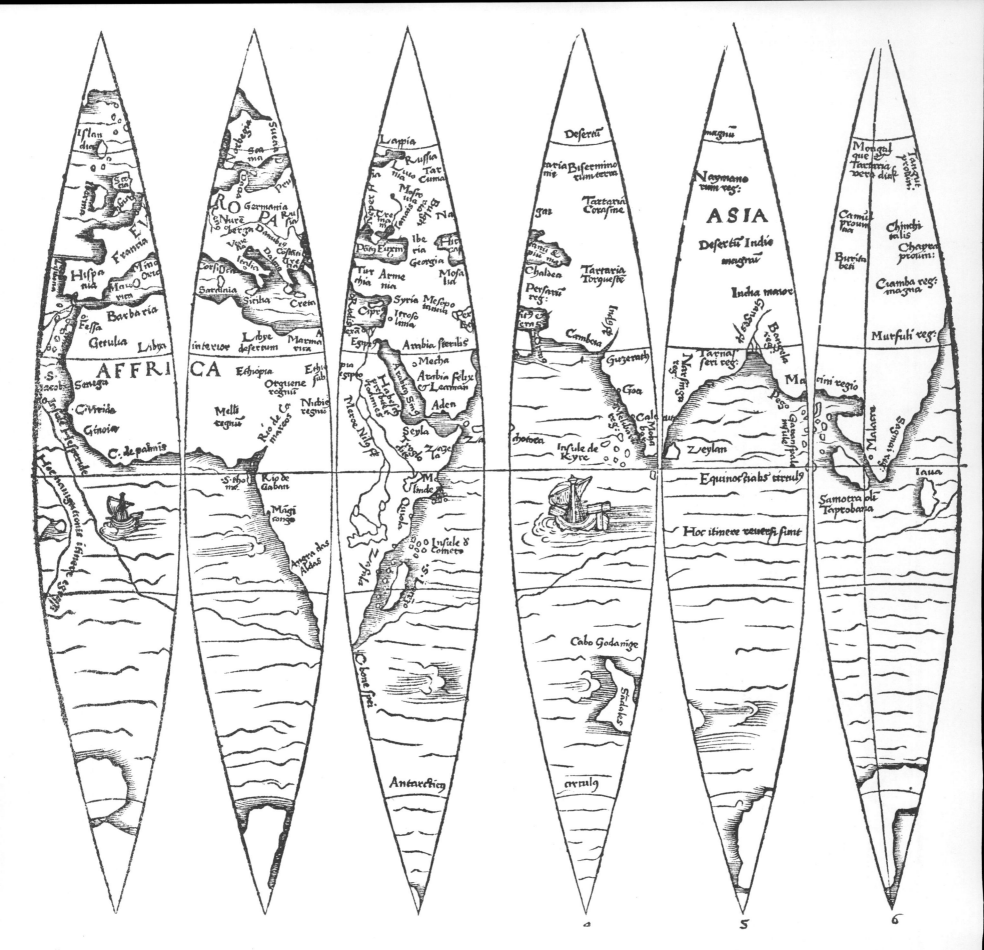

XL. (1)   MAPPA MUNDI AUCTORIS INCERTI, NORIMBERGÆ C. 1540(?)

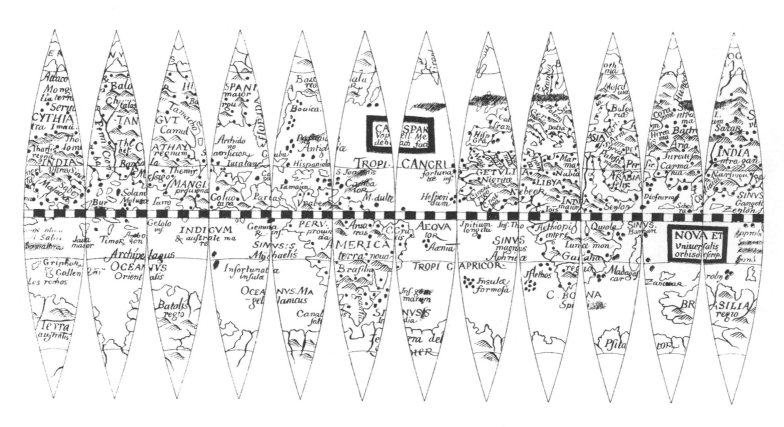

XL. (2)   CASPAR VOPEL 1543.

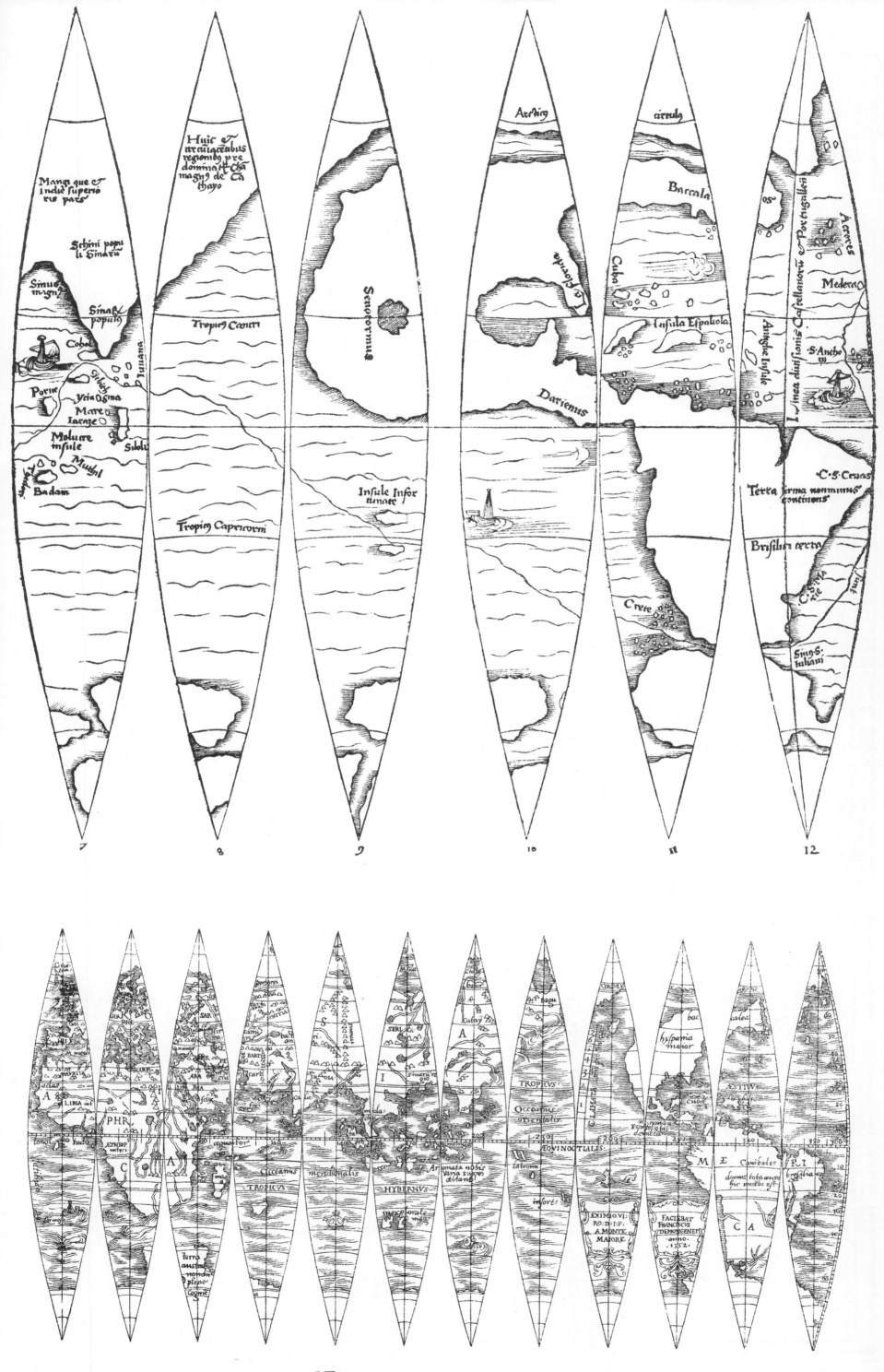

XL. (3) FRANCISCUS DEMONGENET 1552.

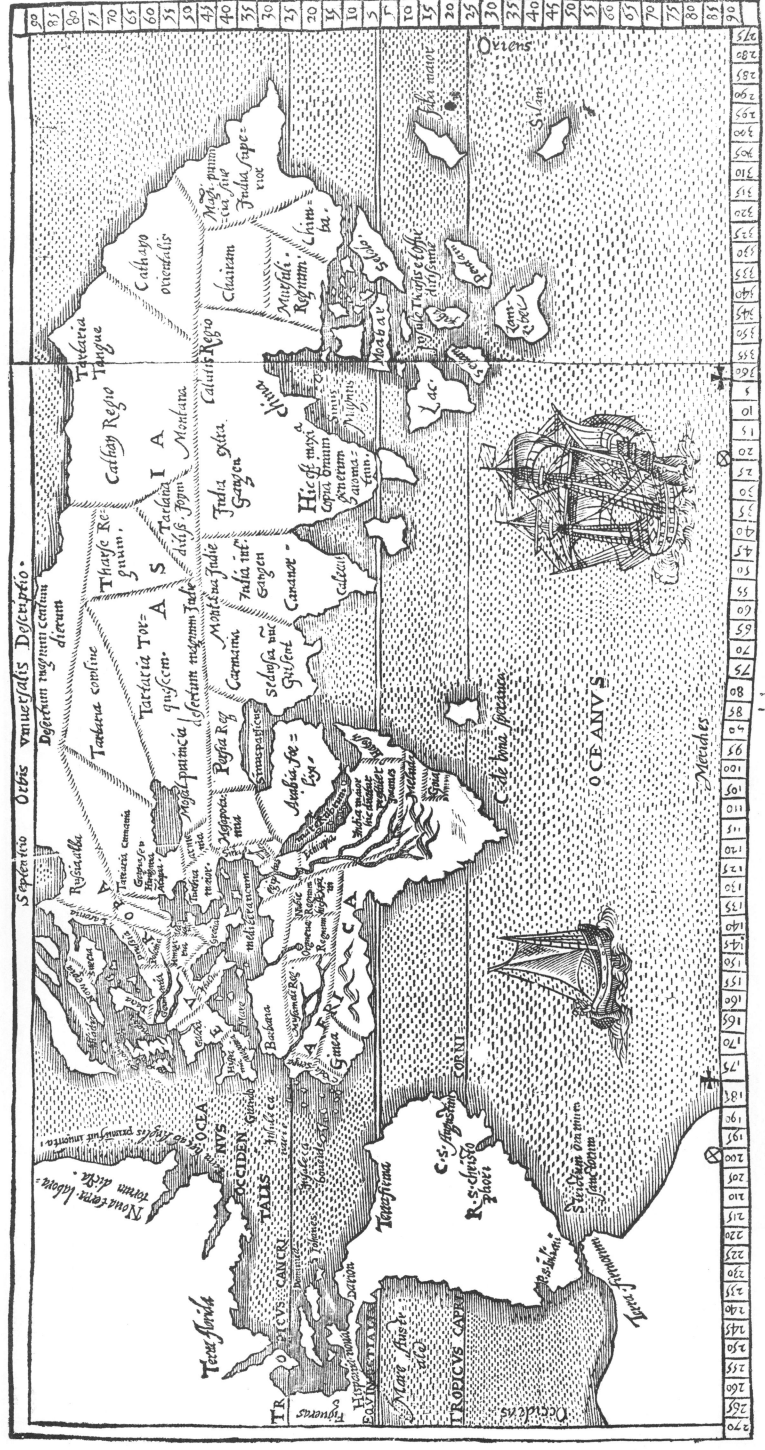

XLI. (1) ROBERT THORNE 1527.

XLI. (2) ORONTIUS FINÆUS 1531

TYPVS COSMOGRAPHI

XLII. SIM. GRYNÆUS, NOVUS ORBIS.
BASILEÆ 1532.

PIPER

MVSCAT

Aquilo seu
Boreas

GARIOFILI

MARE CONGELATVM

SCYTHIA

Tartari Za
lhenses

Scythia intra Imaum

Scythia extra
Imaum

Regnū Cassiæ

Desertum
Belgian

Mare Caspiū

Imaus
mons

Regnum Cu
maniæ

TARTARIA MAGNA
Terra Mongal

ASIA

Regnum
Corasinæ

Cambalu

Regnū Cathay

Mædia

Aria

Emodij mōtes

Regnum
Tharsæ

Cacrius

aldæa.

Persia

Dragiana

Indus
flu.

Ganges
flu.

Regnum
Turquestram

in° Persic

Gedrosia

TROPICVS CANCRI

India extra Gangem

Prouincia Syn

INDIAE
regnum

Ciamba
prouincia
magna

elix

INDIA INTRA
Gangem

Cale
chut

Sinus
Gangetic'

Sinus
aureus

Sinus
magnus

Achiloa
Zaphala

Trap
bana

905

100

110

120

130

140

150

160

170

180

190

200

210

220

230

240

250

260

Sin° Barbaric'

ÆQVATOR

Cherones'

Sinaru
regio

Bubsolang

10

Regnum
Murfuli

Regnum Pego

Idua

Vultuzng

INDIA ab Indo flu. sic appellata, oppidis adeo
exculta dicitur, ut quidam 5000. in ea esse dicant.
Terra est saluberrima, bis in anno metit fruges.
Fert cynnamomū, piper, & calamum aromaticum.
Ebenum arborem sola producit. Psitacum auem &
monoceron bestiam habet. Beryllis, adamantibus,
carbunculis, margaritis, & alijs gemmis preciosis
abundat. Centum & triginta annorum æuum ob
temperatum cœlum quidam agunt. Cultus præci-
puus cum gemmis: alij laneis, alij linteis peplis ue-
stiuntur: pars nudi, pars obscœna tantum amicu-
lati. Niger uulgo corporis color, ex materno utero
sic nati. Potum ex riso & hordeo conficiunt. Aetati
senum prærogatiuā nullam tribuūt, nisi pruden-
tia excellant. Sunt tamen Indorum multæ gentes,
diuersæ forma & lingua, nec eisdem uiuentes moribus.

Regnum
Malach

Zeila'

SCYTHARVM natio primo parua & con-
tempta fuit, sed postea in magnum imperium &
gloriam peruenit, agros ampliens usque ad Ta-
naim fluuiū, à quo Scythia ipsa longo tractu uer-
sus ortum protensa, Imao monte per medium ue-
lut in duas Scythias diuiditur.   Tartaria quæ
& Mongal, maiorem Scythiæ occupans partem:
regio est plurimū montosa, & ubi campestris est,
admixta est glarea harenosa, multis patens deser-
tis. Aër & cœlum intemperatum, tonitrua & ful-
gura in æstate adeo horrenda sæpe fiunt, ut præ
timore homines intereant, Iam calor magnus est,
mox frigus & densißimæ niues cadunt.

Euro auster

VARTOMANVS.

XLIII. GERARDUS MERCATOR 1538.

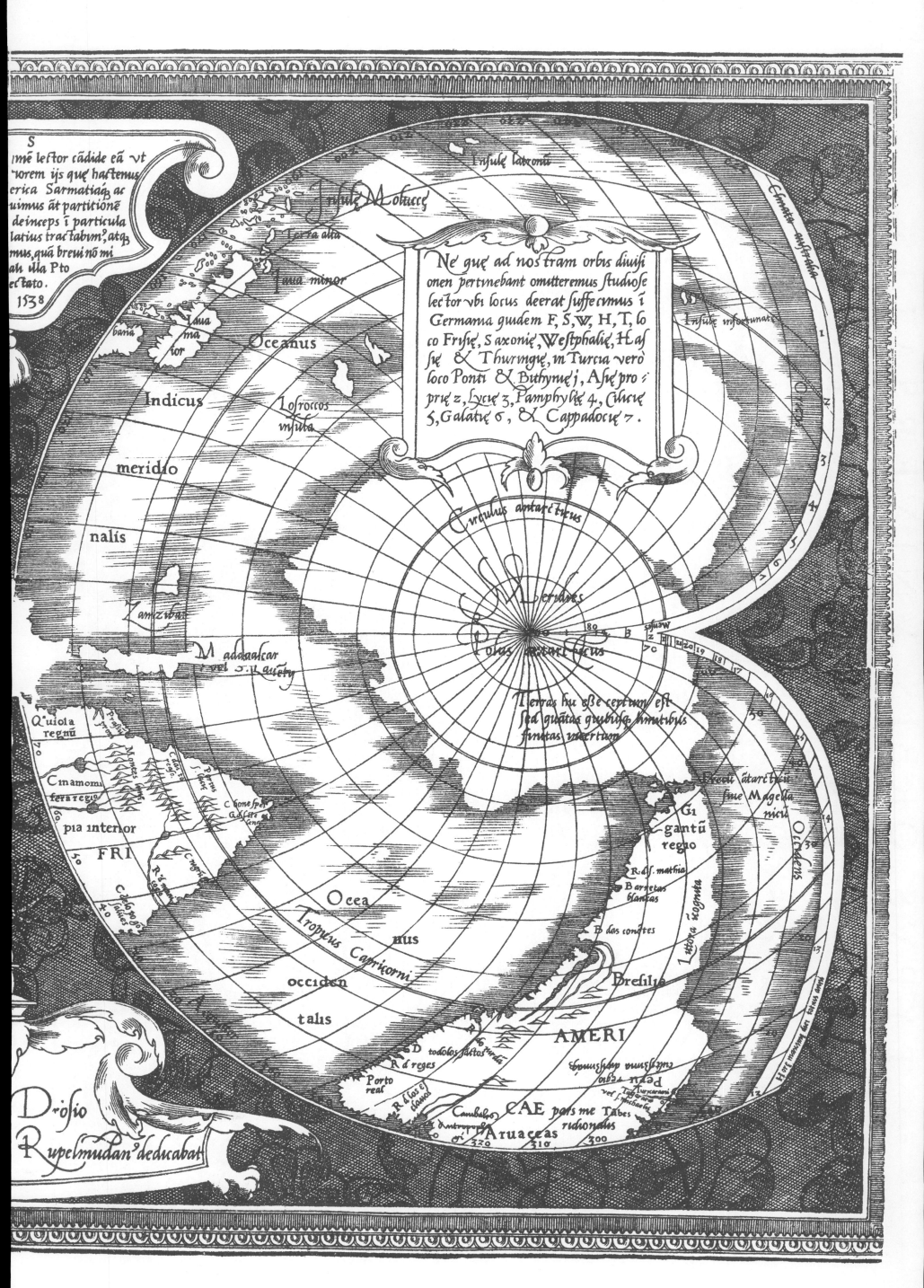

TYPVS ORBIS VNIVERSALIS

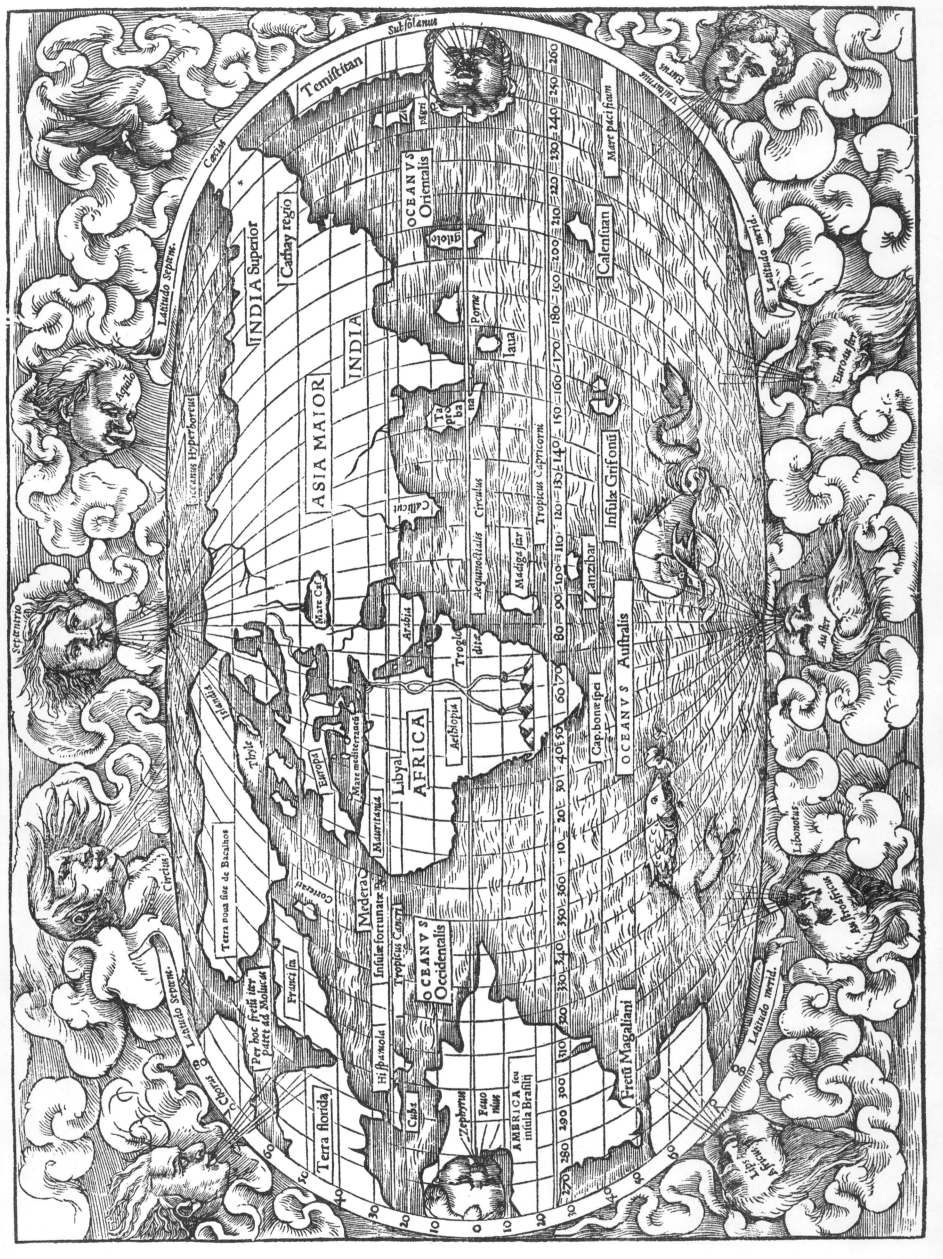

XIIV. (1) PTOLEMÆVS, BASILEÆ. 1540.

VNIVERSALIS COSMOGRAPHIA

SEPTENTRIO

Ordo Climatum   *   *   Gradus Latitudinis

TIGVRI IVE   MERIDIES   .M.D.XLVI.

(2) JOH. HONTERUS, RUDIMENTA COSMOGRAPHICA.

TIGURI 1546.

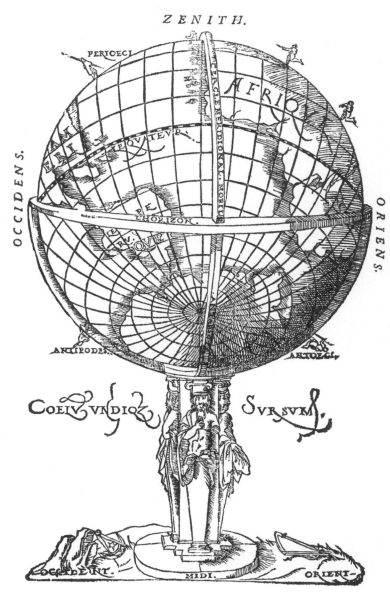

(3) COSMOGRAPHIA PETRI APIANI, PER GEMMAM FRISIUM ILLUSTRATA

PARISIIS 1551.

CHARTA COSMOGRAPHICA, CVM VENTORVM PROPRIA NATVRA ET OPERATIONE.

| Circius North northouest. | SEPTEN | Septentrionalis NORTH. | TRIO. | Aquilo North northest. |

REGIONVM LONGITVDO AD ORTVM.

| Auftroafricus Su fuoueft. | MERI | Auster SV. | DIES. | Euroauster Su fueft. |

XLIV. (4)   COSMOGRAPHIA PETRI APIANI, PER GEMMAM FRISIUM ILLUSTRATA

PARISIIS 1551.

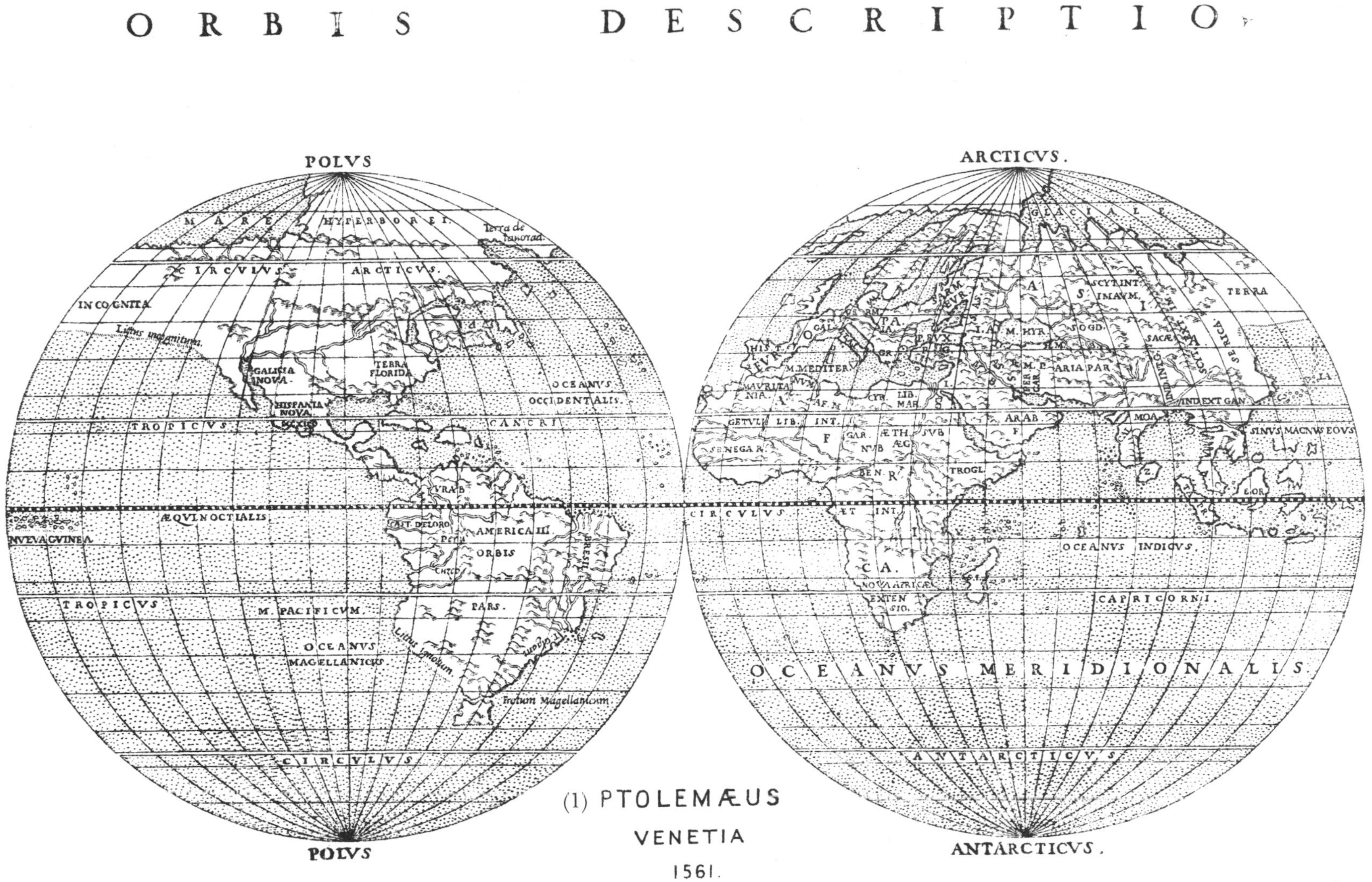

ORBIS DESCRIPTIO

CARTA MARINA NOVA TABVLA

VNIVERSALE NOVO

(1) PTOLEMÆUS VENETIA 1561.

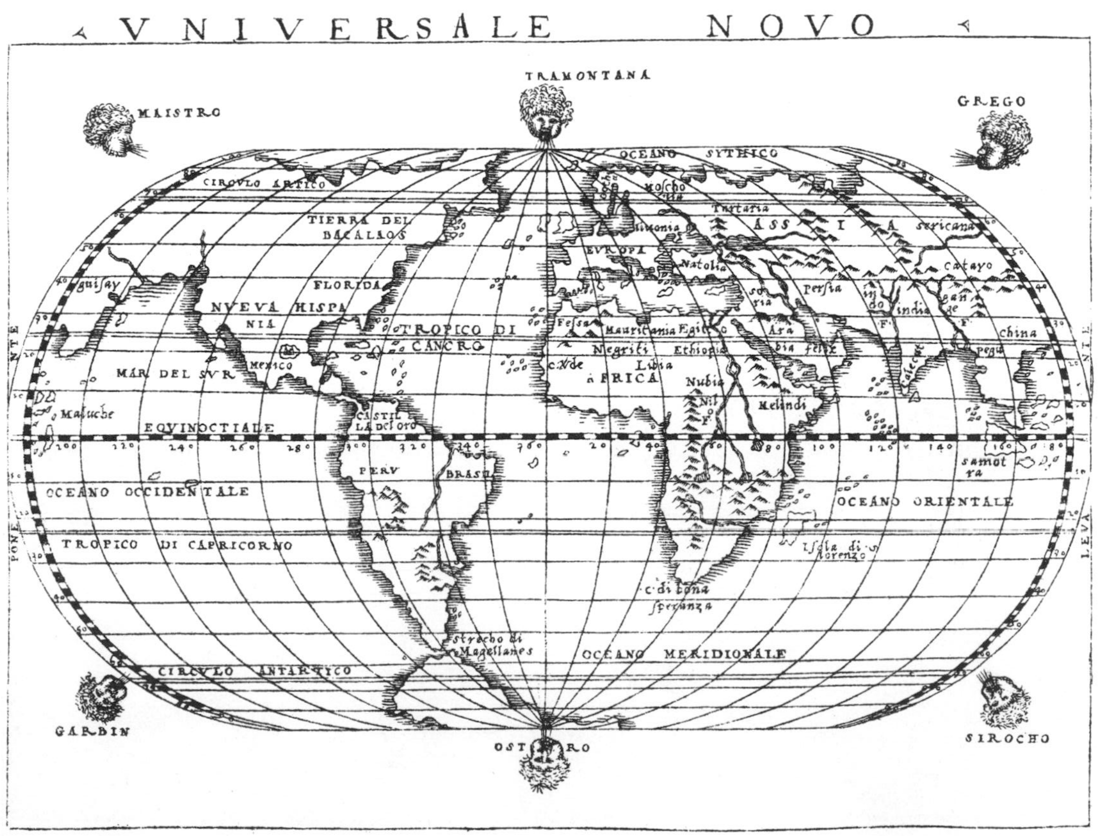

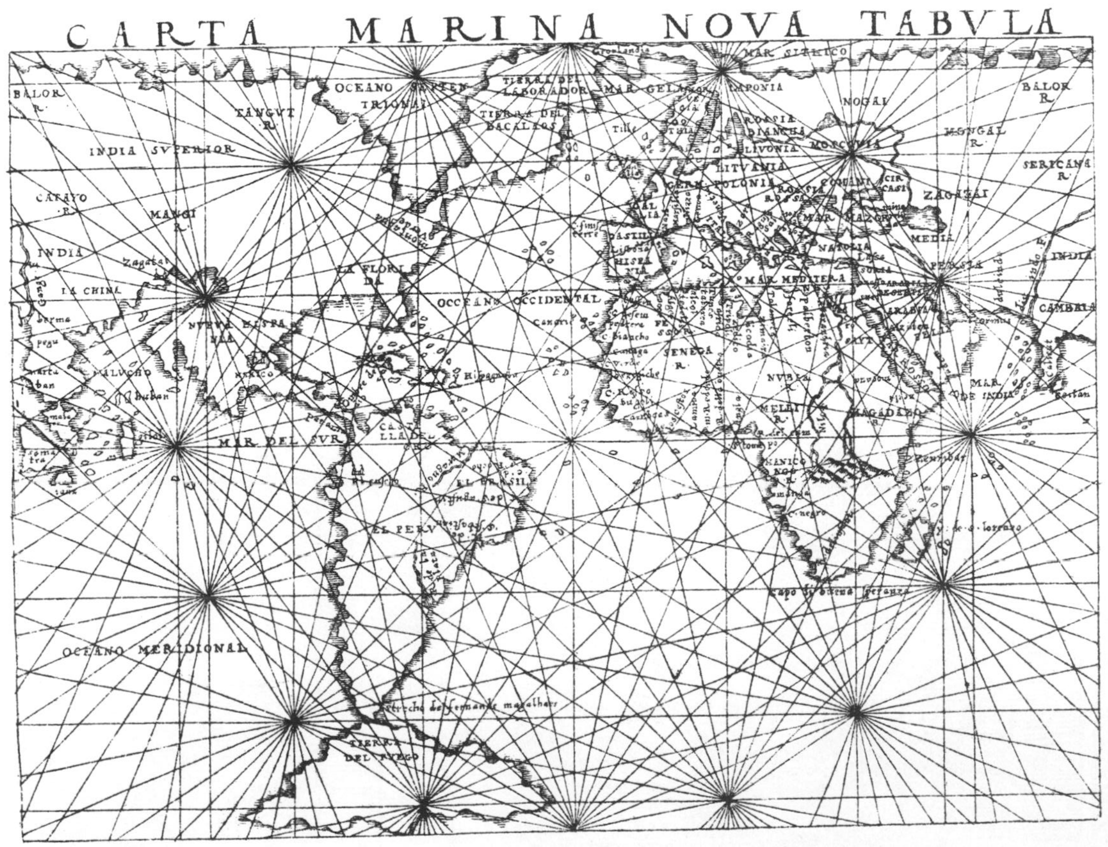

XIV. (23)   LA GEOGRAFIA DI CLAUDIO PTOLEMEO. VENETIA, 1548.

XIV. (4)   HIERONYMO GIRAVA, DOS LIBROS DE COSMOGRAPHIA.

MILAN 1556.

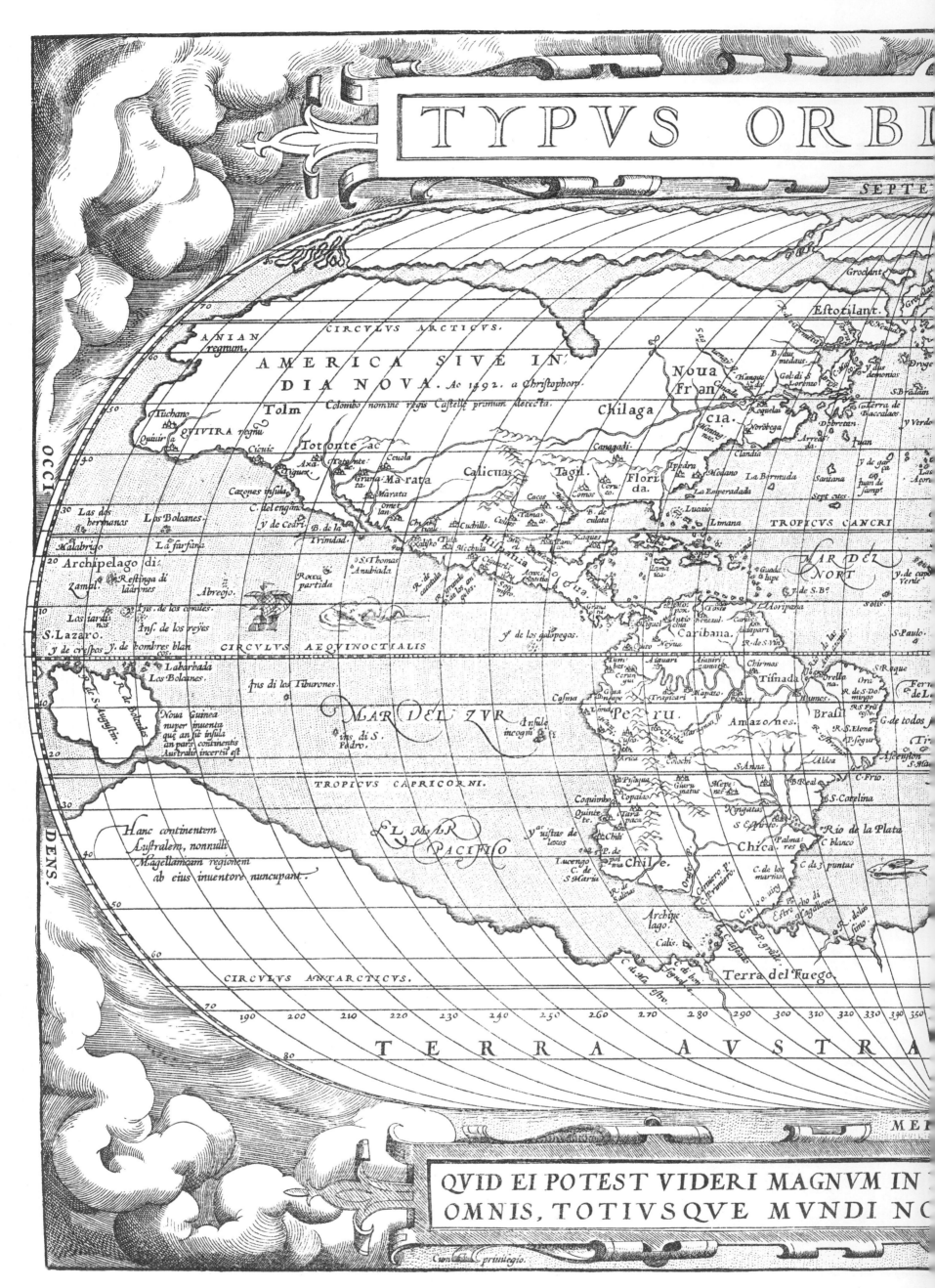

XLVI. ABR. ORTELIUS, THEATRUM ORBIS TERRARUM.
ANTVERPIÆ 1570.

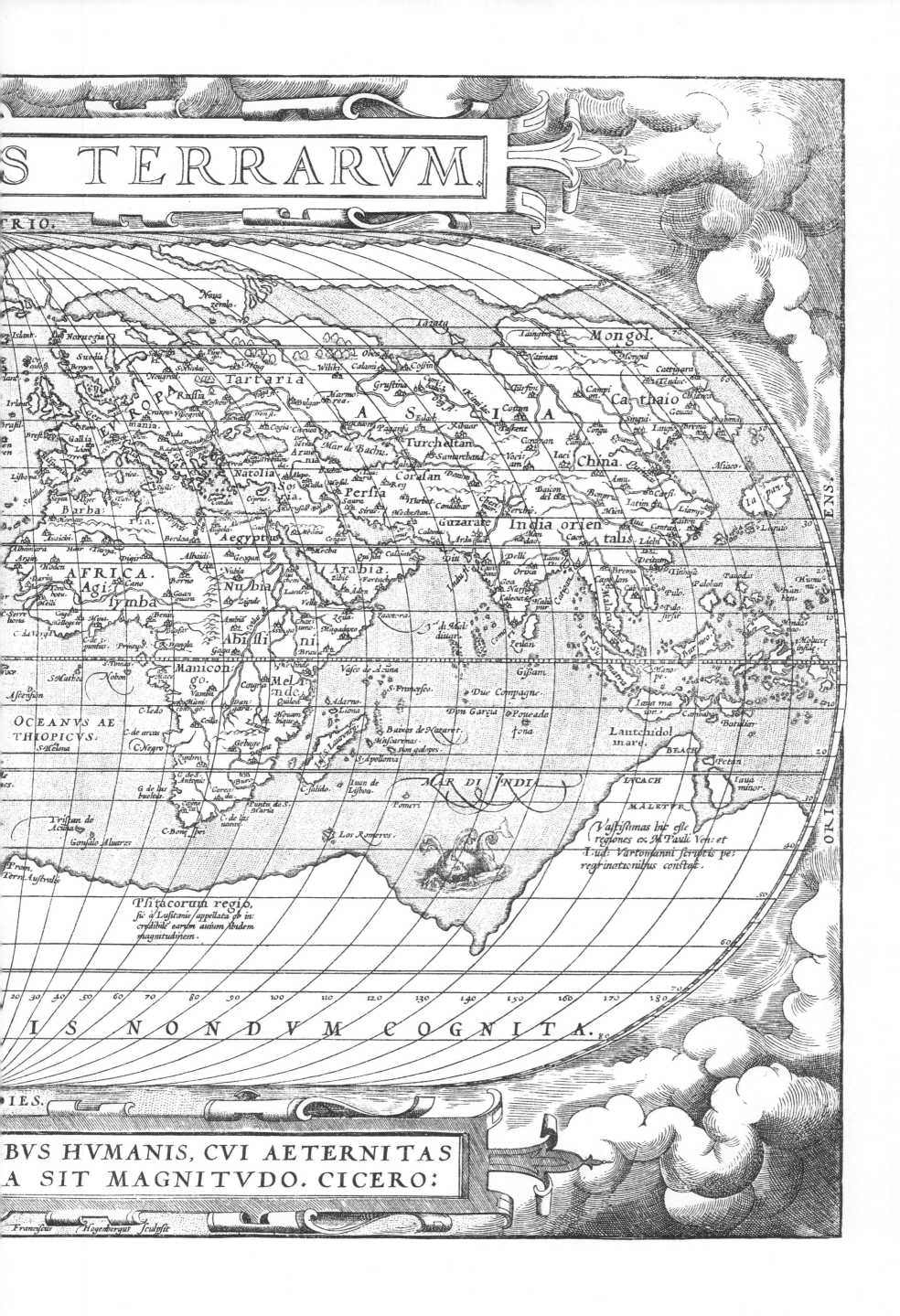

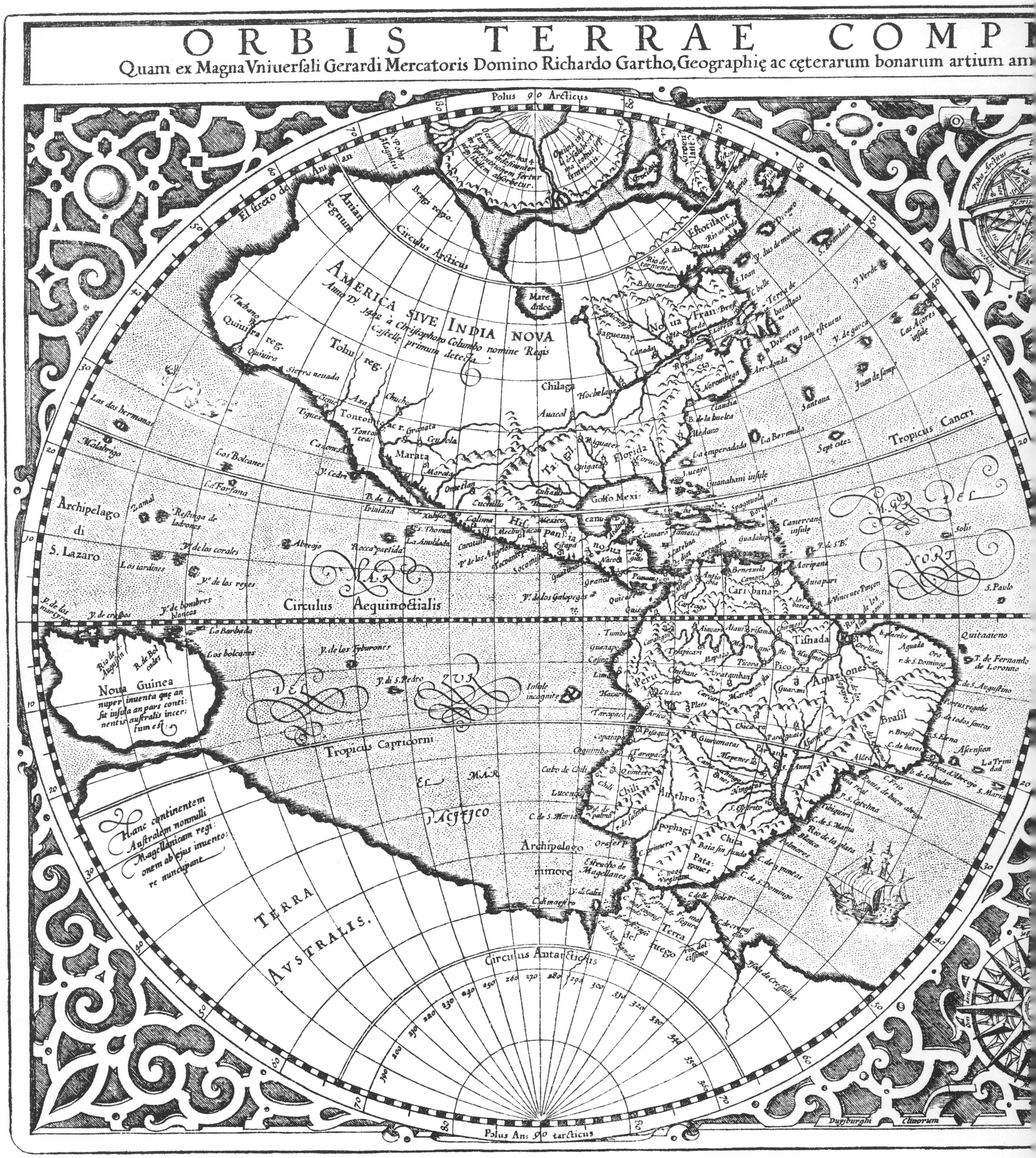

Polus 90 Arcticus

Circulus Arcticus

AMERICA SIVE INDIA NOVA
Anno D: 1492 a Christophoro Columbo nomine Regis
Castellæ primum detecta.

Tropicus Cancri

Circulus Aequinoctialis

MAR DEL ZUR

Nova Guinea nuper inuenta quæ an sit insula an pars continentis australis incertum est.

Hanc continentem Australem nonnulli Magellanicam regionem ab ejus inuentore nuncupant.

Tropicus Capricorni

EL MAR PACIFICO

TERRA AUSTRALIS.

Archipelago minore

Circulus Antarcticus

Polus An: 90 tarcticus

Archipelago di S. Lazaro

Peru

Brasil

Amazones

Terra del fuego

Estrecho de Magellanes

XLVII. MERCATOR 1587.

...NDIOSA DESCRIPTIO

...tori ac fautori summo, in veteris amicitię ac familiaritatis memoriā Rumoldus Mercator fieri curabat A°. M.D.Lxxxvii.

Polus 90 Arcticus

Groenlant

EVROPA

AFRICA

ASIA

India

Tropicus Cancri

Mare mediterraneum

MARE RVBRVM

Aequinoctialis — Aequator.

OCEANVS — MARE DI INDIA

Tropicus Capricorni

Oce... Ae Thiopicvs

Beach prouincia aurifera

Lava maior

Lava minor

C. de Bona Spei

TERRA AVSTRALIS

Circulus Antarcticus

Polus Au: 90 tarcticus

Inter S. Laurentij & Los Romeros insulas vehemens admo: dum est versus ortum & occasum fluxus & refluxus maris.

Psitacorum regio sic'a Lusitanis appellata, ob in credibilem earum auiũ ibidem magnitudinem.

Promontorii Terrę Australis hispano 450 leucis a Caster Bong spei: & Sola a promontorio S. Augustini.

Va(tißimas hic Re regiones ex M. Pauli Veneti & Ludouici Vartomanni scriptis peregrina: tionibus liquido constat.

Typis Aeneis

HEMISPHERIV̄ AB ÆQVINOCTIALI LINEA, AD CIRCVLV̄ POLI ARCTICI.

XLVIII. CORNELIUS DE JUDÆIS, SPECULUM ORBIS TERRÆ.

ANTVERPIÆ 1593.

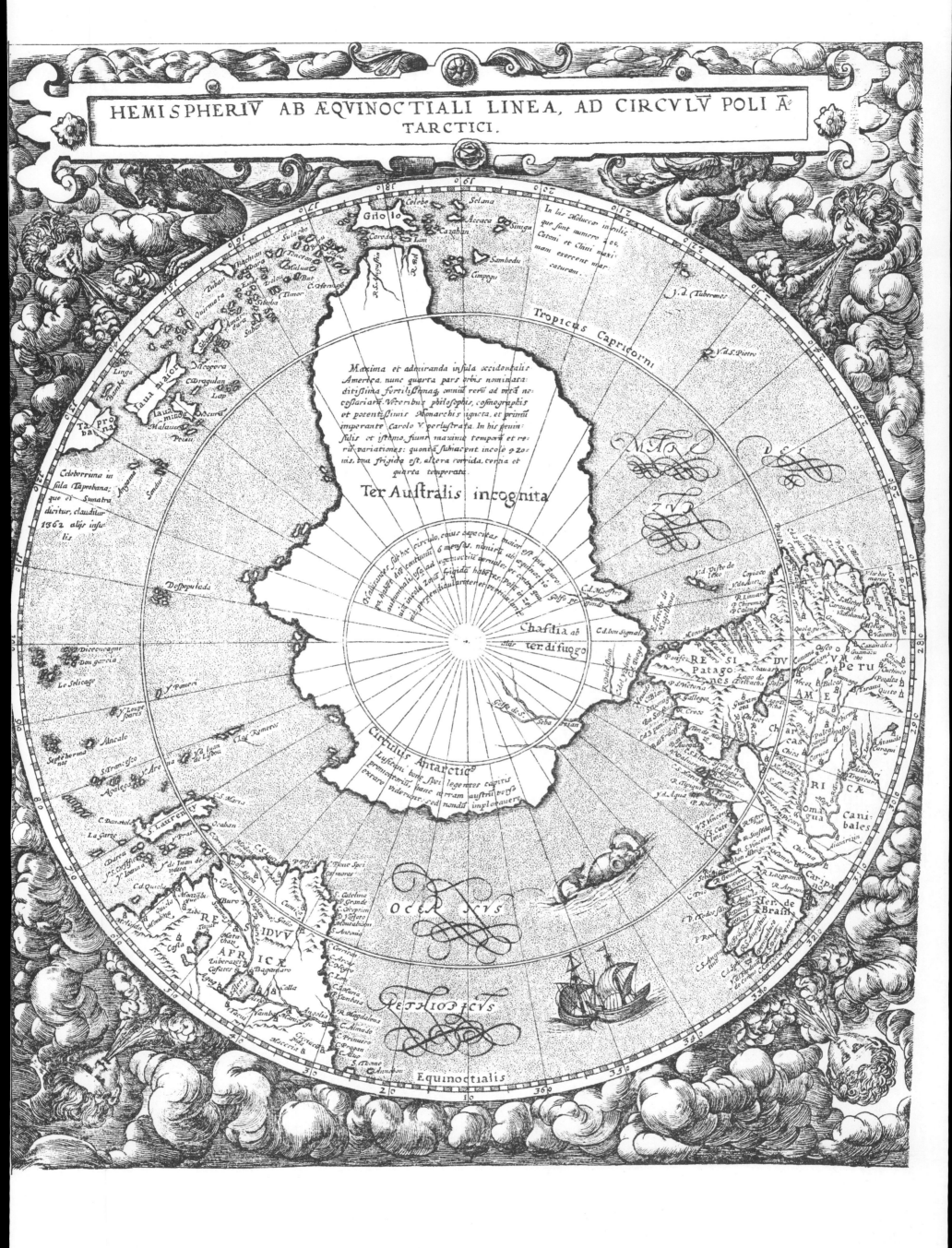

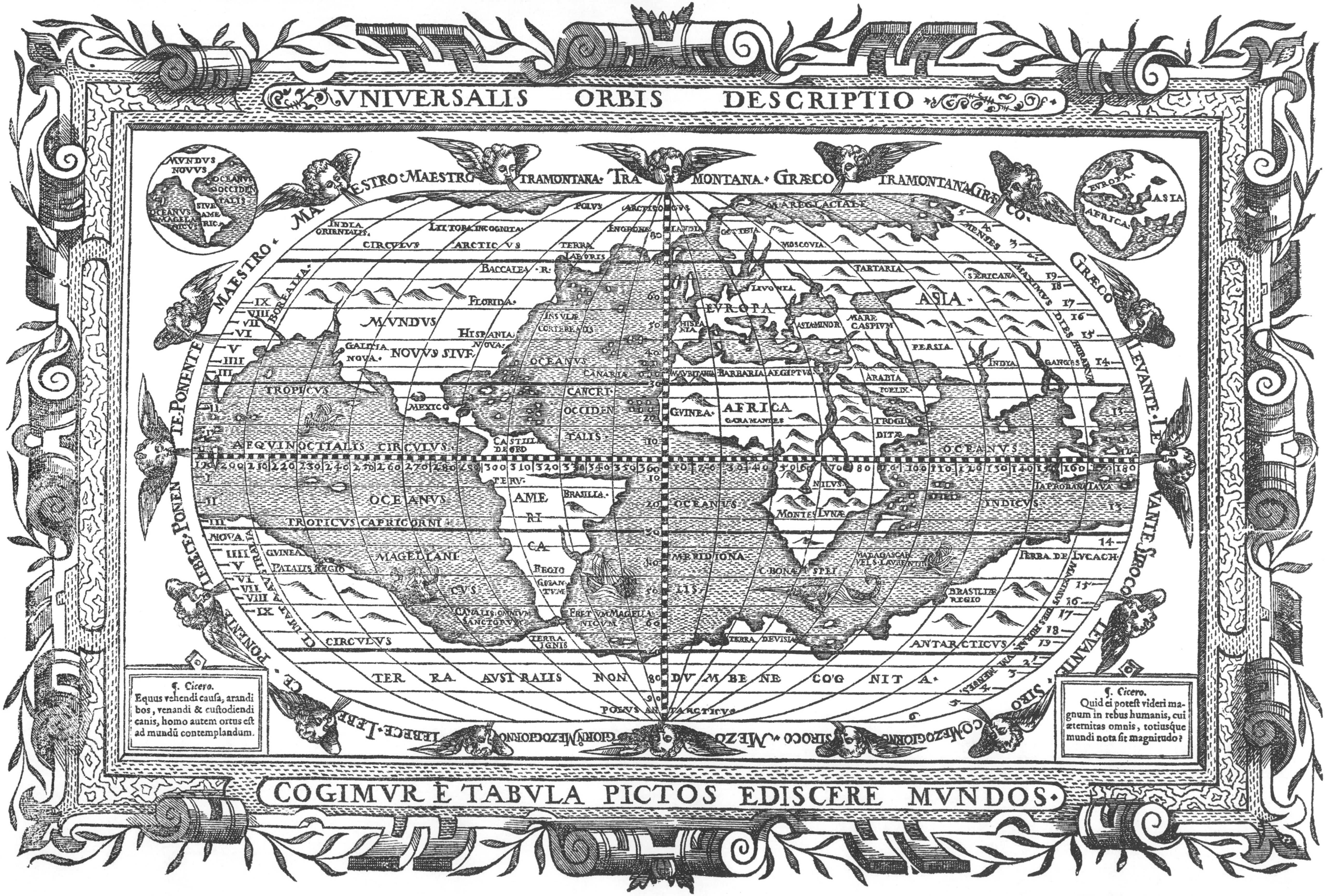

UNIVERSALIS ORBIS DESCRIPTIO

COGIMUR È TABULA PICTOS EDISCERE MUNDOS

TRAMONTANA·GRECO TRAMONTANA· GRECO·LEVANTE·LEVANTE·Siroco LEVANTE·SIRO

MAESTRO·MAESTRO TRAMONTANA MAESTRO COMEZOGIORNO SIROCO MEZO GIORNO MEZOGIORNO

PONENTE LIBECE LIBECE·PONENTE LIBECE·PONEN LEPONENTE MAESTRO

ASIA

AFRICA

AMERI CA

MUNDUS NOVUS

XLIX. (1) JOANNES MYRITIUS, OPUSCULUM GEOGRAPHICUM RARUM.
INGOLSTADII 1590.

(2) L'ISOLE PIU FAMOSE
DEL MONDO.

VENETIA

DESCRITTE DA
THOMASO PORCACCHI.

MDLXXII.

XLIX. (3) MATTHIAS QUADUS, FASCICULUS GEOGRAPHICUS.
KÖLN 1608.

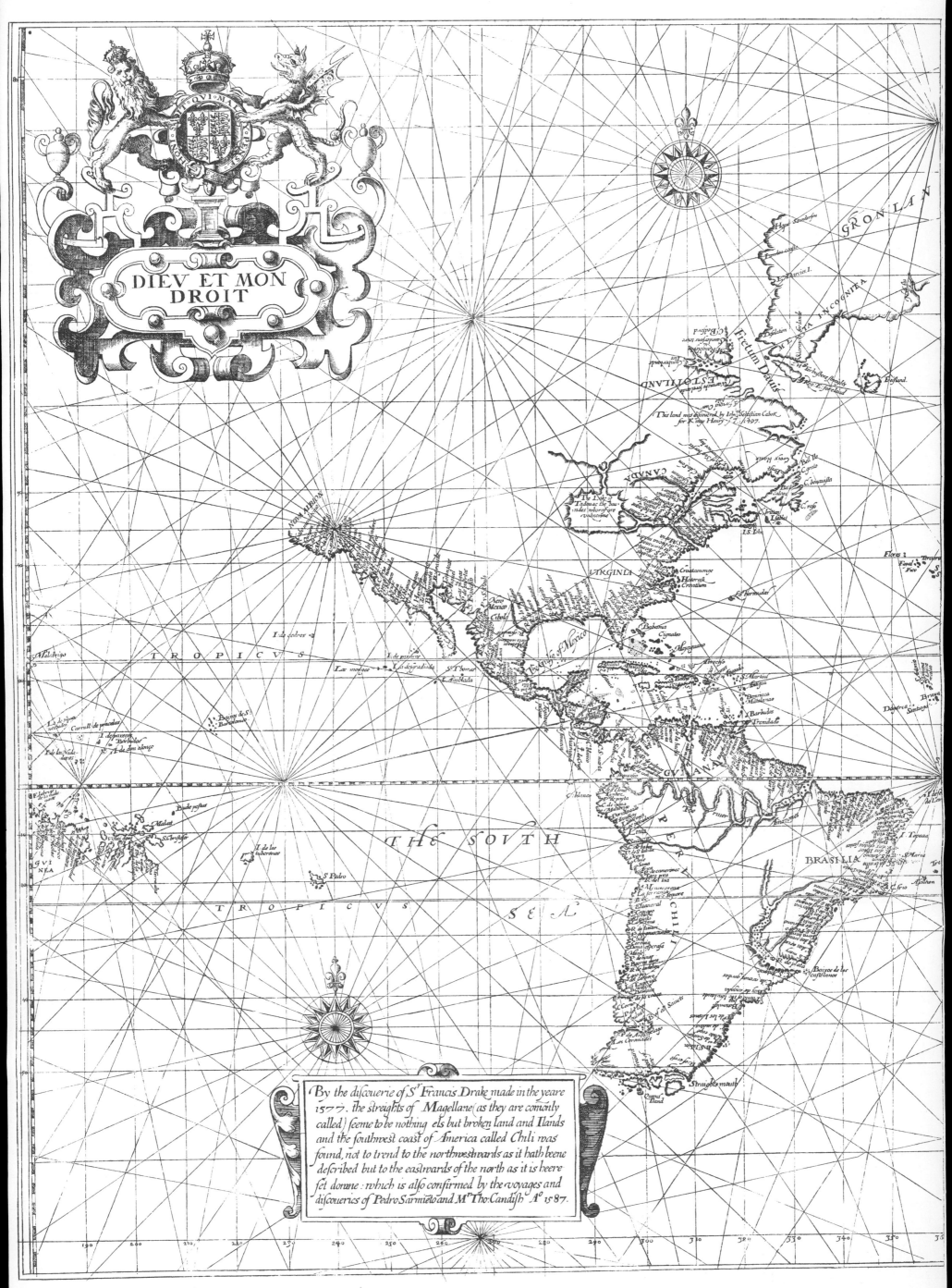

DIEV ET MON DROIT

By the difcouerie of St Francis Drake made in the yeare
1577. the ſtreights of Magellane (as they are comonly
called) ſeeme to be nothing els but broken land and Ilands
and the ſouthweſt coaſt of America called Chili was
found, not to trend to the northweſtwards as it hath beene
deſcribed but to the eaſtwards of the north as it is heere
ſet donne : which is alſo confirmed by the voyages and
diſcoueries of Pedro Sarmiẽto and Mr Tho: Candiſh Aº 1587.

L. RICHARD HAKLUYT, THE PRINCIPAL NAVIGATIONS.
LONDON 1599.

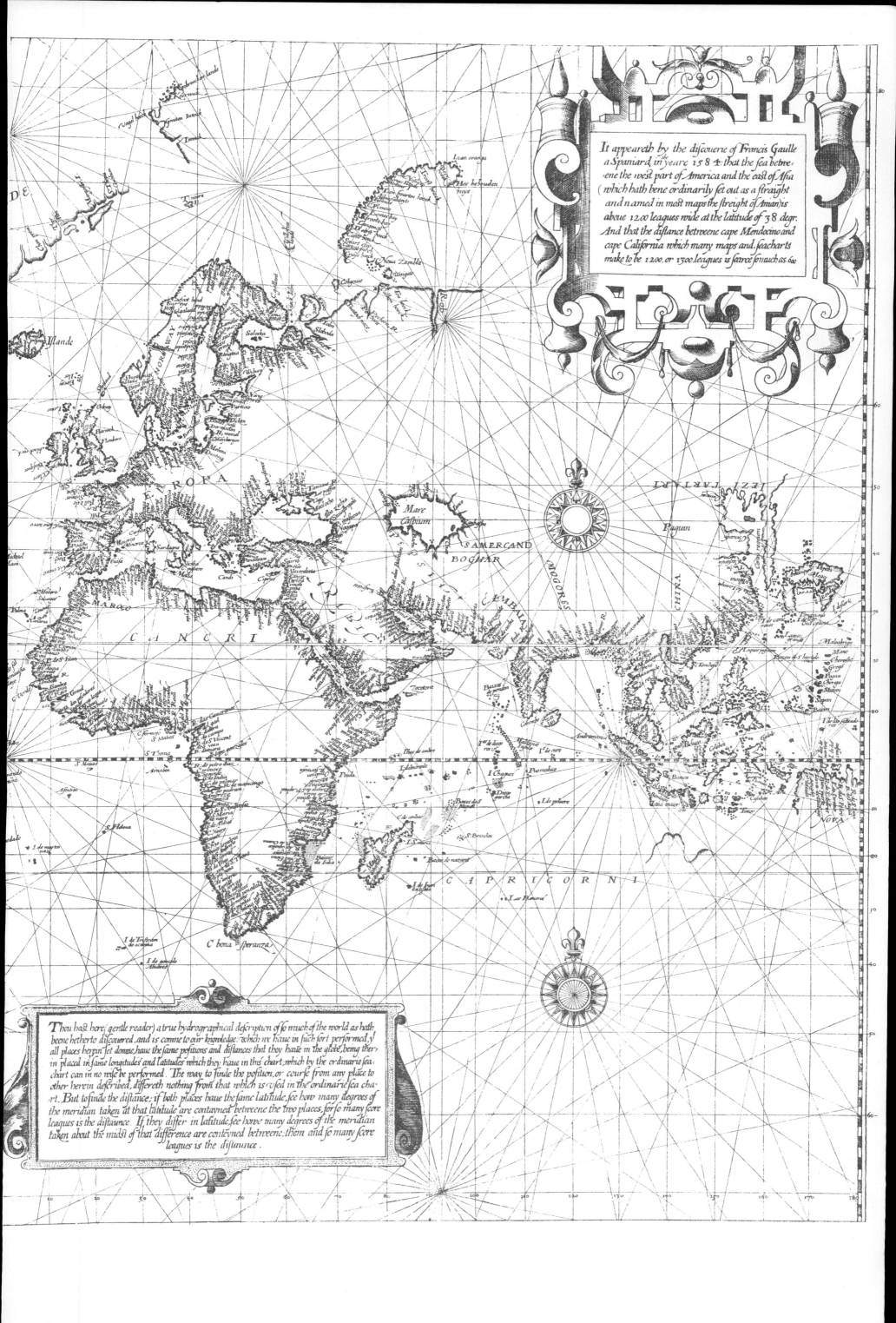

It appeareth by the discouerie of *Francis Gaulle*
a Spaniard, in yeare 1584: that the sea betweene the west part of America and the east of Asia
(which hath bene ordinarily set out as a straight
and named in most maps the streight of Aman) is
aboue 1200 leagues wide at the latitude of 38 degr.
And that the distance betweene cape Mendocino and
cape California which many maps and seacharts
make to be 1200. or 1300 leagues is scarce so much as 600

Thou hast here (gentle reader) a true hydrographical description of so much of the world as hath
beene hetherto discouered, and is conne to our knowledge: which we haue in such sort performed, y¹
all places herein set donne, haue the same positions and distances that they haue in the globe, being ther-
in placed in same longitudes and latitudes which they haue in this chart, which by the ordinarie sea-
chart can in no wise be performed. The way to finde the position, or course from any place to
other herein described, differeth nothing from that which is vsed in the ordinarie sea chart. But to finde the distance; if both places haue the same latitude, see how many degrees of
the meridian taken at that latitude are contayned betweene the two places, for so many score
leagues is the distaunce. If they differ in latitude, see howe many degrees of the meridian
taken about the midst of that difference are contayned betweene them and so many score
leagues is the distaunce.

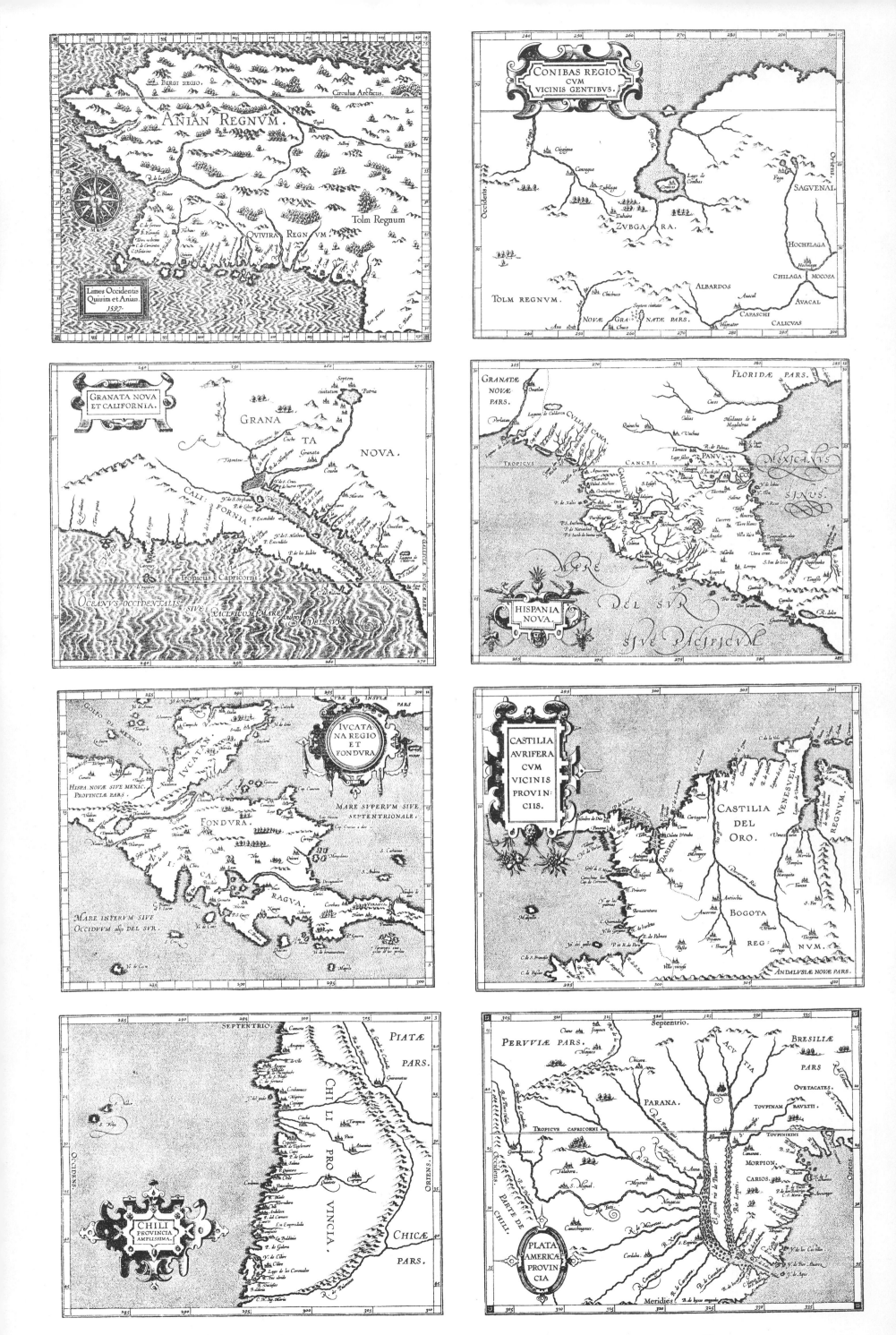

LI.   CORNELIUS WYTFLIET, DESCRIPTIONIS PTOLEMAICÆ AUGMENTUM.   LOVANII 1597.

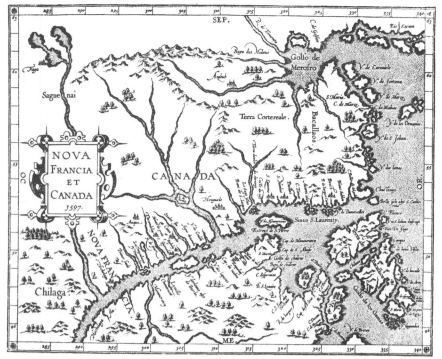

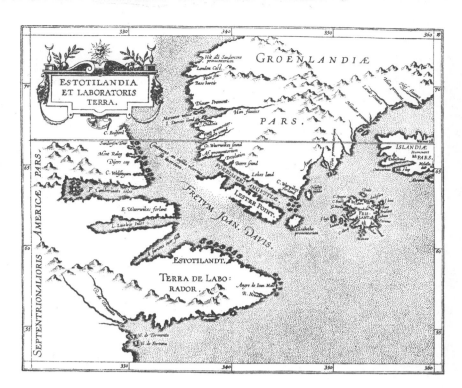

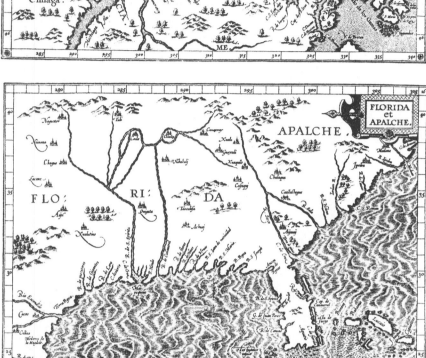

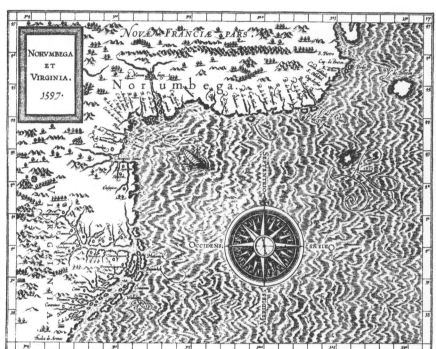

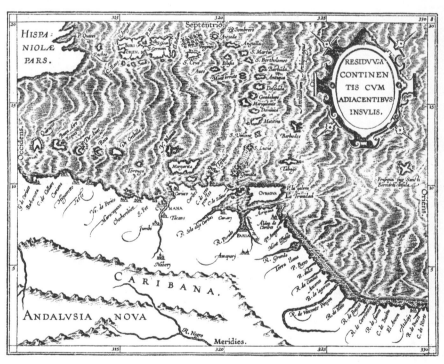

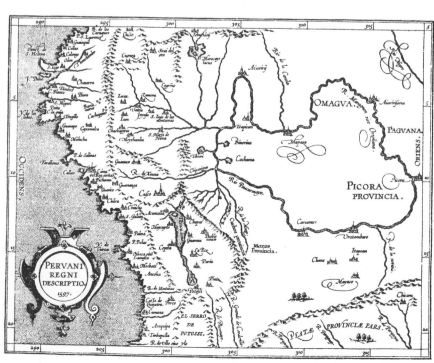

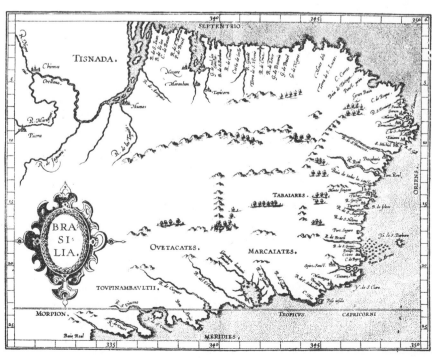